BEGINNER'S GUIDE TO DIGITAL PAINTING IN

CLIP STUDIO PAINT

> Featuring tutorials by expert professional artists

3dtotal Publishing

3dtotalPublishing

Correspondence: publishing@3dtotal.com
Website: store.3dtotal.com

Clip Studio Paint is developed by Celsys, Inc. Clip Studio Paint and Clip Studio are the trademarks or registered trademarks of Celsys, Inc. Find out more about Clip Studio Paint at: www.clipstudio.net/en

First published in the United Kingdom, 2024, by 3dtotal Publishing.

Address: 3dtotal.com Ltd,
29 Foregate Street, Worcester,
WR1 1DS, United Kingdom.

Soft cover ISBN: 978-1-912843-96-1
Printed and bound in China
by C&C Offset Printing Co., Ltd.

Visit **store.3dtotal.com** for a complete list of available book titles.

Editor: Philippa Barker
Designer: Fiona Tarbet
Lead Editor: Samantha Rigby
Lead Designer: Joseph Cartwright
Studio Manager: Simon Morse
Managing Director: Tom Greenway

Front cover artwork by individual artists as credited throughout the book.

Back cover artwork: © Devin Elle Kurtz

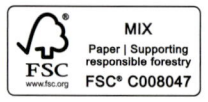

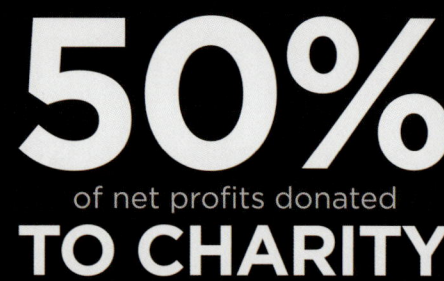

50%
of net profits donated
TO CHARITY

In 2022, 3dtotal Publishing became successful enough to make a pledge to donate **50% of its net profits to charity**. This continues to be possible due to the incredible support from all our customers, employees, and partners. At the time of printing, we have donated over $1.62 million (USD) to charity.

We focus our giving on three charitable areas: **environmental**, **humanitarian**, and **animal welfare**. We use organizations such as Effective Altruism and Founders Pledge to guide who we help within these causes. Some ways of doing good are over 100 times more effective than others, so donating this way hugely increases the impact of our contributions.

See **3dtotal.com/charity** for full details.

Image © Nimrod Villar

Contents

Introduction

BY NIMROD VILLAR

Welcome to this beginner's guide to getting started in Clip Studio Paint (CSP)! If you're reading this, you're undoubtedly interested in embarking on a journey into the world of digital painting. You may be completely new to this, or perhaps you've previously been using a different digital-painting software and want to expand your skills. Whatever your background and experience level, you're in the right place!

Clip Studio Paint is digital painting and drawing software that provides you with the tools to develop illustration projects, concept art, comics, manga, webcomics/webtoons, and animation. Currently available on Windows, MacOS, iPad, iPhone, Android, and Chromebook, it's a highly accessible and versatile software that's well known for its simple menus, wide variety of tools, and useful features to help streamline your workflow and painting process. Popular with amateurs and professionals alike, it has a supportive community of fellow users and creators available to answer questions, plus provides ways to contact Clip Studio Paint's support desk for help.

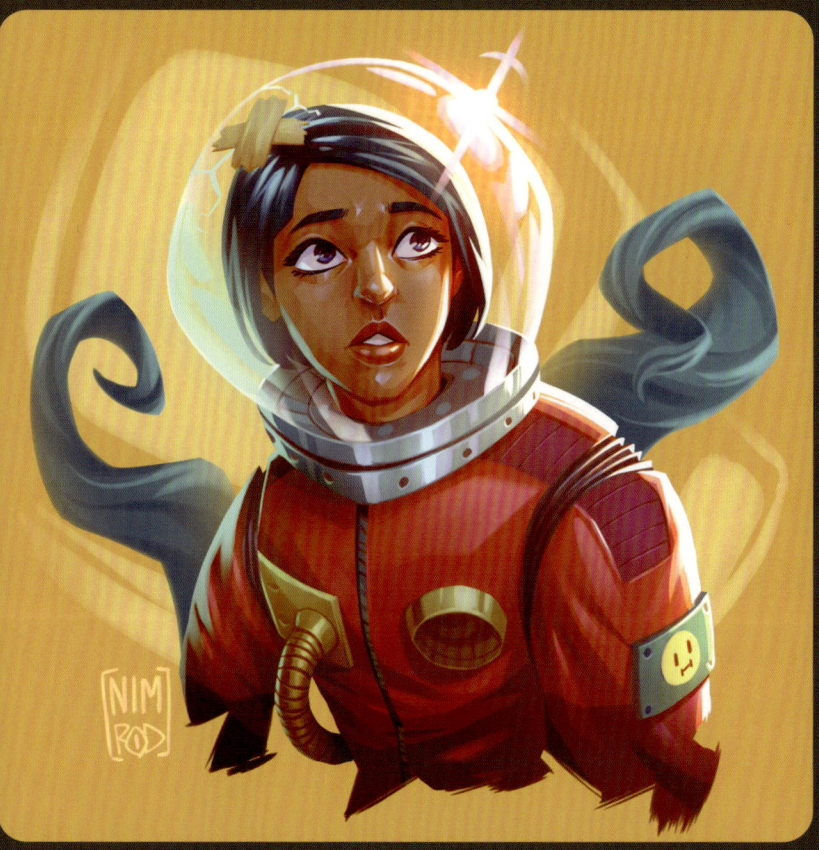

PRACTISE, PRACTISE, PRACTISE!

Practising drawing or painting using traditional media is an excellent way to improve your digital-painting skills. Just like you would warm up before undertaking any physical exercise, creating a series of loose warm-up sketches before you begin a new digital project will help to exercise and prepare your hand.

Clip Studio Paint is divided into two versions, with different pricing and usage models:

CLIP STUDIO PAINT PRO

This version is geared towards creating single-page illustrations, concept art, comics, and manga. It also supports animations of up to twenty-four frames.

CLIP STUDIO PAINT EX

In addition to the features of Clip Studio Paint PRO, this version includes options for creating multi-page projects and more advanced features for professional animations. This book will include some features only found Clip Studio Paint EX.

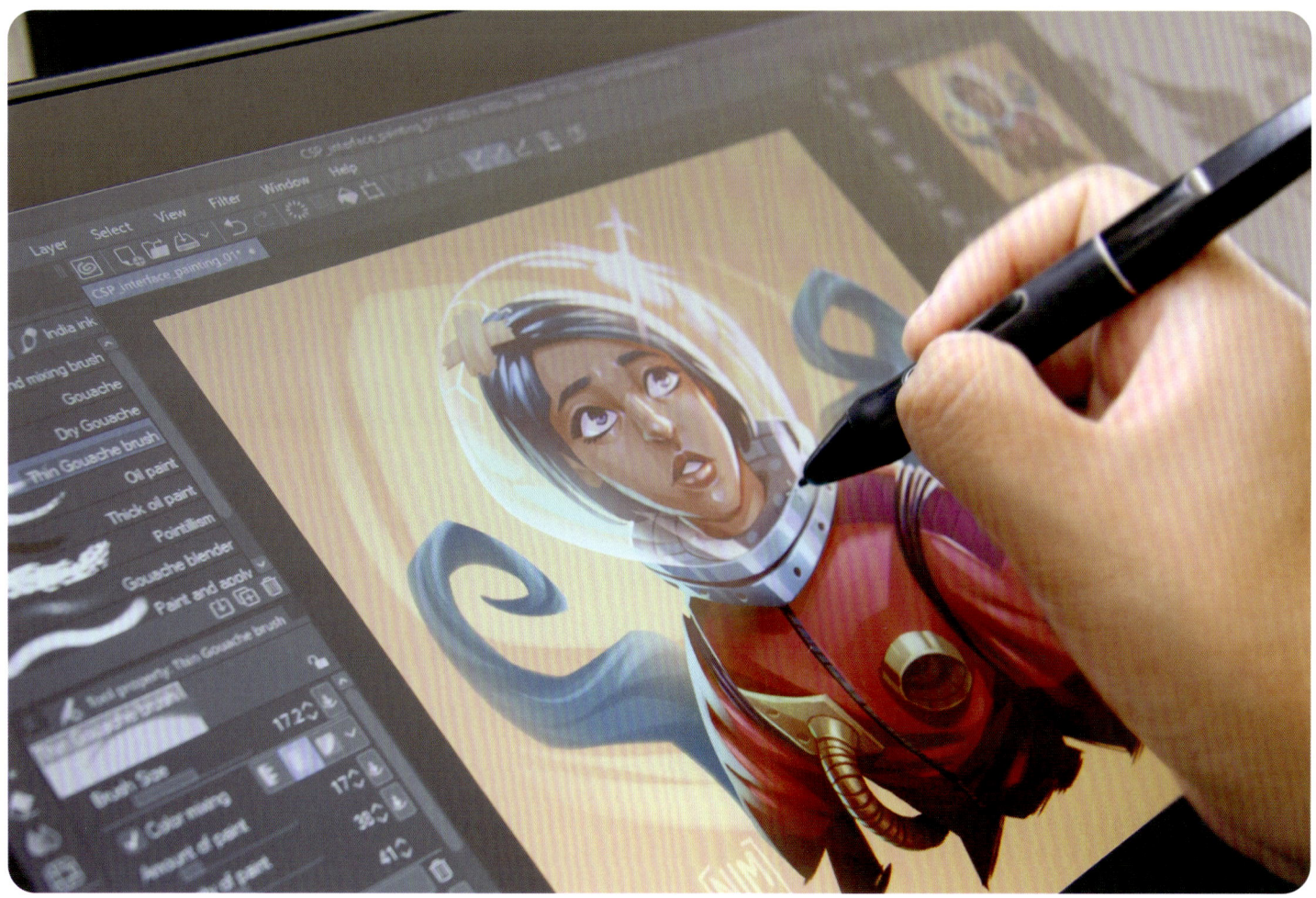

WHAT IS DIGITAL PAINTING?

While traditional painting involves creating artwork using traditional media, such as pencils or watercolours, digital painting utilizes digital software on a computer or compatible device, such as a tablet or smartphone. It also uses hardware, such as a digital tablet and stylus, which closely simulates traditional pencil and paper, though it's also possible to create illustrations using a computer mouse, depending on the desired results or style.

There are many advantages to digital painting, the first being that it allows artists to streamline and speed up processes that are more time-consuming in traditional painting. The wide variety of tools, brushes, special effects, and techniques available in the digital medium are endless, allowing artists to take their artwork to the next level. The wondrous and rather addictive 'undo' feature allows you to undo mistakes or backtrack if you change your mind. Before long, the Ctrl+Z shortcut will become your best friend!

Another significant advantage of working digitally is the ease with which you can edit a painting thanks to the use of layers. Different sections or objects in a painting can be organized on separate layers, which are like transparent sheets stacked one on top of the other to create the final image. You can choose how layers interact with one another. For example, you may wish them to act independently of each other. Painting different elements on different layers allows you to adjust each component without affecting the rest of the illustration. Alternatively, you may want the layers to influence each other, as if layering paint on a canvas.

While creating artwork digitally may seem daunting at first, familiarizing yourself with the basics of Clip Studio Paint and how each menu and tool works is the best place to start. Just like with traditional painting, regular practice is crucial to making progress and creating great artwork in the digital medium.

How to use this book

Working with talented industry professionals, we have created a book aimed at those new to Clip Studio Paint. Whether you're a complete beginner or a learner with some experience, we recommend you begin by reading through the Getting Started chapters. This section of the book starts with chapters on The User Interface, how to Create a New File, and Navigation Tools & Basic Actions, which provide the brief overview you need to get started.

Following on from this, Tools, Sub Tools & Tool Properties; Drawing Tools; Layers; Colour; Selection; Fill & Gradient; Figure Tool; Comics, Manga & Webtoons; Drawing Guides; Tonal Correction; Filters; Materials; and Using Clip Studio Paint on a Tablet will explore the various tools, menus, and palettes CSP has to offer. Read each chapter carefully and take time to experiment with the different tools and techniques to understand how they work.

SUB-TOPIC

SUPPORTING IMAGES

CHAPTER TOPIC

INTRODUCTION

LEARNING OBJECTIVES

IMAGE DESCRIPTION

ARTIST'S TIP BOX

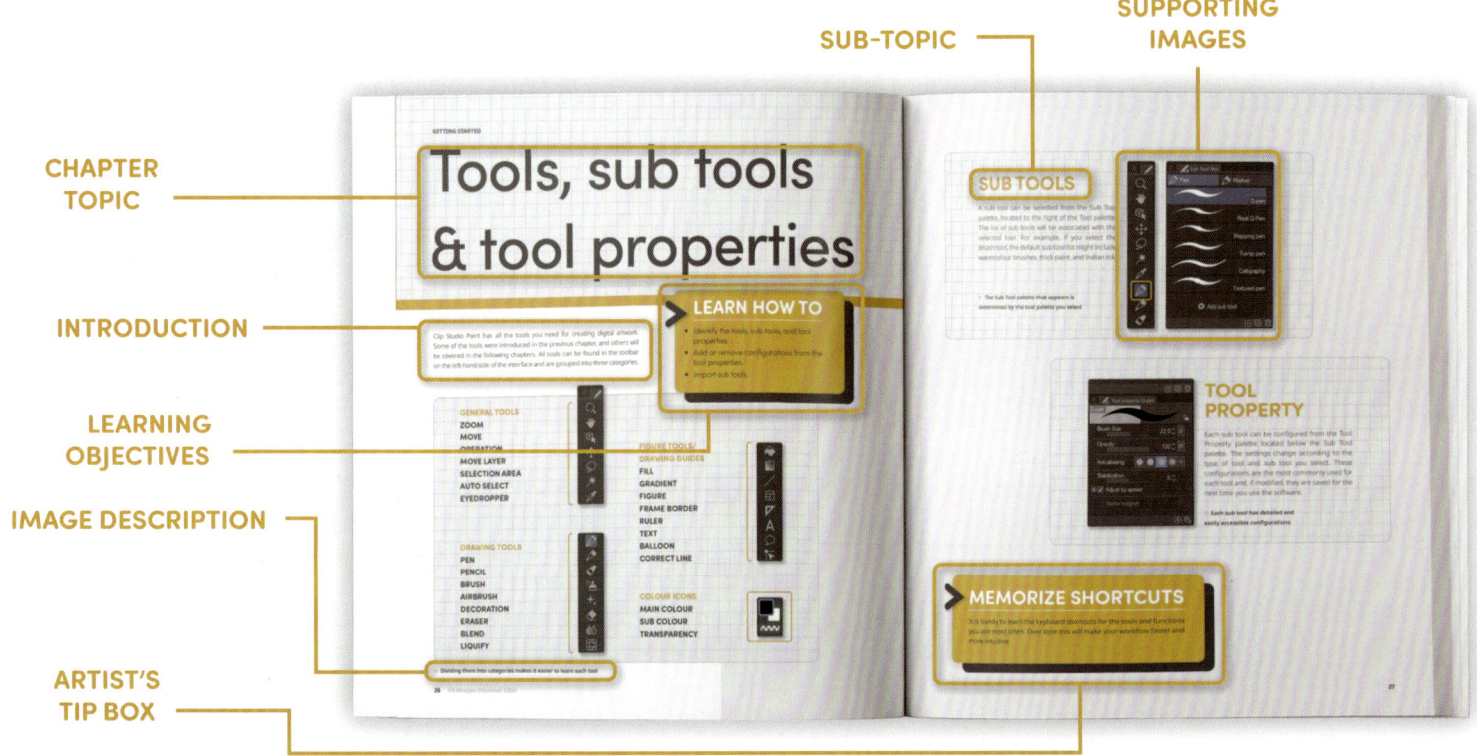

TUTORIAL
TITLE

TUTORIAL
INTRODUCTION

IMAGE
DESCRIPTION

DOWNLOADABLE
RESOURCES ICON

LEARNING
OBJECTIVES

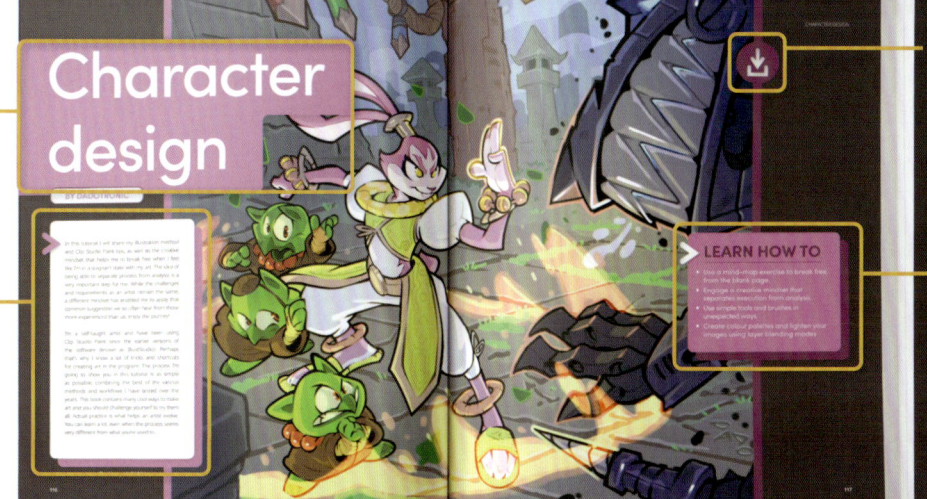

Once you've read the Getting Started chapters and have a good grasp of the basics, work your way through each of the eight Tutorials. Spanning a variety of different themes, styles, and approaches, the tutorials will guide you step-by-step through how to create artwork in Clip Studio Paint. You can complete these in any order you like. There are learning objectives at the start of each tutorial, which list the various skills and creative techniques you will learn as you work through the steps.

You will also find artist's tip boxes scattered throughout the introductory chapters and tutorials. These provide helpful advice and creative insights to aid your learning. At the end of the book is a useful Glossary of terms, which you can refer back to as needed.

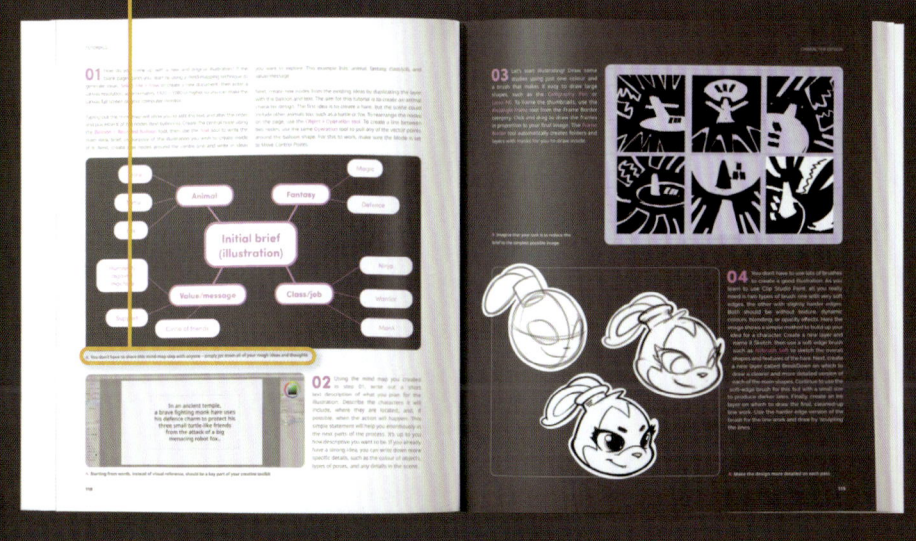

STEP-BY-STEP
INSTRUCTIONS

STEP-BY-STEP
IMAGES

DOWNLOADABLE RESOURCES

The artists featured in this book have supplied a range of downloadable resources to help further your learning. These include custom brushes used in the tutorials, as well as line art and timelapse videos. A full list of downloadable resources can be found on page 257. Make sure to download these before you start the tutorials. Where downloadable resources are available, you will see an arrow icon at the start of the chapter.

Getting started

BY NIMROD VILLAR

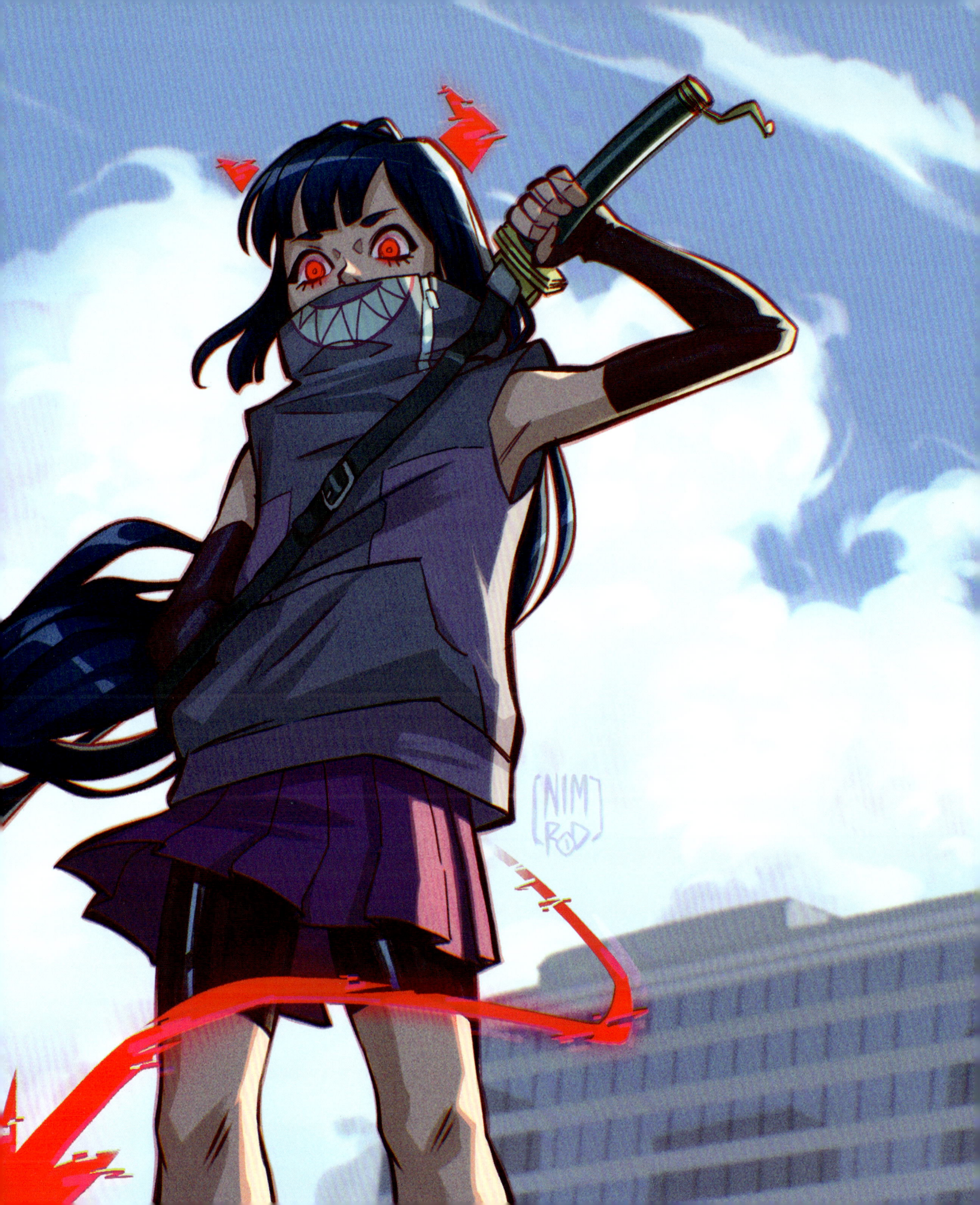

The user interface

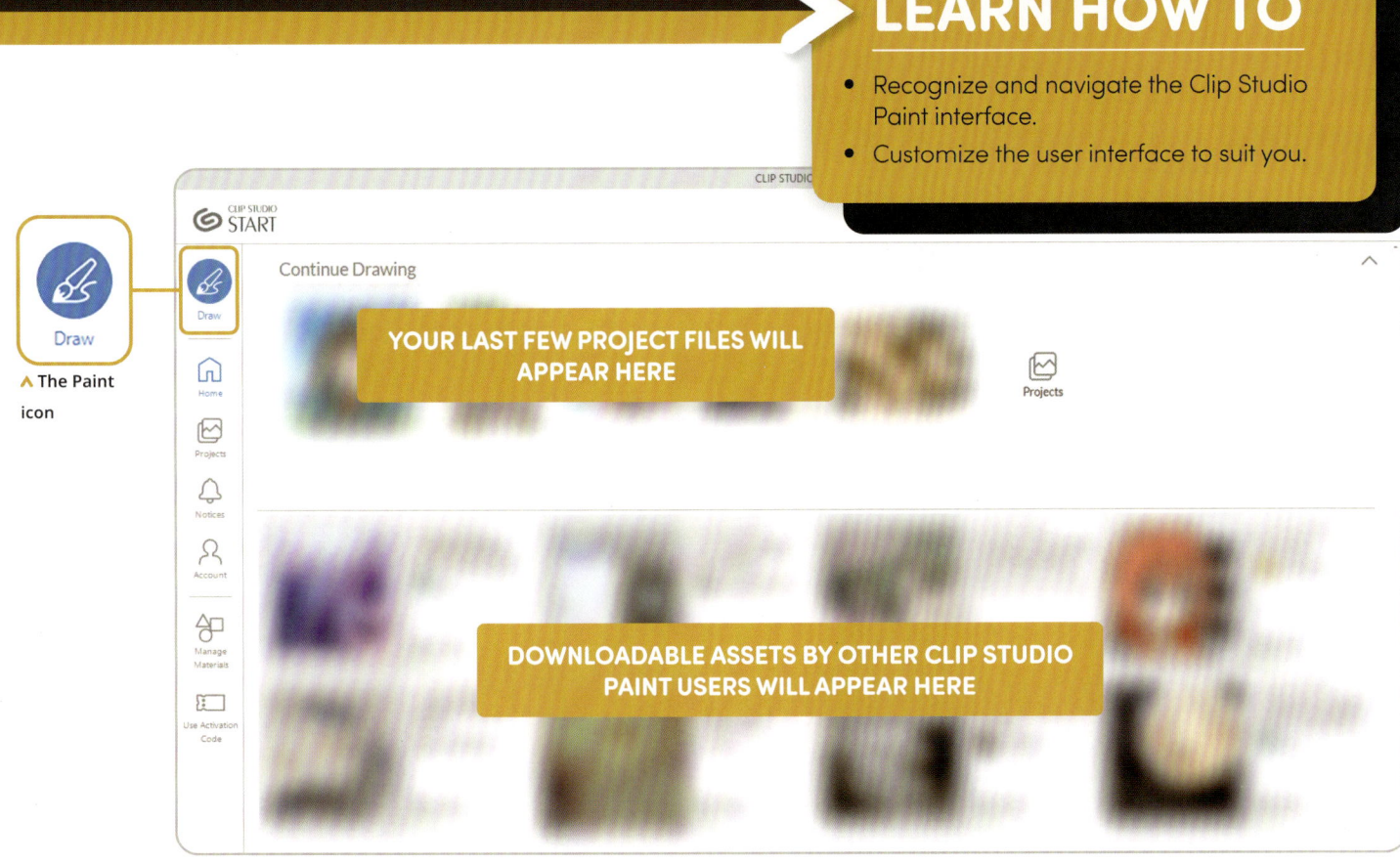

LEARN HOW TO

- Recognize and navigate the Clip Studio Paint interface.
- Customize the user interface to suit you.

^ The Paint icon

YOUR LAST FEW PROJECT FILES WILL APPEAR HERE

DOWNLOADABLE ASSETS BY OTHER CLIP STUDIO PAINT USERS WILL APPEAR HERE

If you find the Clip Studio Paint user interface a little overwhelming to begin with, you're not alone. It has many different elements and menus on display, but it is designed so that everything is easily accessible. This chapter will explore where the various tools and menus are located and the function of each.

To launch Clip Studio Paint, select the Paint icon located in the top-left corner of the Clip Studio Launcher. Clip Studio Launcher is the central platform of Clip Studio Paint where, in addition to launching CSP, you can find notifications, program updates, contests, and help sections.

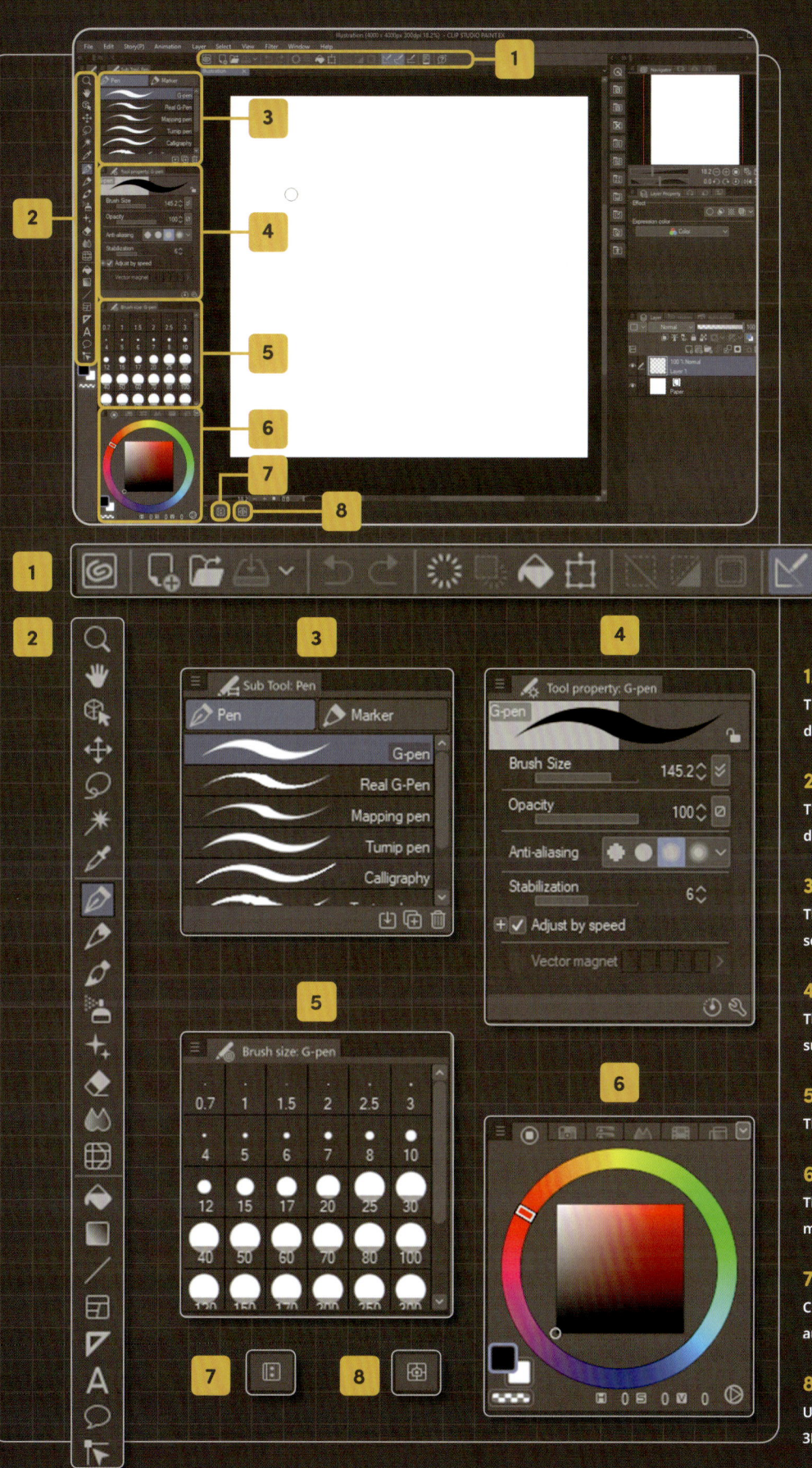

1. COMMAND BAR

This offers options to create, open, and save a document, plus other functions.

2. TOOL PALETTE

This palette contains tools for the canvas, such as drawing, painting, selection, text, and more.

3. SUB-TOOL PALETTE

This holds a list of sub tools based on the tool selected from the main tool palette.

4. TOOL PROPERTY PALETTE

This represents the properties of the selected sub tool.

5. BRUSH SIZE PALETTE

This palette provides quick selections for brush size.

6. COLOUR PALETTES

This contains various types of colour selection and mixing palettes.

7. TIMELINE PALETTE

Clicking the icon will display the the animation timeline.

8. ALL SIDES VIEW PALETTE

Used to visualize various angles of the selected 3D material.

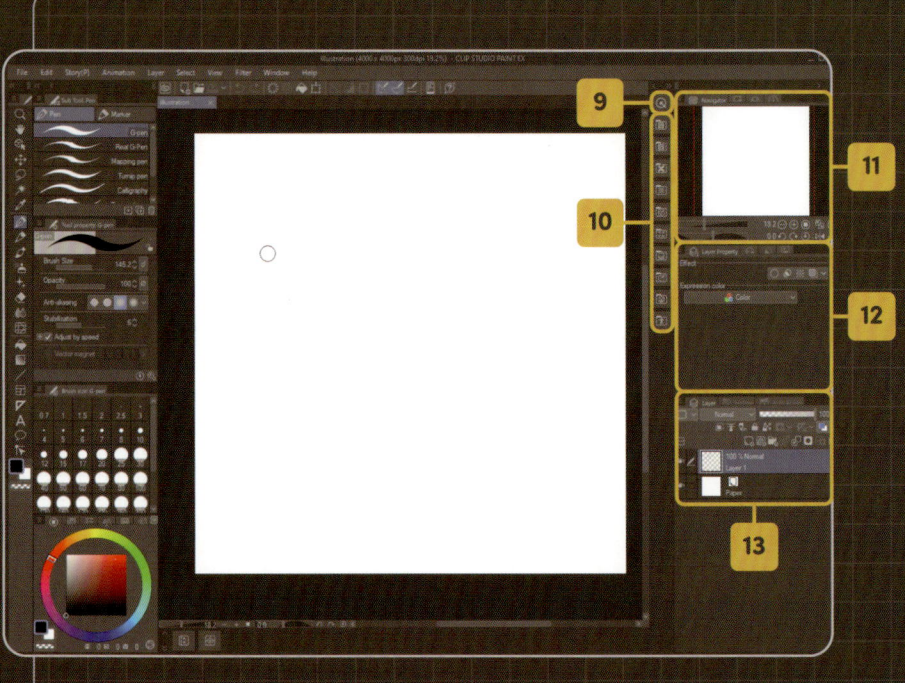

ARTIST'S TIP

At first it may seem like there are lots of icons and menus on the interface, but as you spend more time using the program, you will soon become familiar with them and their functions. It's not necessary to memorize every single one of the options and functions before you start. Just use what's necessary for the projects you want to create.

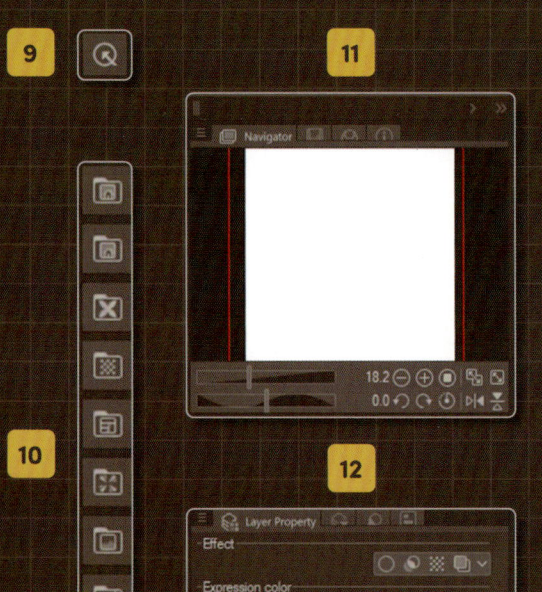

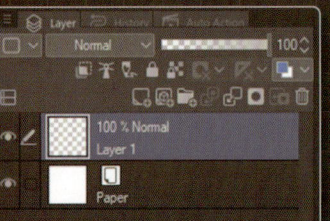

9. QUICK ACCESS PALETTE

This contains quick access functions and tools. It can be customized to contain your most frequently used tools.

10. MATERIAL PALETTE

In this folder group, materials and assets can be organized for use. It also stores materials and assets downloaded from Clip Studio Assets (found in Clip Studio Launcher).

11. NAVIGATOR PALETTE

This palette features a canvas thumbnail for reference. It also includes functions including zoom, horizontal and vertical canvas inversion, and canvas rotation, among others.

12. LAYER PROPERTY PALETTE

This contains properties of the selected layer.

13. LAYER PALETTE

In this palette, you can manage the layers of your project and other related functions.

HOW TO ADJUST THE USER INTERFACE?

Over time you will find that there are some palettes you use more often, and some you rarely seek out. Clip Studio Paint allows you to customize the user interface, moving or hiding them to suit your needs.

The group of side palettes each have three buttons at the top, which allow you to hide them or change their width.

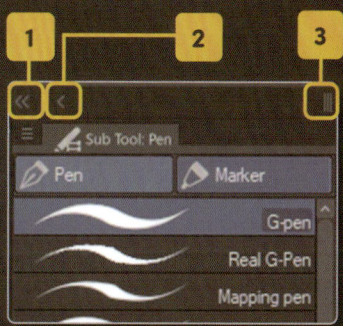

1. HIDE THE SIDE PALETTE

2. HIDE THE SIDE PALETTE, BUT STILL SHOW THE ICON

3. CLICK AND DRAG SIDEWAYS TO CHANGE THE WIDTH OF THE PALETTE

ᐯ Palettes can also be closed or added from the drop-down Window menu

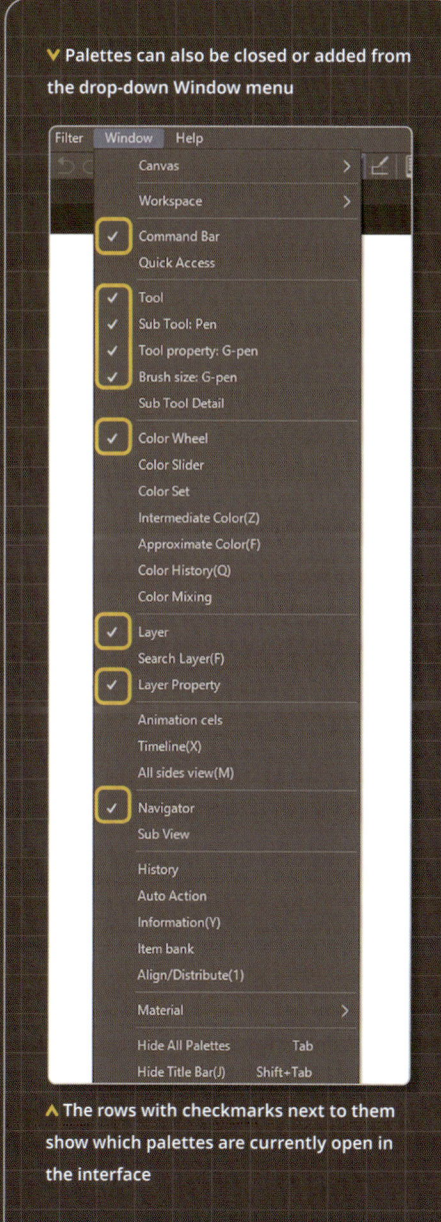

ᐱ The rows with checkmarks next to them show which palettes are currently open in the interface

ᐯ Each palette can be detached and placed anywhere on the interface; they can also be moved to a palette group or created as a side palette

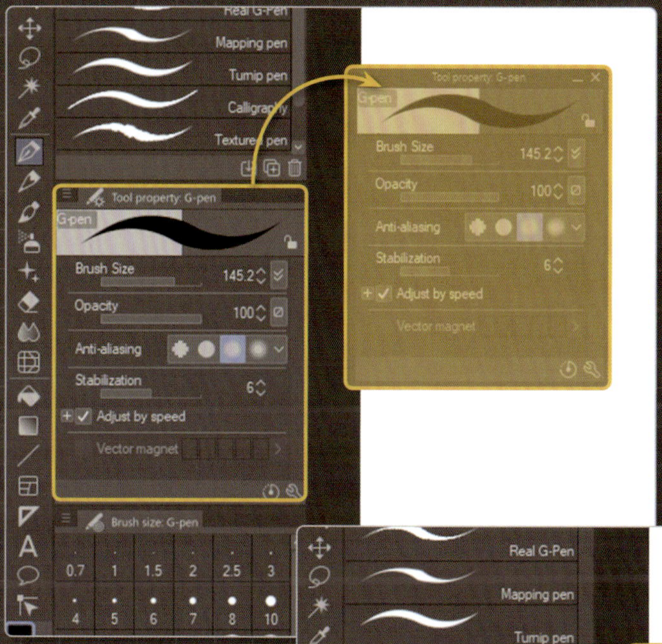

ᐱ Click and hold the button at the top of the palette, then drag it to the desired location

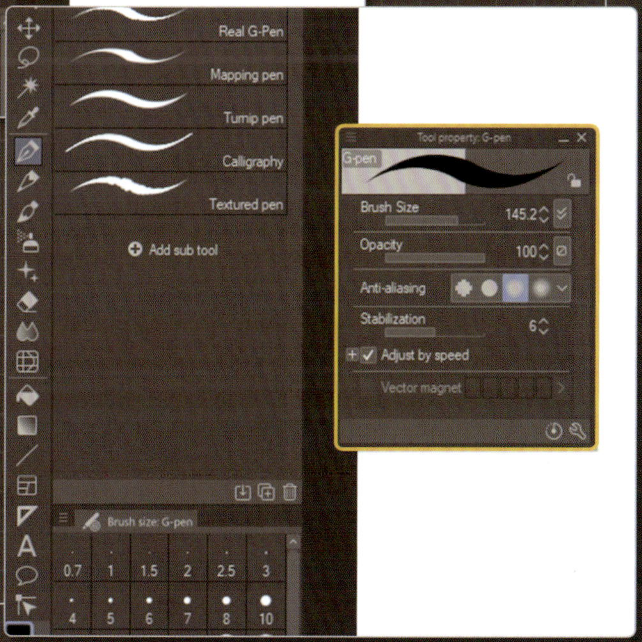

Create a new file

Clip Studio Paint has everything you need for creating many different kinds of projects, including illustration, concept art, comics, webtoons, manga, and animation. The category of artwork you plan to make will dictate the type of file you need to create.

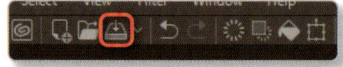

< To create a new file, select the New icon in the Command bar

LEARN HOW TO

- Create a new document.
- Understand the different types of documents you can create.
- Modify the canvas (resolution and size).
- Save and choose the save file type.

TYPES OF FILE

ILLUSTRATION

This file type is the most commonly used for projects like illustration or concept art. You can configure the canvas size, resolution, colour expression, and paper colour. Additionally, templates and preset size formats can also be used.

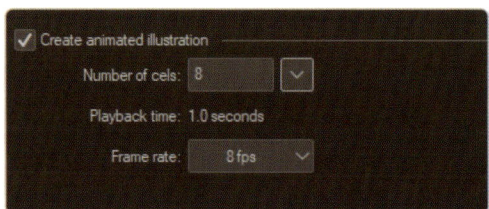

∧ The Create Animated Illustration option allows you to create short animations – this can be used to create animated stickers or GIFs

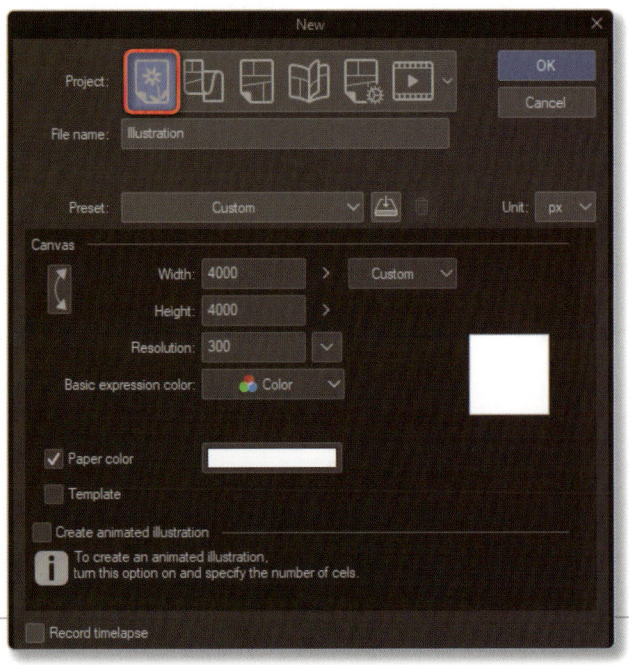

WEBTOON

This modern comic format is read vertically on electronic devices. As well as the basic settings and size, there are additional settings that allow you to select the file save path and add split pages.

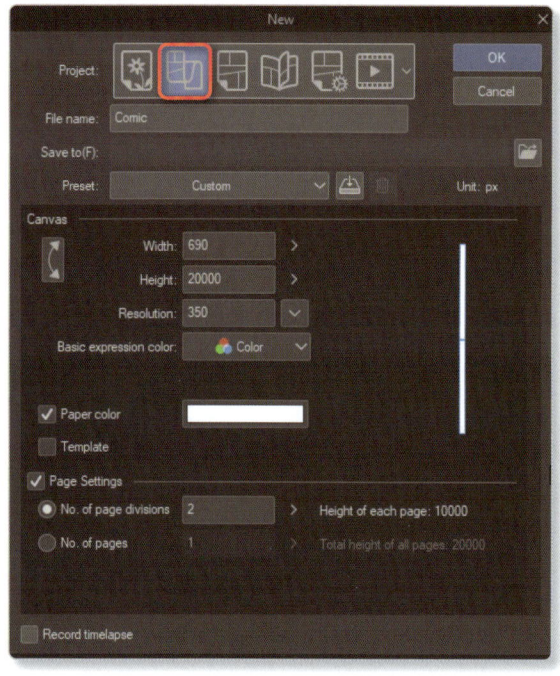

∧ The webtoon file type provides additional settings, such as dividing the page into frames

COMIC

This format is for classic comics or manga, whether single or multiple pages. (Settings related to multi-pages are only available in Clip Studio Paint EX.) It allows you to configure the file with the trim size, comic bleed width, number of pages, binding direction (left is used for American-style comics and right is used for manga), the option to add a cover, plus more.

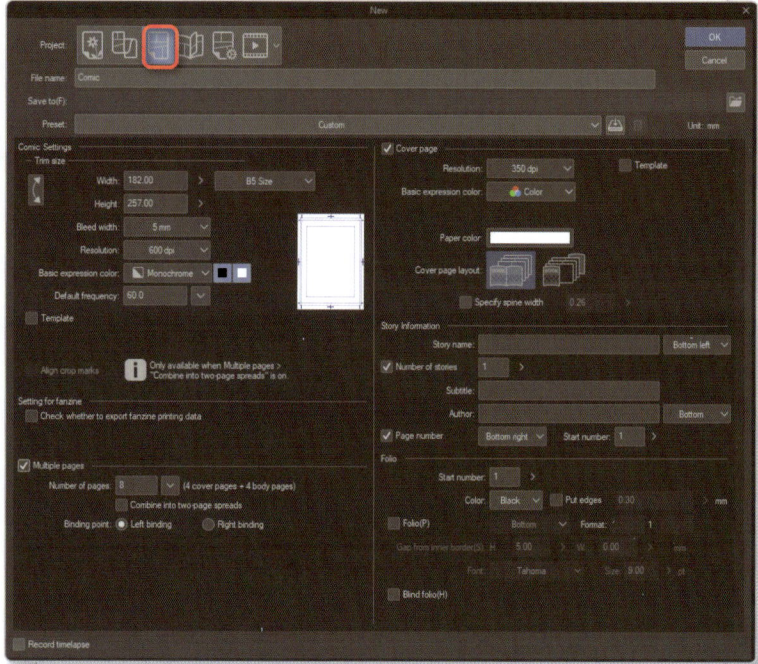

∧ The classic comic format comes with numerous options, including setting the trim size and bleed width

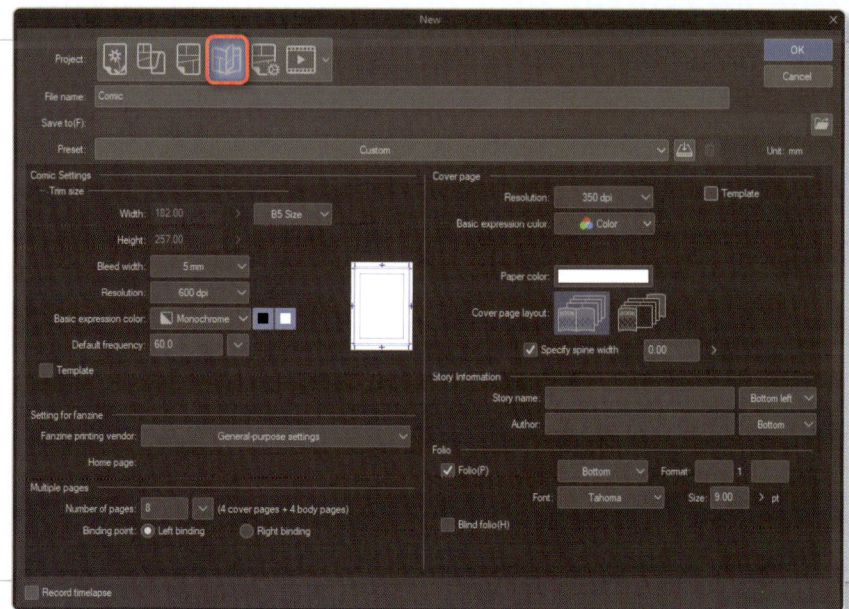

PRINTING A FANZINE

Unlike the comic format, this file type is for multi-page comics with simplified preset settings and size options. (The fanzine category and settings related to multi-page projects are only available in Clip Studio Paint EX.)

‹ This simplified version of the comic file format provides fewer options for customization

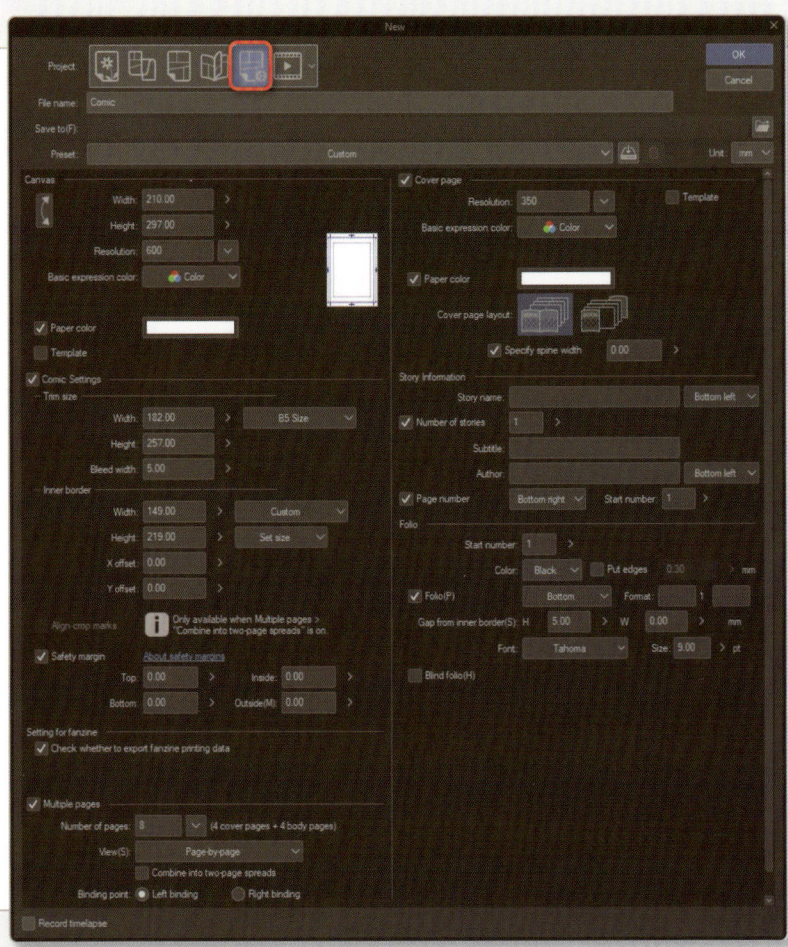

SHOW ALL COMIC SETTINGS

This format includes all of the configurations, size details, and printing options for a single or multi-page comic project.

> Choose this version of the comic format if you want maximum flexibility and choice when customizing the file

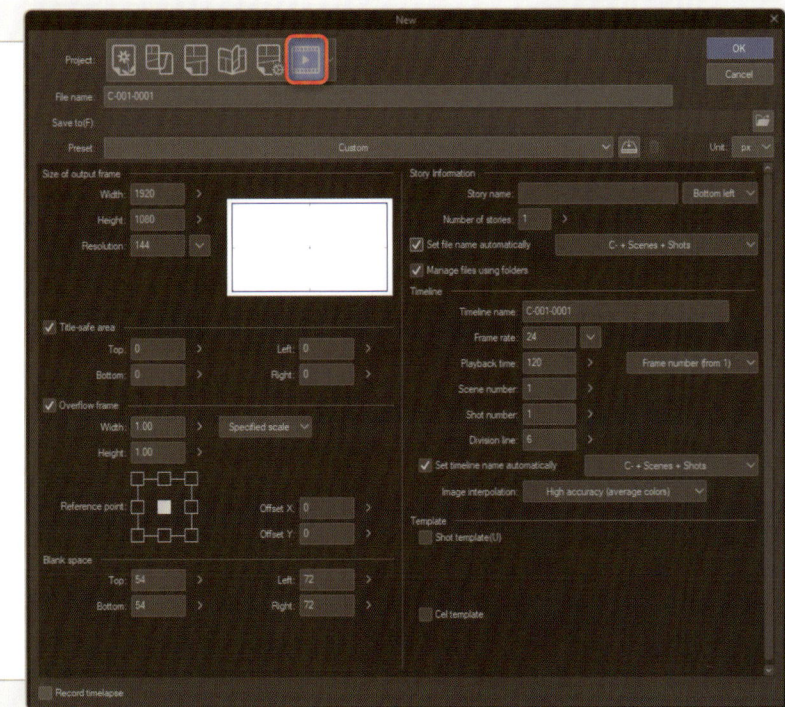

ANIMATION

When creating a new animation project, you can configure the blank space area, output frame, and other timeline details, such as the frame rate.

< The animation format contains various options, including frame rate and timeline preferences

RECORD TIMELAPSE

Clip Studio Paint's timelapse feature allows you to record your painting process and export it as a video. To use this function, enable the Record Timelapse option when creating a new file.

When you finish a project, you can export the video by selecting File > Timelapse > Export Timelapse. The Export Timelapse window allows you choose the duration, size or resolution, aspect ratio, and the option to add a watermark. The exported video file format is mp4. You can then share this video on your social platforms or online portfolio, or simply watch it back to track your progress and see how you can improve your workflow.

^ When creating a new file, make sure to turn on Record Timelapse at the bottom of the window

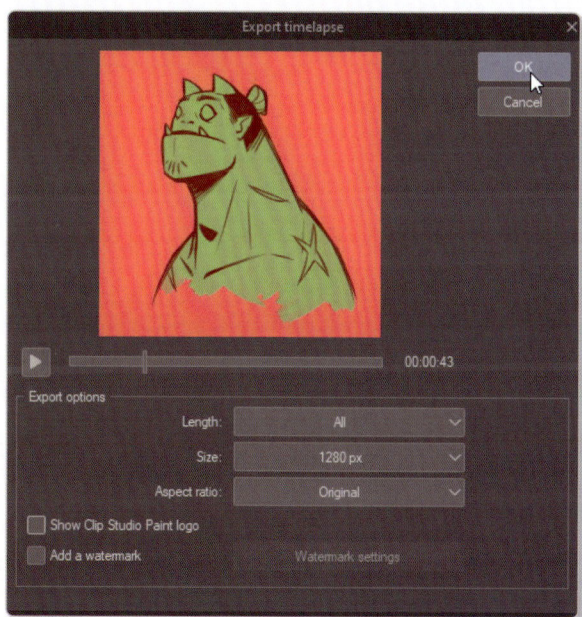

^ After choosing the video length, size, and ratio, you can export the timelapse recording to share with others

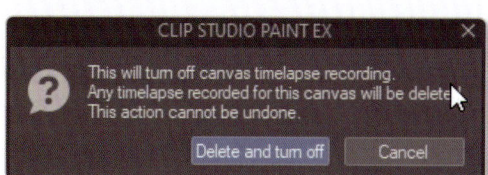

^ Turn off recording to reduce your project's file size

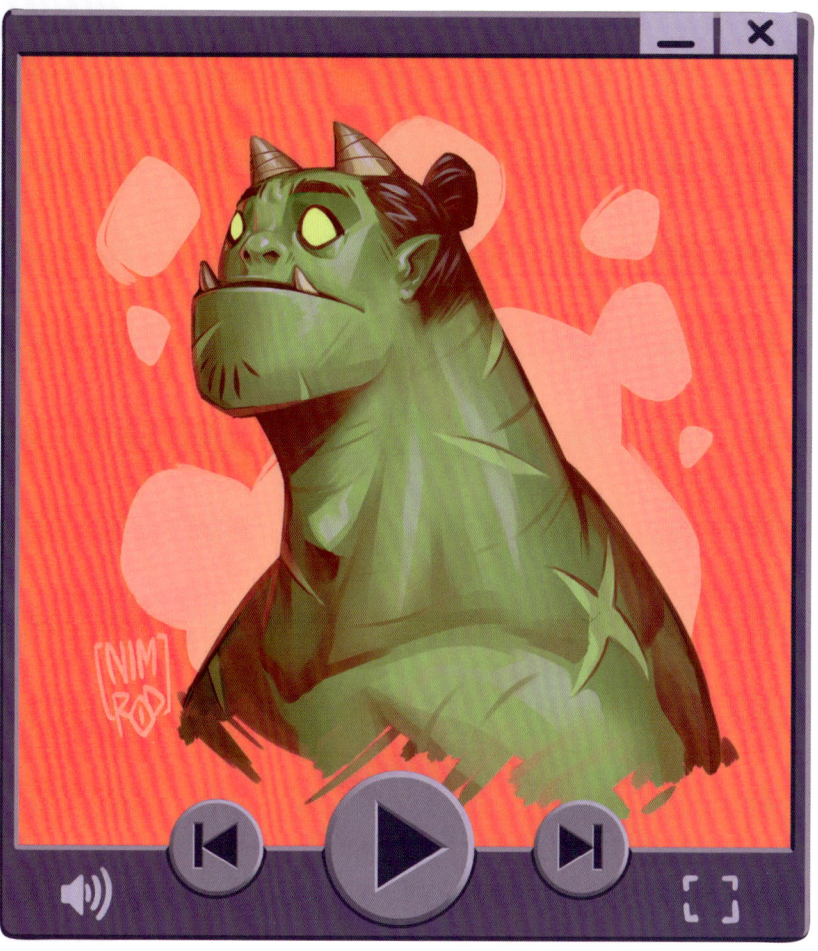

^ The timelapse recording tool allows you to watch the creation of an artwork from start to finish

Files that have the timelapse feature activated will increase in size the longer you work on a project. The longer the painting process, the larger the file will be. If you have already exported your video and want to reduce the size of the project file, you can delete the recording by deselecting the Record Timelapse option. When a warning message appears, choose the Delete and Turn Off option.

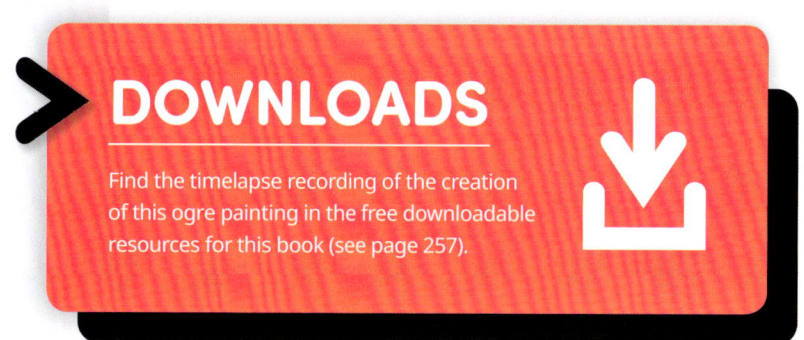

DOWNLOADS

Find the timelapse recording of the creation of this ogre painting in the free downloadable resources for this book (see page 257).

MODIFY CANVAS SETTINGS

After creating the canvas, you are still able to modify the size, resolution, and scale settings whenever you like.

CHANGE IMAGE RESOLUTION

Select Edit > Change Image Resolution to bring up resolution options. Modify the Resolution and Unit options as desired. The 300 px value in this image is a resolution of 300 dpi (dots per inch).

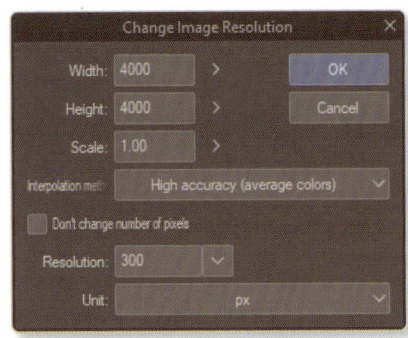

∧ Select Change Image Resolution to adjust the size and resolution of your canvas

CHANGE CANVAS SIZE

To change the size of the canvas without altering the proportions of the project's content, select Edit > Change Canvas Size. This will allow you to crop part of the canvas or expand it.

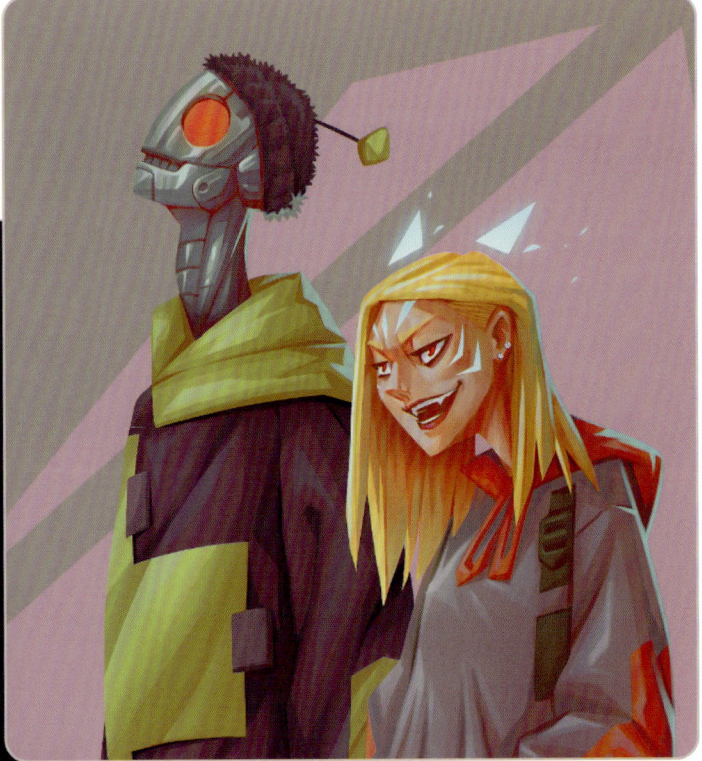

∧ The resized canvas

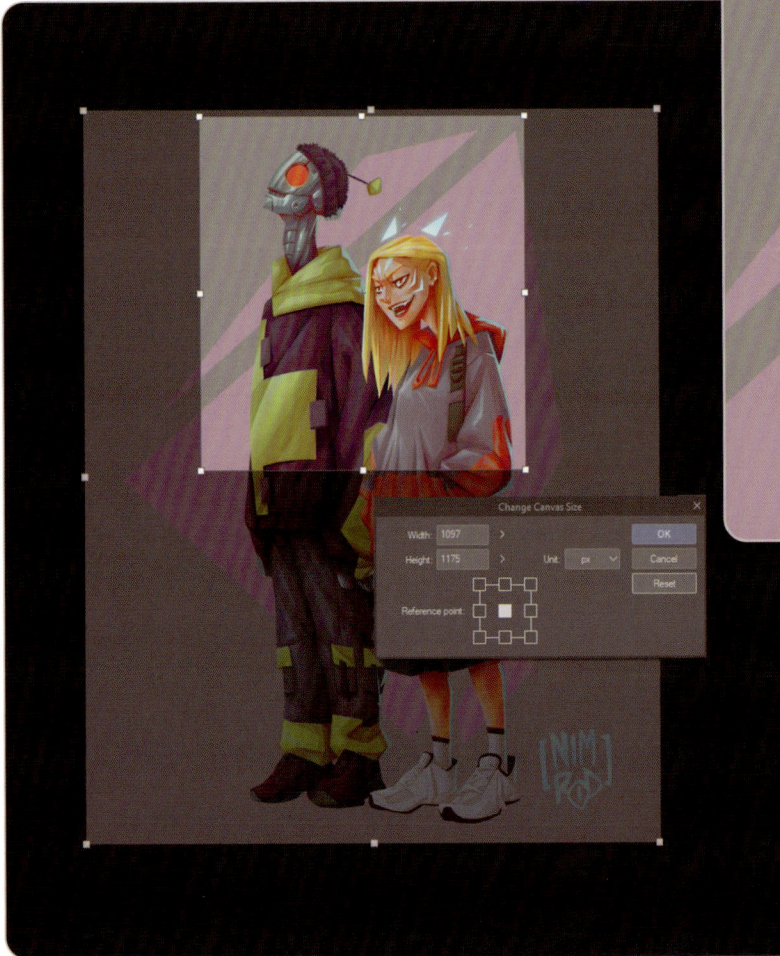

There are two ways to change the canvas size. The first is by dragging one of the handles that will appear at the edges of the canvas. If you hold down the Shift key while dragging one of the points, this will maintain the original proportion of the image. The second way to change the canvas size is to adjust the numbers in the width and height boxes.

< Change the canvas size, while maintaining the original proportions, by dragging a handle at the same time as holding down Shift

SAVE A FILE

Select the Save button in the Command bar to save the file you're working on. Another way to do this is by selecting File > Save. The keyboard shortcut to save the file is Ctrl+S. Alternatively, select File > Save As to save your file in a different format or another folder location.

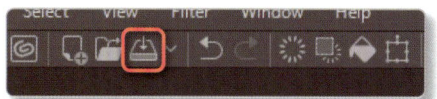

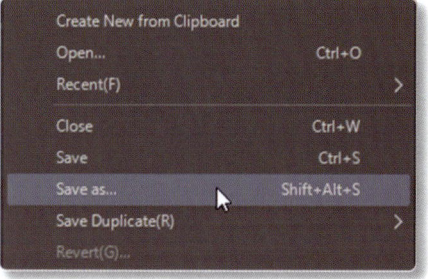

∧ There are numerous ways to save a file – use the one that works for you

FILE SUPPORT

Clip Studio Paint provides various file-saving formats, ranging from the native format (Clip Studio Format) to the well-known Photoshop format (PSD).

CLIP STUDIO FORMAT (*. CLIP)
This is the exclusive format of Clip Studio Paint.

JPEG & PNG
These are compressed image formats that are commonly used for sharing artwork on digital platforms. The PNG format allows transparency if you want to export an image with a transparent background.

BMP, TIFF & TARGA
These are uncompressed image formats. The export quality in these formats is excellent if you want to retain detail in your images. The disadvantages are they are not widely used, not all programs can open them, and the file size is much larger.

PSD & PSB
Photoshop Document (PSD) and Photoshop Big (PSB) are editable formats in Photoshop and other programs that support this format. When you save in this format, you retain layer information for further editing.

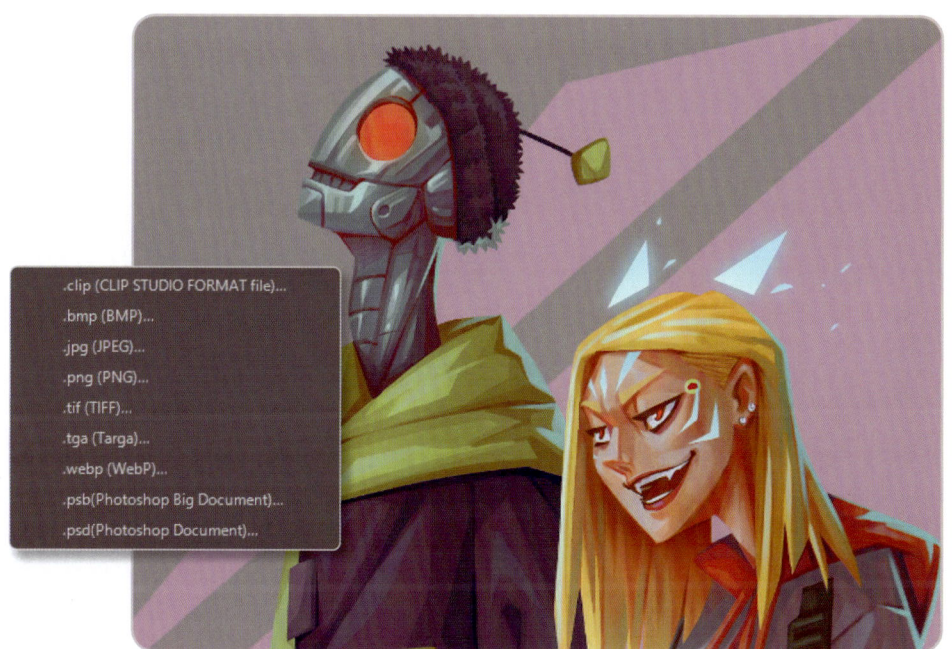

UNIQUE FEATURES

If you want to save your file from the native format (*. clip) to another format, some unique features of the software will be affected in the final format. For example, if you have the Record Timelapse option enabled, it will disappear when you save to a format other than Clip Studio Paint.

Navigation tools & basic actions

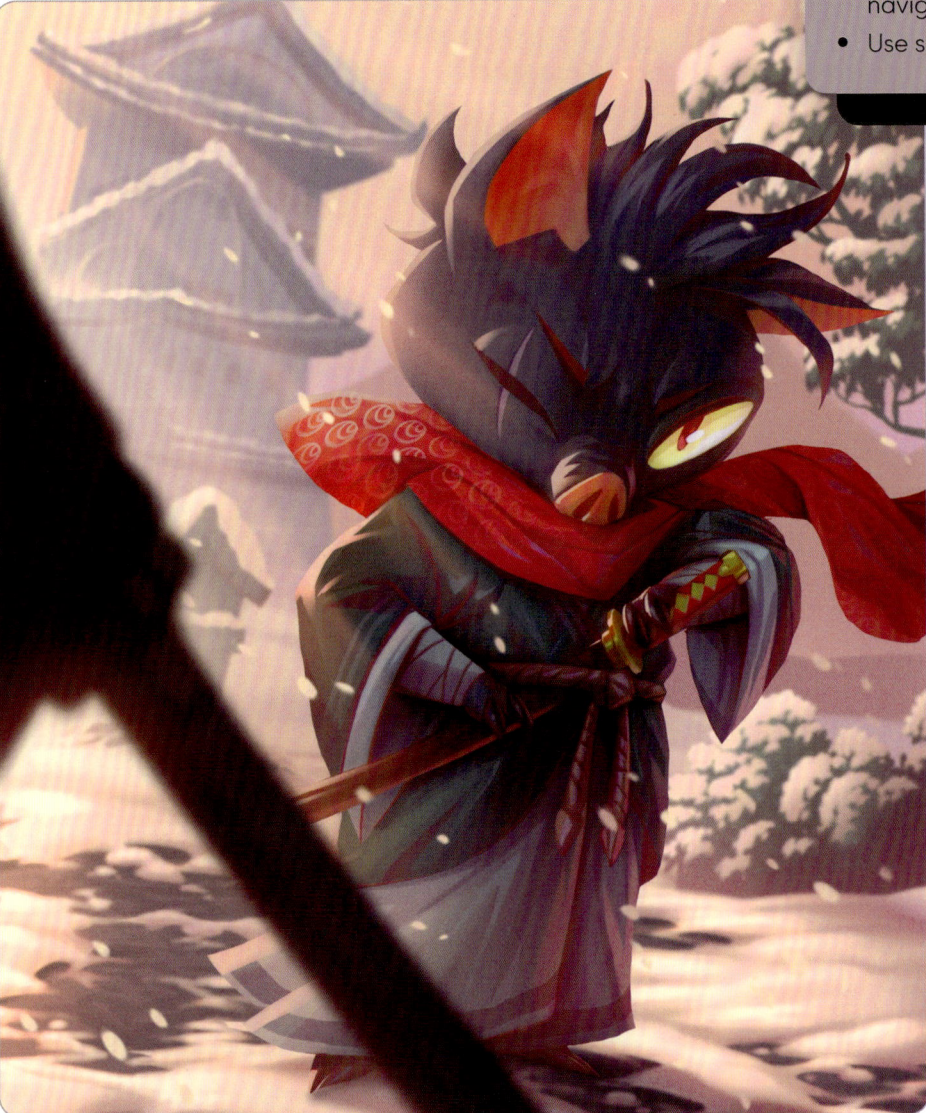

Before you begin creating artwork in Clip Studio Paint, there are several basic actions you need to know in order to confidently navigate your way around the interface. Taking the time to familiarize yourself with them now will make the painting process quicker and easier. These actions will be performed so often, they will soon become second nature to you. Most of these actions and tools can be found on the Navigator palette on the right-hand side of the interface.

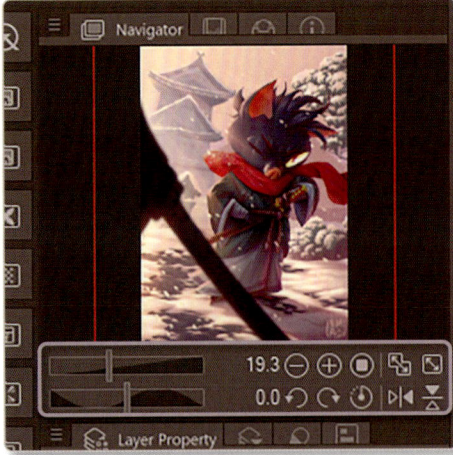

⌃ **The Navigator palette also provides an overview of the whole canvas**

ZOOM IN, ZOOM OUT

The Zoom tool on the Navigator palette allows you to zoom in or out of the canvas, depending on whether you wish to view an area of a painting close up, or view the canvas as a whole.

➢ **The red frame represents the canvas limit when zooming in**

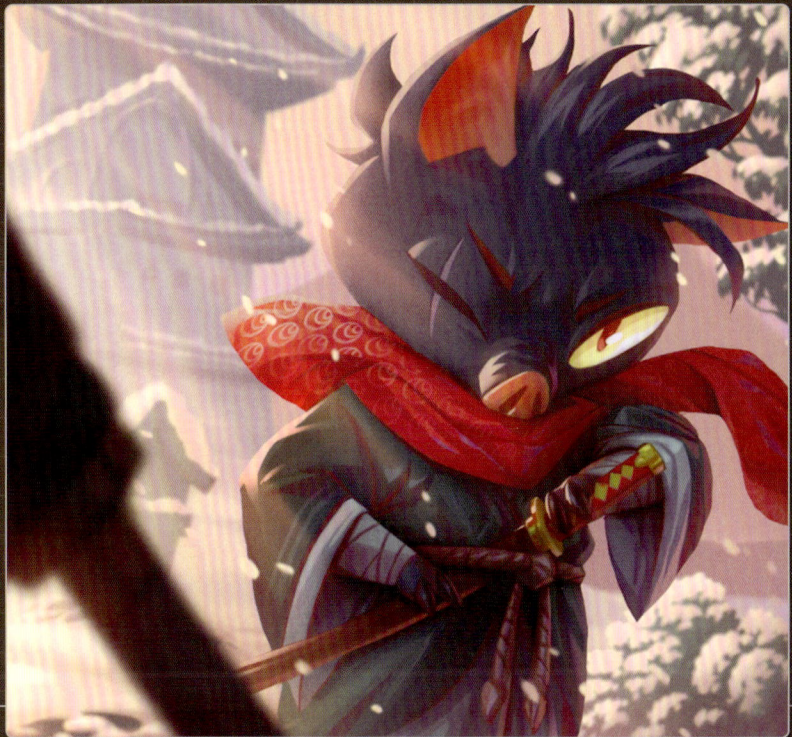

1. ZOOM BAR
2. ZOOM OUT
3. ZOOM IN
4. RESET ZOOM

Another way to zoom in or out is by using the toolbar on the left-hand side of the interface. Simply hold down the left mouse button and drag the mouse over the area you want to zoom in or out on.

SHORTCUTS

- To select the Zoom tool, press the Z key on the keyboard.

- To zoom in, use **Ctrl + plus** key (+). To zoom out, use **Ctrl + minus** key (-).

- To zoom without selecting the Zoom tool, simply roll the mouse wheel forwards to zoom in or backwards to zoom out.

- To temporarily use the Zoom tool, hold **Ctrl + space bar**, then drag the left mouse button over the area you want to zoom in or out on. The Zoom tool will deactivate when you release the keyboard keys.

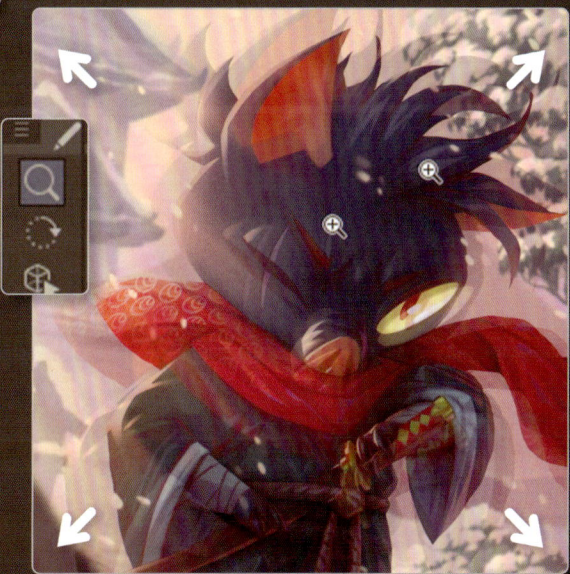

⌃ Drag the mouse over the area of the artwork you wish to zoom in or out on

21

ROTATE CANVAS

The Rotate tool is useful for directing brushstrokes to an area that's difficult to paint when the canvas is in its normal position. There are various methods you can use to rotate the canvas from the Navigator palette.

> The Navigator palette offers various methods to rotate the canvas

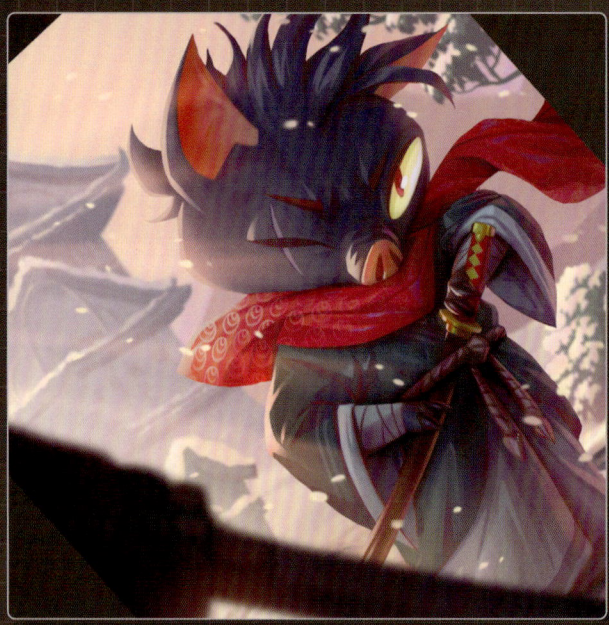

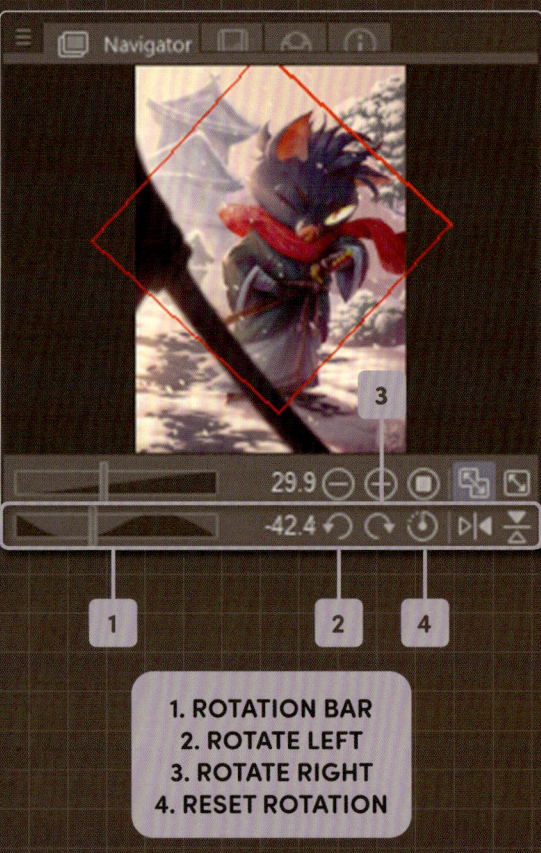

1. ROTATION BAR
2. ROTATE LEFT
3. ROTATE RIGHT
4. RESET ROTATION

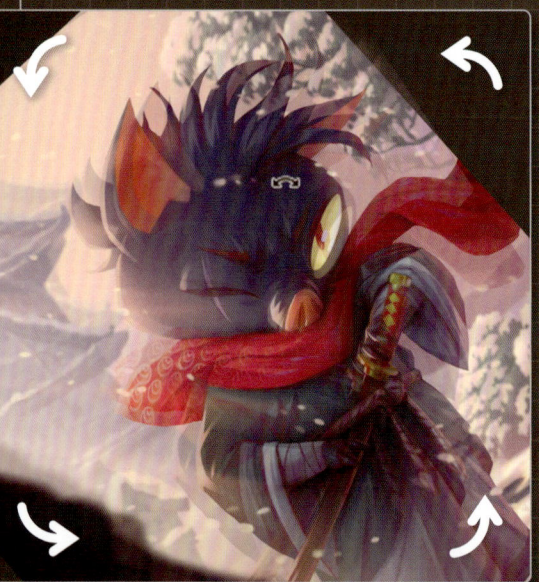

< Rotate the canvas to your desired angle by holding the left mouse button and dragging

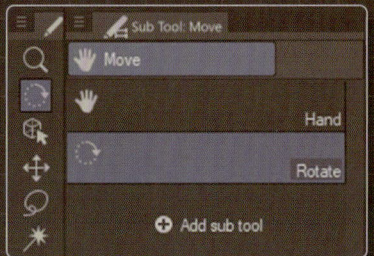

Another way to rotate the canvas is by selecting the Rotation tool on the left-hand side of the interface. Holding down the left mouse button while dragging it will enable you to rotate the canvas to the desired angle.

SHORTCUTS

- To select the Rotate tool, press the R key on the keyboard.

- Pressing the Left Shift key while rotating will rotate the canvas in fixed increments.

CANVAS SCROLLING

Use the Navigator palette's thumbnail view to move around the canvas. The cursor will switch to a hand, enabling you to click and drag to move the canvas to wherever you want.

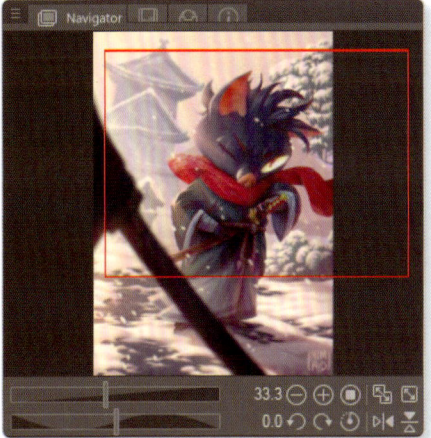

∧ Navigate around the canvas using the thumbnail view, found in the Navigator palette

Another way to scroll through the canvas is by selecting the Hand tool located on the left-hand side of the interface. As before, simply click and drag to move around the canvas.

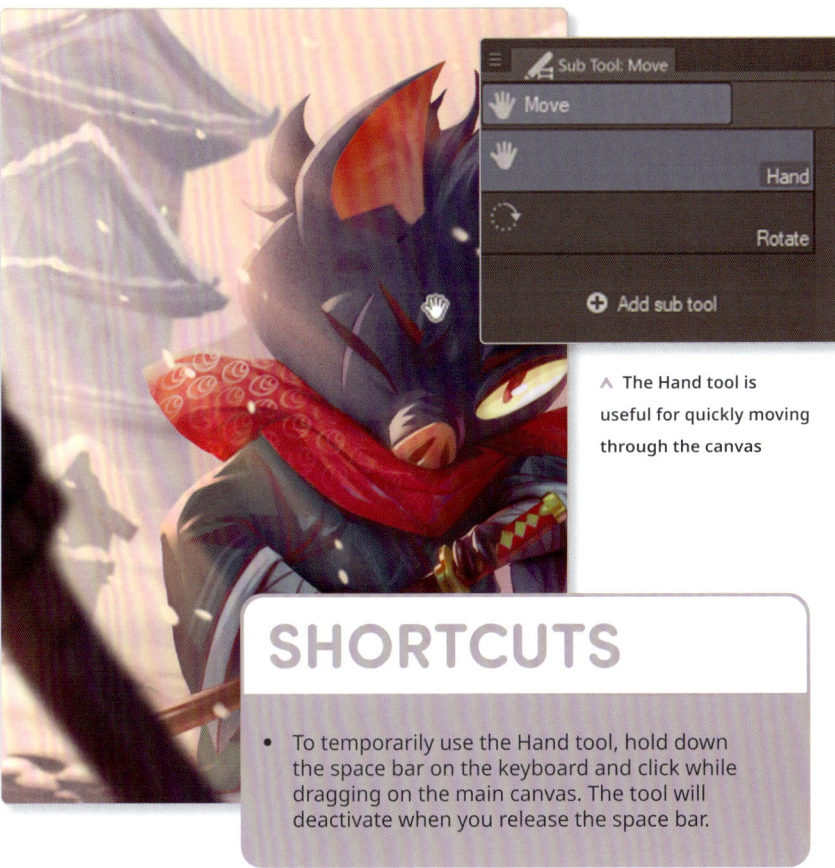

∧ The Hand tool is useful for quickly moving through the canvas

SHORTCUTS

- To temporarily use the Hand tool, hold down the space bar on the keyboard and click while dragging on the main canvas. The tool will deactivate when you release the space bar.

FLIP CANVAS

Flipping the canvas is a technique used to check progress midway through a painting. Viewing the artwork flipped helps you to spot errors (such as incorrect proportions or perspective) that may go unnoticed when the canvas is not flipped. This function can be used horizontally or vertically.

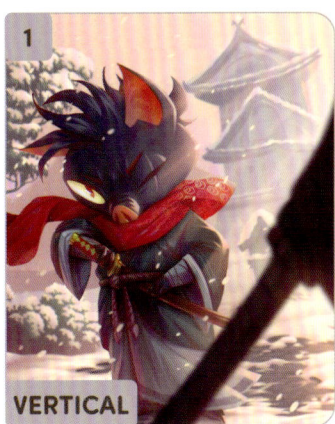

VERTICAL

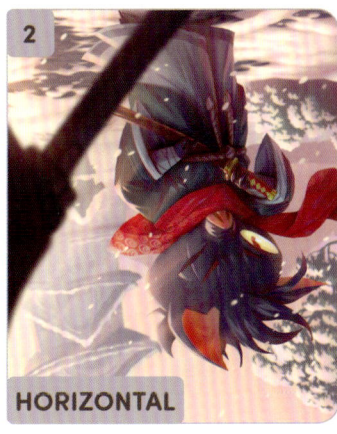

HORIZONTAL

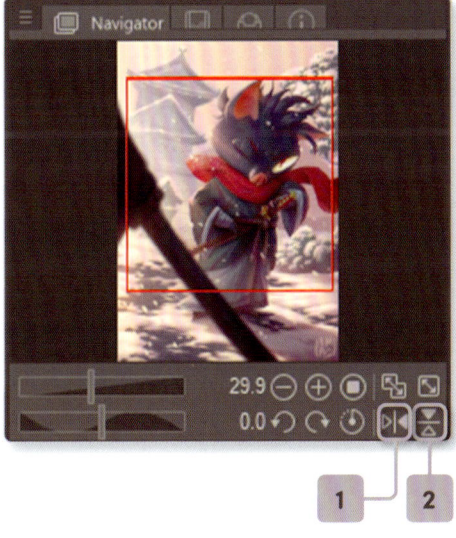

< Flipping the canvas vertically and horizontally can help you to spot mistakes more easily

Tools, sub tools & tool properties

Clip Studio Paint has all the tools you need for creating digital artwork. Some of the tools were introduced in the previous chapter, and others will be covered in the following chapters. All tools can be found in the toolbar on the left-hand side of the interface and are grouped into four categories.

GENERAL TOOLS

ZOOM
MOVE
OPERATION
MOVE LAYER
SELECTION AREA
AUTO SELECT
EYEDROPPER

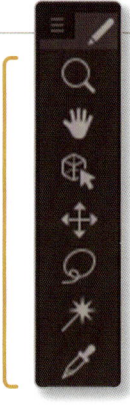

FIGURE TOOLS/ DRAWING GUIDES

FILL
GRADIENT
FIGURE
FRAME BORDER
RULER
TEXT
BALLOON
CORRECT LINE

DRAWING TOOLS

PEN
PENCIL
BRUSH
AIRBRUSH
DECORATION
ERASER
BLEND
LIQUIFY

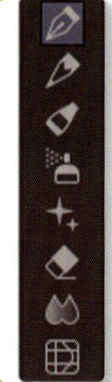

COLOUR ICONS

MAIN COLOUR
SUB COLOUR
TRANSPARENCY

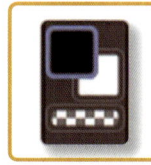

∧ Dividing them into categories makes it easier to learn each tool

SUB TOOLS

A sub tool can be selected from the Sub Tool palette, located to the right of the Tool palette. The list of sub tools will be associated with the selected tool. For example, if you select the Brush tool, the default sub tool list might include watercolour brushes, thick paint, and Indian ink.

 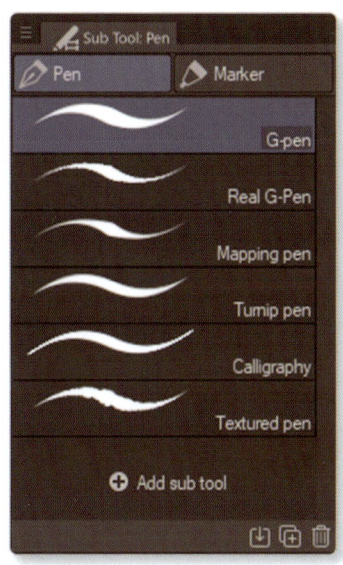

> **The Sub Tool palette that appears is determined by the tool palette you select**

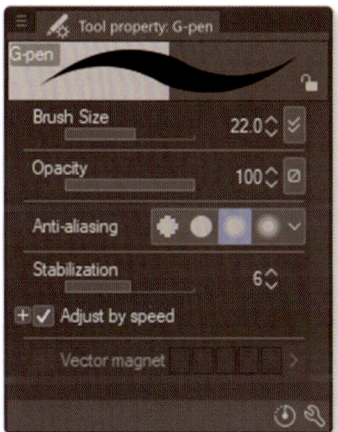

TOOL PROPERTY

Each sub tool can be configured from the Tool Property palette located below the Sub Tool palette. The settings change according to the type of tool and sub tool you select. These configurations are the most commonly used for each tool and, if modified, they are saved for the next time you use the software.

< **Each sub tool has detailed and easily accessible configurations**

MEMORIZE SHORTCUTS

It is handy to learn the keyboard shortcuts for the tools and functions you use most often. Over time this will make your workflow faster and more intuitive.

CUSTOMIZE SUB TOOL SETTINGS

The Sub Tool Detail palette allows you to customize all settings. Select the wrench icon at the bottom right of the Tool Property palette to access it.

To make a configuration visible in the Tool Property palette, click on the box to the left of each one. When the eye icon appears, it means the setting is visible. Take some time to experiment with the various settings each sub tool offers, toggling on the options you might need to access regularly.

If you want to revert to the original sub-tool configurations, simply click on Reset Selected Sub Tool to Default Settings located at the bottom right.

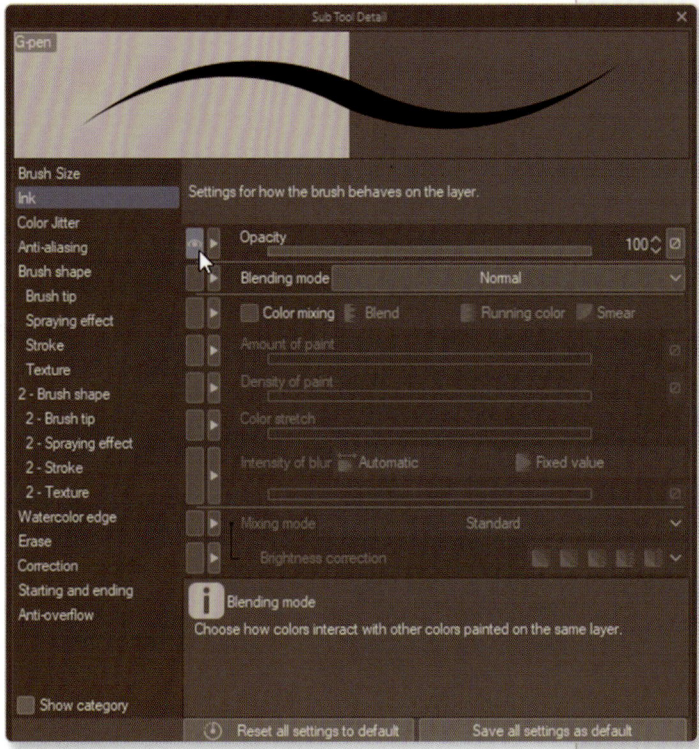

∧ Customize the various settings in the Sub Tool Detail palette

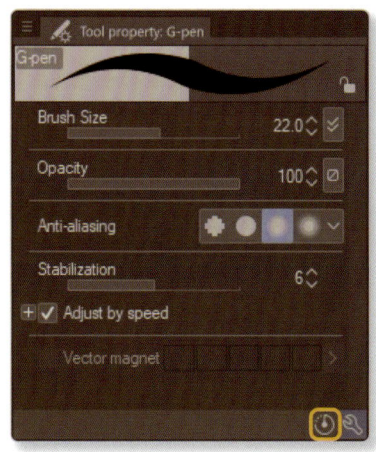

‹ Select Reset Selected Sub Tool to Default Settings to undo any changes made to the sub-tool configurations

EXPLORE & EXPERIMENT

Experimenting with other tools, especially brushes, in addition to Clip Studio Paint's default tools, can help you to discover new ways to develop your art and even accelerate your workflow.

IMPORT SUB TOOL

If you find that the default sub tools don't allow you to fully explore the style you're trying to create in your artwork, or if you want to continue experimenting with even more options, Clip Studio Paint allows you to import sub tools. Click on the Add Sub Tool button, located at the bottom of the sub tool list. A window will appear where you can search for the sub tools you have downloaded from Clip Studio Assets. (Clip Studio Assets will be covered in Materials on page 109).

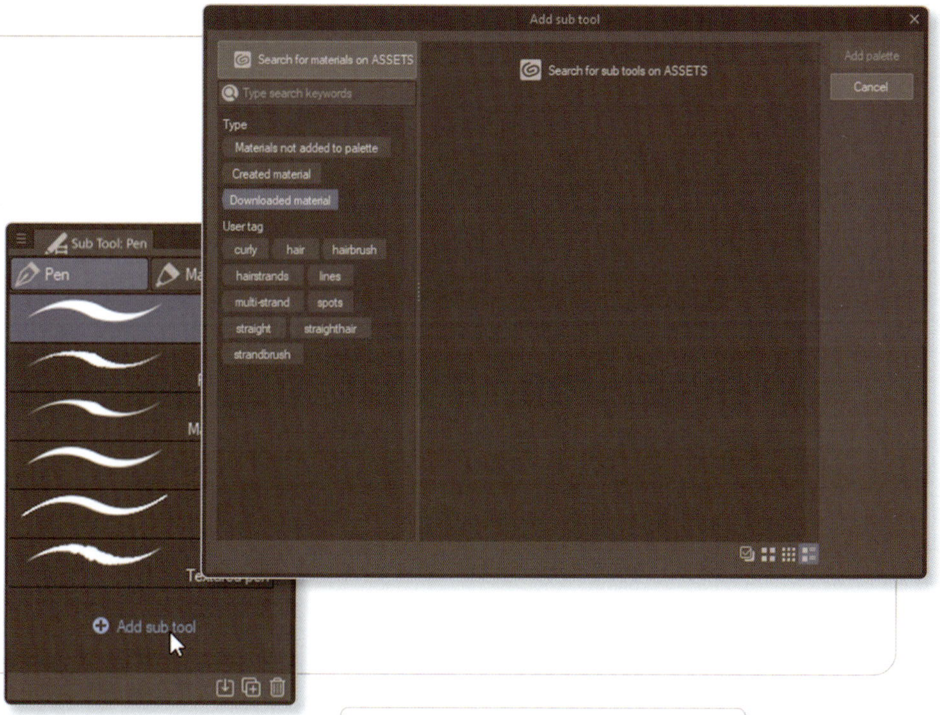

> **Import sub tools downloaded from Clip Studio Assets**

The Sub Tool Palette menu provides another way to add sub tools. Click on the Import Sub Tool option, then locate the sub-tool file you want to import.

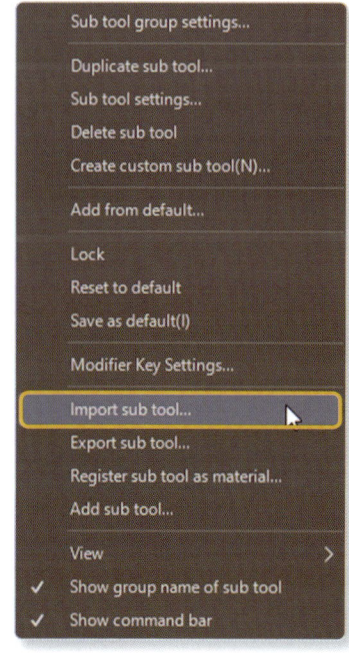

∧ **Import sub tools using the Sub Tool palette menu**

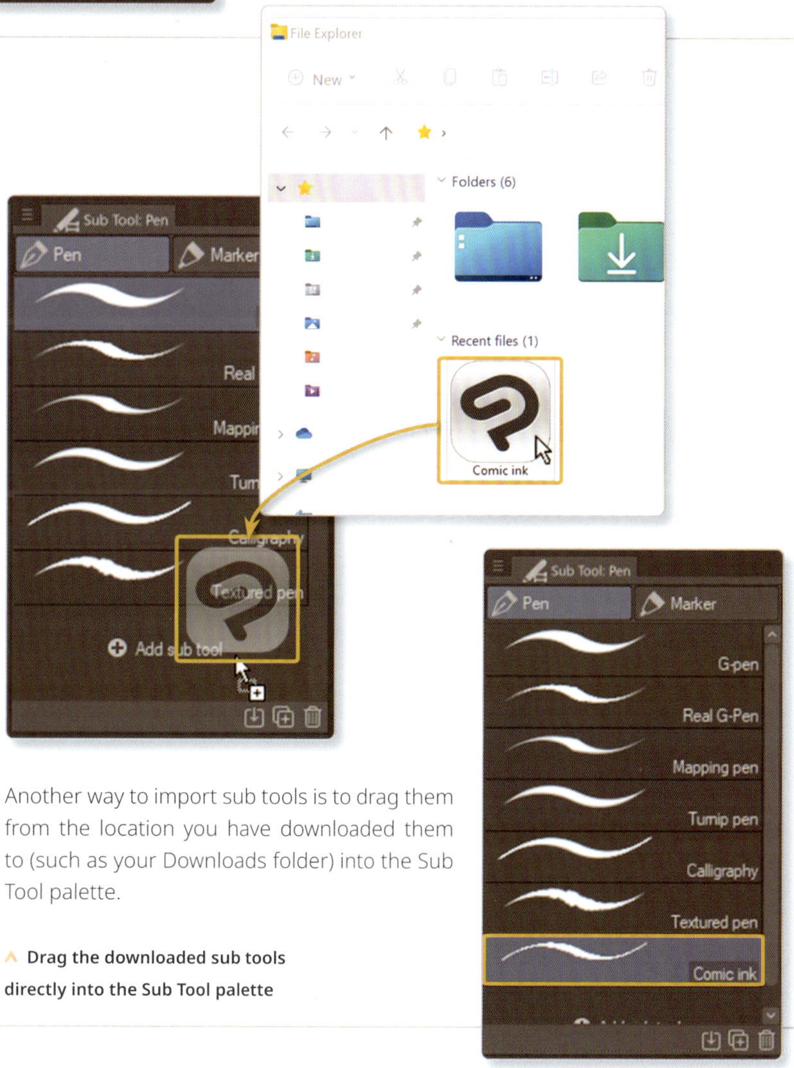

Another way to import sub tools is to drag them from the location you have downloaded them to (such as your Downloads folder) into the Sub Tool palette.

∧ **Drag the downloaded sub tools directly into the Sub Tool palette**

Drawing tools

> ## LEARN HOW TO

- Identify the drawing tools and how they work on the canvas.
- Use the general settings in the Tool Property palette.

Clip Studio Paint offers a wide variety of default brushes, pencils, and pens, allowing artists to create a range of different styles and techniques. This chapter will cover the various types of drawing tools and their configurations.

Drawing tools are the second group on the Tool palette, located on the left side of the interface. They are a group of tools used for drawing and painting, including brushes, pens, erasers, special-effect tools, and more.

PALETTES

The palettes associated with the drawing tools are the Sub Tool palette, Tool Property palette, and Brush Size palette, which are all located to the right of the Tool palette.

SUB TOOL PALETTE
This is a palette of brush sub-tool groups associated with the selected tool.

TOOL PROPERTY PALETTE
With this palette, you can configure the properties of the selected brush, such as size, opacity, density, and more.

BRUSH SIZE PALETTE
This palette provides quick shortcuts to preset brush sizes.

> **Palettes associated with the Tool palette are located to the right of it for easy access**

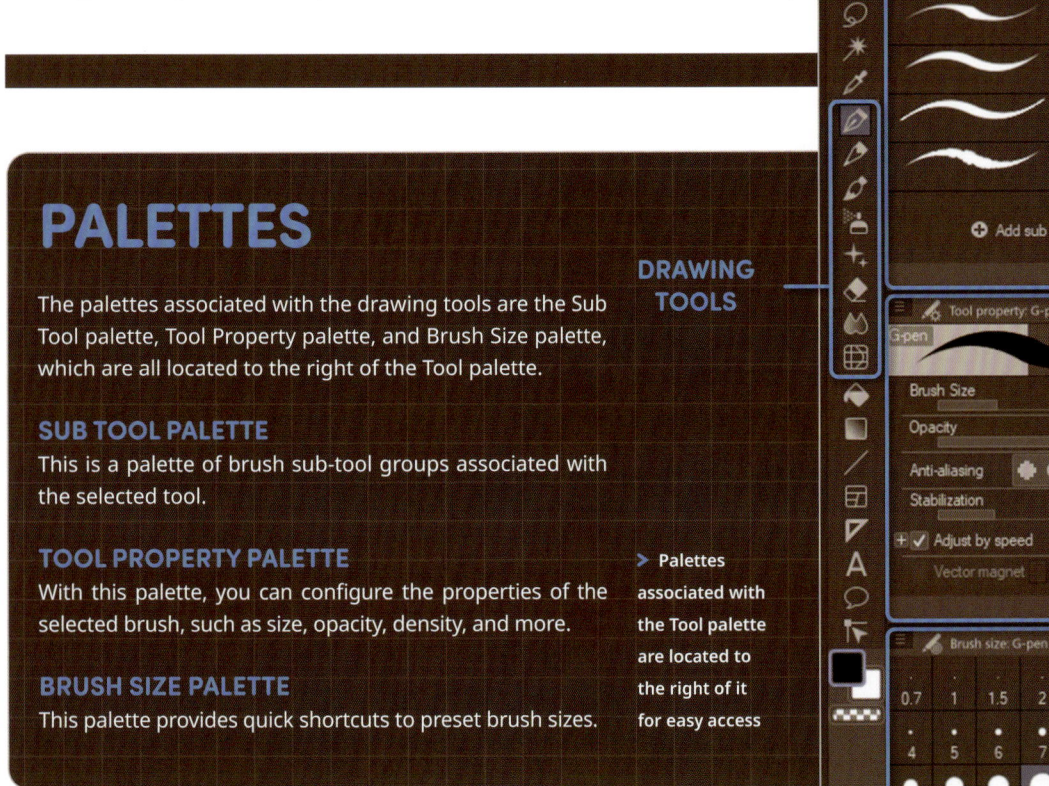
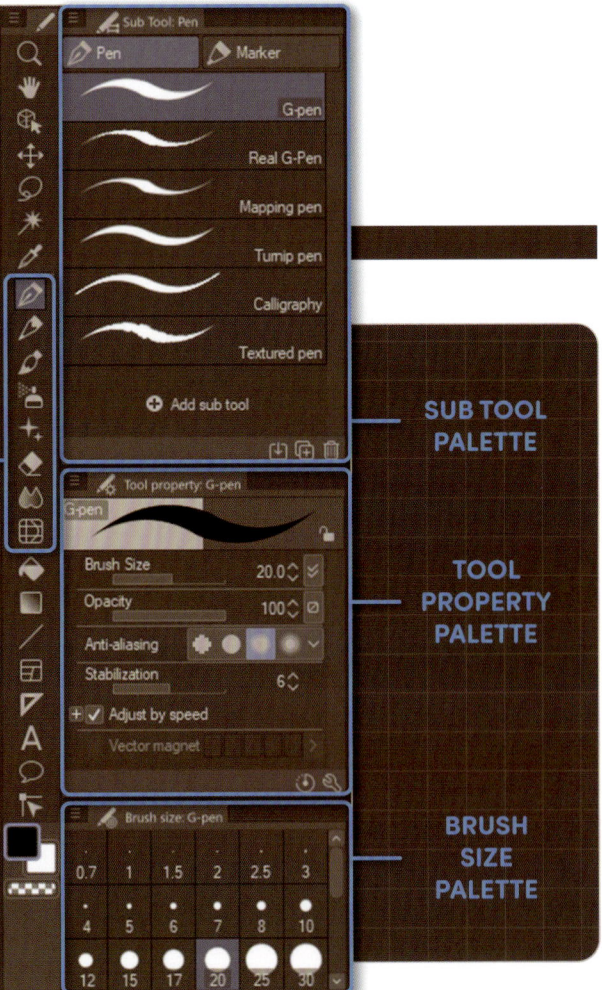

DRAWING TOOLS

SUB TOOL PALETTE

TOOL PROPERTY PALETTE

BRUSH SIZE PALETTE

DRAWING TOOLS

PEN

This group of tools is divided into two sub-groups. These simulate traditional pens with varying thicknesses (Pen) and markers with a uniform thickness (Marker).

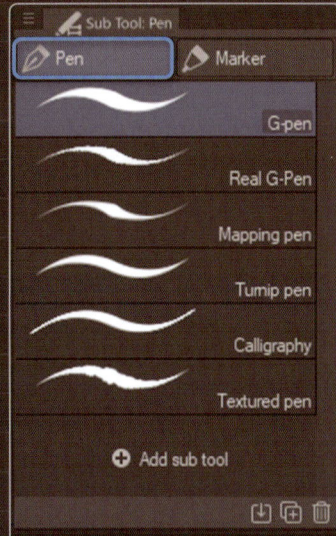

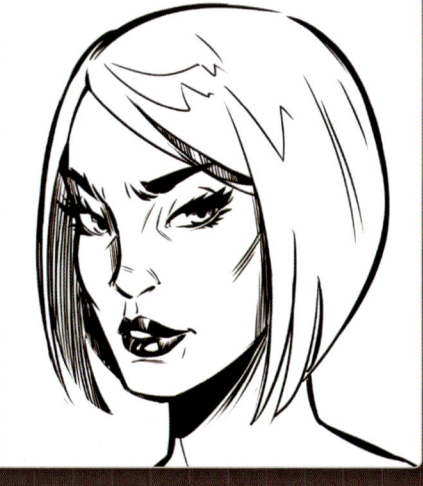

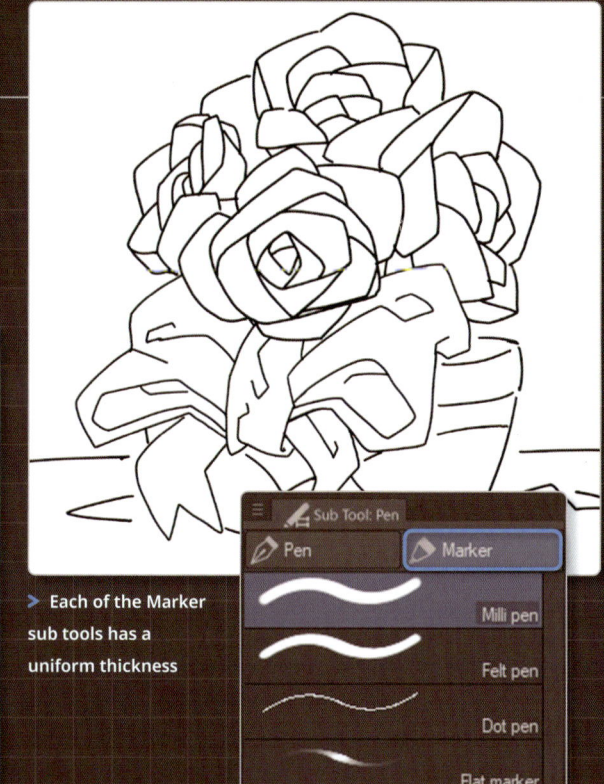

> Each of the Marker sub tools has a uniform thickness

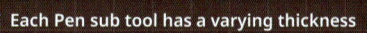
< Each Pen sub tool has a varying thickness

PENCIL

Under the Pencil tool you will find both Pencil and Pastel sub tools. These are ideal for conveying the essence of a traditional drawing or sketch.

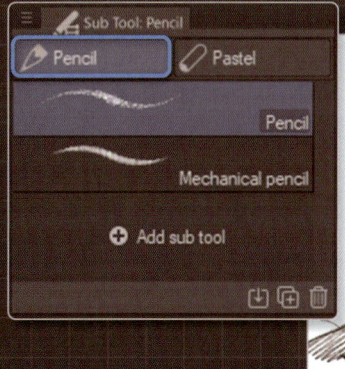

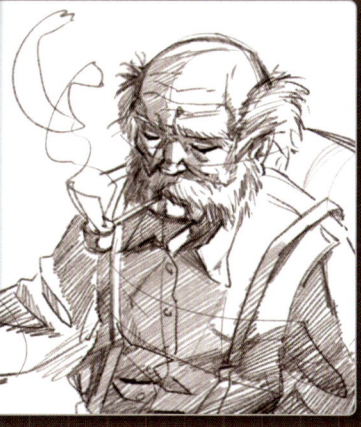

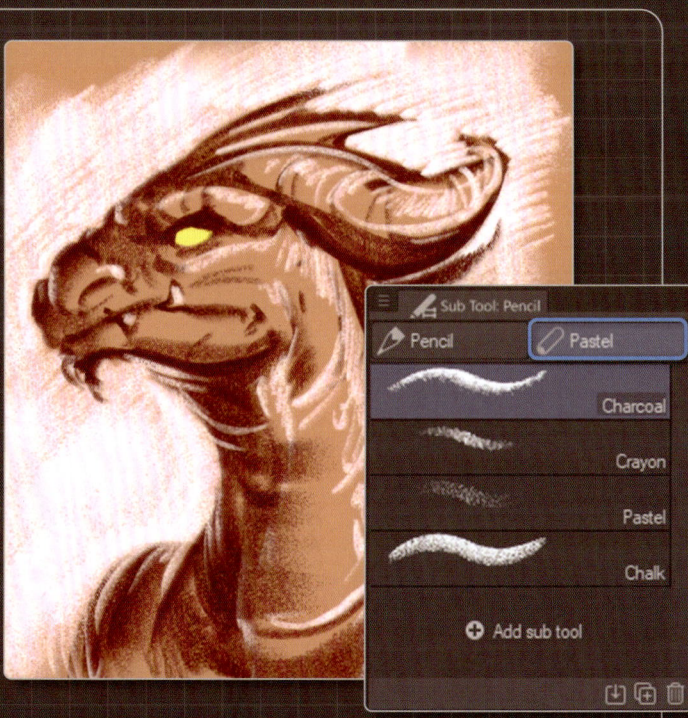

∧ The sketchy nature of the Pencil tool can replicate the look of traditional media

BRUSH

The three sub-groups of the Brush tool are Watercolour, Thick Paint, and India Ink. There are further brush options within each of these, allowing you to create a multitude of different types of brushstroke.

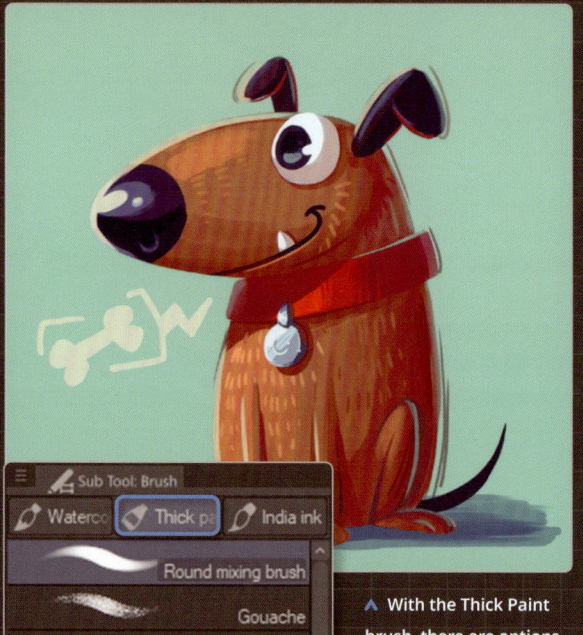

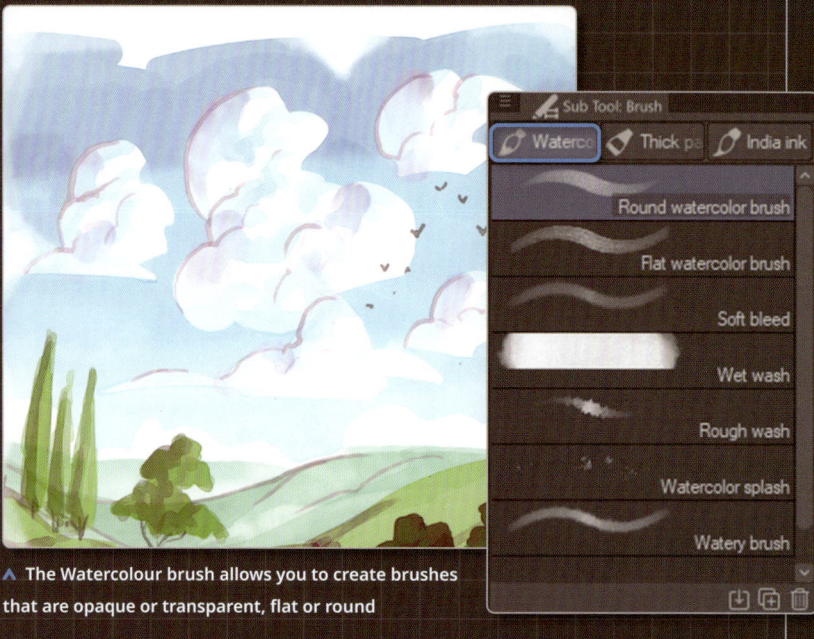

∧ The Watercolour brush allows you to create brushes that are opaque or transparent, flat or round

∧ With the Thick Paint brush, there are options for brushstrokes in the style of oil and gouache

> The India Ink brush enables you to create brushstrokes that are dry or watery, smooth or rough

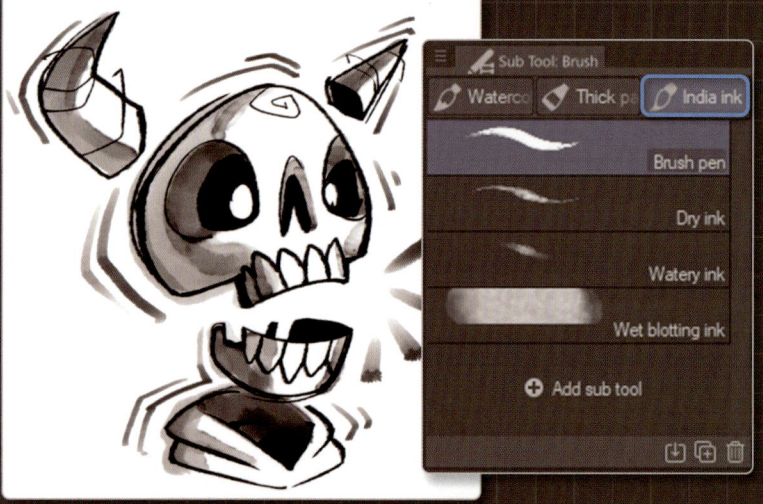

AIRBRUSH

This group of brush tools can be used as an airbrush or for blending.

> The Airbrush tool has various options, of which Soft is a popular choice

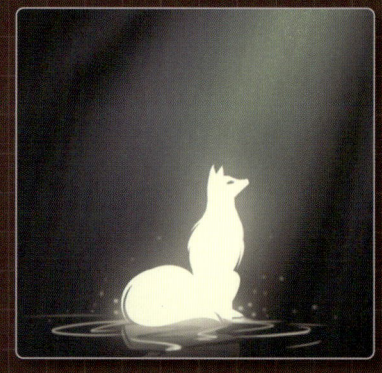

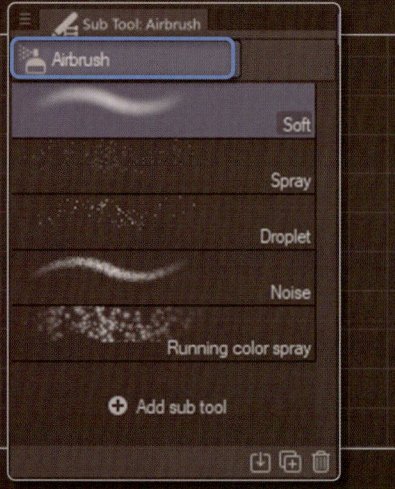

DECORATION

The Decoration group has five categories of brushes: Effect, Background, Clothing, Hatching, and Ruled Lines. These allow you to draw a wide variety of different types of patterns. They are also useful because they can simplify and speed up effects that are more difficult and time-consuming to do manually.

> The Decoration brushes allow you to quickly create various patterns, shapes, and effects

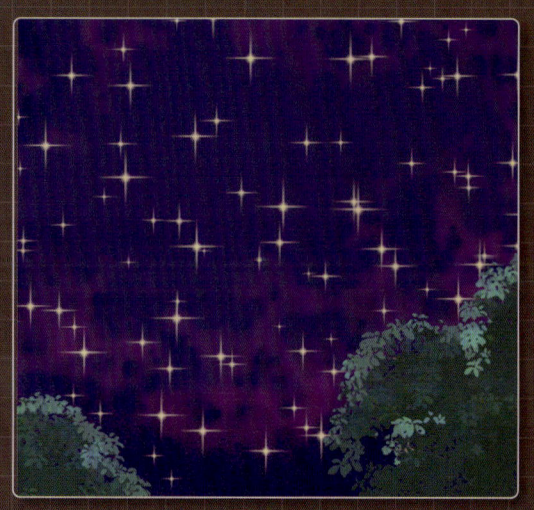

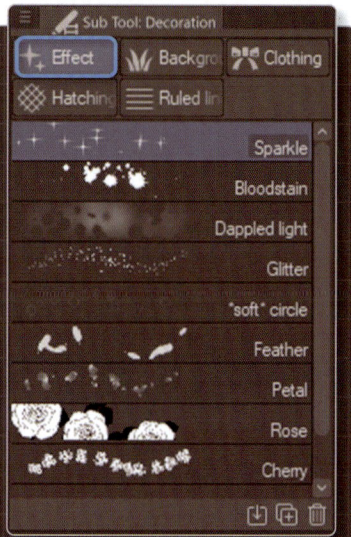

BLEND

This group doesn't add colour, but contains brushes that allow you to drag or blend existing colours on the canvas instead.

> The Blend brush is used for blending colours rather than adding additional colour

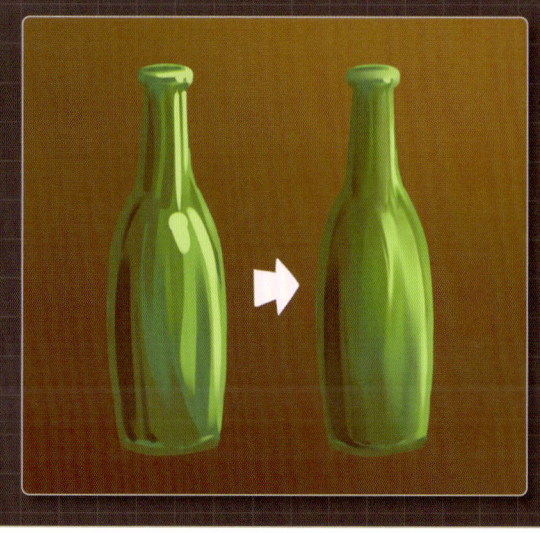

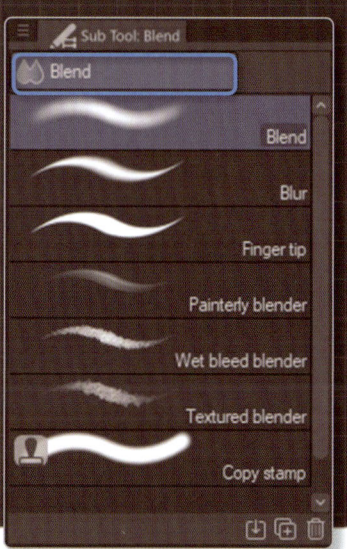

LIQUIFY

This tool allows you to use your stylus to push or drag on the contents of your canvas, changing their size and shape.

> The Liquify tool can be used to warp, distort, and alter what's on the canvas

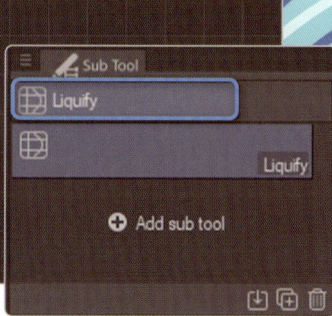

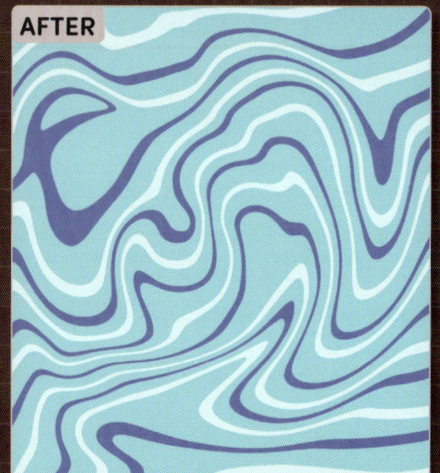

ERASER

This group of tools allows you to erase any part of the painting by drawing over it.

Another erasing method is to use the Opacity function located in the colour icons at the bottom of the Tool palette. When activated, your brush will function as an eraser. This is useful when you want the erasing stroke to have the same texture or shape as your brush.

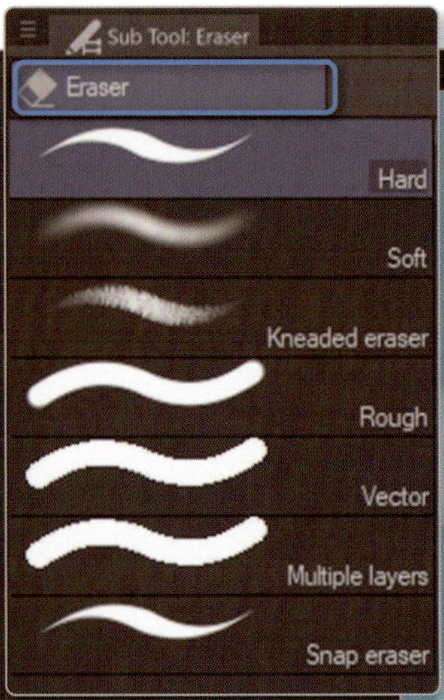

Sub Tool: Eraser

Eraser

Hard

Soft

Kneaded eraser

Rough

Vector

Multiple layers

Snap eraser

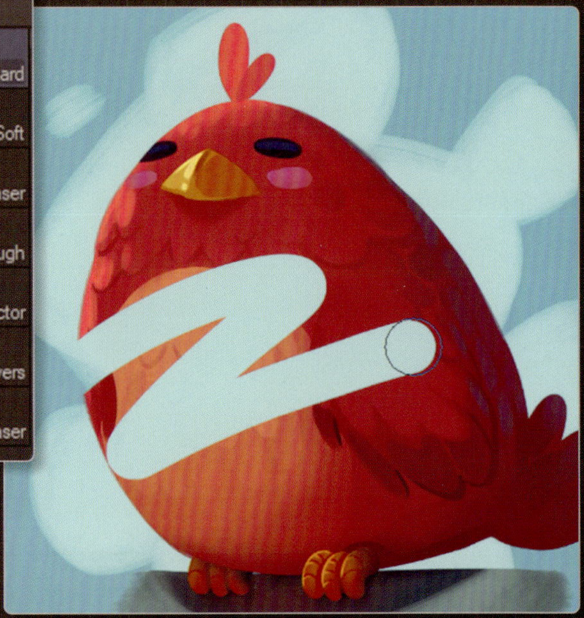

∧ The eraser comes with various options, such as a hard or soft stroke

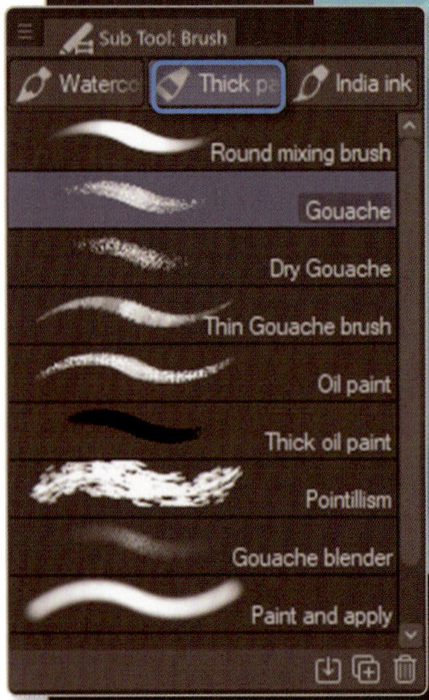

Sub Tool: Brush

Waterco | Thick pa | India ink

Round mixing brush

Gouache

Dry Gouache

Thin Gouache brush

Oil paint

Thick oil paint

Pointillism

Gouache blender

Paint and apply

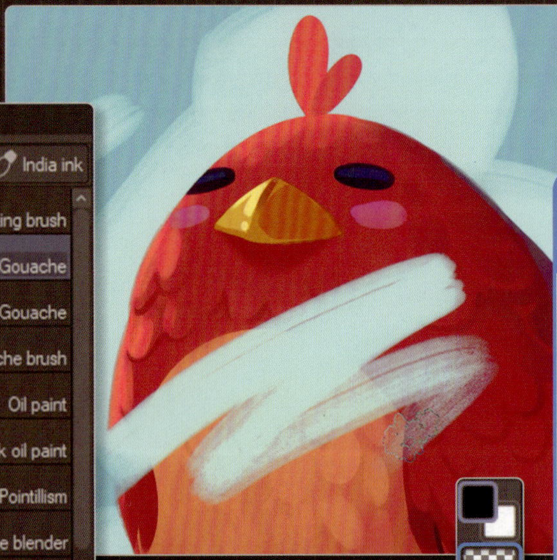

∧ Select the Opacity icon from the left-hand Tool palette to erase with the same stroke as your current brush

SHORTCUTS

- P key: Pen > Pencil

- B key: Brush > Airbrush > Decoration

- J key: Blend > Liquify

Some keyboard shortcuts toggle through more than one tool. For example, if you press the B key, this activates the Brush tool. But if you then press the B key again, the Airbrush tool will be activated. The Decoration tool will be activated if you press the B key for the third time. Pressing the B key for the fourth time will revert to selecting the Brush tool.

COLOUR MIXING

Activating this setting allows you to blend the colour of the brushstroke with the colour currently on the canvas. This is useful for generating new colours on the canvas while painting.

> The Colour Mixing setting is similar to the effect you create when painting with traditional media

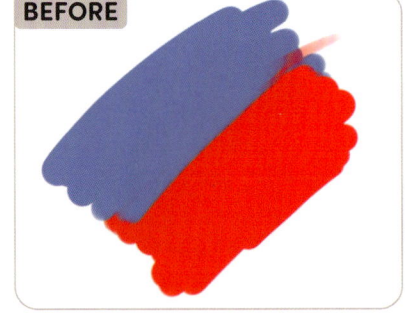

BEFORE

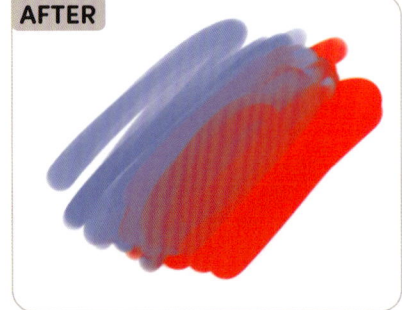

AFTER

TOOL PROPERTIES

Sub tools from the Drawing tools groups may not all have the same settings in the Tool Property palette, but the vast majority offer general options for brush size and opacity settings. As explained on page 26, you can add new settings to the Tool Property palette from the Sub Tool Detail palette.

The configuration groups or categories of the Sub Tool Detail palette are:

BRUSH SIZE
This group has settings for brush size.

INK
This includes settings for brush opacity, blending modes, colour mixing, paint density, and more.

COLOUR JITTER
These settings allow you to add variation to the brush tip's colour or randomization to the colour of a whole stroke.

ANTI-ALIASING
This group has settings for controlling brush smoothing.

BRUSH SHAPE
The settings in this group allow you to change the shape of the selected brush to a preset one, or to add the selected brush to the preset brush list.

BRUSH TIP
This category includes settings for the brush tip, such as thickness, hardness, shape, direction, and so on.

SPRAYING EFFECT
This group allows you to enable various brush spraying and scattering effects.

STROKE
These settings control the amount of space between brush tips, plus other settings related to the brushstroke.

TEXTURE
In this category, you can apply a texture to your strokes.

DUAL BRUSH
These configurations are applied to create a second brush linked to the main brush.

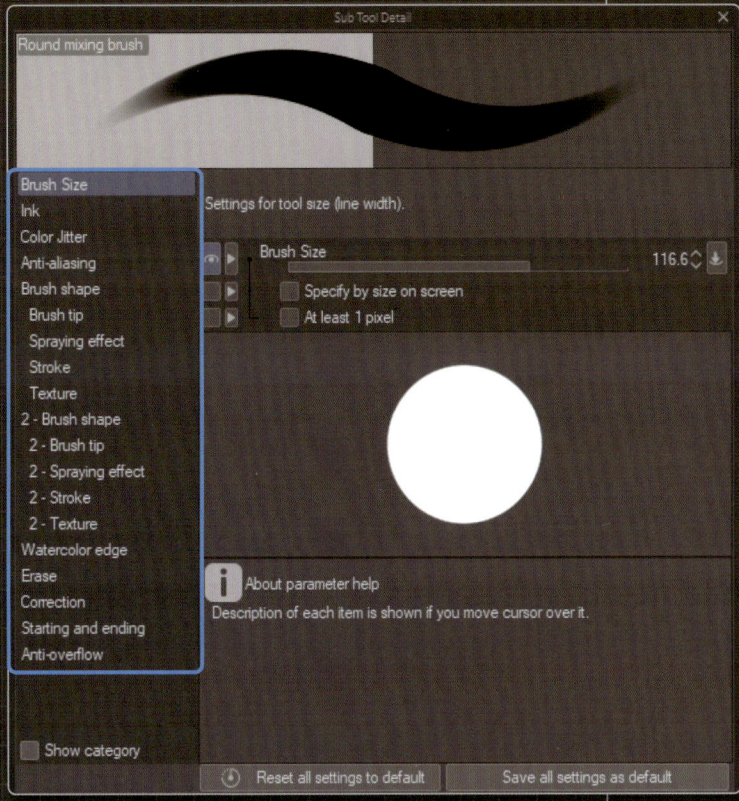

∧ Settings and categories will change, or will show as unavailable, depending on your chosen tool

WATERCOLOUR EDGE
These settings allow you to simulate watercolour-like opacity on the edges of your brushstrokes.

ERASER
These settings are only enabled for eraser-type brushes. The available settings include vector erasing and erasing on all layers.

CORRECTION
These settings modify the software's reaction to your stylus. You can find stabilization settings, speed adjustments, post-correction, and more.

STARTING AND ENDING
In this group, you can define effects at the beginning and end of brushstrokes.

ANTI-OVERFLOW
Enabling this option ensures your brushstrokes don't overflow outside the lines or layer you have set as a reference.

Layers

Layers are a basic yet important tool in digital painting. They are like transparent sheets or panes stacked on top of each other, on which you can paint or add information on the canvas. Each layer exists independently of the others, so what you do on one layer won't affect the others. This allows you to make different actions and brushstrokes without fear of messing up or losing what is already on the canvas. This provides great freedom and flexibility when experimenting during the creative process.

Layers can be arranged and grouped in a way that lets you keep your artwork ordered. You can resize them, reposition them, and change tones in specific areas. What's more, whether a layer is stacked above or below another layer will determine how it interacts with it. Unlike traditional media, painting on layers gives you effective control over each element, allowing you to make adjustments at any stage in the painting process.

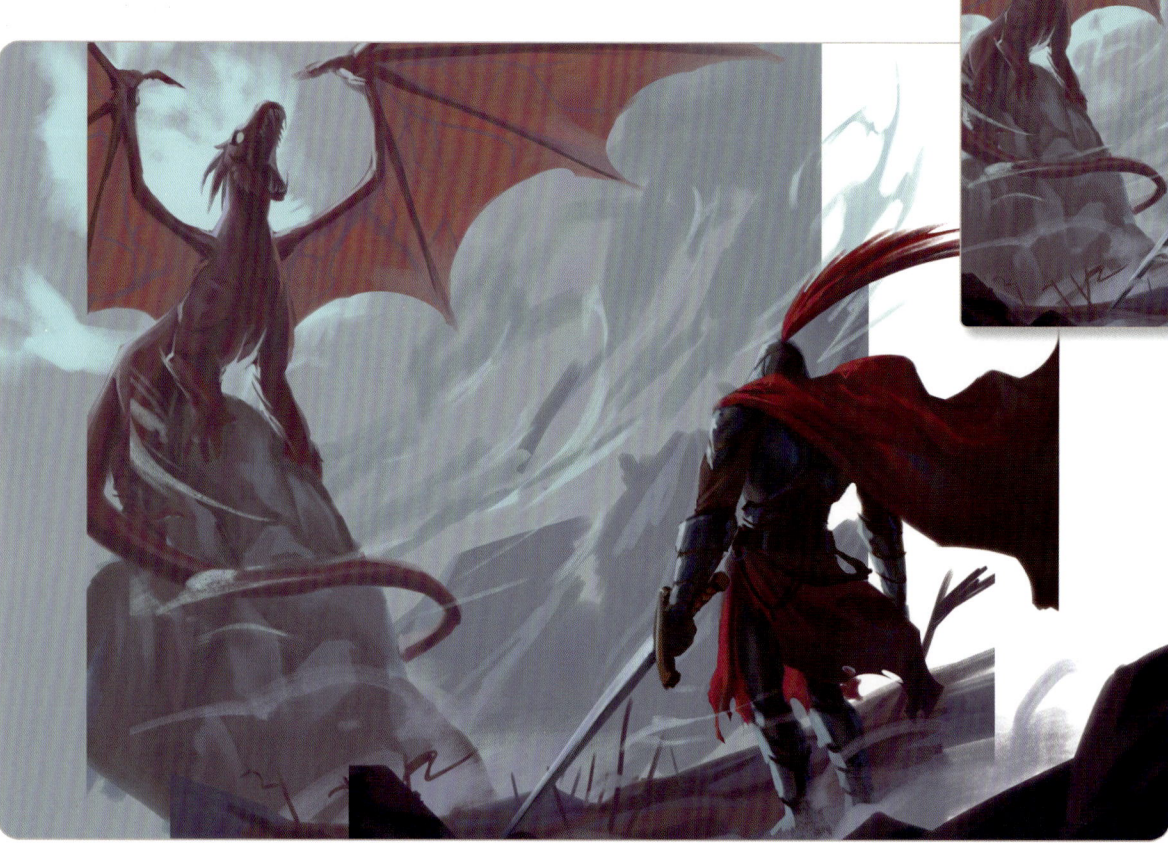

‹ What you paint on a higher layer will cover what is on the layers below

LAYER PALETTE

The creation, organization, and editing of layers is all done from the Layer palette. This shows you the number of layers you have and how they are organized. The Layer palette has various parts and functions.

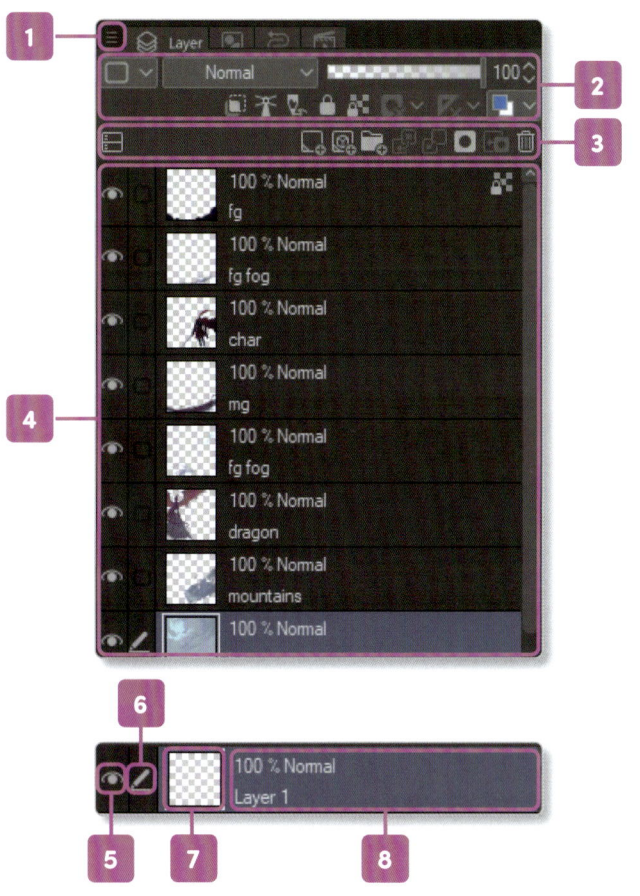

1. LAYER LIST

This list allows you to organize, group, or deactivate layers.

2. LAYER COMMAND BAR

This bar has functions for creating and modifying layers.

3. LAYER PROPERTY BAR

This bar allows you to modify the properties of the selected layer.

4. LAYER PALETTE MENU

Clicking on this will display a list with various layer functions.

5. VISIBILITY FUNCTION

This hides or keeps the layer visible. When it's visible, an icon of an open eye is shown. When it's hidden, the icon disappears.

6. LAYER STATUS FUNCTION

This indicates whether the layer is active or selected. When the layer is active, a pencil icon is displayed. When you select one or more additional layers, these layers will show a checkmark icon.

7. LAYER THUMBNAIL

This displays a miniature of the layer's content. You can make a selection of the entire layer by holding down the Ctrl key and clicking it.

8. LAYER DETAILS

This shows the name of the layer plus other information, such as opacity or blending mode. To change the name of a layer, double-click on the current layer name and modify it.

BASIC OPERATIONS

LAYER CREATION

Clip Studio Paint allows you to create different types of layers. One of the most used is the Raster layer, which can be created by selecting the New Raster Layer icon in the Layer palette. You can also create a new layer from the Layer menu. If you want to configure the layer name, expression colour, and blending mode right from the start, select New Layer

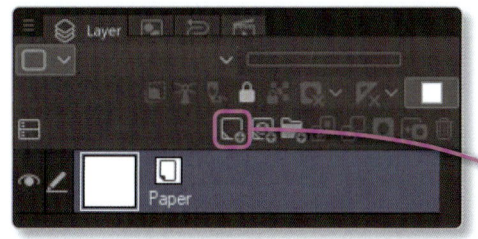

> Raster Layer from the Layer menu. You will see a small window with those settings. You can also hold down the Alt key while clicking the New Raster Layer icon in the Layer palette. The keyboard shortcut to quickly create a new Raster layer is Shift+Ctrl+N.

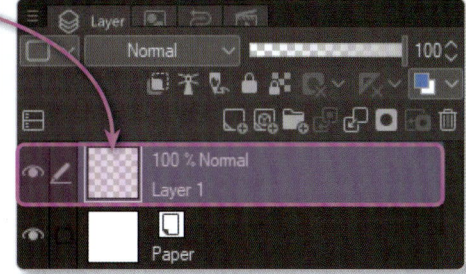

∧ There are various ways to create a new Raster layer

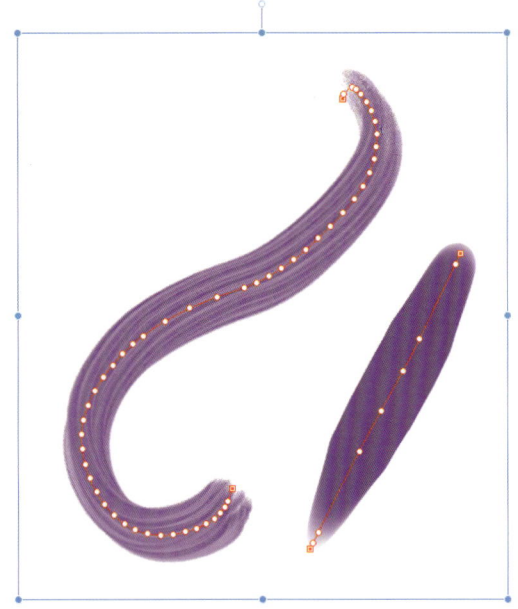

THE VECTOR LAYER

The Vector layer is another commonly used option. It converts everything added or painted onto it into vector lines, maintaining good quality if the image needs to be scaled. Each stroke can be modified using the Object sub tool or the Correct Line tool and its sub tools.

< **Anything painted onto a Vector layer is converted into vector lines**

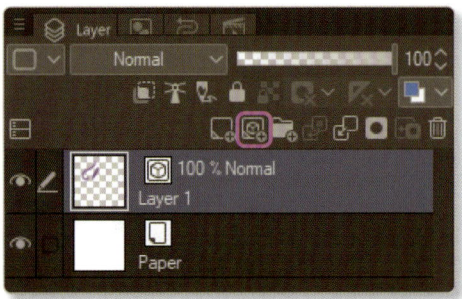

DUPLICATE & DELETE LAYERS

To duplicate one or several layers, right-click on the layer you want to duplicate in the Layer palette, then select the Duplicate Layer button. Alternatively, go to Layer > Duplicate Layer.

There are also two even quicker ways to duplicate layers. The first is to hold down the Alt key while clicking and dragging the layer up or down, then release the mouse. The second is to press Ctrl+C to copy the layer and then Ctrl+V to paste.

To delete a layer, click the Delete Layer button in the Layer palette or use Layer > Delete Layer. Pressing the Del key will not delete the layer, only its content.

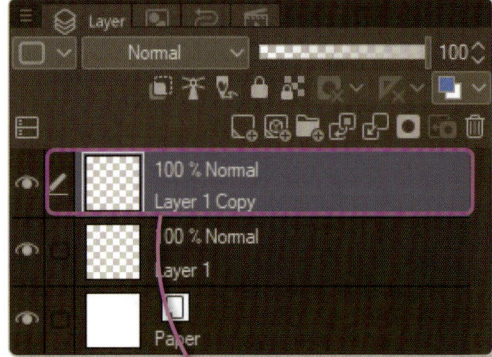

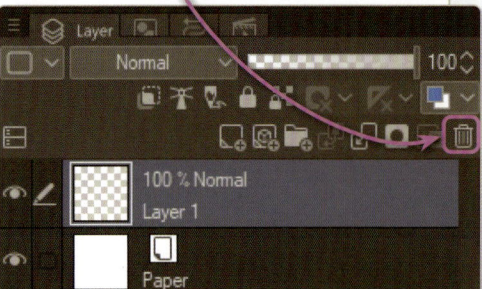

▲ **Click the trash can icon to delete the selected layer**

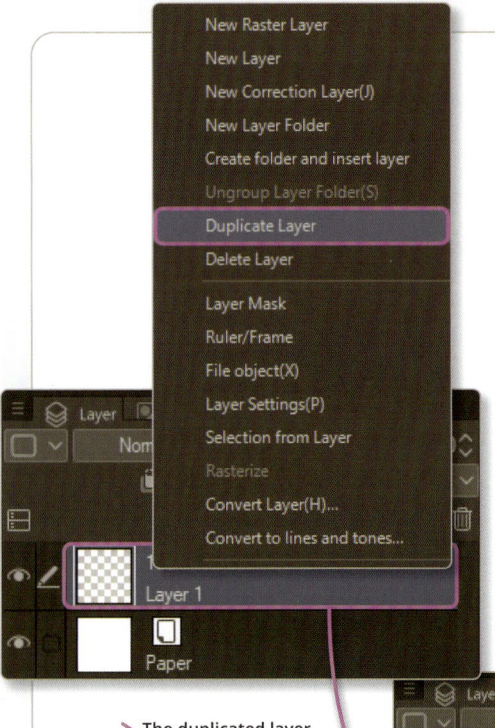

> **The duplicated layer will have same name as the original layer, but with 'Copy' added to it**

LOCKING LAYERS

When a layer is locked, you won't be able to draw or edit anything on it. This setting is useful to prevent accidentally editing a layer, or simply to isolate it from other layers. Click the Lock Layer icon in the Layer palette to activate this setting. You can also lock a layer by selecting Layer > Layer Settings > Lock Layer. When a layer is locked, a padlock icon will appear next to it.

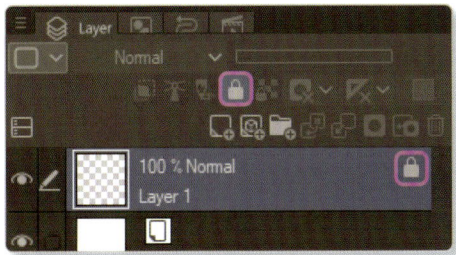

∧ The padlock icon shows the layer is locked

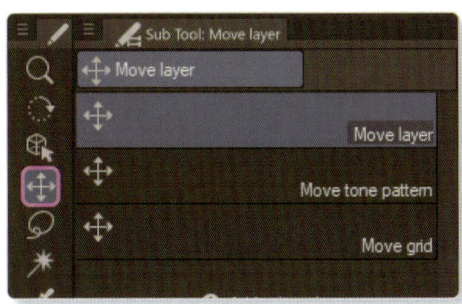

∧ The Move Layer tool lets you move layers when you need to tweak the composition

MOVING A LAYER

When creating digital paintings, you will find that you need to move layers to modify or enhance the composition of your artwork. The Move Layer tool, located in the Tool palette, can be used to do this. Simply left-click and drag on the canvas to move the contents of the active layer. You can also click and drag while holding Ctrl for the same effect, or hold Shift while dragging to move the layer in set increments.

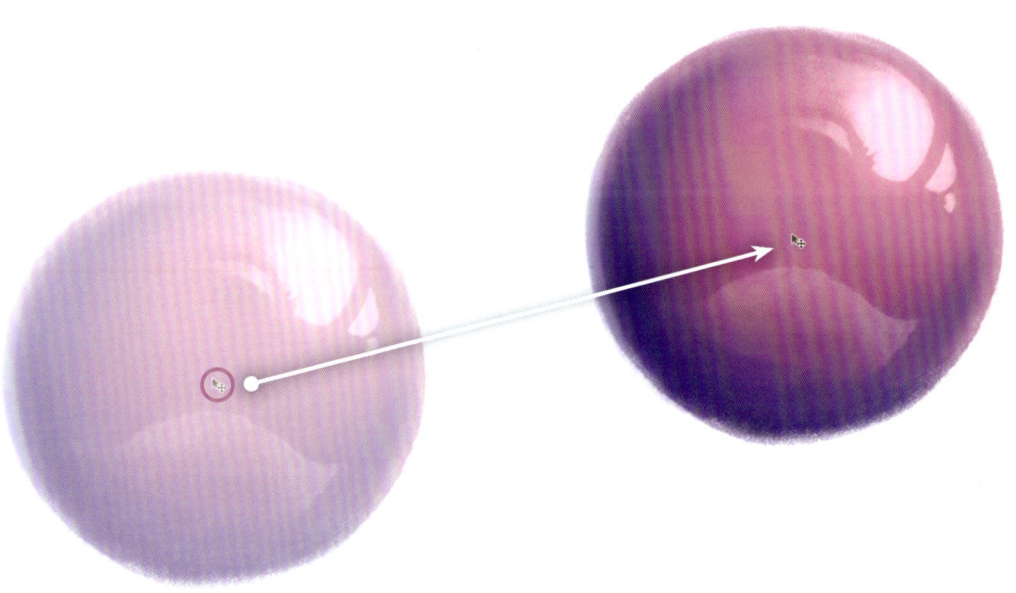

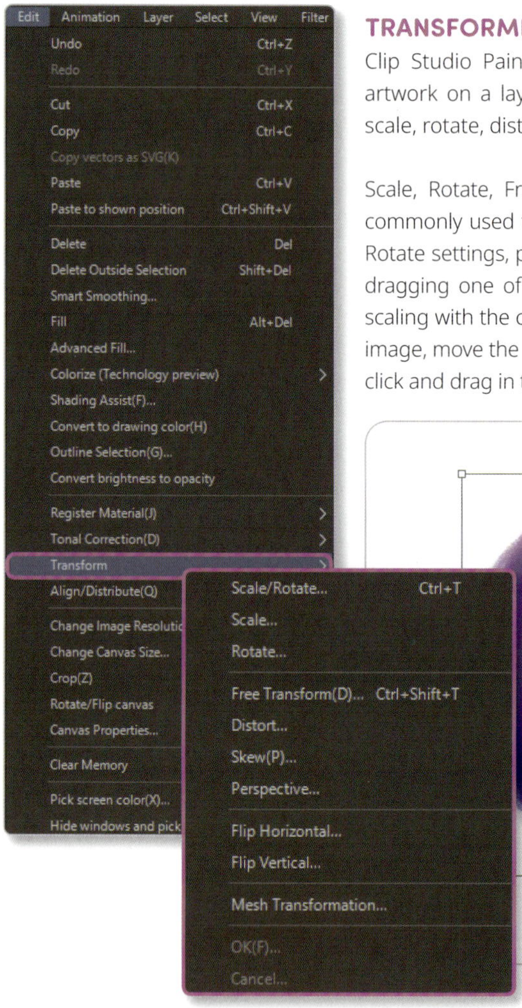

| Edit | Animation | Layer | Select | View | Filter |

Undo	Ctrl+Z
Redo	Ctrl+Y
Cut	Ctrl+X
Copy	Ctrl+C
Copy vectors as SVG(K)	
Paste	Ctrl+V
Paste to shown position	Ctrl+Shift+V
Delete	Del
Delete Outside Selection	Shift+Del
Smart Smoothing...	
Fill	Alt+Del
Advanced Fill...	
Colorize (Technology preview)	>
Shading Assist(F)...	
Convert to drawing color(H)	
Outline Selection(G)...	
Convert brightness to opacity	
Register Material(J)	>
Tonal Correction(D)	>
Transform	>
Align/Distribute(Q)	
Change Image Resolution	
Change Canvas Size...	
Crop(Z)	
Rotate/Flip canvas	
Canvas Properties...	
Clear Memory	
Pick screen color(X)...	
Hide windows and pick	

Scale/Rotate...	Ctrl+T
Scale...	
Rotate...	
Free Transform(D)...	Ctrl+Shift+T
Distort...	
Skew(P)...	
Perspective...	
Flip Horizontal...	
Flip Vertical...	
Mesh Transformation...	
OK(F)...	
Cancel...	

TRANSFORMING

Clip Studio Paint's transformation settings can make it easier to edit artwork on a layer. Selecting Edit > Transform will bring up options to scale, rotate, distort, flip, and more.

Scale, Rotate, Free Transform, and Mesh Transformation are the most commonly used transformation settings. To quickly activate the Scale and Rotate settings, press Ctrl+T. You can then scale the image by clicking and dragging one of the surrounding handles. Hold down the Alt key while scaling with the cursor to scale from the centre of the image. To rotate the image, move the cursor outside the box until a curved arrow appears, then click and drag in the desired direction.

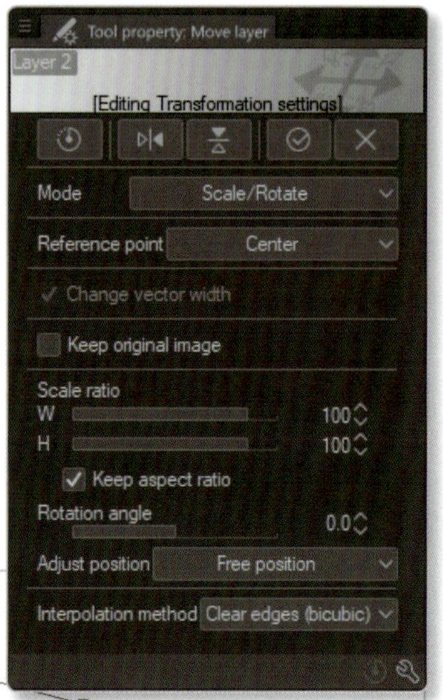

Tool property: Move layer

Layer 2

[Editing Transformation settings]

| Mode | Scale/Rotate |
| Reference point | Center |

✓ Change vector width

☐ Keep original image

Scale ratio
| W | 100 |
| H | 100 |

✓ Keep aspect ratio

Rotation angle   0.0

| Adjust position | Free position |

| Interpolation method | Clear edges (bicubic) |

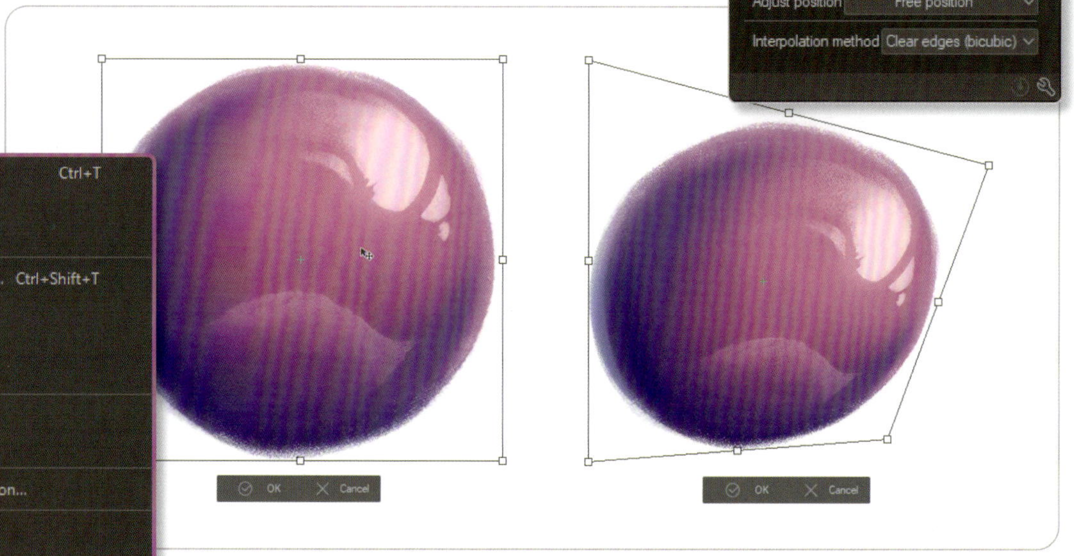

< The Transform tool offers numerous options for altering a layer

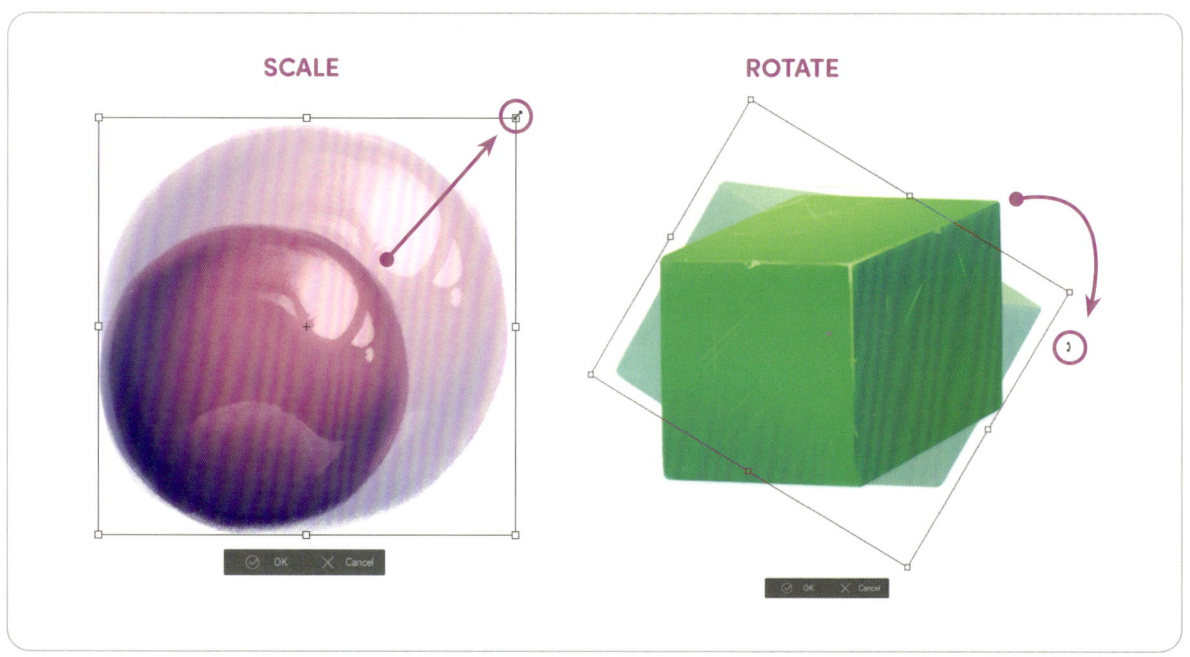

SCALE

ROTATE

< Click and drag to scale or rotate a layer

To use the Free Transform setting, hold down the **Ctrl** key and drag on one of the square handles to distort or skew the object in the desired direction.

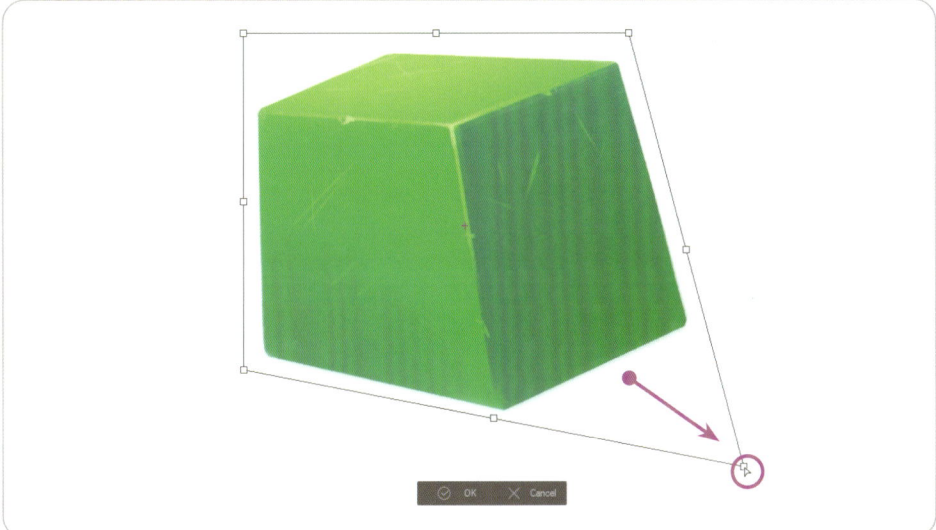

> Free Transform offers more freedom in how you can alter an image

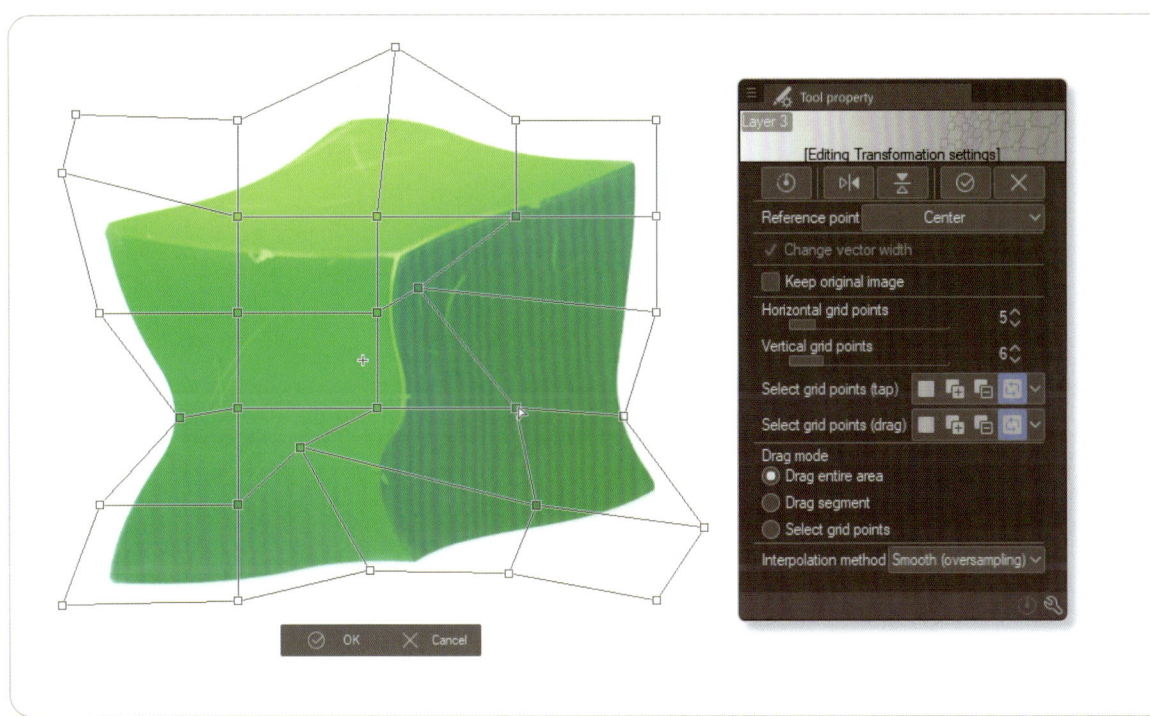

< Mesh Transformation provides greater flexibility and detail when adjusting a layer

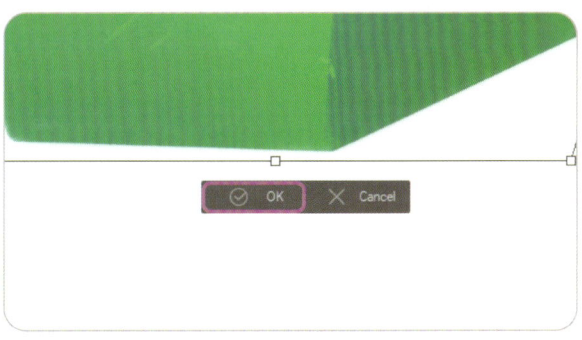

> Choose whether to finalize the adjustments made, or undo them

Mesh Transformation allows you to modify a layer with more detail by generating more handles to adjust. To enable this option, select **Edit > Transform > Mesh Transformation**. From the Tool Property palette, you can modify the number of lattice points or the rotation centre.

After modifying a layer, select the OK button to finalize the edit, or Cancel if you don't want it to be applied. You can also press the **Enter** key to accept the transformation, or the **Esc** key to cancel it.

LOCK TRANSPARENT PIXELS

This useful setting allows you to lock the transparency of the layer and only edit the opaque pixels on it. To activate it, click the Lock Transparent Pixels button in the Layer palette. Alternatively, you can select Layer > Layer settings > Lock transparent pixels.

> A small padlock icon on a small opacity grid will appear next to the layer when the transparent pixels are locked

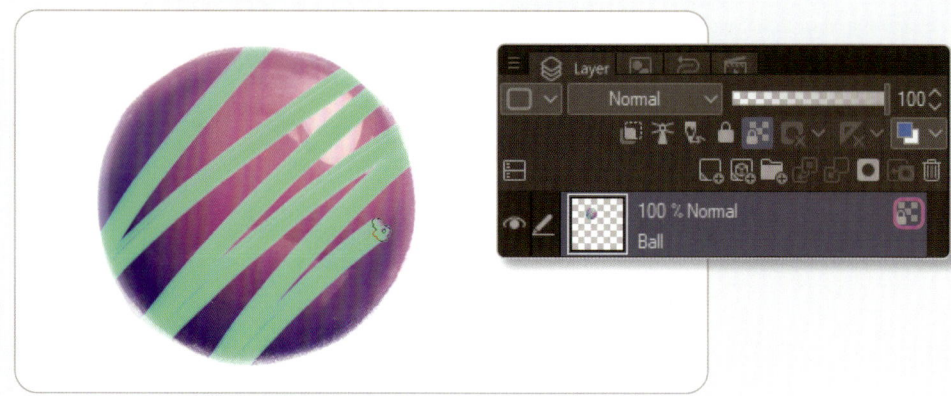

CLIP TO LAYER BELOW

When you activate this setting, the selected layer's visible area is confined to the drawn area of the layer directly below it. This is useful if you want to paint on one layer while staying within the confines of another. Click the Clip to Layer Below button in the Layer palette to activate it. A pink bar will appear next to the layer thumbnail when this function is activated.

< Clipping a layer to the layer below restricts the area in which you can paint to what is already painted on the lower layer

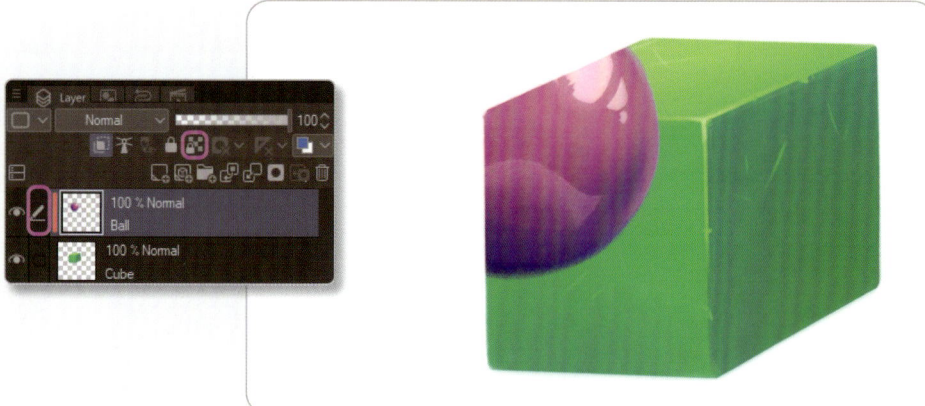

SET AS REFERENCE LAYER

This setting converts the selected layer into a reference for other layers. It enables you to draw and fill on other layers while only referring to the specific reference layer or layers. Filling an element without interference from other line layers in this way is very useful. They can also help to sort a painting into different stages – line art on one layer, filled block colours on the next, and so on.

Click the lighthouse icon in the Layer palette to activate this setting. Alternatively you can select Layer > Layer settings > Set as Reference Layer, or right-click on the layer thumbnail and select the Set as Reference Layer icon (the lighthouse).

> Set a layer as a reference layer when you want what you draw on top to be informed by this specific layer

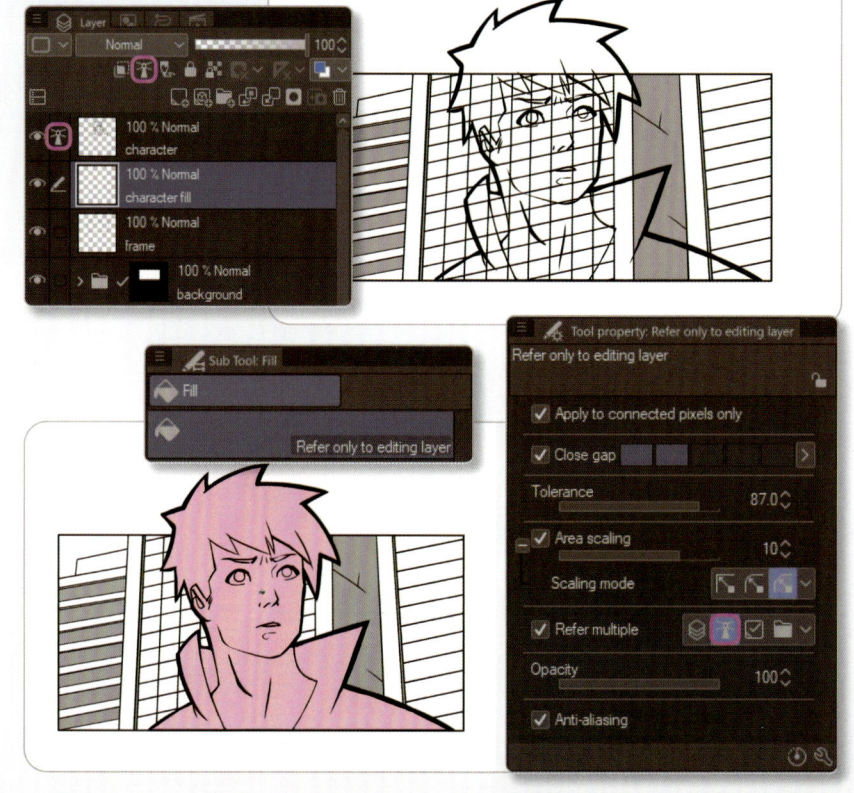

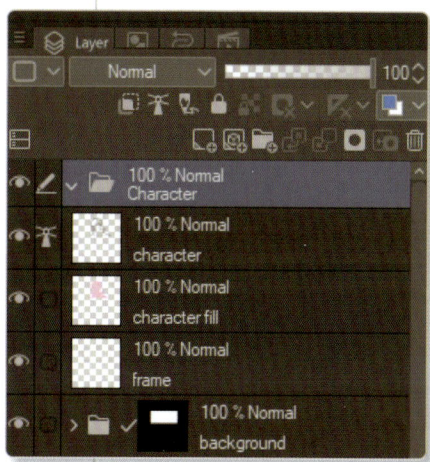
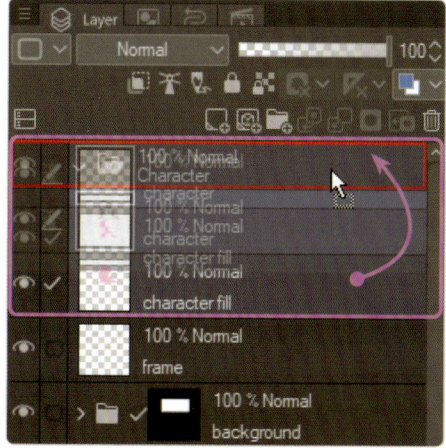
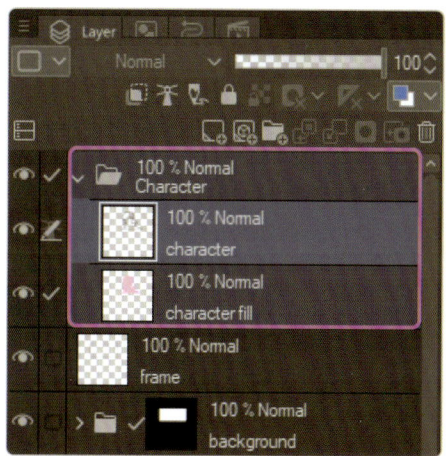

∧ **Make a folder and drag layers into them to create more organized groups**

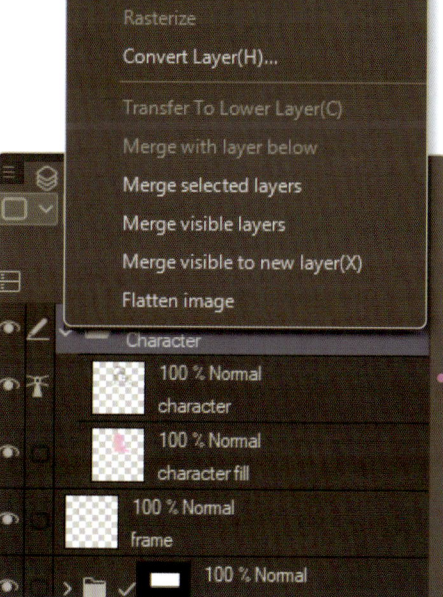

LAYER FOLDER

This tool is very useful for grouping layers and maintaining better organization in your layer list. To create a folder, click on the New Layer Folder icon in the Layer palette (the folder with the small plus sign). Alternatively you can select Layer > New Layer Folder. To add layers to the folder, select a layer, then click, hold, and drag it into the folder. You can also create a folder with the layers grouped by right-clicking and selecting the option Create Folder and Insert Layer.

To delete the folder but not its layers, right-click on it and choose Ungroup Layer Folder. If you press the Del key on the folder containing the layers, only the *contents* of each layer will be deleted.

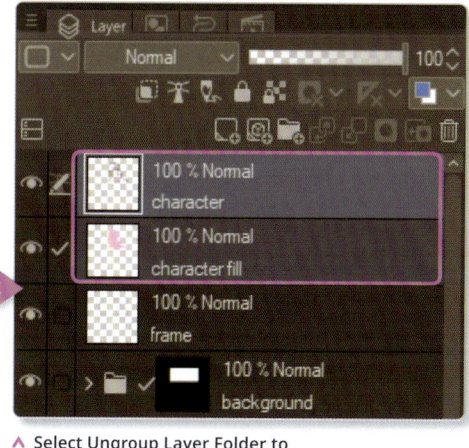

∧ **Select Ungroup Layer Folder to delete a folder but not its layers**

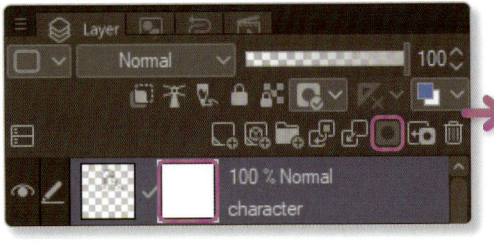

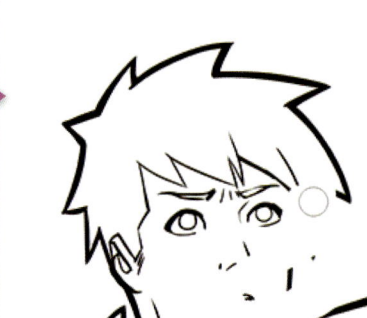

◁ Layer masks can be used to conceal parts of a layer without erasing them

LAYER MASKS

These are very useful if you want to hide parts of the layer without erasing them. To create a layer mask, select the layer and then click on the Create Layer Mask icon in the Layer palette. The mask will display next to the layer thumbnail. You can hide areas by clicking on the mask thumbnail and using an eraser. Another way to create a layer mask is by selecting Layer > Layer Mask > Mask Selection.

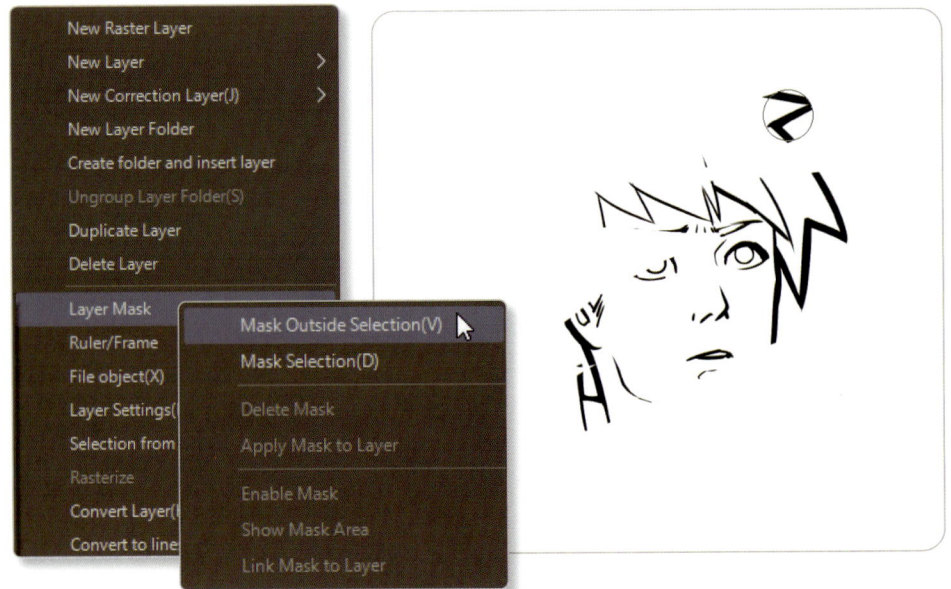

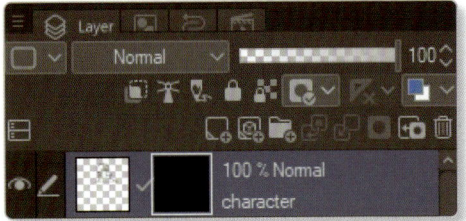

Another type of mask you can create is the Mask Outside Selection, which is the opposite of the Mask Selection. When you create this type of mask, the entire layer's content will be hidden. You can then use a brush to make it visible. Activate this by selecting Layer > Layer Mask > Mask Outside Selection. Alternatively you can hold the Alt key and click the Create Layer Mask icon in the Layer palette.

▲ The Mask Outside Selection option hides a layer's content, which can be revealed with a brush

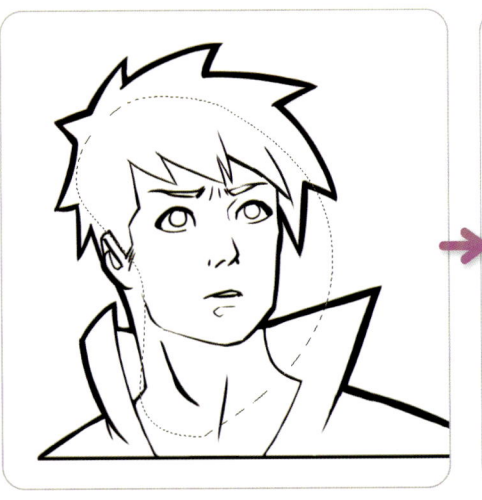

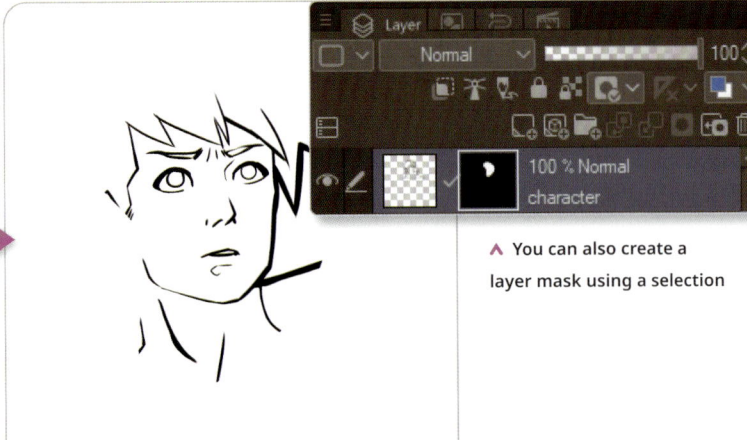

▲ You can also create a layer mask using a selection

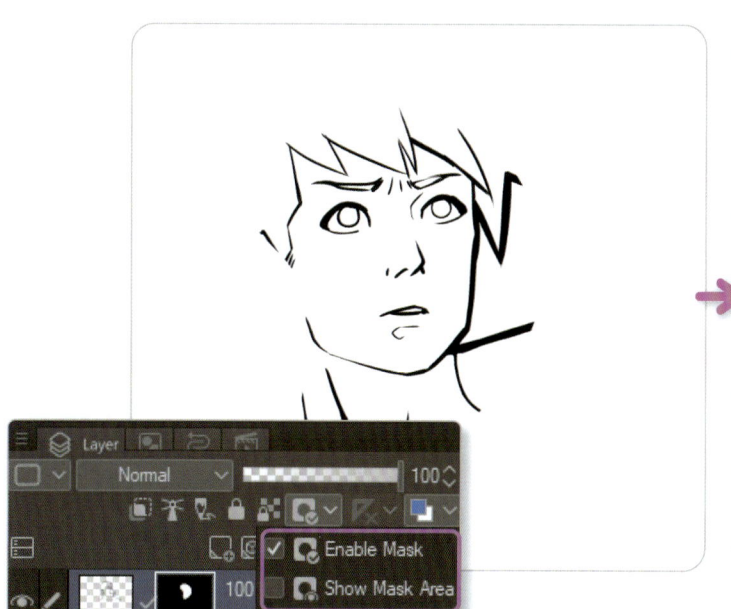

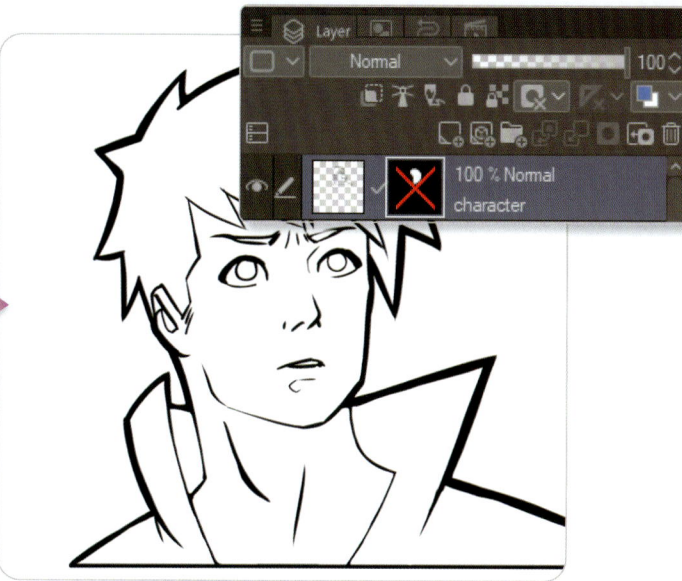

You can temporarily disable the mask and show the mask area on the Layer palette. You can also do this by right-clicking on the mask and disabling the Enable Mask option, or by enabling the Show Mask Area option. These can be accessed from Layer > Layer Mask submenu.

When you create a layer mask, the mask and the layer are always linked. This means that when you move the layer or apply transformations, the mask will also move or be modified accordingly. You can unlink the layer and the mask by clicking the chain icon. This link will be disabled, allowing you to edit and move the layer without affecting the mask.

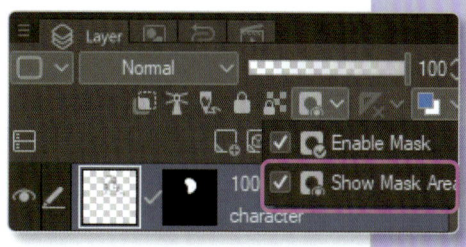

∧ There are multiple ways to disable a layer mask; a disabled mask will have red cross through it

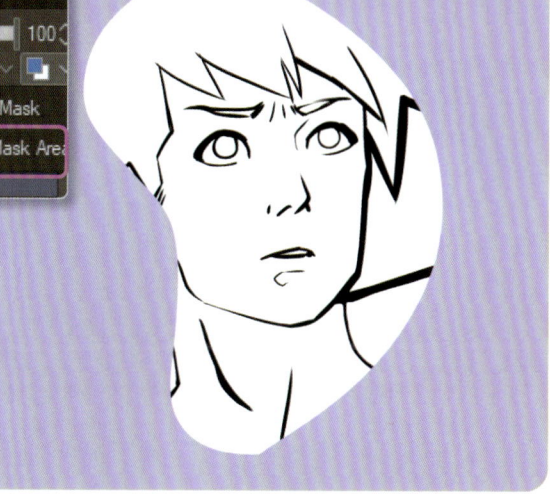

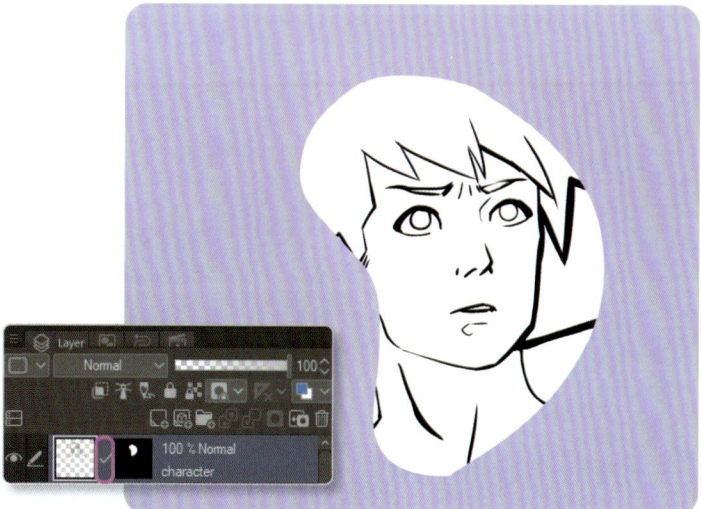

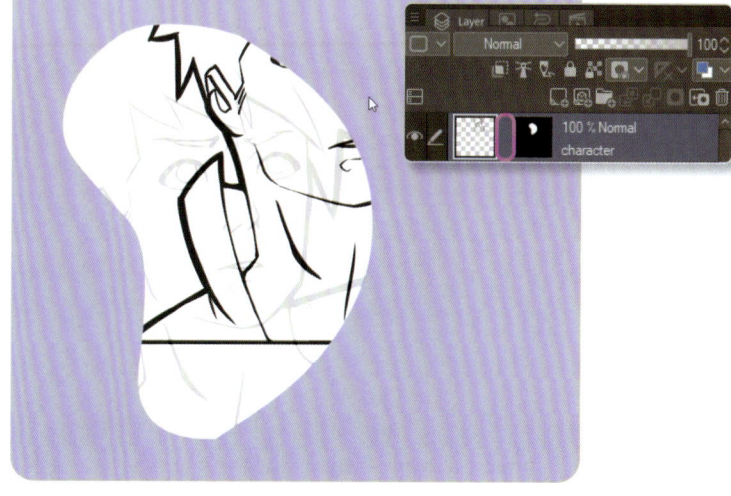

∧ The mask and the layer are always linked

43

BLENDING MODES

Blending modes, also known as blending options or layer blending modes, are settings that control how colours on a layer interact with the colours on the layers beneath it. They offer various effects and styles to apply to your layer. You can find this setting in the Layer palette. Clicking on it will display a list of the different blending modes on offer. You can also enable blending modes for many of the drawing tools by opening the Sub Tool Detail palette, going to Ink, and selecting the Blending Mode option.

In the following images, yellow and purple stripes have been added on a separate layer over the right-hand side of the image to help demonstrate how each blending mode works. Artists often use yellow to paint highlights and purple to paint shadows.

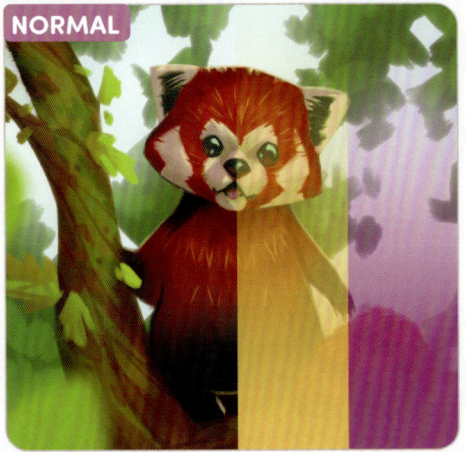

NORMAL

∧ **This is the default mode in which the layer is originally set, without any applied effects.**

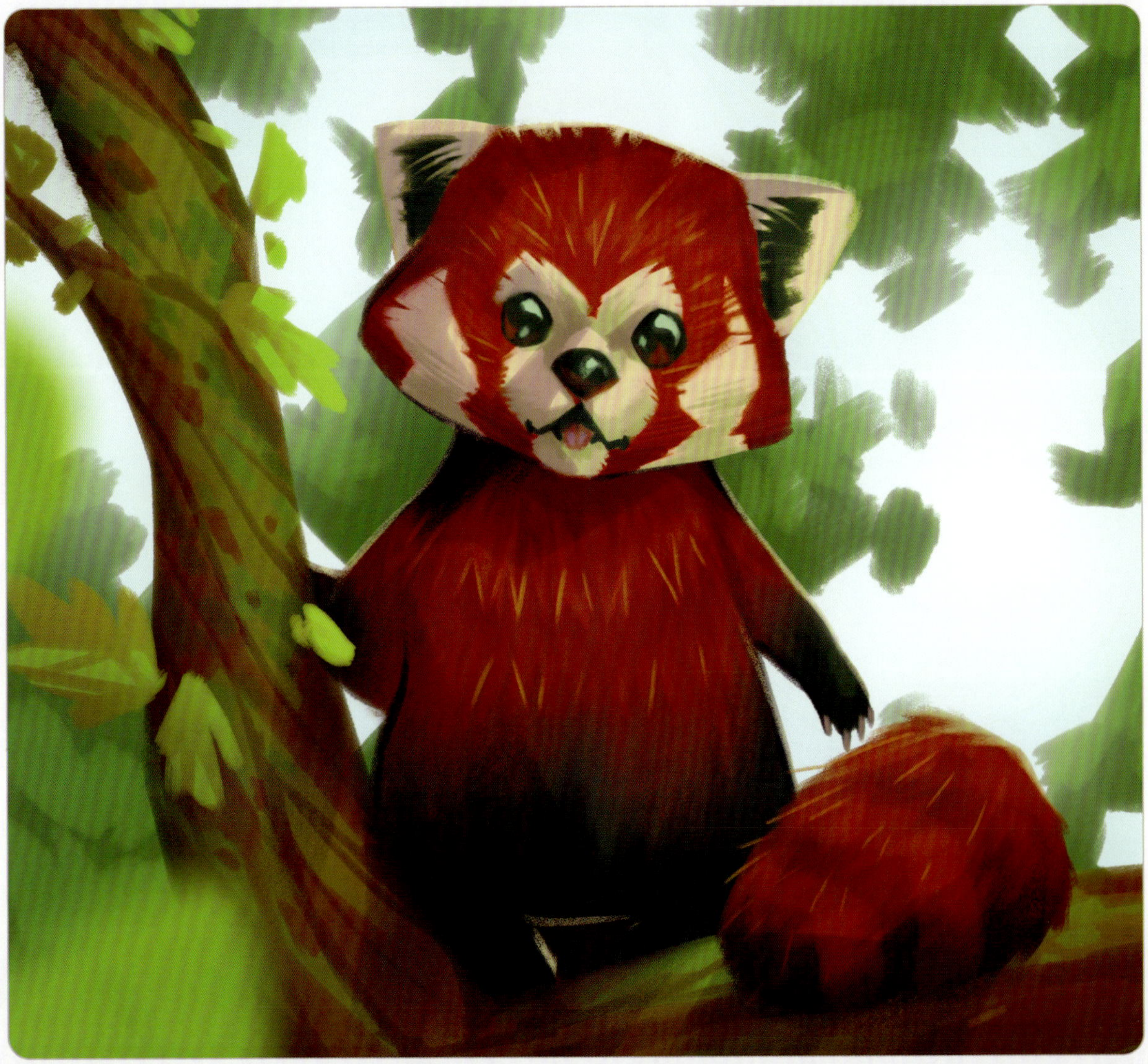

> **Original image with no blending modes applied**

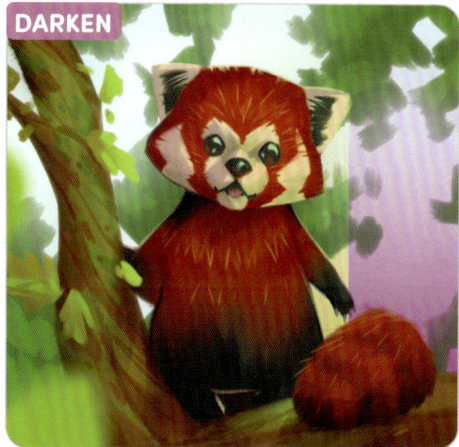

DARKEN

∧ Darken highlights the dark tones of the layer and darkens the entire image. It intensifies shadows for greater contrast in the image.

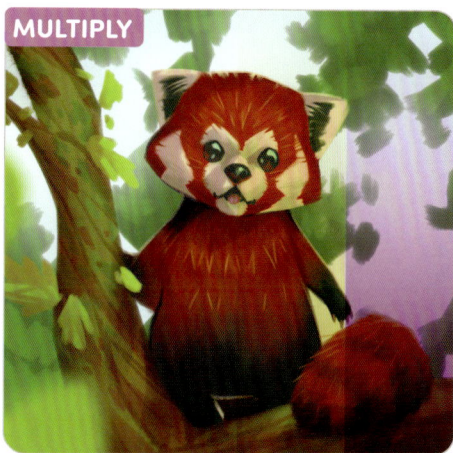

MULTIPLY

∧ This mode multiplies the layer's colours, resulting in a darker colour. It helps to deepen shadows and increase overall contrast in the image.

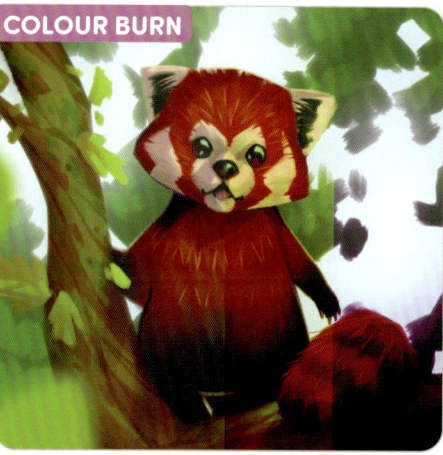

COLOUR BURN

∧ This mode intensifies the contrast of the layer's dark colours, generating a dramatic effect in the illustration. Use it moderately to avoid losing details when applied.

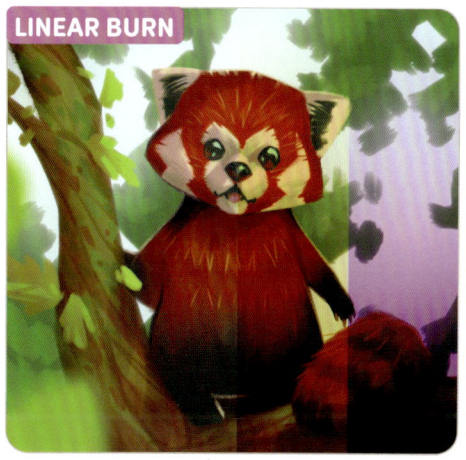

LINEAR BURN

∧ Similar to Colour Burn, this mode helps to intensify and increase contrast, but more linearly, highlighting dark colours.

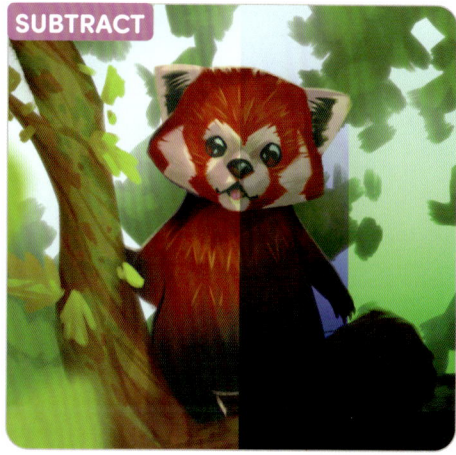

SUBTRACT

∧ Subtract allows you to create and modify colour values to achieve fading effects. It's very useful for correcting the colour balance of the image.

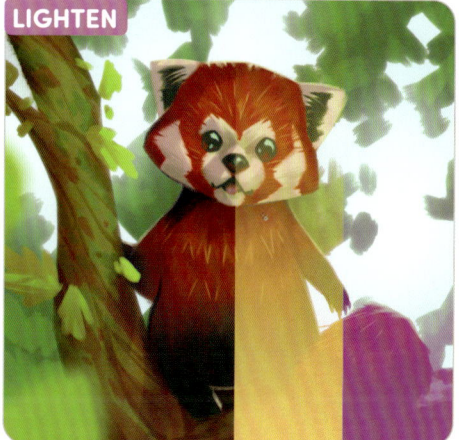

LIGHTEN

∧ This displays and enhances the lightest colour in a layer. Used to create luminosity, it will highlight illuminated areas and increase the image's brightness.

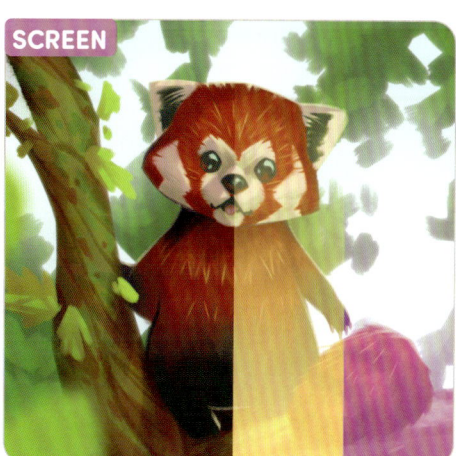

SCREEN

∧ This blending mode creates a soft appearance and luminous brightening effects.

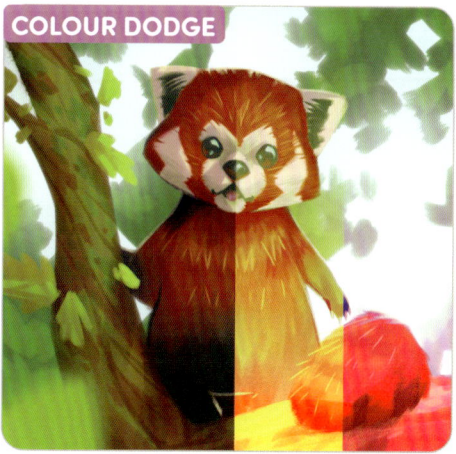

COLOUR DODGE

∧ This lightens an image, especially in dark and midtone areas, creating an effect that intensifies colours.

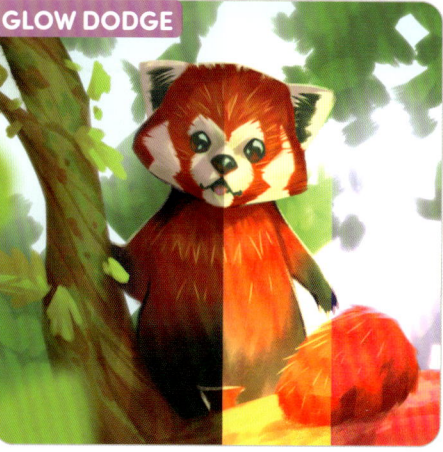

GLOW DODGE

∧ This mode creates an intense glow that highlights bright areas.

ADD

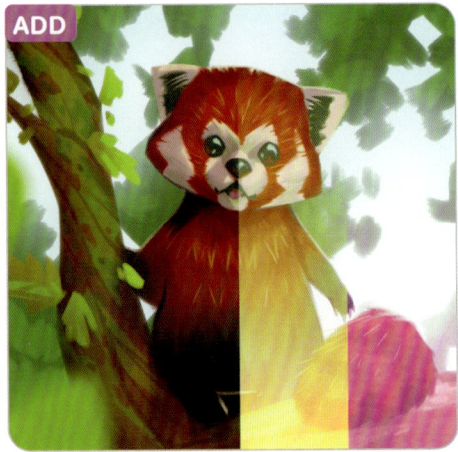

⌃ This adds the blended layer's values to the colours below, resulting in a brighter and more saturated image.

ADD (GLOW)

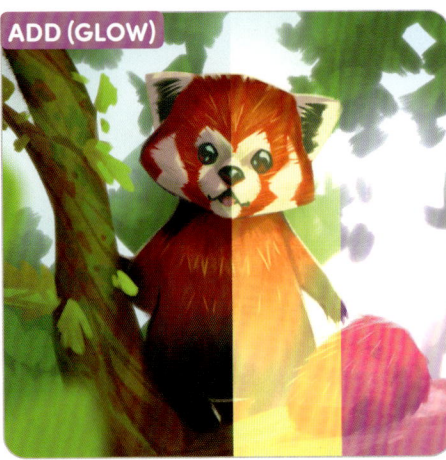

⌃ This blending mode creates a more intense glowing effect than Add, intensifying light and brightness.

OVERLAY

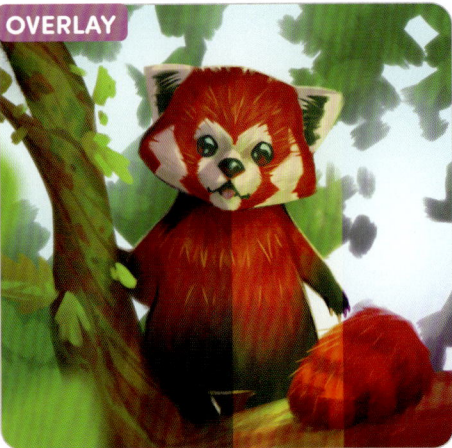

⌃ Overlay combines colours based on contrast, intensifying the lights and darks below.

SOFT LIGHT

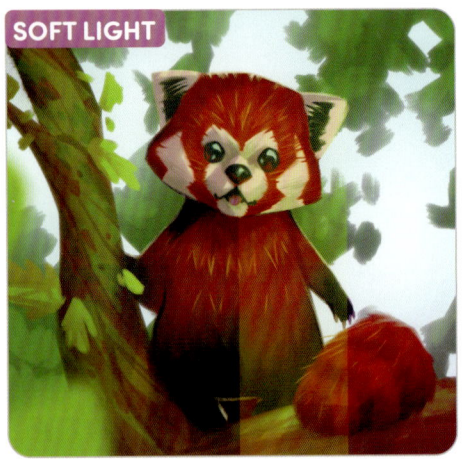

⌃ This mode subtly softens the image's colours, creating a gentle lighting effect.

HARD LIGHT

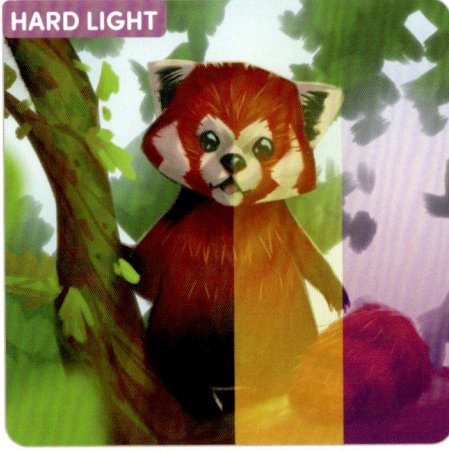

⌃ Hard Light intensifies and combines colours with greater contrast, resulting in deep, vivid colours.

DIFFERENCE

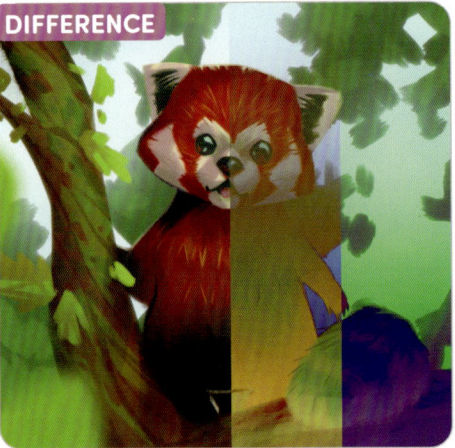

⌃ This mode 'subtracts' the blended layer's colours from the colours below.

VIVID LIGHT

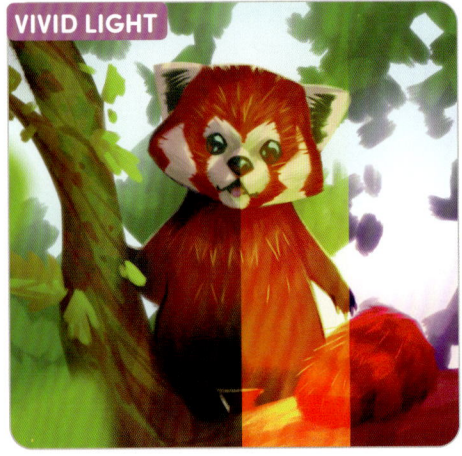

⌃ This mode intensifies the values below, depending on the values of the blended layer.

LINEAR LIGHT

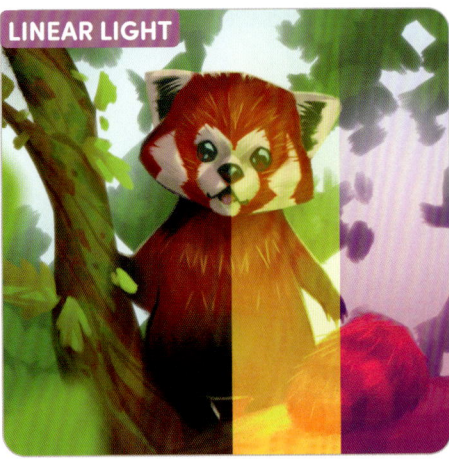

⌃ This applies a mix of Linear Dodge and Linear Burn for an intense effect.

PIN LIGHT

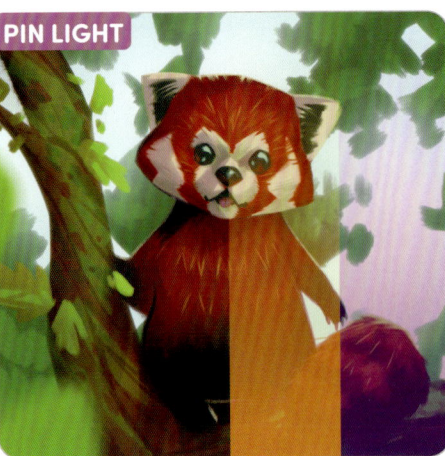

⌃ This mode modifies the colours below if they are darker than the blended colour.

HARD MIX

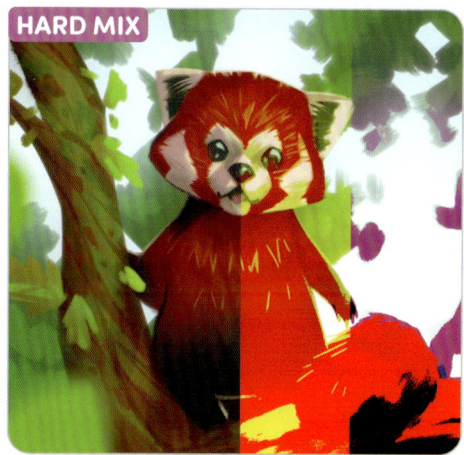

⌃ Hard Mix dramatically reduces the displayed colours of the image, creating an intense, simplified result.

EXCLUSION

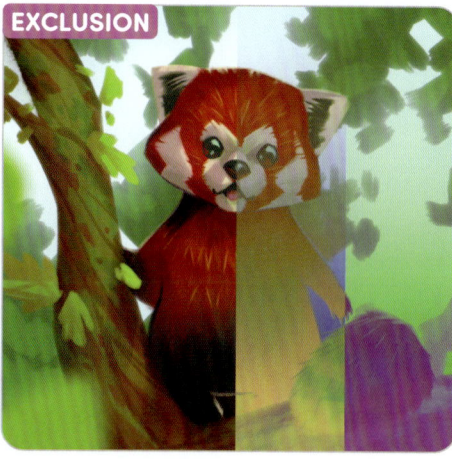

⌃ This mode works similarly to Difference, but creates a less intense effect.

DARKER COLOUR

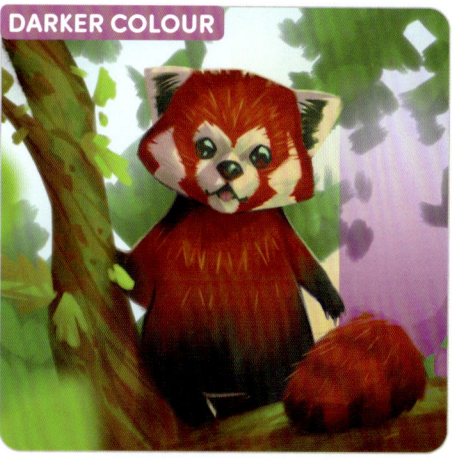

⌃ This mode applies the colour of the blended layer if it's darker than the colour below.

LIGHTER COLOUR

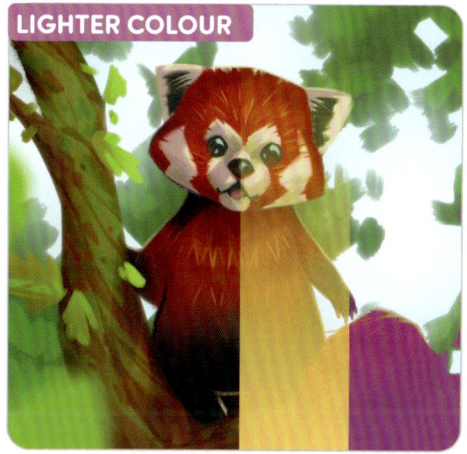

⌃ This mode applies the colour of the blended layer if it's lighter than the colour below.

DIVIDE

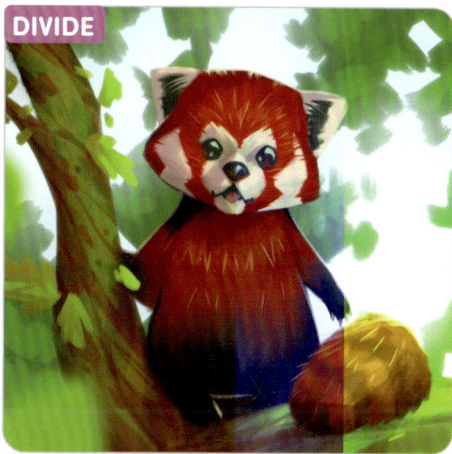

⌃ This mode divides the colour values of the layers, applying a glowing effect to the image.

HUE

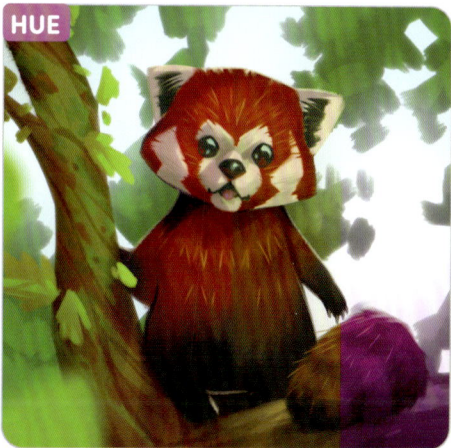

⌃ Hue adjusts the image's hue while preserving the saturation and brightness.

SATURATION

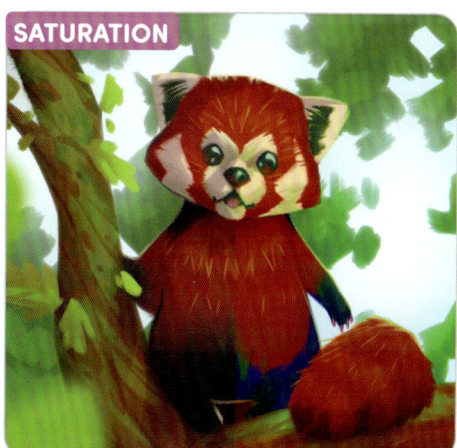

⌃ This mode applies the saturation of the blended layer to the colours below.

COLOUR

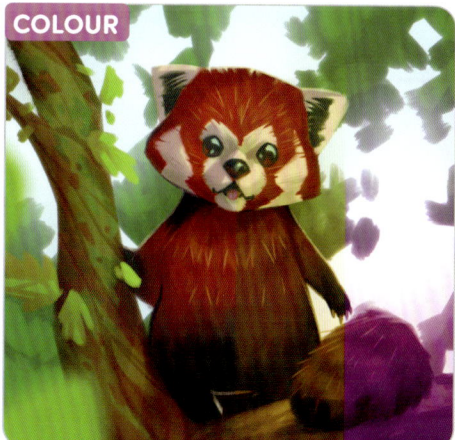

⌃ This mode applies the colour of the blended layer while preserving the saturation and brightness of the colours below.

BRIGHTNESS

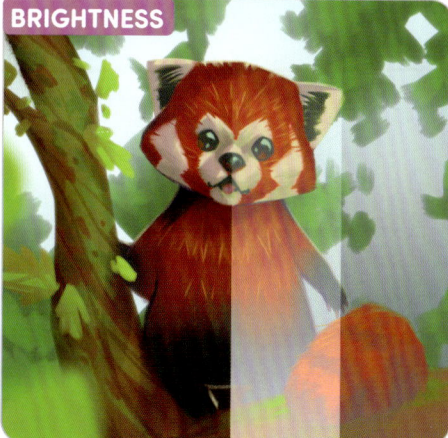

⌃ This mode applies the brightness of the blended layer while preserving the hue and saturation of the colours below.

LAYER PROPERTY PALETTE

Located on the right side of the interface above the Layer palette, the Layer Property palette contains a variety of useful settings that can be applied to a layer. There are options to change the colour expression, as well as to add effects to the layer. Some settings will change, or more/fewer will be offered, depending on the type of layer selected. An advantage is that various configurations can be activated or deactivated without altering the layer.

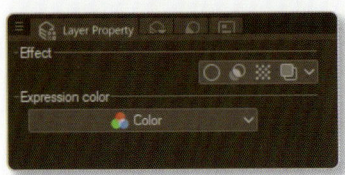

< The Layer Property palette provides options to change certain properties and configurations of a layer

EFFECT

This setting has four effects that you can apply to the selected layer.

BORDER EFFECT

This adds an outline around the drawing on the layer. You can modify the thickness, colour, and type of outline (for example, edge or watercolour edge).

EXTRACT LINES

This effect extracts the lines and dark areas from the layer, giving it the appearance of a line drawing. (This tool is for Clip Studio Paint EX only.)

TONE

Tone converts the selected layer into screentones. Further details are mentioned in the Comics, Manga & Webtoons chapter on page 74.

LAYER COLOUR

This effect allows you to change the colour of a layer. You can modify the main colour of the dark tones and add a sub colour to modify the white areas of the layer.

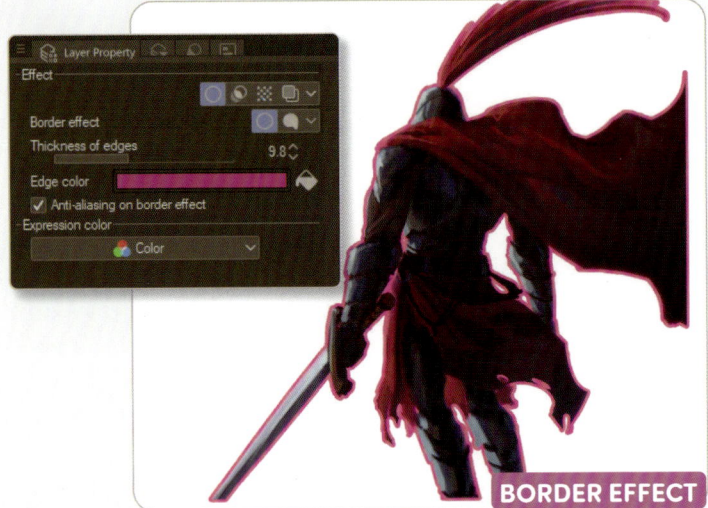

BORDER EFFECT

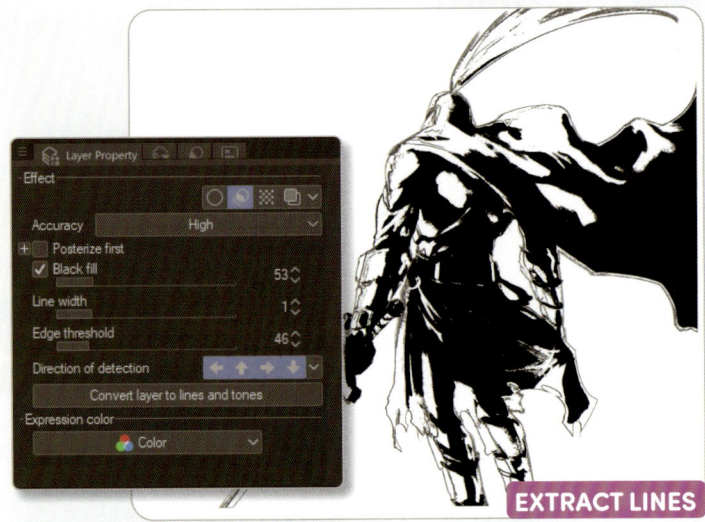

EXTRACT LINES

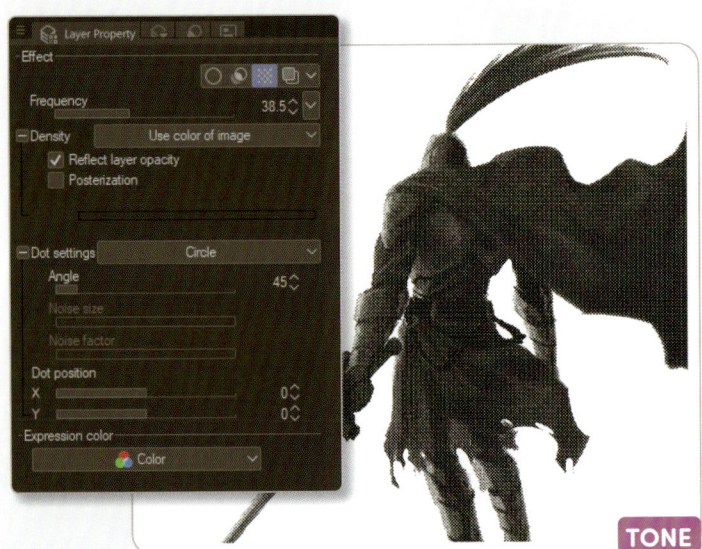

TONE

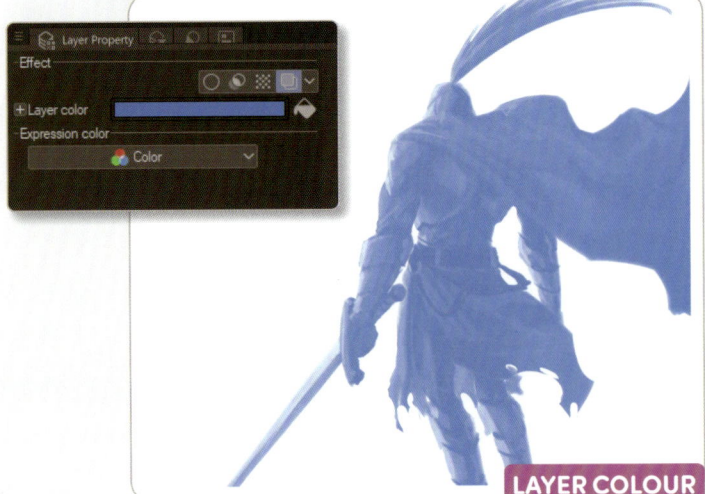

LAYER COLOUR

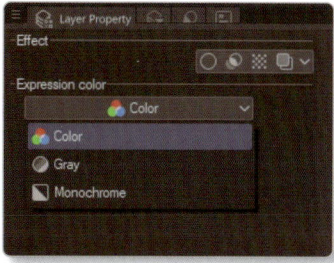

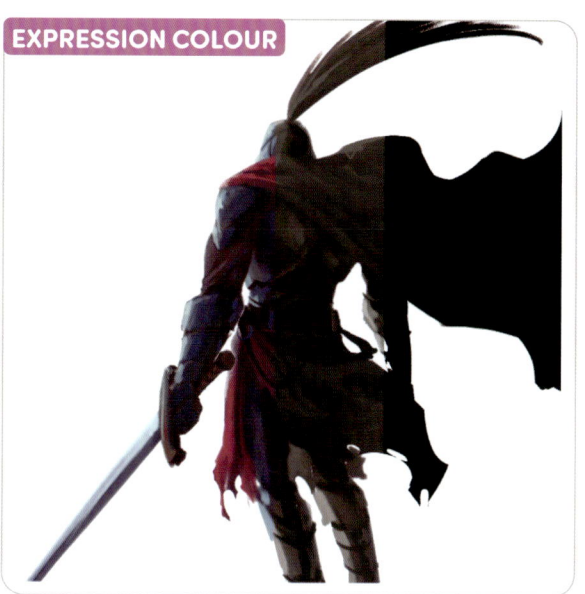

EXPRESSION COLOUR

EXPRESSION COLOUR

This setting allows you to modify the colour expression of the layer. The types of colour expressions are: Colour, Grey, and Monochrome.

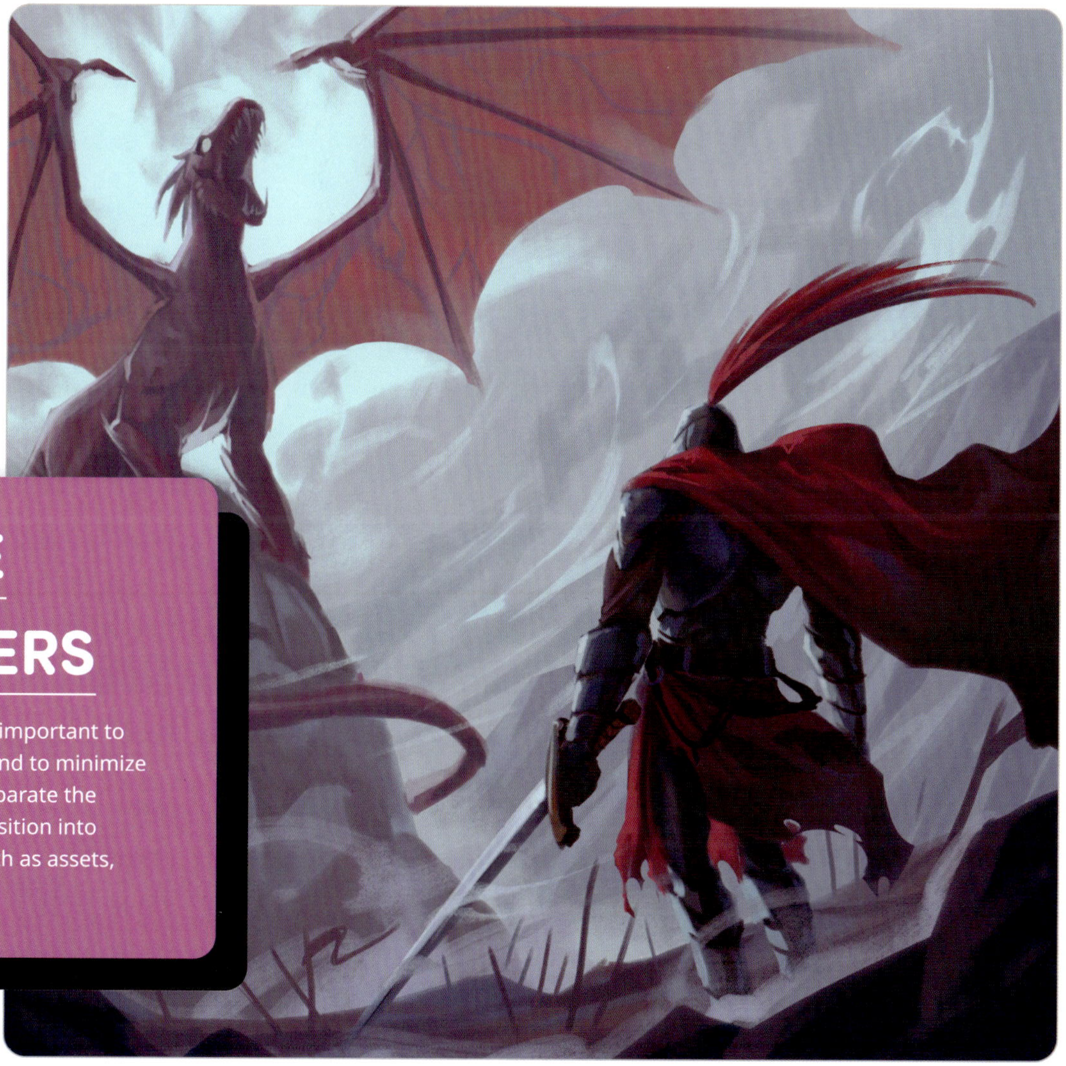

ORGANIZE
YOUR LAYERS

When working with layers, it's important to maintain some kind of order and to minimize the number of layers used. Separate the different elements of a composition into separate folders (page 41), such as assets, characters, and backgrounds.

Colour

> **LEARN HOW TO**

- Understand and use the Colour Selection icons in the Tool palette.
- Use the Eyedropper tool.
- Familiarize yourself with the Colour palettes.

The influence of colour in painting is fundamental for enriching your artwork. Not only can a colour palette convey mood and emotion, but it can also change the whole dynamic or look of a composition. Clip Studio Paint provides a variety of palettes and tools that will help you to find the ideal colours for your digital-painting projects.

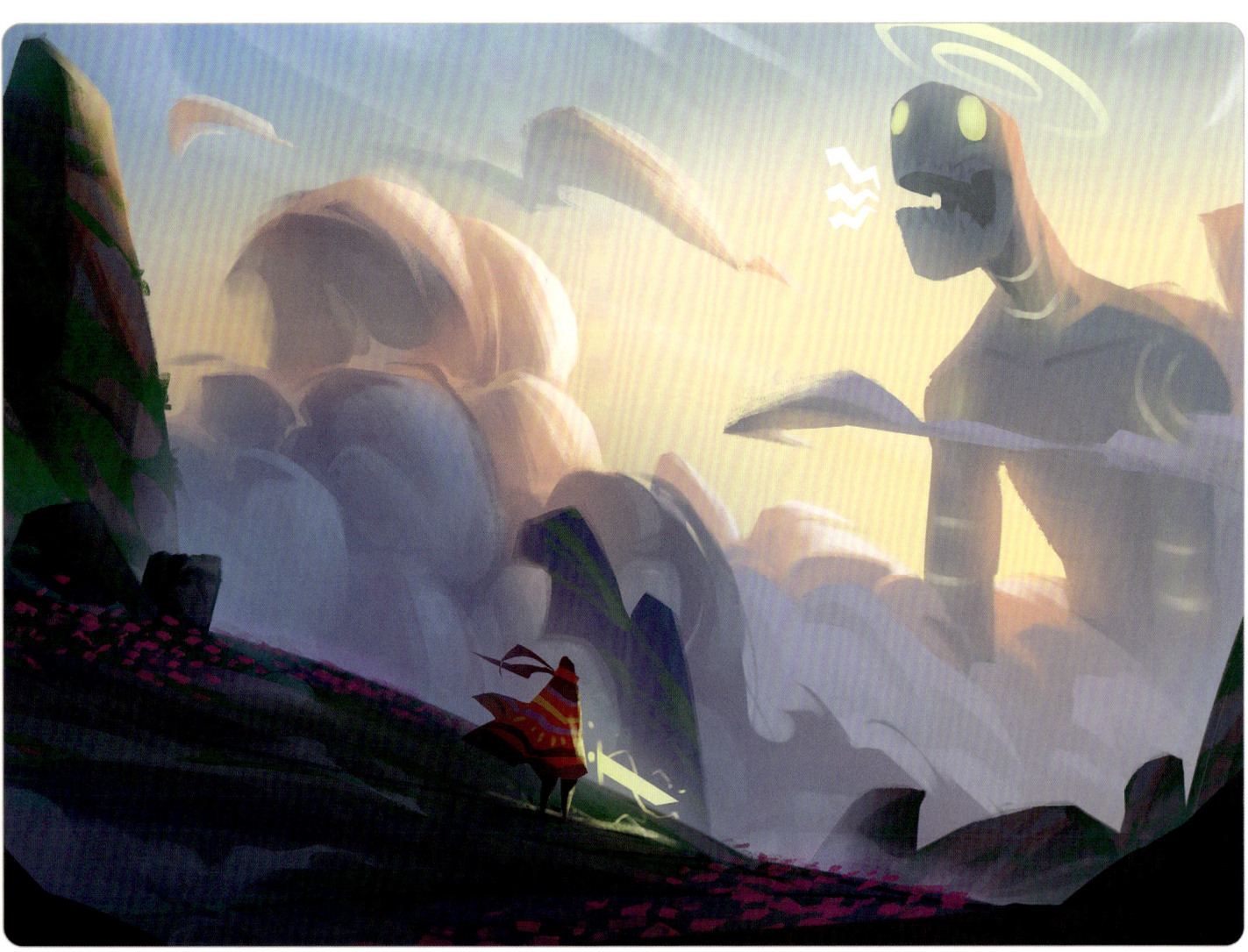

COLOUR PICKER

1. MAIN COLOUR

This function allows you to choose the drawing's main colour. When you double-click this icon, you can change the colour in the Colour Settings dialog box.

2. SECONDARY COLOUR

This function allows you to choose the drawing's secondary colour. As with the main colour, you can change the colour in the Colour Settings dialog box.

3. TRANSPARENT COLOUR

Unlike the previous two, choosing this colour function allows you to use your brush, or any drawing tool, to apply transparency. This can also be understood as erasing any area with the same brush used for painting.

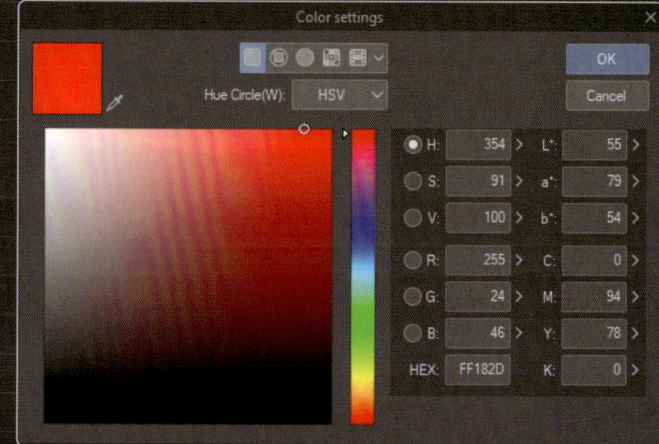

▲ The colour picker, found on the tool palette, is divided into three parts

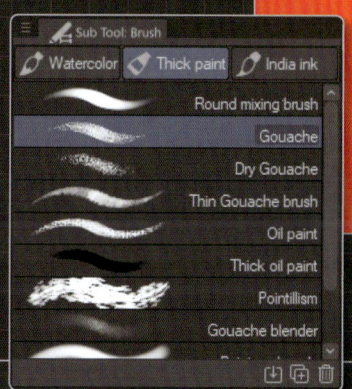

▲ The Transparent colour function is useful if you want to erase with the same texture as the brush you painted the colour with

The Colour settings window lets you choose the colour manually, or input it numerically in the colour model of your choice. The colour models you can use are: HSV (hue, saturation, value), RGB (red, green, blue), LAB (lightness, red & green, blue & yellow), CMYK (cyan, magenta, yellow, key-black), and HEX (hexadecimal colour code).

> When you select or type in one colour model, the other colour models will automatically change their code to the colour you set

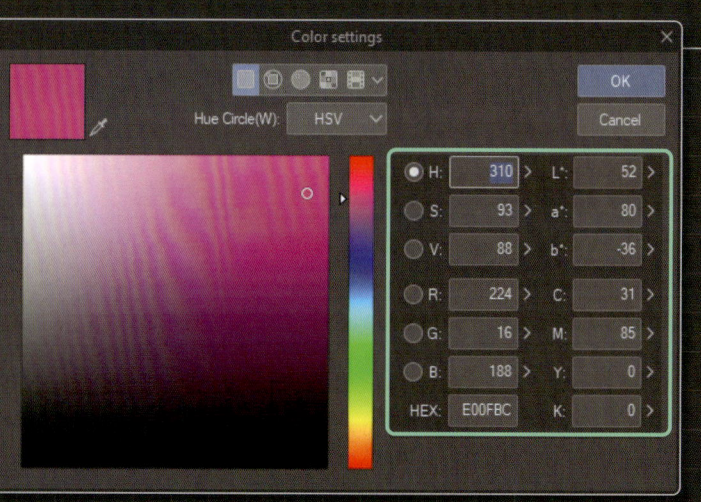

EYEDROPPER TOOL

Located in the Tool palette on the left-hand side of the interface, the Eyedropper tool allows you to select a colour from an element and apply it to the artwork. After selecting the Eyedropper tool, simply click on the specific colour you want to work with and a ring divided in two will appear. The top half is the newly selected colour, and the lower half shows the previous colour that will be replaced. This is a quick way of selecting a colour when painting or making colour adjustments.

The Eyedropper tool has two sub tools, each with a slightly different function:

PICK DISPLAYED COLOUR
Allows you to select any colour from your canvas's active layers.

PICK COLOUR FROM LAYER
Allows you to select any colour from the layer you have selected in your Layer palette.

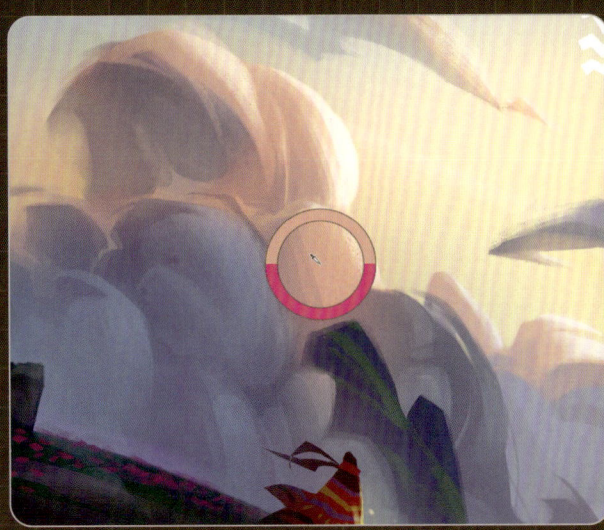

< The lower half of the ring shows the previous colour, while the top half of the ring displays the new colour that will replace it

> You can modify the settings of both sub tools in the Tool Property palette

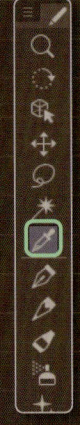
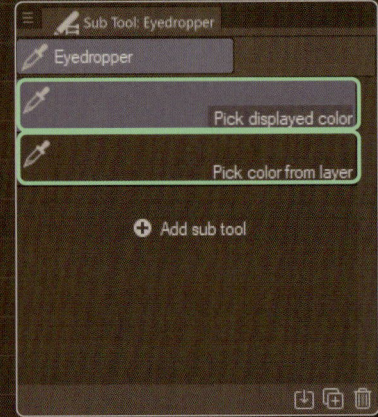

SHORTCUTS

- To select the Eyedropper tool, press the I key.

- To quickly enable the Eyedropper tool, hold down the Alt key. The tool will deactivate when you release the key.

PALETTES

Clip Studio Paint features a variety of palettes for mixing or generating the desired colour. These palettes are grouped together and located in the lower left of the interface, and displays the Colour Wheel view by default. You can also enable a palette from the Window menu.

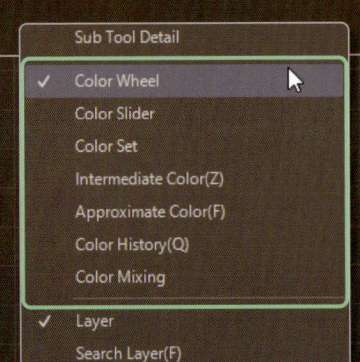

< The check mark shows which palette is active

COLOUR WHEEL PALETTE

The Colour Wheel palette is one of the most useful palettes for selecting colours. Presented as a circular spectrum, it offers a wide variety of colours composed of primary and secondary colours, organized according to their hue or tone. The palette also shows the HSV or HLS colour space.

> To switch to the HLS colour space, click the icon in the bottom right of the palette

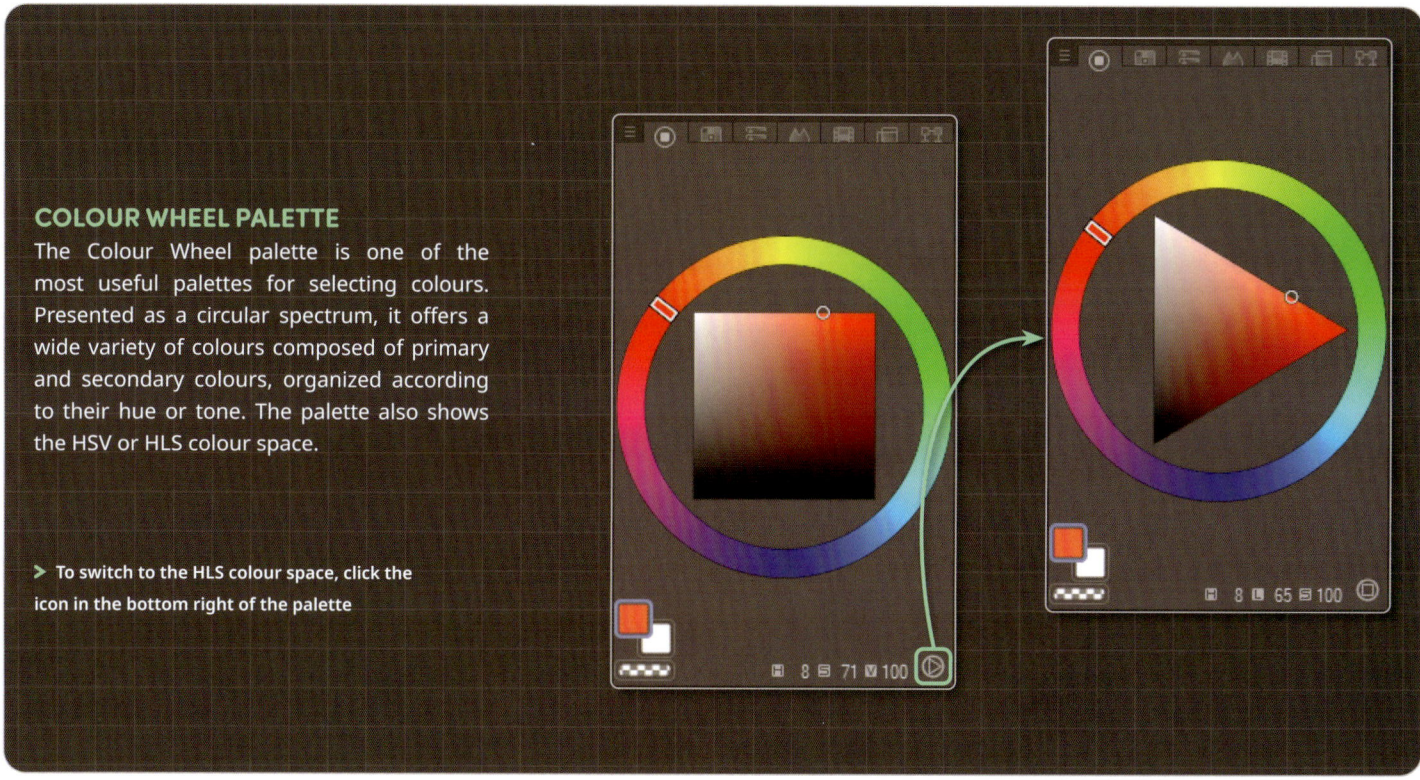

COLOUR SET PALETTE

This palette allows you to store and quickly access the exact custom colours you select or create for your artwork.

ⱽ Here you can store your favourite custom colours for quick access

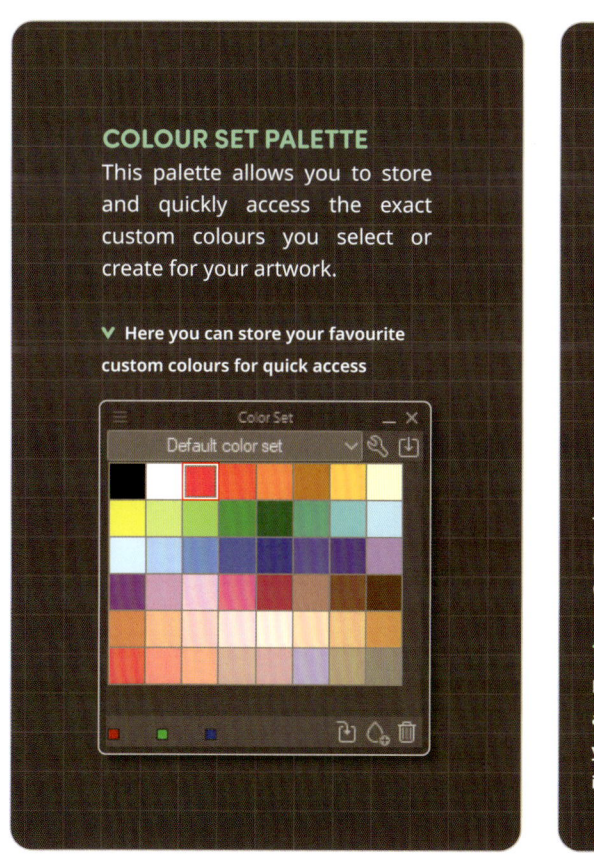

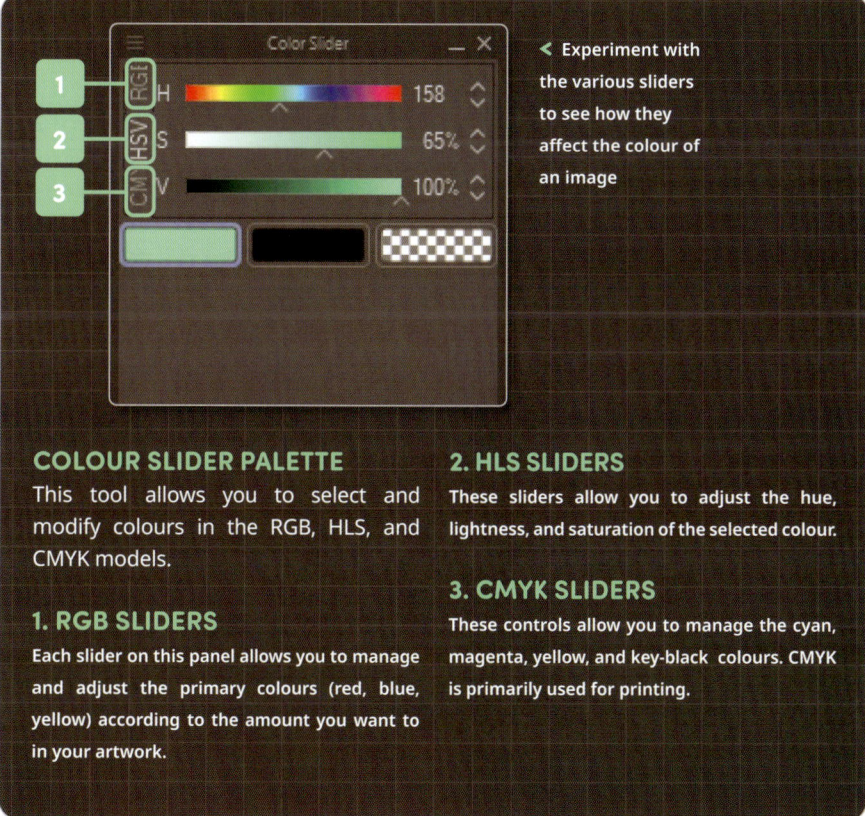

< Experiment with the various sliders to see how they affect the colour of an image

COLOUR SLIDER PALETTE

This tool allows you to select and modify colours in the RGB, HLS, and CMYK models.

1. RGB SLIDERS

Each slider on this panel allows you to manage and adjust the primary colours (red, blue, yellow) according to the amount you want to in your artwork.

2. HLS SLIDERS

These sliders allow you to adjust the hue, lightness, and saturation of the selected colour.

3. CMYK SLIDERS

These controls allow you to manage the cyan, magenta, yellow, and key-black colours. CMYK is primarily used for printing.

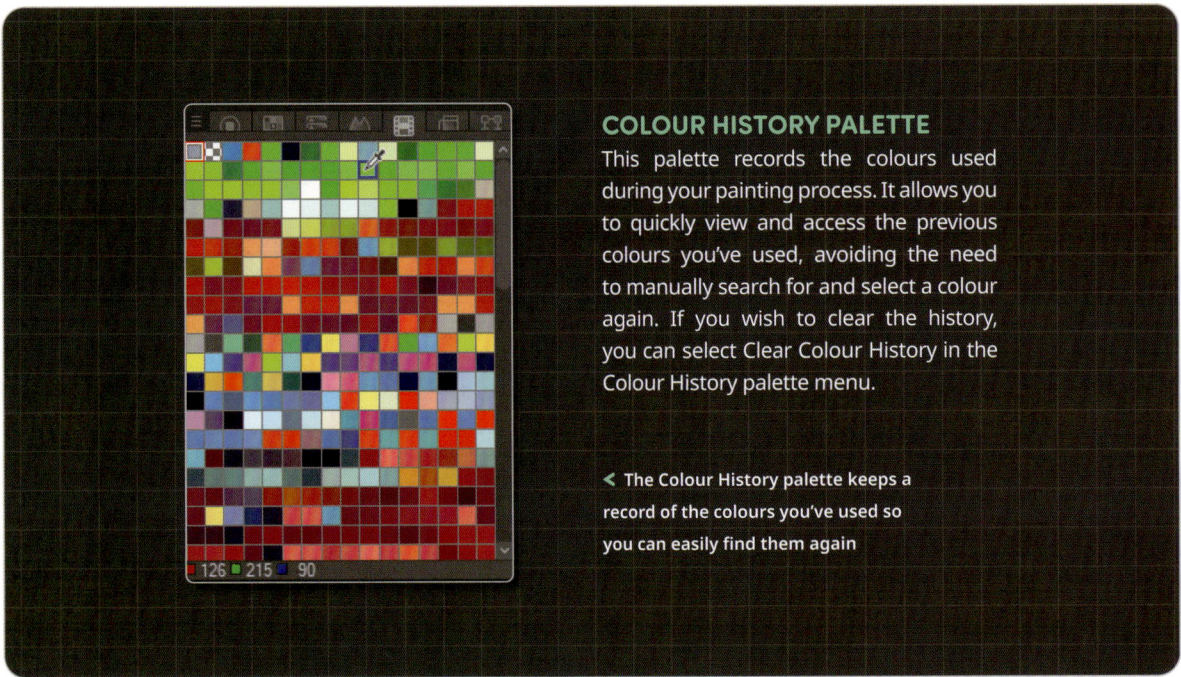

COLOUR HISTORY PALETTE

This palette records the colours used during your painting process. It allows you to quickly view and access the previous colours you've used, avoiding the need to manually search for and select a colour again. If you wish to clear the history, you can select Clear Colour History in the Colour History palette menu.

< The Colour History palette keeps a record of the colours you've used so you can easily find them again

APPROXIMATE COLOUR PALETTE

This palette allows you to select colours close to a specific colour you have previously selected.

∨ Find varying shades similar to a previously used colour using the Approximate Colour palette

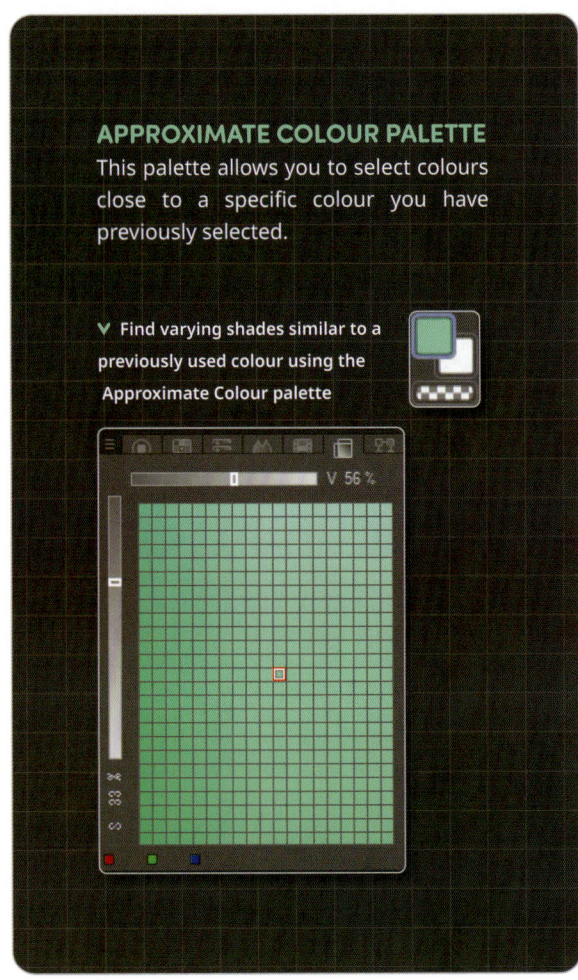

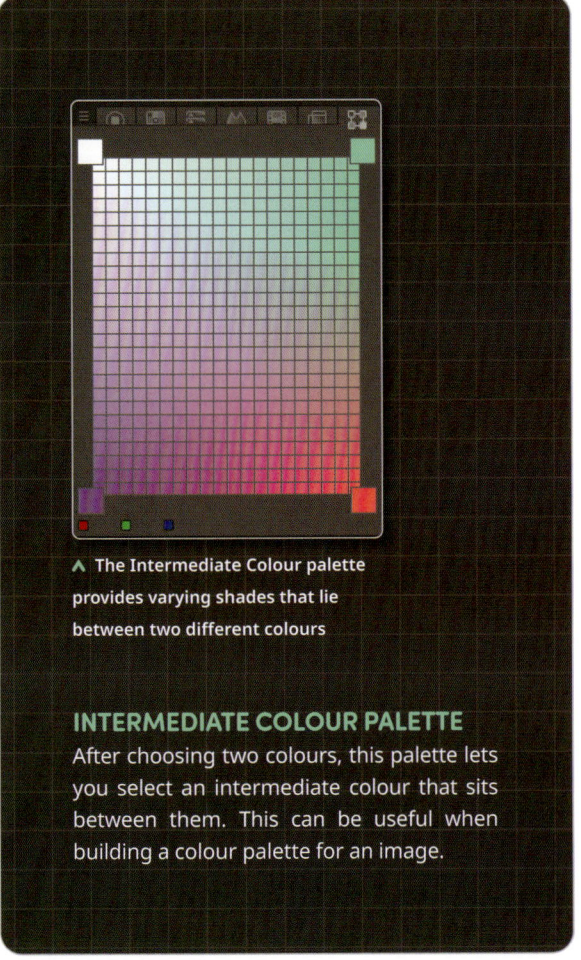

∧ The Intermediate Colour palette provides varying shades that lie between two different colours

INTERMEDIATE COLOUR PALETTE

After choosing two colours, this palette lets you select an intermediate colour that sits between them. This can be useful when building a colour palette for an image.

EXPERIMENT WITH COLOUR

Try using the Eyedropper tool and Brush tool to create different colours. For example, start by painting a secondary colour over a base colour, press Alt to activate the Eyedropper tool, and select the colour produced from the blend of the base and secondary colours. Repeat the same process to select another colour generated from the blend of the selected colour and the other colours.

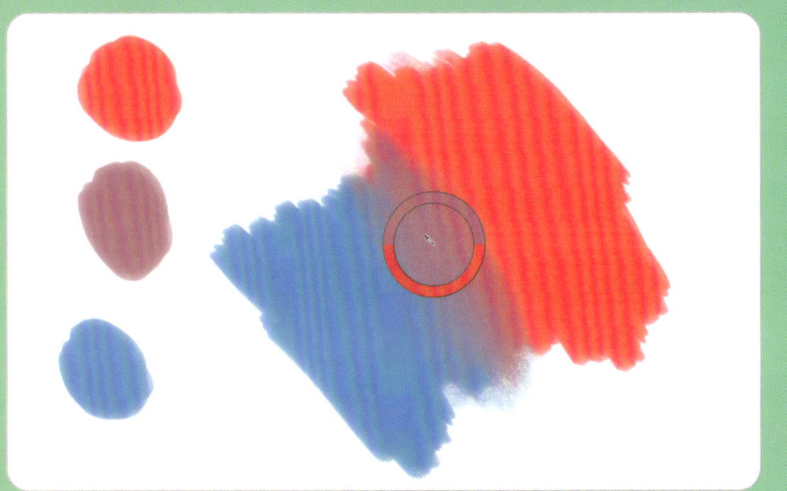

COLOUR MIXING PALETTE

This palette allows you to simulate colour mixing on a physical paint palette. By selecting combinations of two or more colours, you can mix and blend them together to create new digital tones.

1. COLOUR AREA

This displays the current colour you have selected. You can change the colour by clicking on the icon.

2. CLEAR

This clears the drawing area.

3. UNDO & REDO

4. BRUSH SIZE

5. USE SAME SUB TOOL AS CANVAS

6. BRUSH

7. BLEND

Activating this function blends the colours.

8. EYEDROPPER

Used to select colours from the canvas.

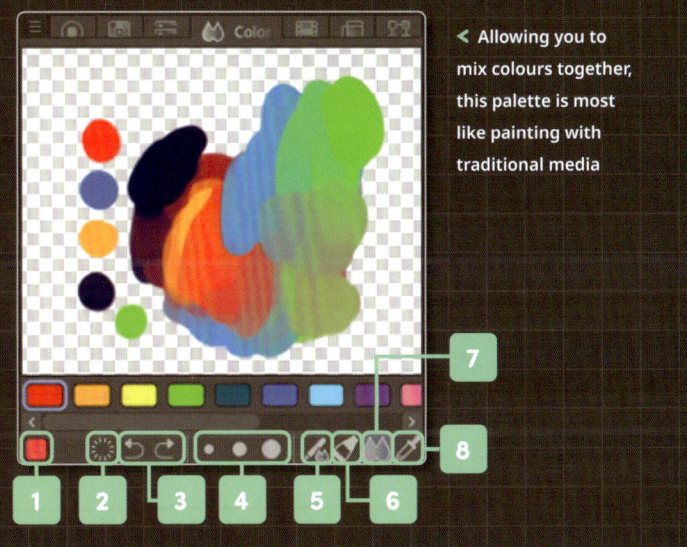

< Allowing you to mix colours together, this palette is most like painting with traditional media

SHORTCUTS

- Hold down the Alt key to temporarily activate the Eyedropper tool in the Colour Mixing palette.

- Use the mouse wheel to zoom in on the drawing area.

- Use the space bar to pan around the drawing area.

- Hold down Ctrl+Alt, then click and drag to change the brush size in the Colour Mixing palette.

Selection

The Selection tool allows you to establish which areas of an artwork you wish to paint or manipulate. Clip Studio Paint's Selection tools are varied and can be useful for numerous different situations. The basic functions of selection can be found in the Selection menu, located at the top of the interface.

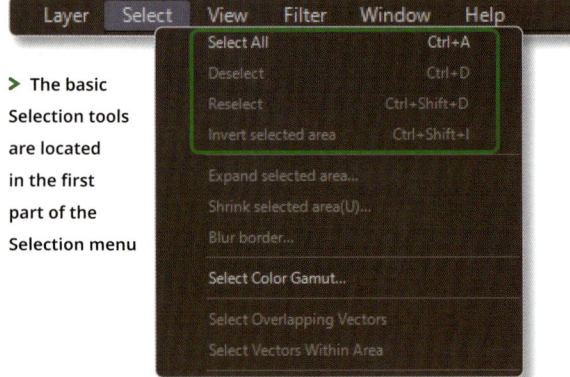

> The basic Selection tools are located in the first part of the Selection menu

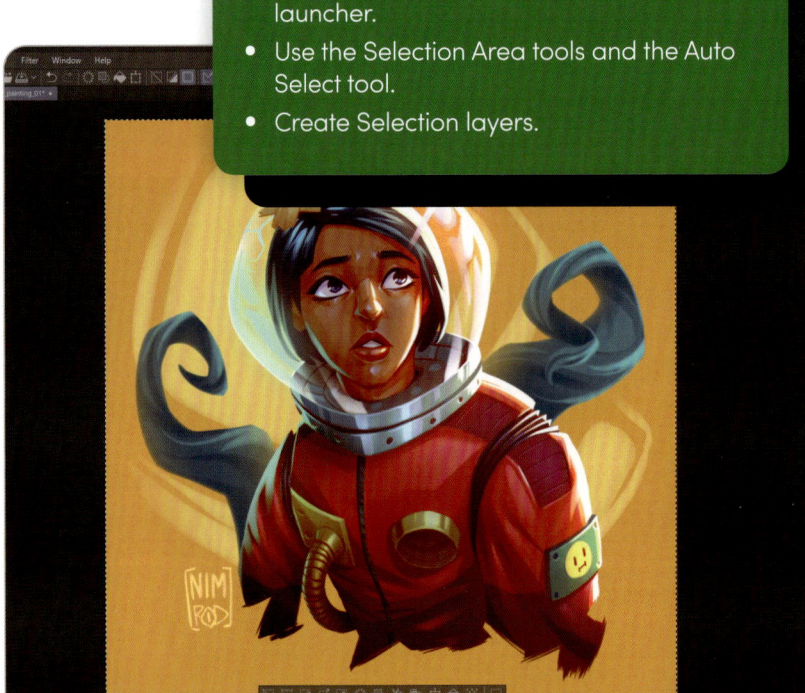

∧ Press Ctrl+A to select everything on the canvas

SELECTION MENU

SELECT ALL
This function selects everything on the canvas. The keyboard shortcut is **Ctrl+A.**

DESELECT
This function deselects your current selection. The keyboard shortcut is **Ctrl+D.**

RESELECT
This function reselects the last deselection. The keyboard shortcut is **Ctrl+Shift+D.**

INVERT SELECTED AREA
This function inverts the selected area, selecting the area not touched in the previous selection. The last selection becomes a deselection. The keyboard shortcut is **Ctrl+Shift+I.**

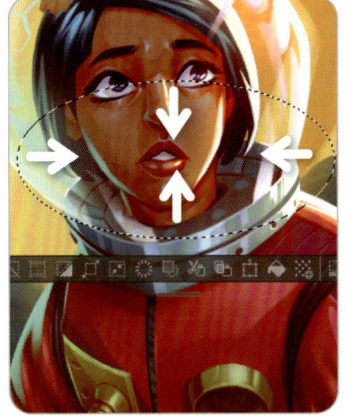
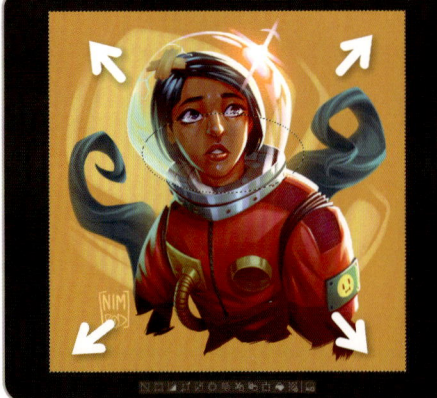

∧ Press Ctrl+Shift+I to invert the selected area

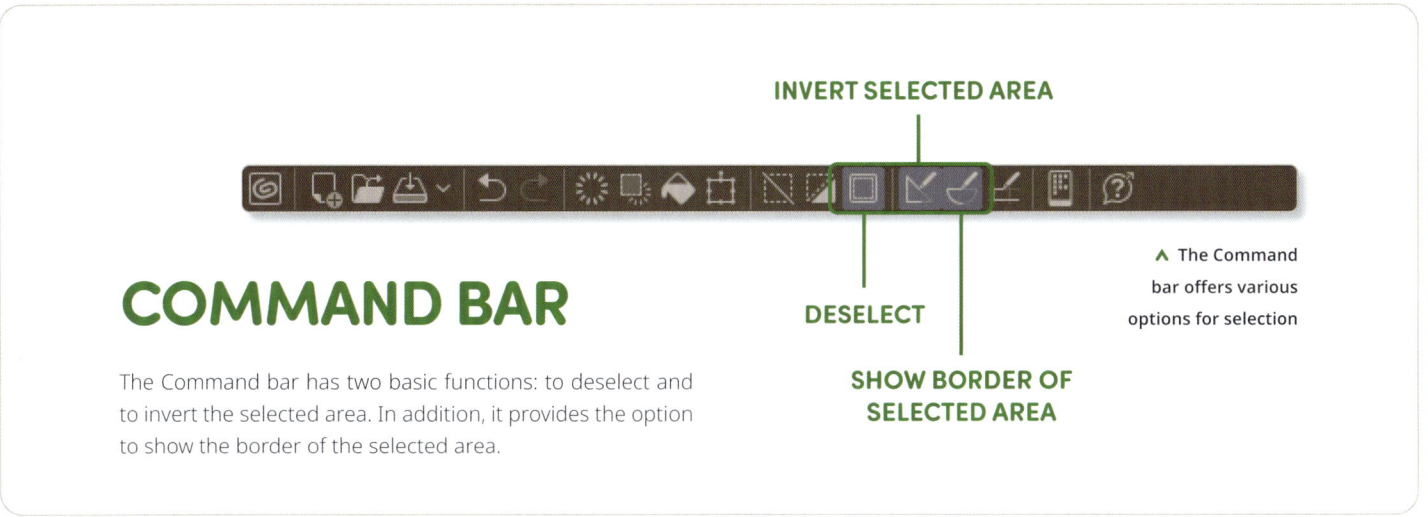

INVERT SELECTED AREA

DESELECT

**SHOW BORDER OF
SELECTED AREA**

∧ The Command
bar offers various
options for selection

COMMAND BAR

The Command bar has two basic functions: to deselect and to invert the selected area. In addition, it provides the option to show the border of the selected area.

SELECTION LAUNCHER

When you select an area on the canvas, a bar with various icons will automatically appear. This contains some of the basic selection actions already covered, as well as new selection functions that can be applied to your selection, such as Fill and Scale/Rotate. You can hide the Selection Launcher by selecting View > Selection Launcher.

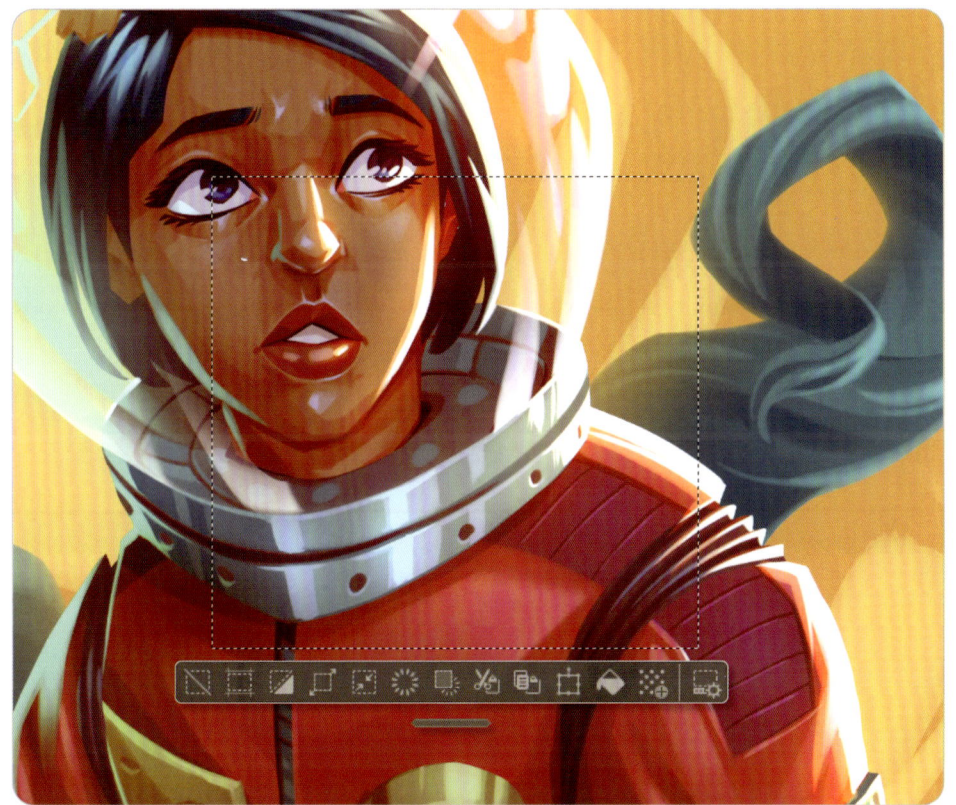

∧ The Selection Launcher bar appears below the selected area

1. DESELECT

Deselect the selected area.

2. CROP

Crop the current canvas size to the size of the selection.

3. INVERT SELECTED AREA

4. EXPAND SELECTED AREA

A small window will appear where you can define the size in pixels and type of expansion.

5. SHRINK SELECTED AREA

Similar to Expand Selected Area, a window will appear where you can define the size in pixels and the type of area reduction.

6. DELETE

This removes everything within the selected area. You can also find this function in the Command bar. Another way to delete is by pressing the **Del** key.

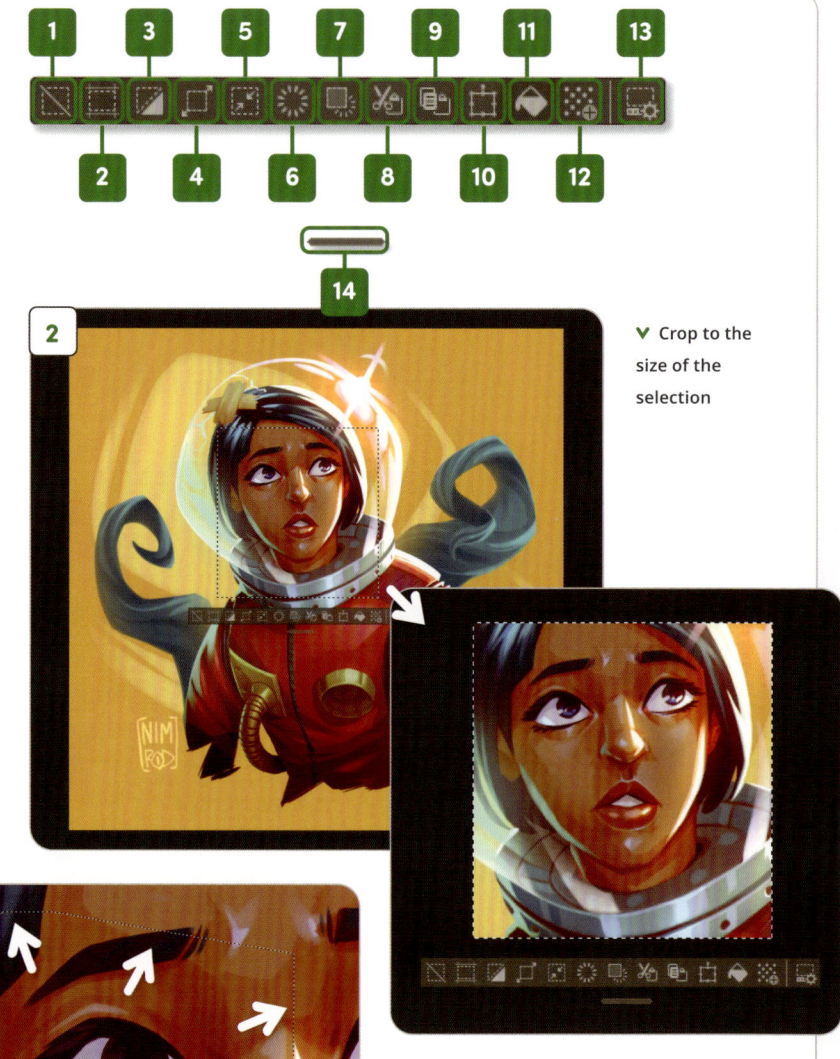

∨ Crop to the size of the selection

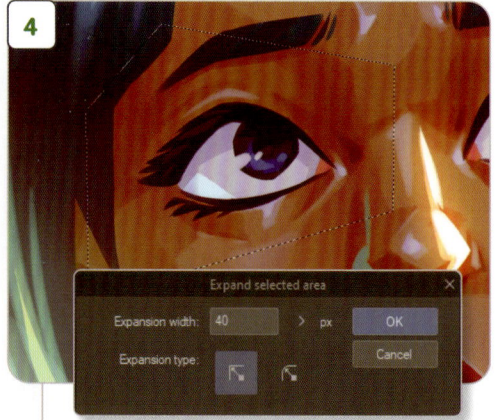

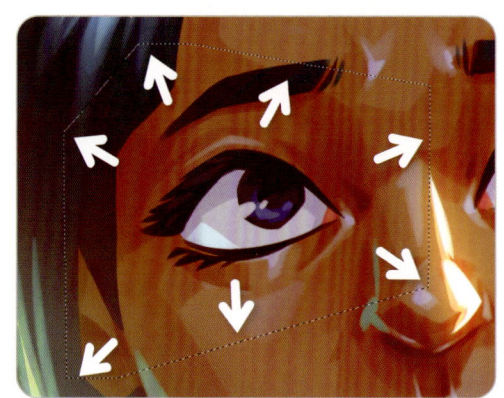

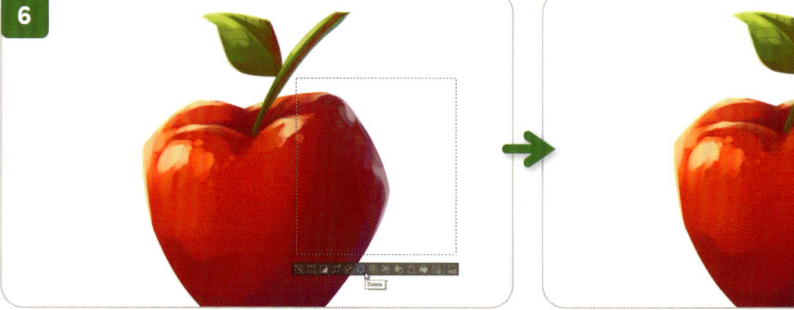

› Delete will remove the selection

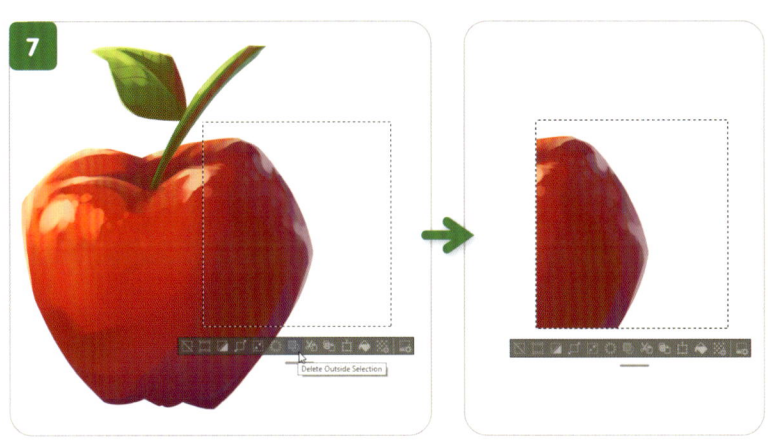

▲ Delete outside selection

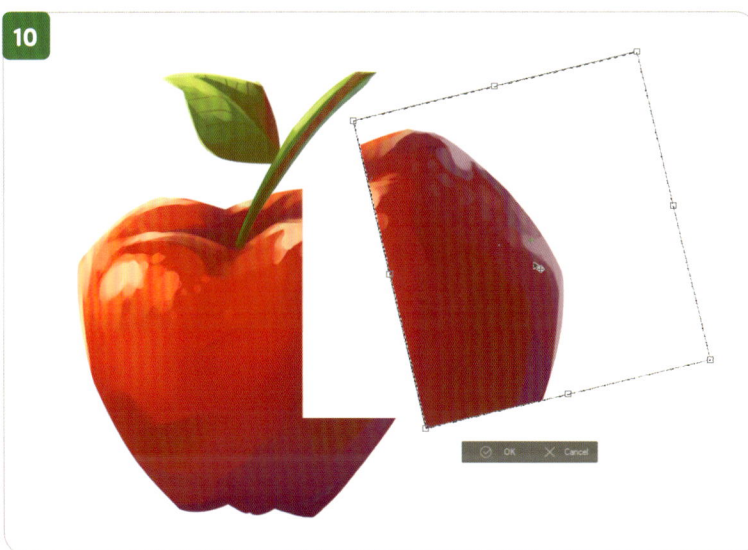

▲ Transform the interior of the selected area

7. DELETE OUTSIDE SELECTION

This function deletes everything outside of the selected area. This action can also be found in the Command bar.

8. CUT AND PASTE

This cuts the interior of the selected area and pastes it onto a new layer.

9. COPY AND PASTE

This option copies the interior of the selected area and pastes it onto a new layer.

10. SCALE/ROTATE

Also found in the Command bar, the Transformation function allows you to move, scale, and rotate the interior of the selected area. (This is also explored in **Layers**, page 38.)

11. FILL

This action fills the selected area with colour. The function can also be found in the Command bar.

12. NEW TONE

This creates a new selection tone layer and will be further explored on page 80.

13. SELECTION LAUNCHER SETTINGS

This option allows you to customize the Selection Launcher.

14. MOVE SELECTION LAUNCHER

Located at the bottom of the bar, this lets you move the Selection Launcher to any part of the canvas. Simply use the cursor to click and drag it to the desired location.

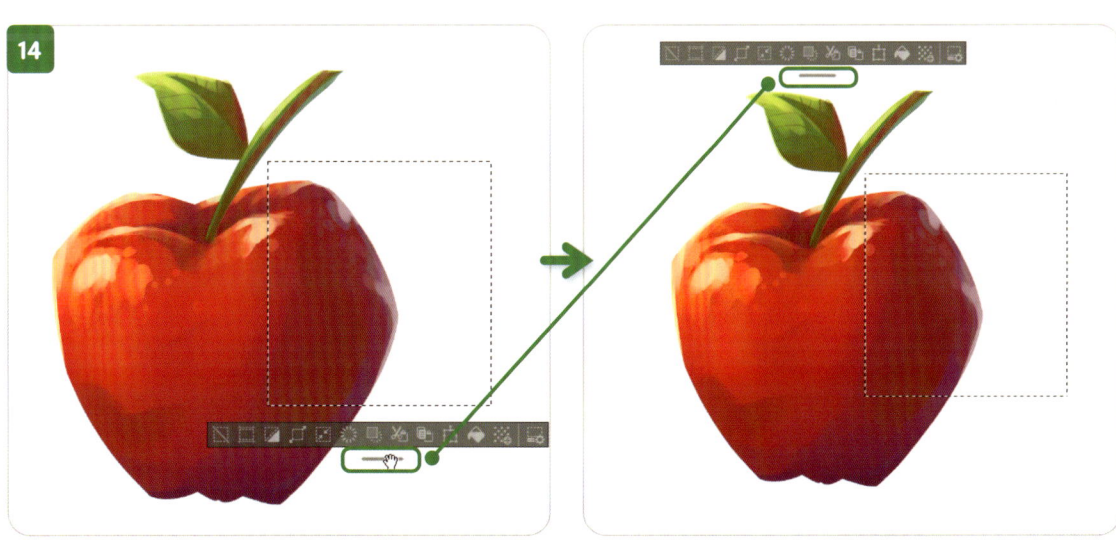

◀ Move the Selection Launcher to where you would like it to sit on the interface

SELECTION AREA TOOL

Located in the Tool palette on the left-hand side of the interface, the Selection Area tool allows you to select areas in various shapes. Sub tools include Rectangle, Ellipse, Lasso, and Selection Pen. The keyboard shortcut is the **M** key.

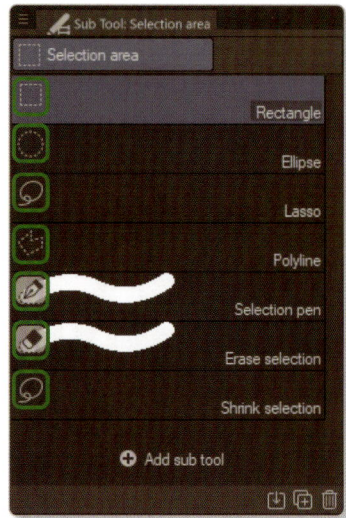

∧ The Selection Area tool offers various sub tools for different ways to make a selection

RECTANGLE

This sub tool enables you to create a rectangular selection. Use the cursor to click and drag to shape the size of the selection.

> **Create rectangular selections using the Rectangular sub tool**

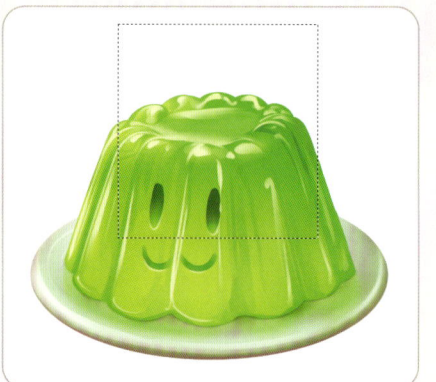

ELLIPSE

This allows you to create an elliptical selection. Simply click and drag the cursor to shape the selection.

> **Use the Ellipse sub tool to create elliptical selections**

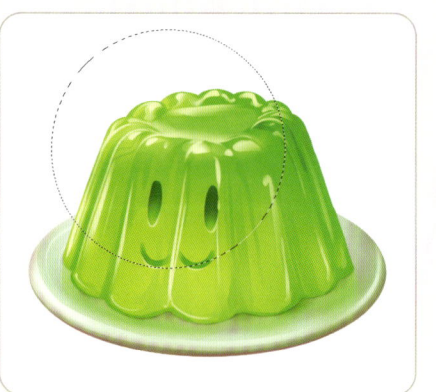

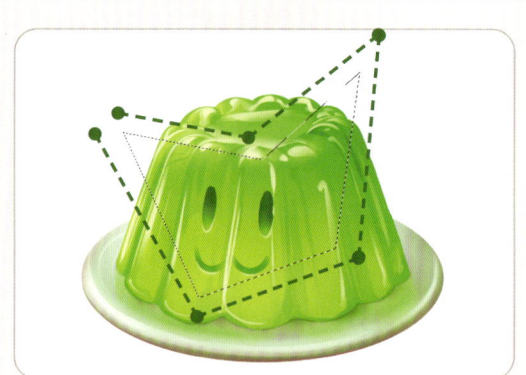

∧ The Polyline sub tool lets you create a selection made up of straight lines

POLYLINE

This sub tool allows you to create a selection with continuous straight segments. Start by clicking to set a starting point, then drag the cursor, click where the next segment will be, and repeat this process until clicking on the segment's origin point to close up the shape.

LASSO

The Lasso sub tool enables you to make a selection by drawing around it.

> **Draw around the area you wish to select using the Lasso sub tool**

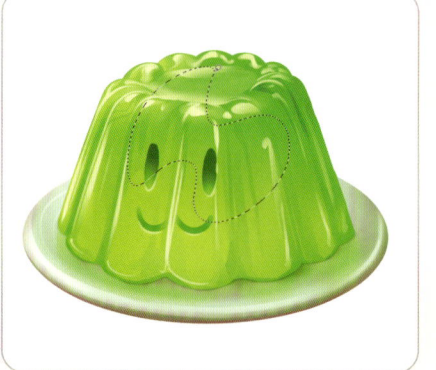

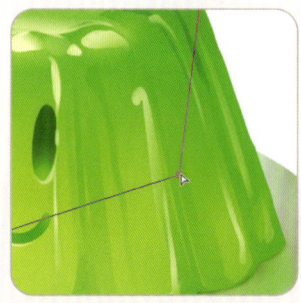

< **Before closing your selection, you can reposition the handles by clicking and dragging them while holding the Ctrl key**

SELECTION PEN

This sub tool lets you make a selection by drawing the area you want to select, as if it were a brush. You can alter the brush size, density, stabilization, or anti-aliasing in the Tool Property palette.

> **Draw the area you want to select using the Selection Pen**

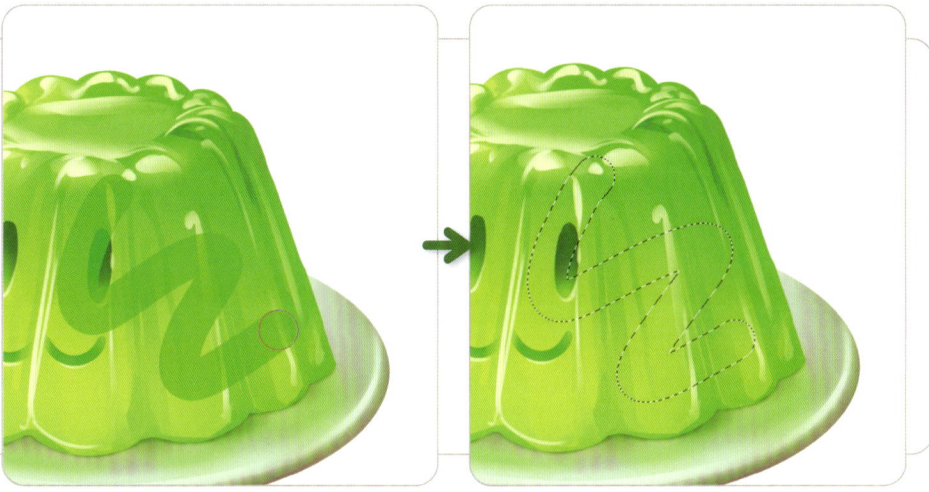

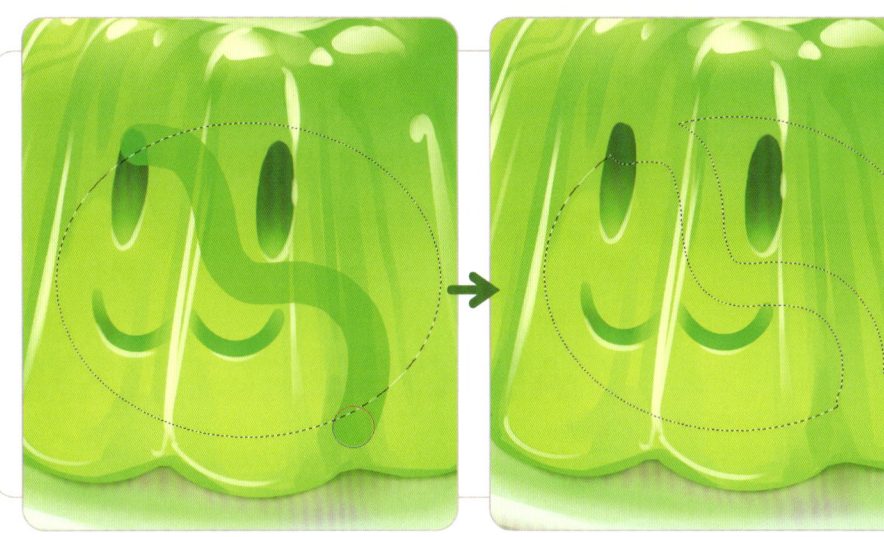

ERASE SELECTION

This sub tool lets you erase part of a selection, as if it were the eraser tool.

< Here a random line across the selected ellipse is being erased from the selection

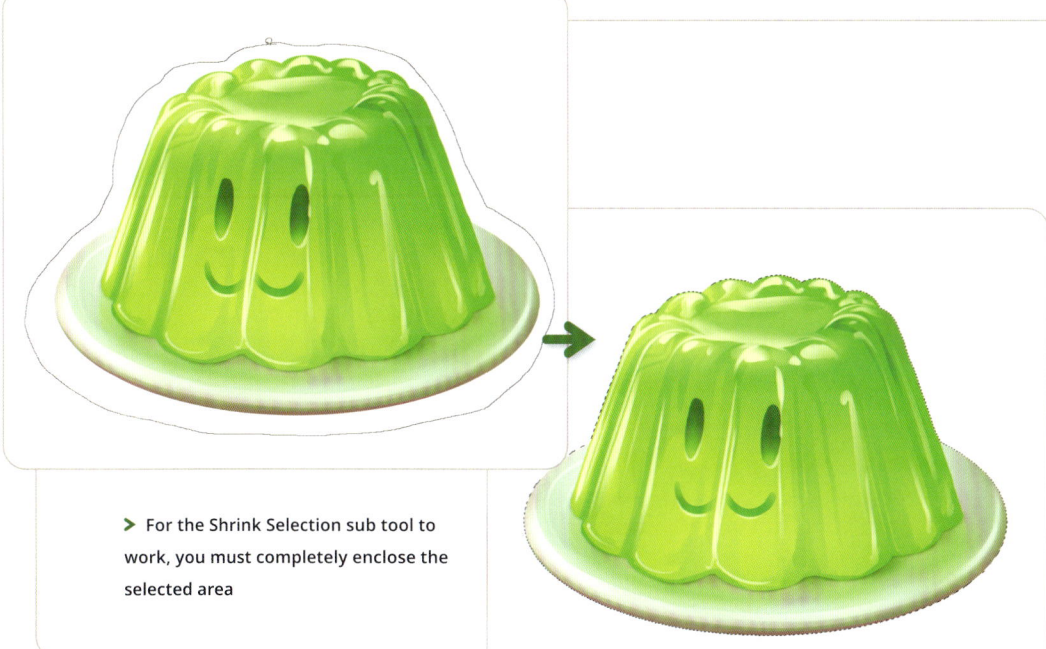

SHRINK SELECTION

This works in a similar way to the Lasso tool. When you surround a drawn area, it adjusts to the shape of the selected area. This area must have closed strokes. It also works on groups of multiple layers.

> **For the Shrink Selection sub tool to work, you must completely enclose the selected area**

SELECTION MODE

Each Selection sub tool comes in four selection modes. Each one differs in the way it makes the selection. These selection modes can be found in the Tool Property palette.

NEW SELECTION

This option creates a new selection. If you create a selection for the second time, the first selection will disappear.

ADD TO SELECTION

This adds a new selection to a previously created selection. You can also activate this function by holding down the **Shift** key.

REMOVE FROM SELECTION

This action erases the selected area. Another way to activate this is by holding down the **Alt** key.

SELECT FROM SELECTION

This option selects the intersection of a new selection and an existing one. Another way to activate this is by holding down the **Shift** and **Alt** keys.

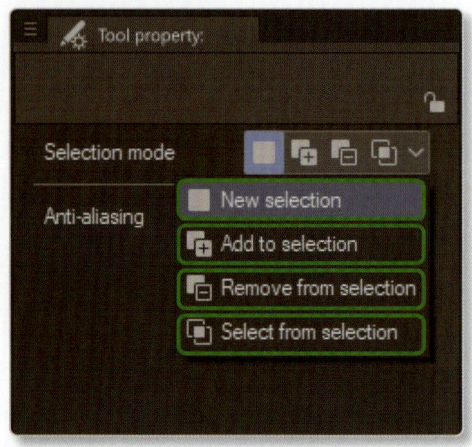

▲ **Expand the menu to see the name of each selection mode**

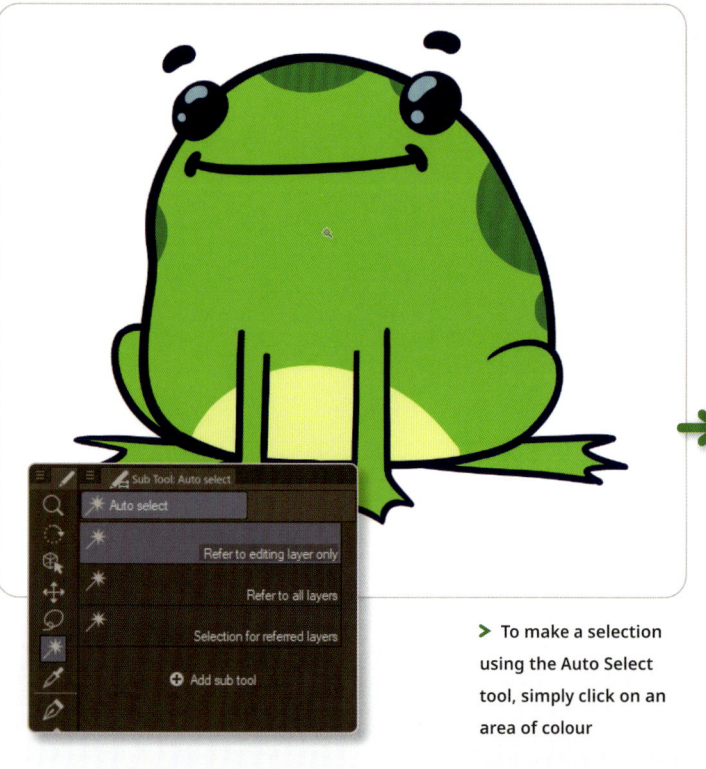

> To make a selection using the Auto Select tool, simply click on an area of colour

AUTO SELECT TOOL

The Auto Select tool works by selecting areas of the same colour. You can add more than one selection by holding down the **Shift** key, or remove more than one selection by holding down the **Alt** key. The keyboard shortcut for this tool is the **W** key.

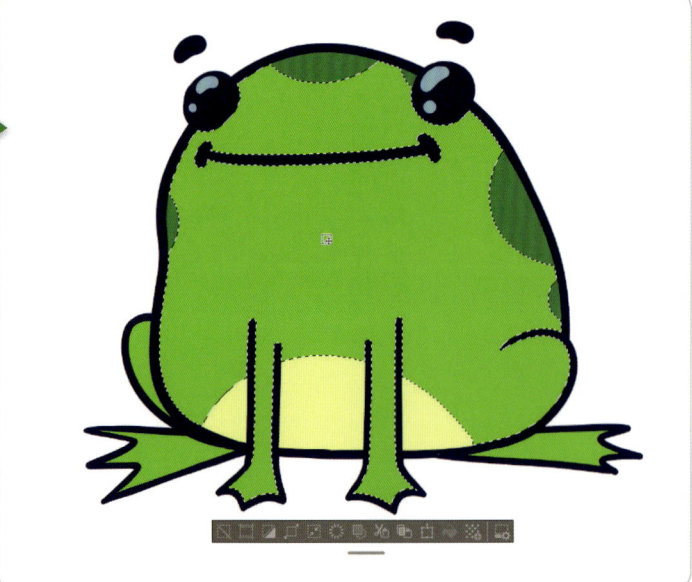

SELECTION LAYER

Once you've selected an area of your work, you may want to save that selection to a new layer to use later. Selection Layers allow you to do this easily. With your selection made, simply click **Select > Convert to Selection Layer**.

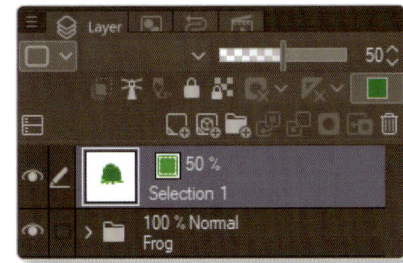

< In this example, the selected outline of the frog has been moved to the base colour layer

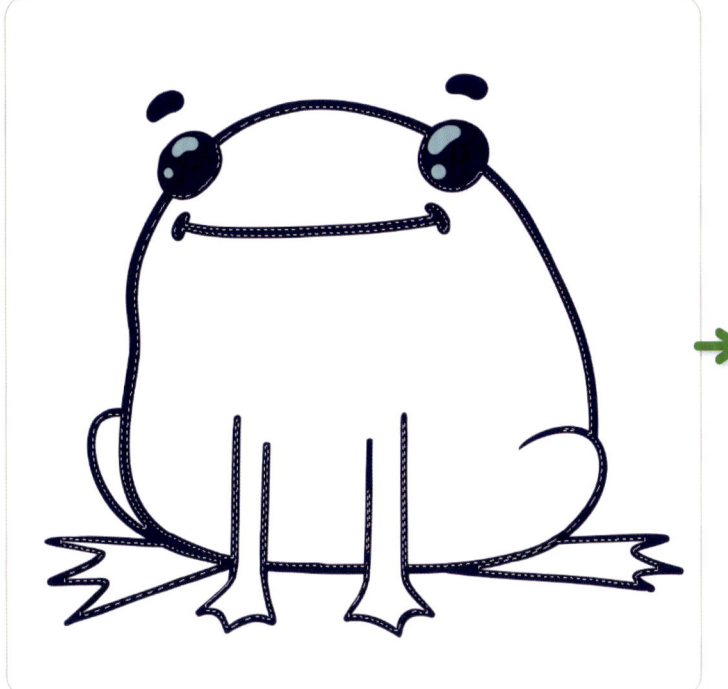

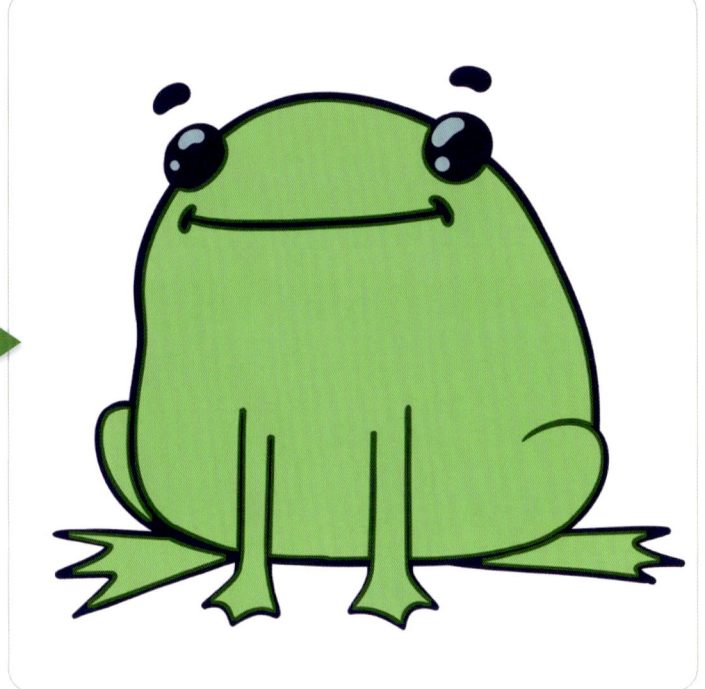

SELECTING LAYERS

To quickly make a selection of everything on a specific layer, hold down the **Ctrl** key and click the layer icon you want to select.

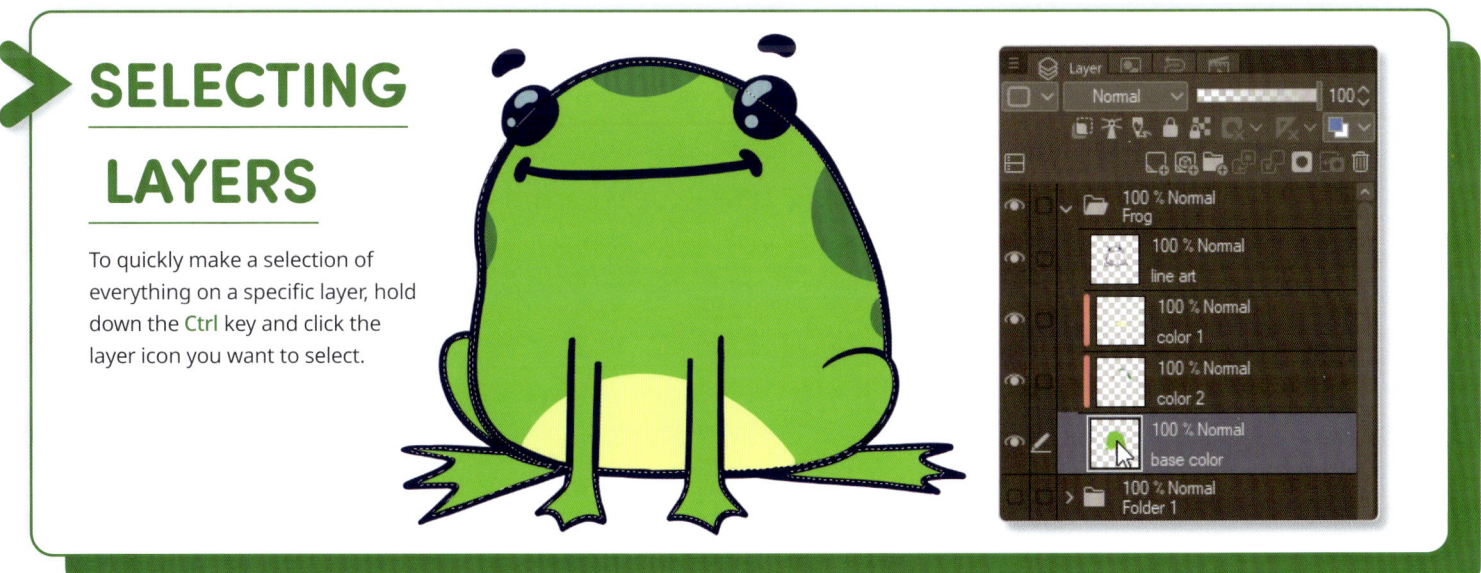

Fill & gradient

The Fill and Gradient tools assist the process of applying uniform painting or gradient painting styles. These are located in the Tools palette on the left-hand side of the interface.

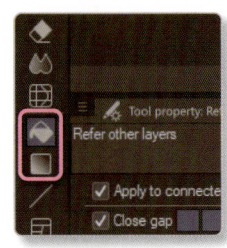

> **The Fill and Gradient tool**
icons in the Tools palette

> LEARN HOW TO

- Use the Fill tool and its sub tools.
- Use the Gradient tool and its sub tools.

FILL

The Fill tool enables you to fill a free area, painted area, or area bordered by lines, with colour. It offers the following sub tools.

REFER ONLY TO EDITING LAYER
This sub tool isolates the fill to the layer you are currently working on.

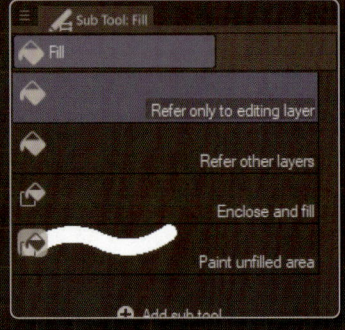

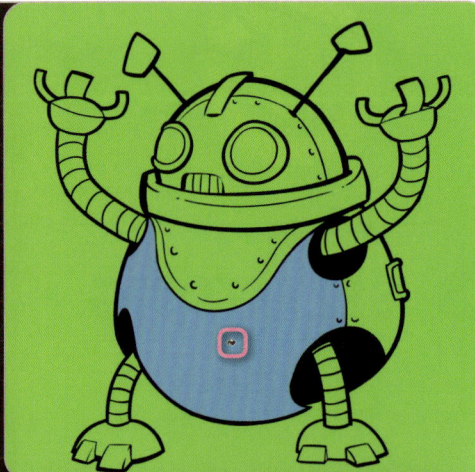

> PAINTING PROCESS

When painting an artwork – whether a character or full scene – the painting process usually starts with the base colours to help establish the colour palette. Additional layers of colour and detail are then applied to harmonize with the base colours. The Fill tool is useful for this, saving you from needing to paint in each base colour manually.

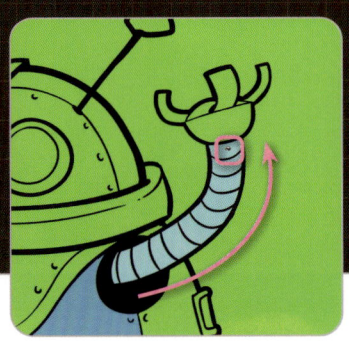

ʌ Click to fill the area surrounded by the lines with the selected colour

< Click and drag the Fill tool over multiple areas to fill them with colour

REFER OTHER LAYERS

This sub tool fills colour according to what's visible on other layers. This is useful if you need your colour to follow the shapes or boundaries on another layer, while still keeping it on its own layer.

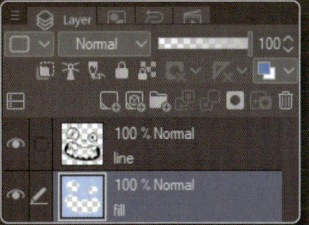

< This Fill sub tool allows you to keep colour and line work layers separate

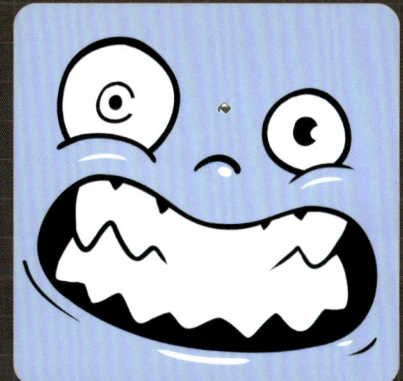

SHORTCUTS

- Press the G key once to activate the Fill tool.

- Press Alt+Delete to fill the selected area with colour. If nothing is selected, the entire canvas will fill with colour.

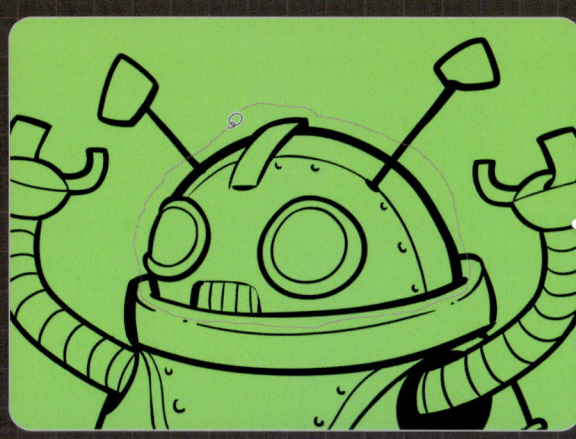

^ Draw and encircle the area you want to fill with colour

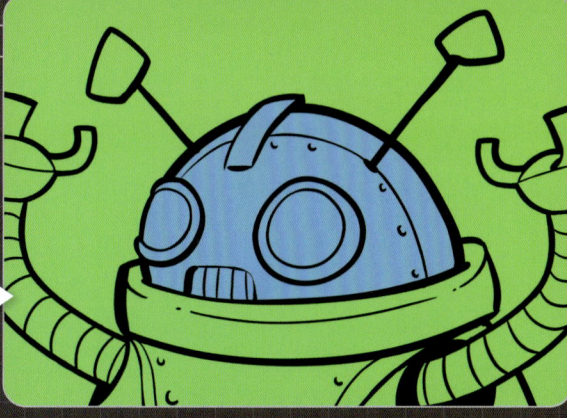

ENCLOSE & FILL

This sub tool can be used to quickly fill selected areas by manually surrounding them. Similar to the previous sub tool, this also uses all layers in the project as a reference.

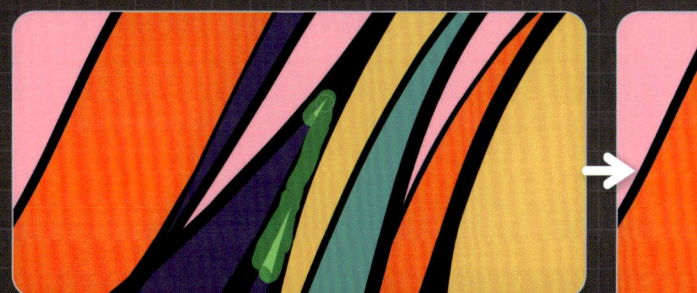

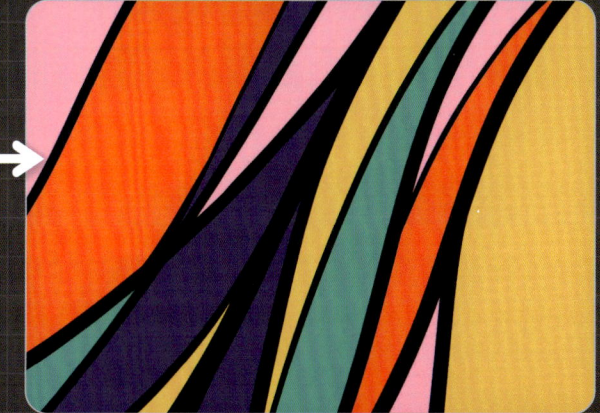

PAINT UNFILLED AREA

This sub tool is perfect for reaching those small areas that are difficult to fill with colour. Simply click and drag over any areas that were previously missed or forgotten about.

^ Click and drag over small areas to fill them with colour

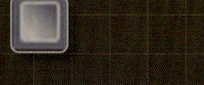

GRADIENT

This tool can be used to create colour gradients. Simply click and drag the pen on the canvas in the direction you wish to place the gradient.

> The Gradient tool contains various sub tools and preset variations that can be modified

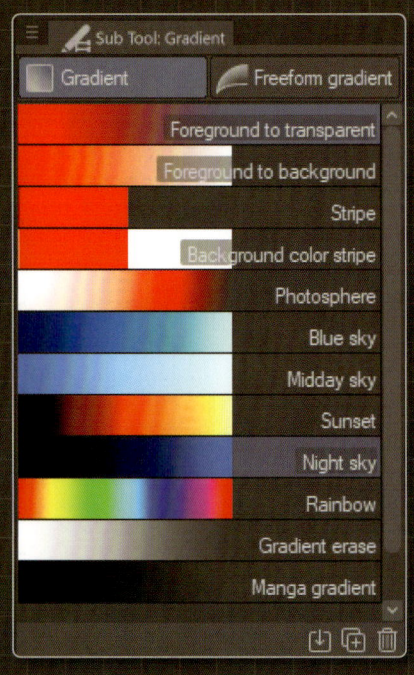

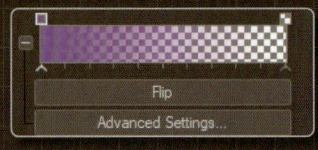

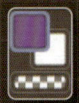

∧ The Gradient bar lets you choose colours, add more colours, and invert the colours

GRADIENT BAR

The Gradient bar is one of the main settings on the Tool Property palette. It provides the option to specify the gradient colours, as well as add more colours. It also allows you to invert the gradient colours by selecting the Flip button below the bar. To specify a gradient colour, click on one of the small squares at the top of the Gradient bar. Another way to modify the colours is in the Edit Gradient window when you click the Advanced Settings button.

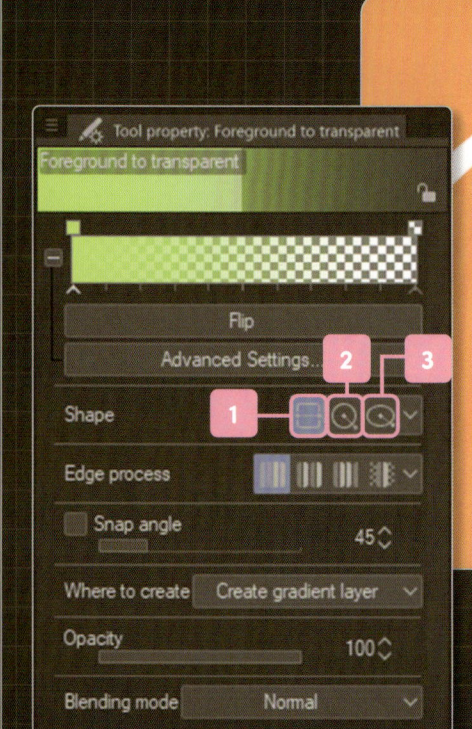

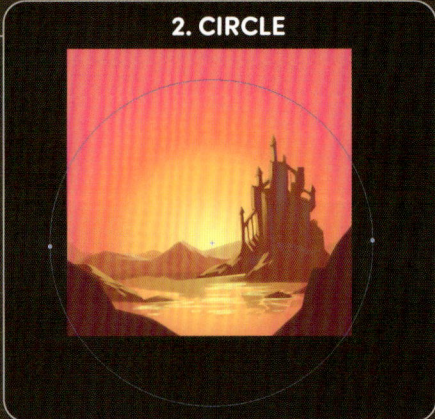

2. CIRCLE

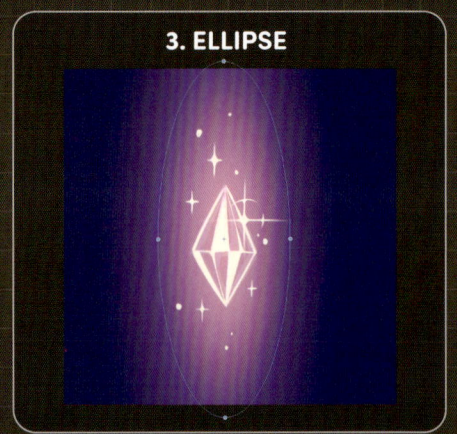

3. ELLIPSE

1. STRAIGHT LINE

∧ Gradients can be created in various different shapes

SHAPE

The Shape setting offers three ways to define the shape of the gradient: straight line, circle, and ellipse.

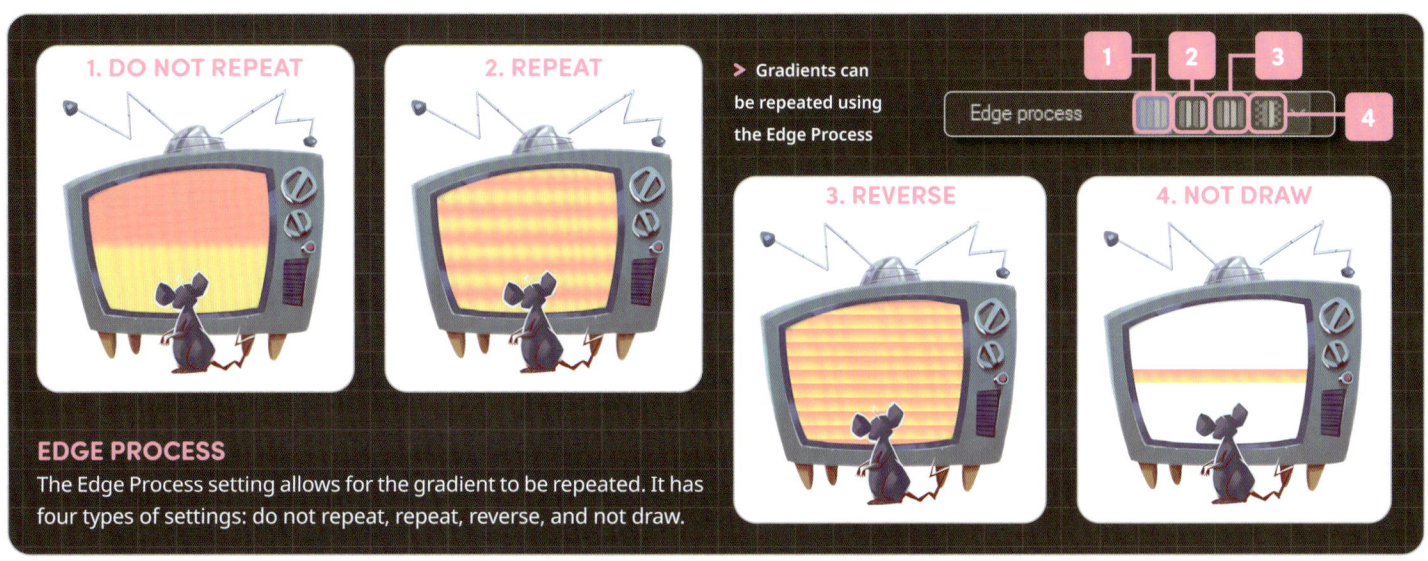

EDGE PROCESS

The Edge Process setting allows for the gradient to be repeated. It has four types of settings: do not repeat, repeat, reverse, and not draw.

> Gradients can be repeated using the Edge Process

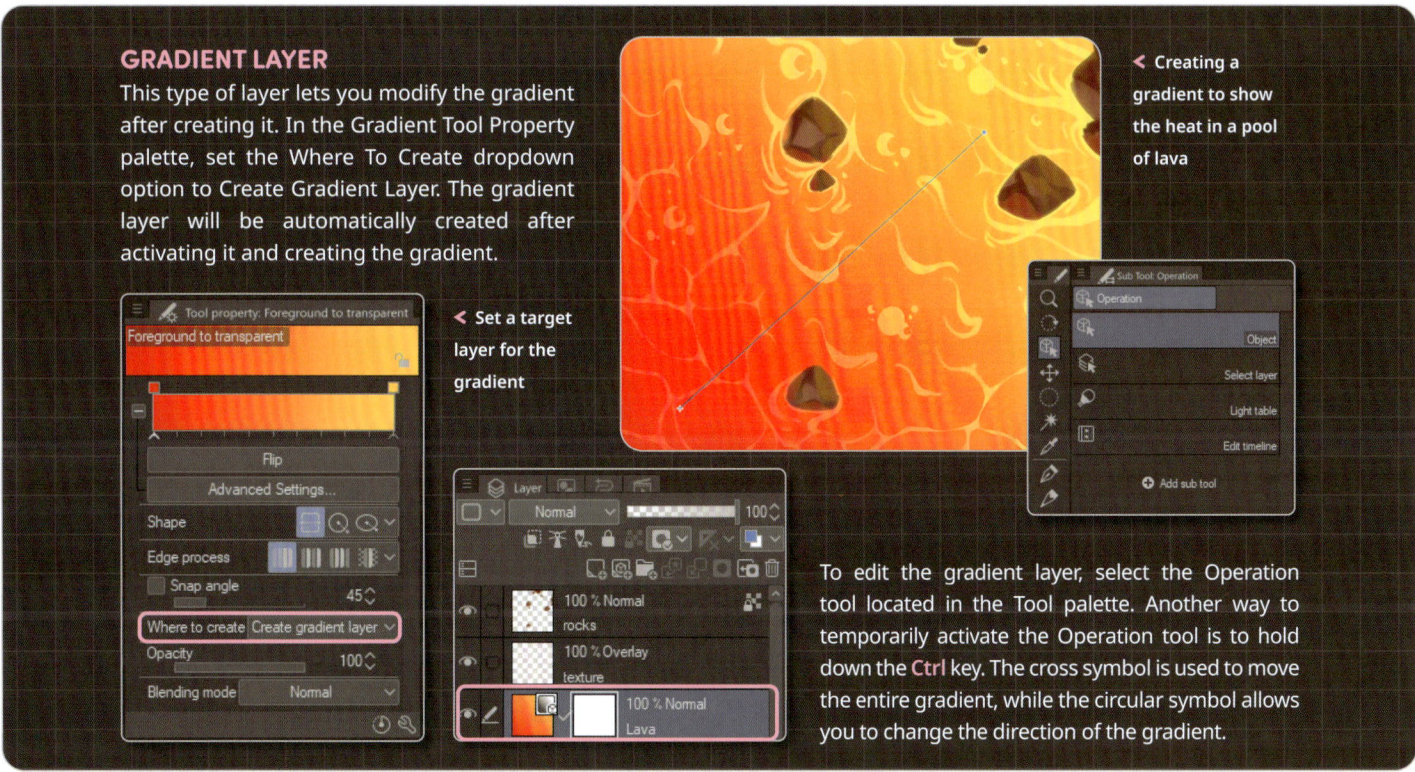

GRADIENT LAYER

This type of layer lets you modify the gradient after creating it. In the Gradient Tool Property palette, set the Where To Create dropdown option to Create Gradient Layer. The gradient layer will be automatically created after activating it and creating the gradient.

< Creating a gradient to show the heat in a pool of lava

< Set a target layer for the gradient

To edit the gradient layer, select the Operation tool located in the Tool palette. Another way to temporarily activate the Operation tool is to hold down the **Ctrl** key. The cross symbol is used to move the entire gradient, while the circular symbol allows you to change the direction of the gradient.

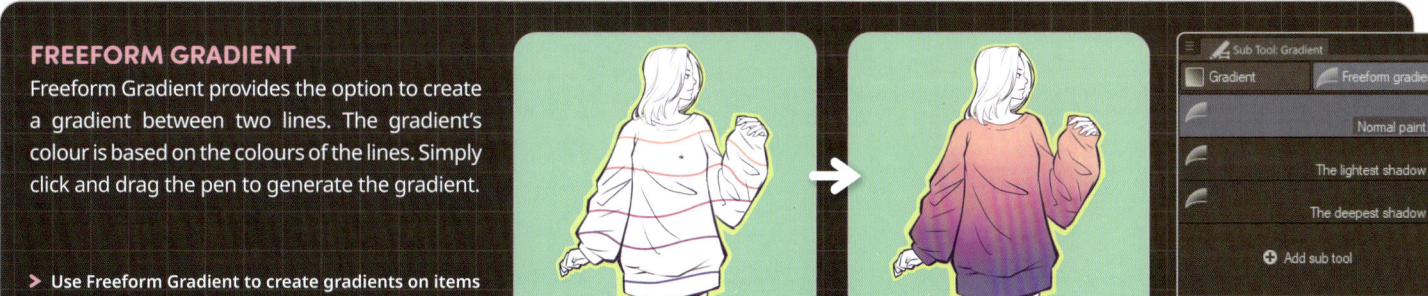

FREEFORM GRADIENT

Freeform Gradient provides the option to create a gradient between two lines. The gradient's colour is based on the colours of the lines. Simply click and drag the pen to generate the gradient.

> Use Freeform Gradient to create gradients on items of clothing

Figure tool

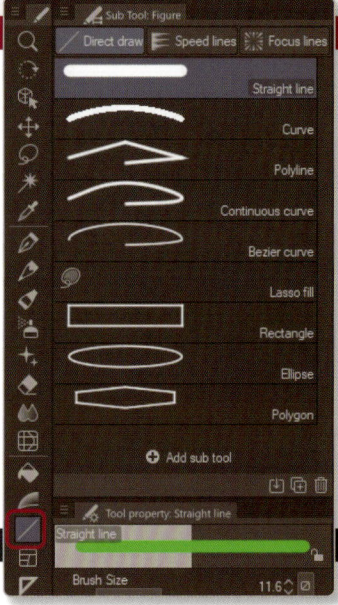

LEARN HOW TO

- Recognize and use the Figure tool and its sub tools.

Some artwork may require you to draw geometric elements or straight lines, but these can be difficult to draw freehand. The Figure tool is useful as it can be used to create neat lines, curves, and shapes.

> **The Figure tool is the straight-line icon on the Tool palette on the left-hand side of the interface**

DIRECT DRAW

The first category, Direct Draw, has sub tools for creating lines, curves, and various shapes.

STRAIGHT LINE

This sub tool allows you to create straight lines. Simply click and drag in the direction you want the line to be. The line will be created when you release the click.

> **Drawing a perfectly straight line to complement a portrait**

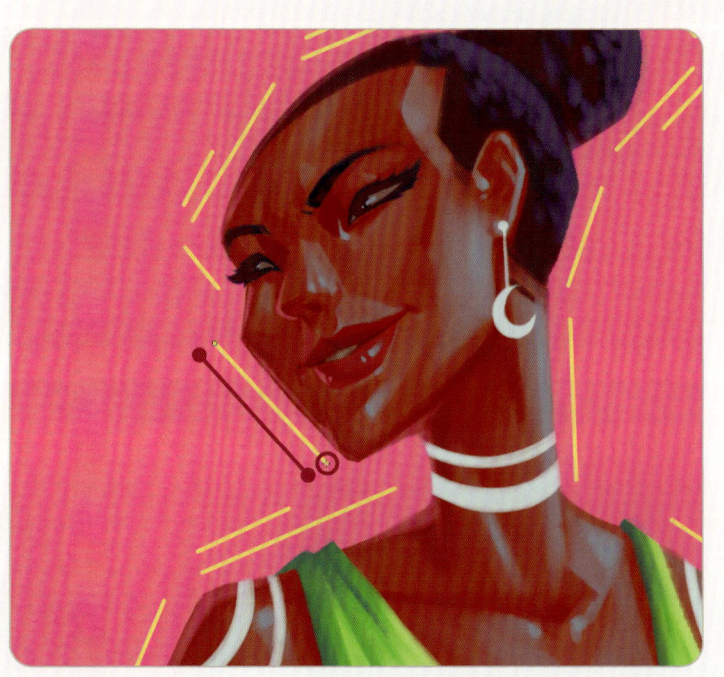

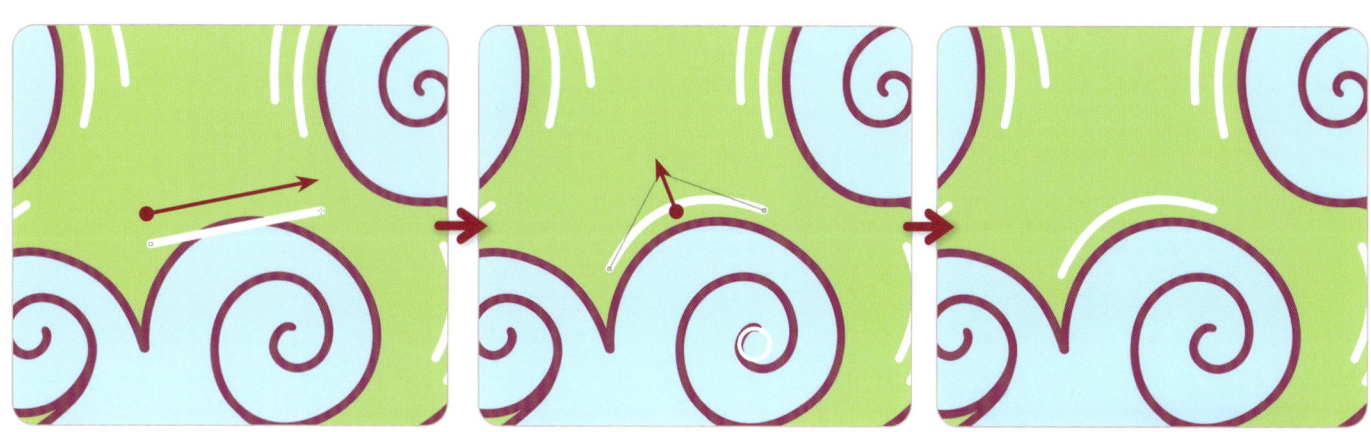

CURVE

This allows you to create curved lines. Click and drag, then release the click. The cursor will then allow you to control the curvature. Once you're happy with the curve of the line, click again, and the curve will be created.

∧ **Drawing curved lines with the Curve sub tool**

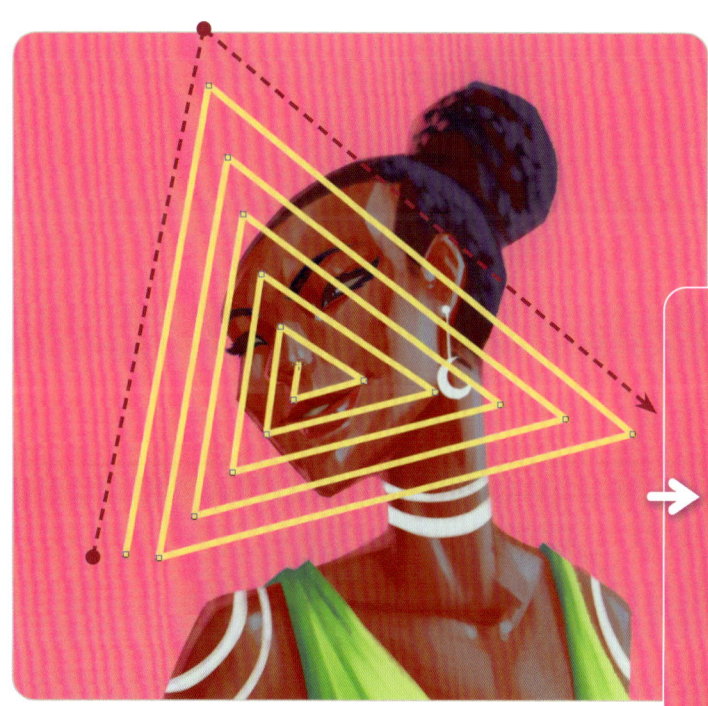

∧ **Drawing continuous straight lines to create a dynamic background**

POLYLINE

The Polyline sub tool allows you to create any shape using continuous straight lines. Start by clicking at the point where you want the line to begin, then continue to generate more lines by moving the cursor and clicking where you want each line to finish. Double-click to finish, or close up the polyline by clicking on the start point.

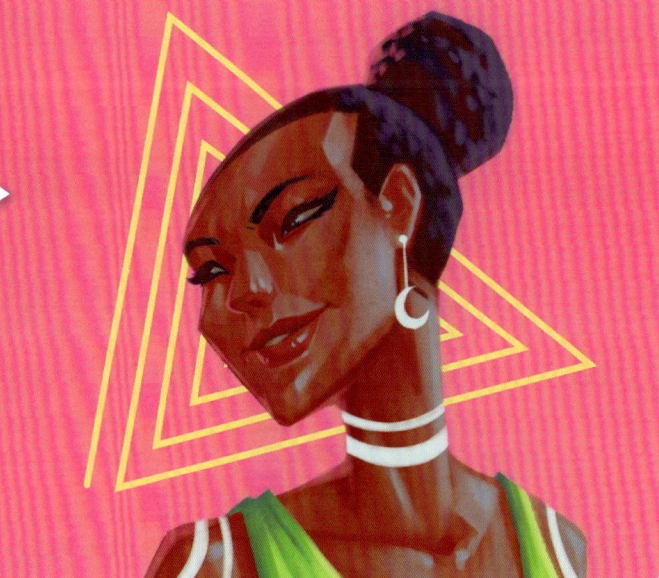

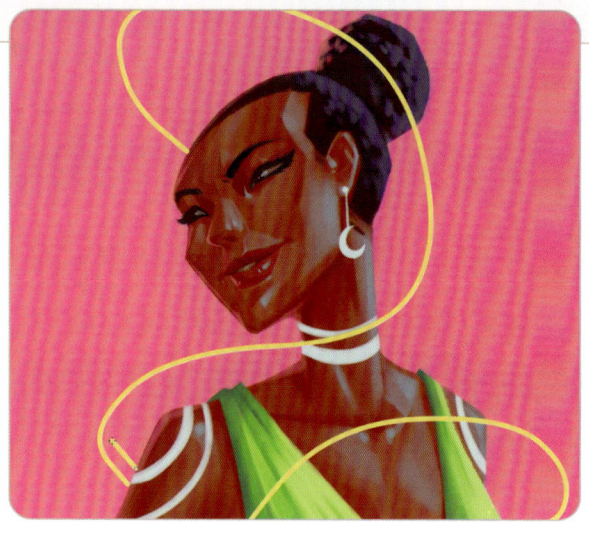
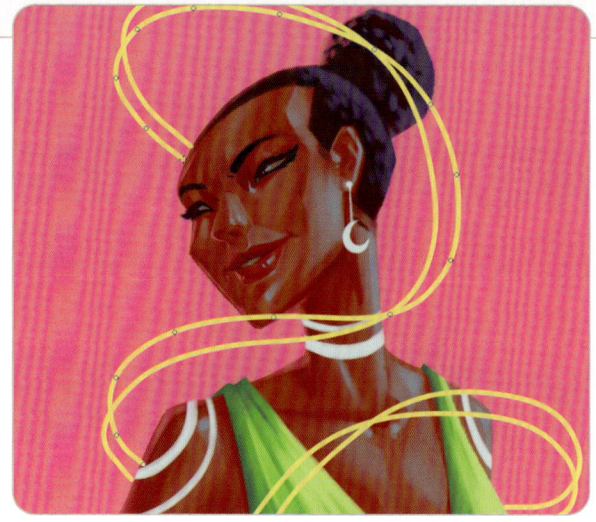
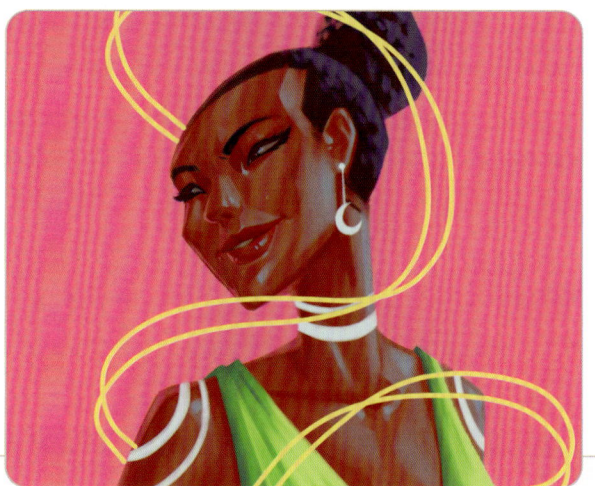

CONTINUOUS CURVE

This sub tool you lets you create continuous curves. Click where you want the line to begin, then drag the cursor while clicking where you want to generate more curves along the same line. Double-click to finish, or close up the curve by clicking back on the original point.

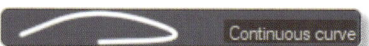

∧ **Using the Continuous Curve sub tool**

to create movement in an image

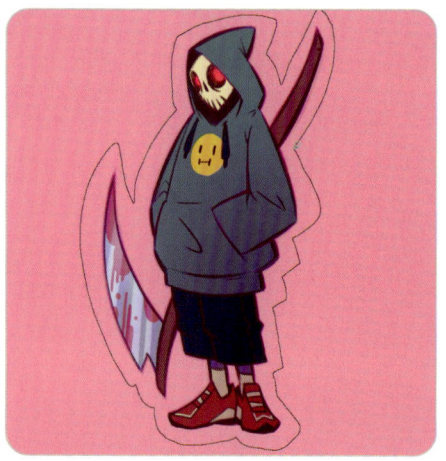
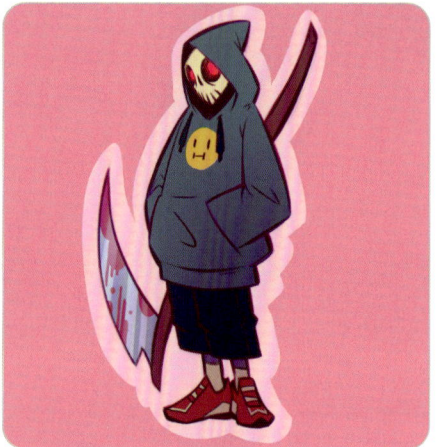
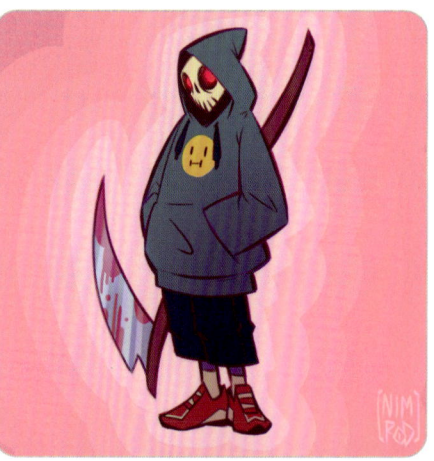

LASSO FILL

This creates a colour fill based on the area you trace with the Lasso tool. To use Lasso Fill, click and drag the brush without releasing it, drawing the area you want to be filled with colour.

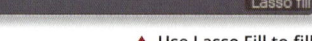

∧ **Use Lasso Fill to fill**

an area with colour

BEZIER CURVE

This sub tool allows you to create continuous straight or curved vector lines and modify them after creation. It's essential that they're created in a vector layer so they can be edited. Click at the starting point, then click anywhere else and drag to create a curve, or click anywhere else to create a straight line.

Use the Object sub tool, located in the Operation tools group, to modify the vector curves or lines. You can also activate this tool by holding down the **Ctrl** key.

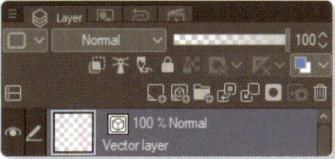

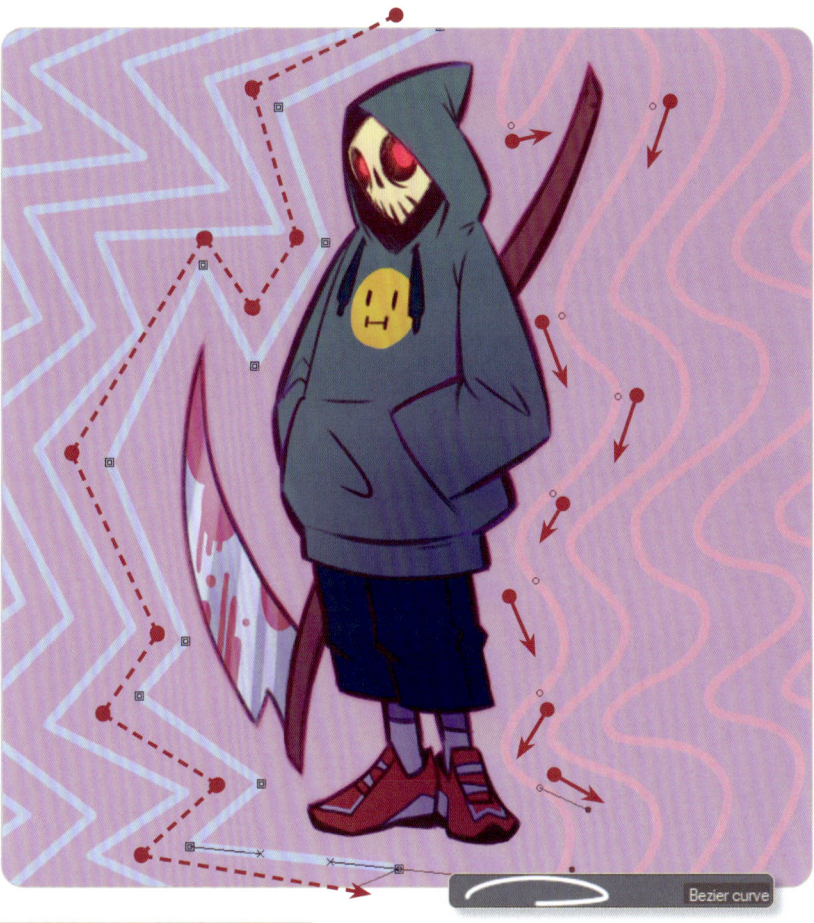

∧ Here the background is created using continuous straight and curved vector lines

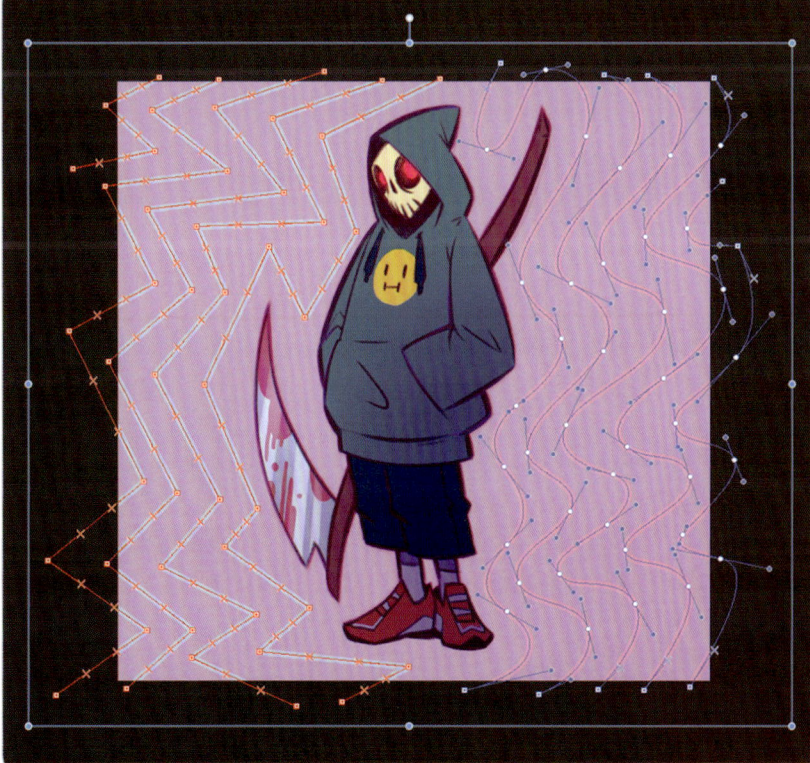

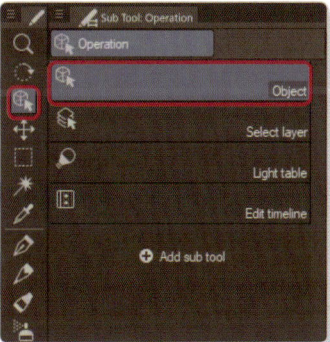

∧ Modify the vector lines using the Object sub tool

71

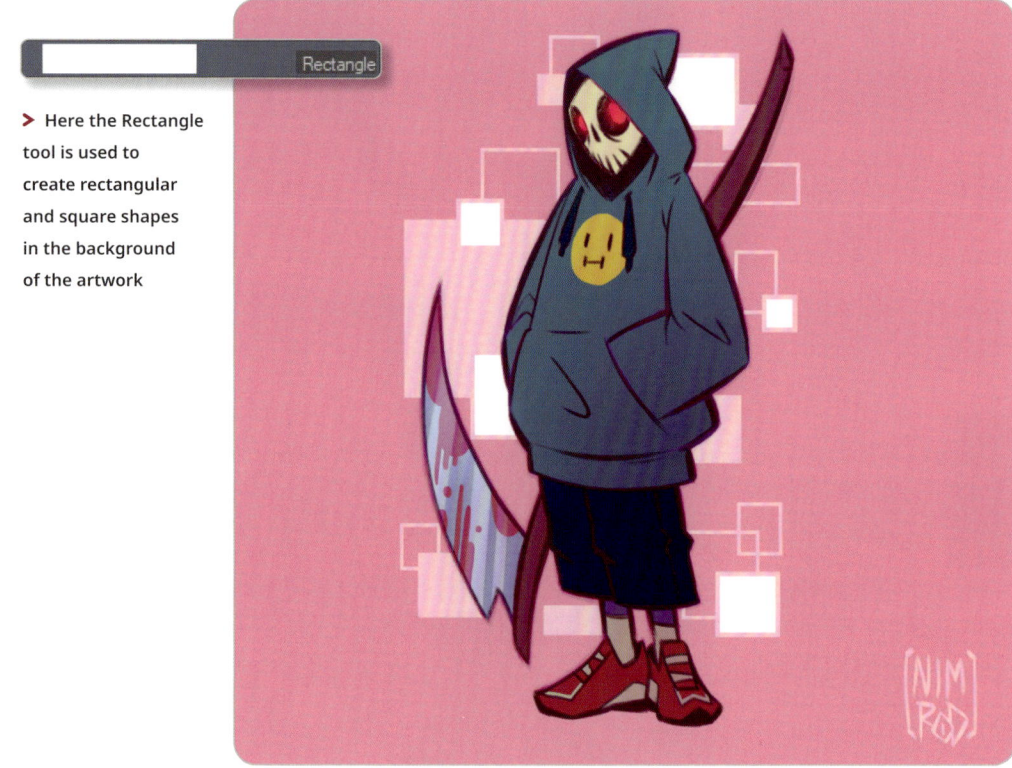

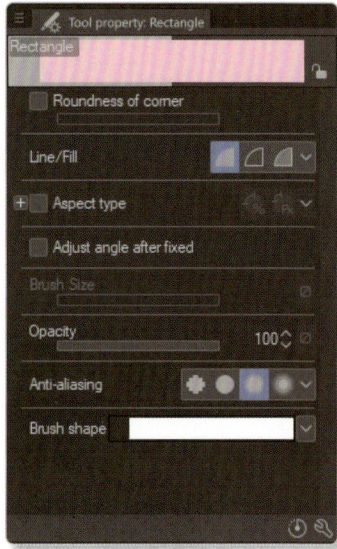

> **Here the Rectangle tool is used to create rectangular and square shapes in the background of the artwork**

RECTANGLE

This sub tool is used to create rectangle or square shapes. Modify the stroke, fill, and corner roundness in the Tool Property palette.

ELLIPSE

The Ellipse sub tool creates circle or oval shapes. You can then edit the stroke and fill colour in the Tool property palette. To create a circle, click and drag the cursor to set the size, then release the click to set the orientation. If you want to create a perfect circle, hold down the **Shift** key while dragging and clicking to set the size of the circle.

> **Produce a variety of different oval and circle shapes using the Ellipse tool**

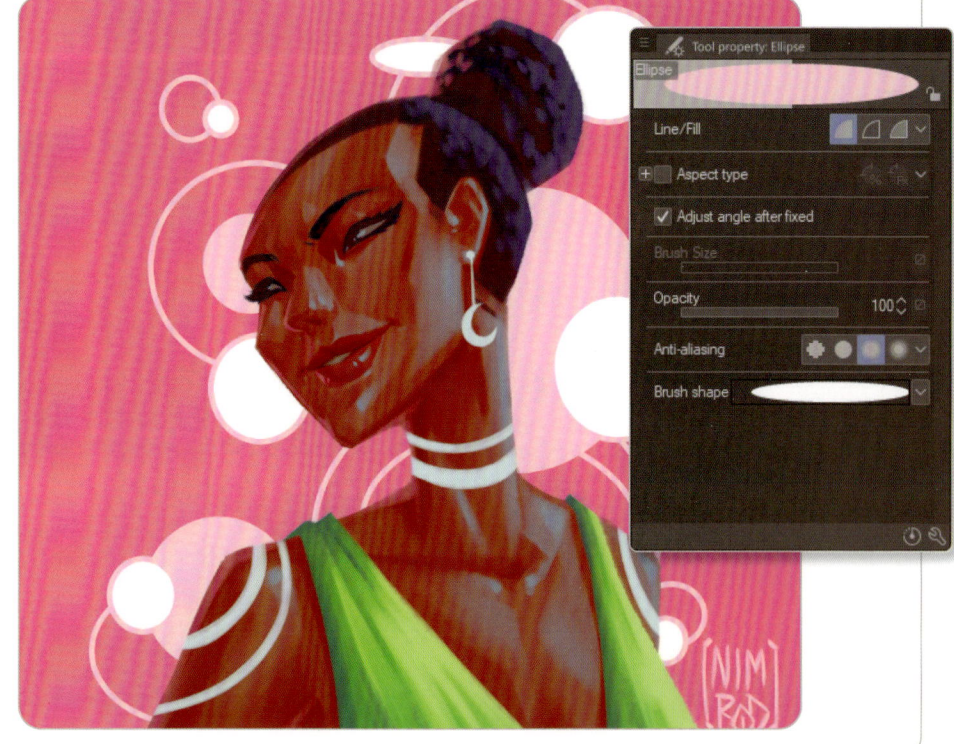

> Create multi-sided shapes using the Polygon sub tool

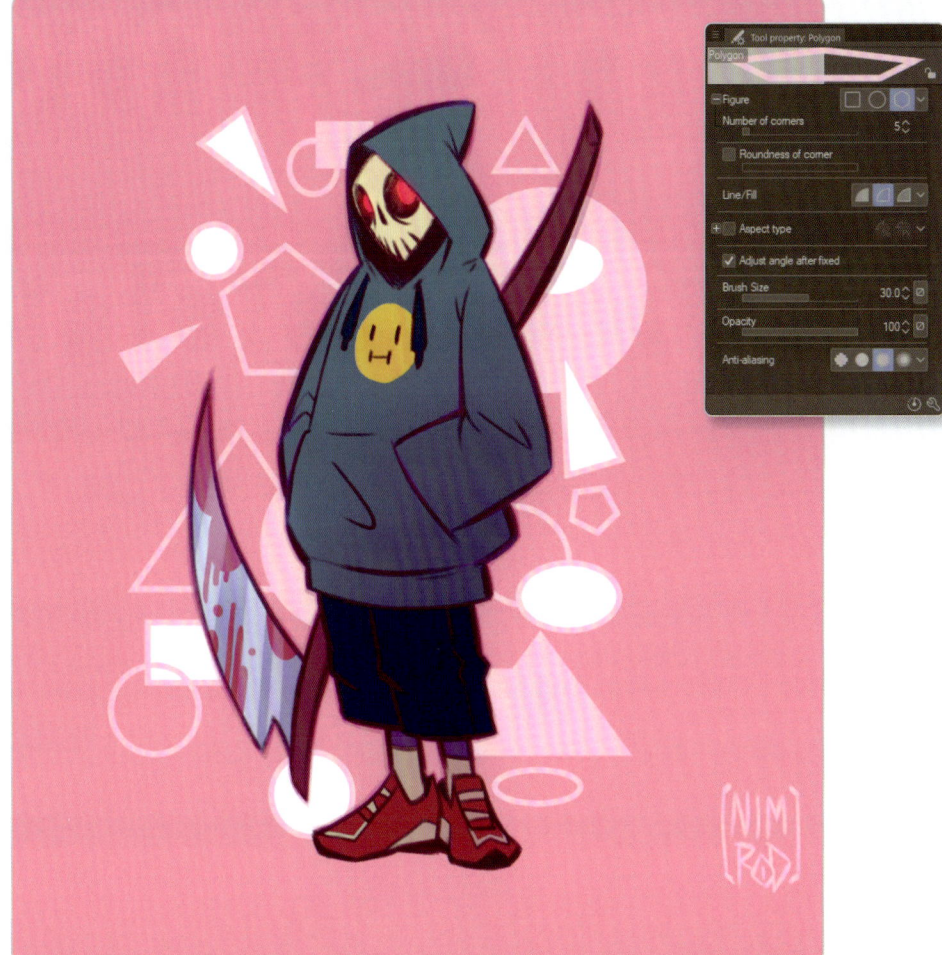

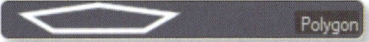

POLYGON

The Polygon sub tool lets you create shapes with three, four, or more vertices (points where two lines meet). You can modify the number of vertices – as well as the stroke, fill, and corner roundness – in the Tool Property palette. To create a polygon, click and drag to define the size, use the cursor to define the orientation, and click to finish.

If you want to create a perfect square or polygon, hold down the **Shift** key while dragging and clicking to set the shape's size.

MODIFY
UNIFORM LINES

When drawing shapes, the stroke line is uniform by default. If drawn using brushes, you may find that the lines blend poorly with the overall artwork. Try using the Eraser tool to erase some of the edges – this will give a line a slightly hand-drawn effect. You can also change the brush shape settings in the Sub Tool Detail palette to make it blend better with the artwork.

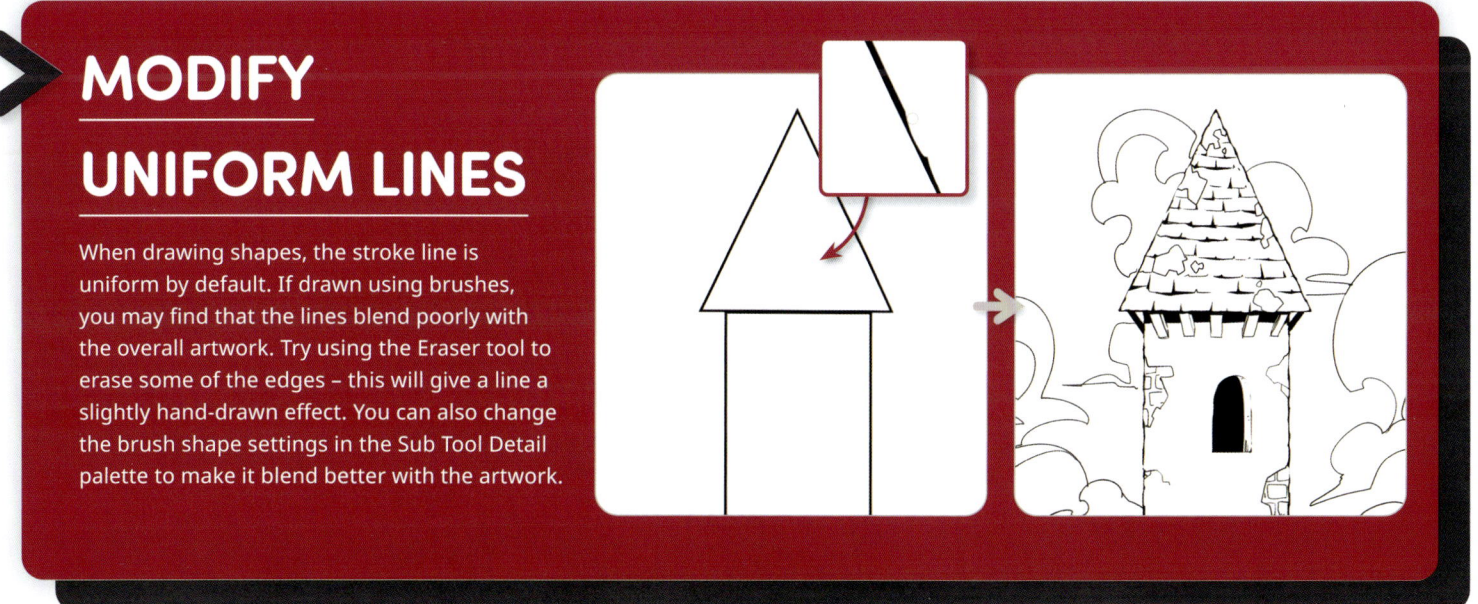

Comics, manga & webtoons

Clip Studio Paint provides a variety of user-friendly tools for creating comics, manga, and webtoons. From Text Balloon tools to special Screentone layers, it offers all the resources you could need to complete your project. The following pages will cover the main tools and how to use them. You can find additional instruction on how to create a page from a comic on page 220 and a manga webtoon on page 238.

LEARN HOW TO

- Explore the main tools used in creating comics, manga, and webtoons.
- Create screentone layers and modify them.

HATCHING & SAND PATTERN BRUSHES

These special brushes have an irregular scattering effect and can be used to create cross-hatching-style strokes. They are much quicker and easier to use than drawing cross-hatching by hand. Their sub tool settings can be found in the Decoration tool located in the Tool palette.

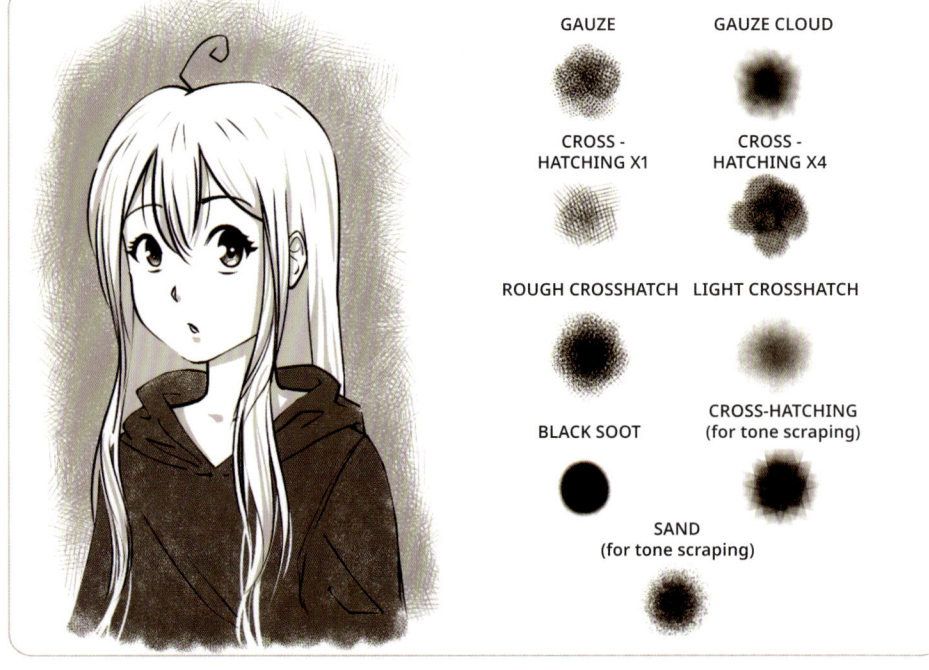

GAUZE

GAUZE CLOUD

CROSS - HATCHING X1

CROSS - HATCHING X4

ROUGH CROSSHATCH

LIGHT CROSSHATCH

BLACK SOOT

CROSS-HATCHING
(for tone scraping)

SAND
(for tone scraping)

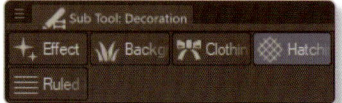

∧ Hatching and Sand Pattern brushes save time when shading an image

FIGURE TOOL

The Figure tool, partially covered in the previous chapter, has two line categories that can be used to quickly add effects to your comic, manga, and webtoon projects.

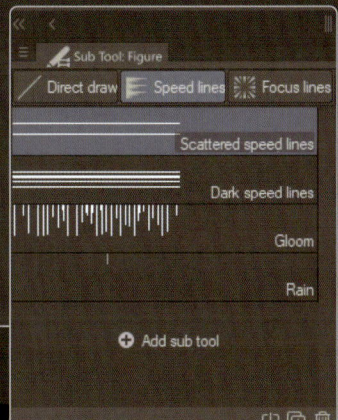

< The Figure tool offers options for speed lines and focus lines to create various effects

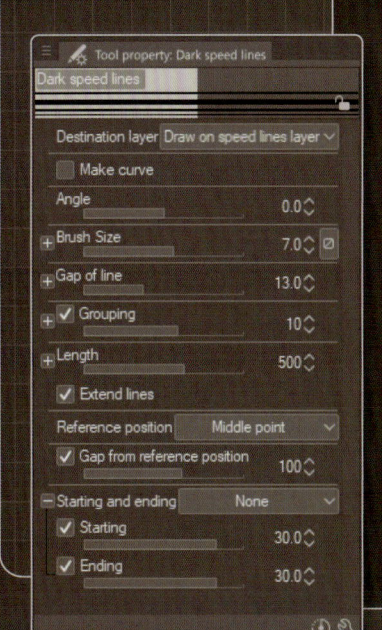

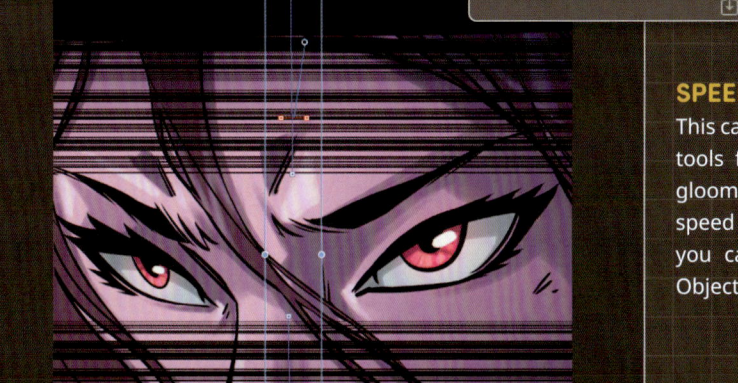

SPEED LINES

This category contains a list of sub tools for creating speed effects, gloom, or rain. If you create speed lines on an editable layer, you can modify them using an Object sub tool.

< These lines help to create a dark and tense mood, and draw focus to the character's gaze

FOCUS LINES

This category has sub tools to add special effect lines that focus the viewer's attention on a certain element in an image. They can be modified if you create them on an editable layer.

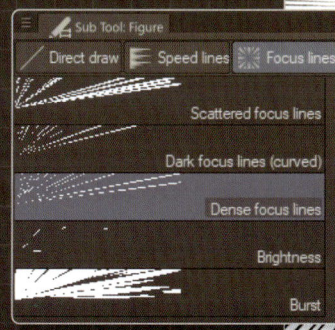

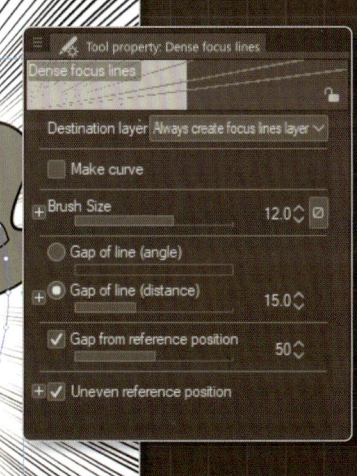

< Here the focus lines direct the viewer's attention to the central character

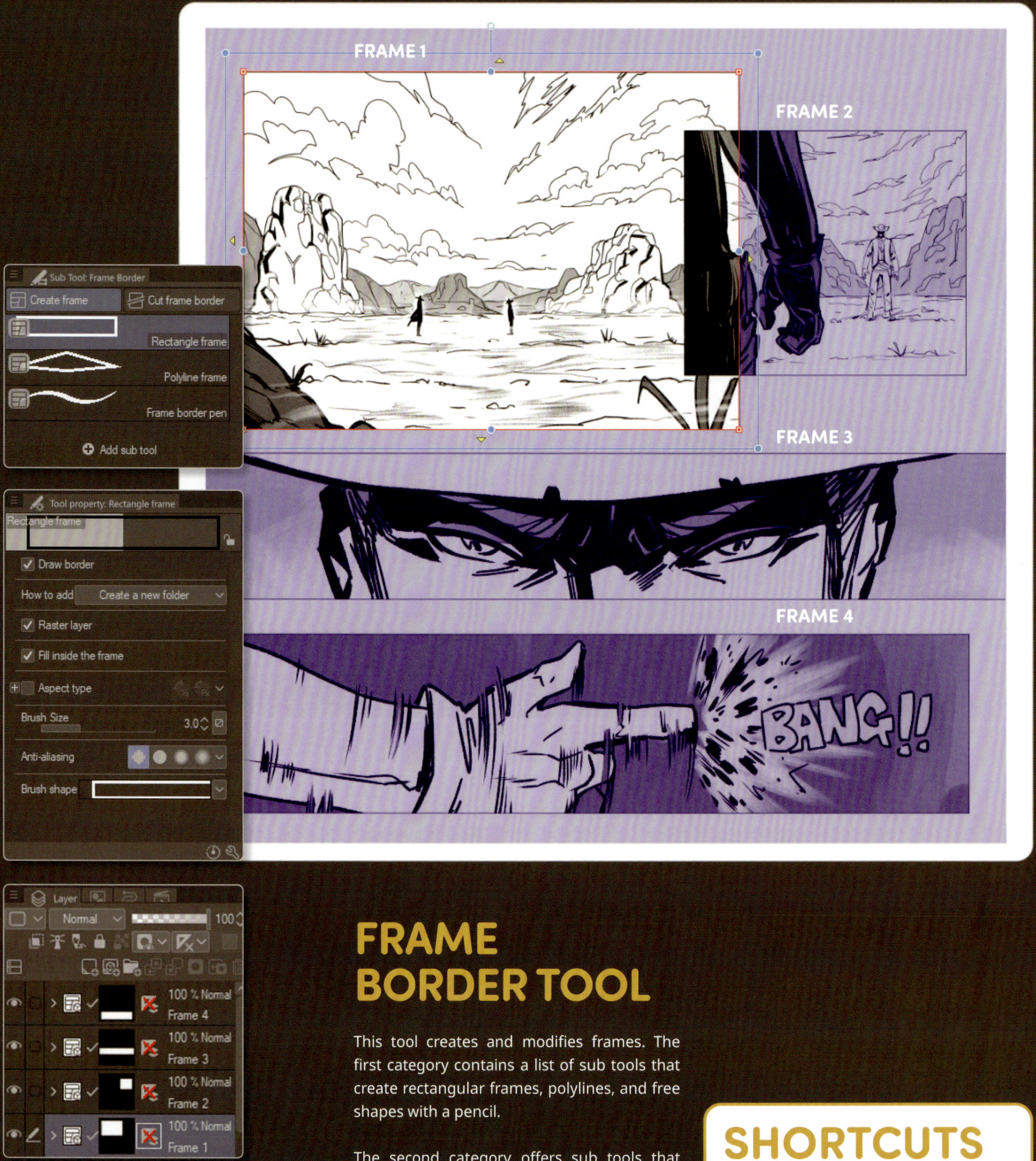

FRAME 1

FRAME 2

FRAME 3

FRAME 4

BANG!!

Sub Tool: Frame Border

Create frame	Cut frame border

Rectangle frame

Polyline frame

Frame border pen

⊕ Add sub tool

Tool property: Rectangle frame

Rectangle frame

☑ Draw border

How to add — Create a new folder

☑ Raster layer

☑ Fill inside the frame

⊞ ☐ Aspect type

Brush Size — 3.0

Anti-aliasing

Brush shape

Layer

Normal — 100

			100 % Normal Frame 4
			100 % Normal Frame 3
			100 % Normal Frame 2
			100 % Normal Frame 1

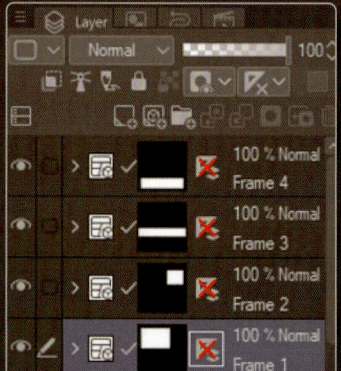

∧ **Use the Frame Border tool to create frames/panels in various different shapes**

FRAME
BORDER TOOL

This tool creates and modifies frames. The first category contains a list of sub tools that create rectangular frames, polylines, and free shapes with a pencil.

The second category offers sub tools that allow you to divide the frame folder into different folders, or divide the frame border and keep the division in the same folder.

SHORTCUTS

- Press the **U** key to activate the tool.

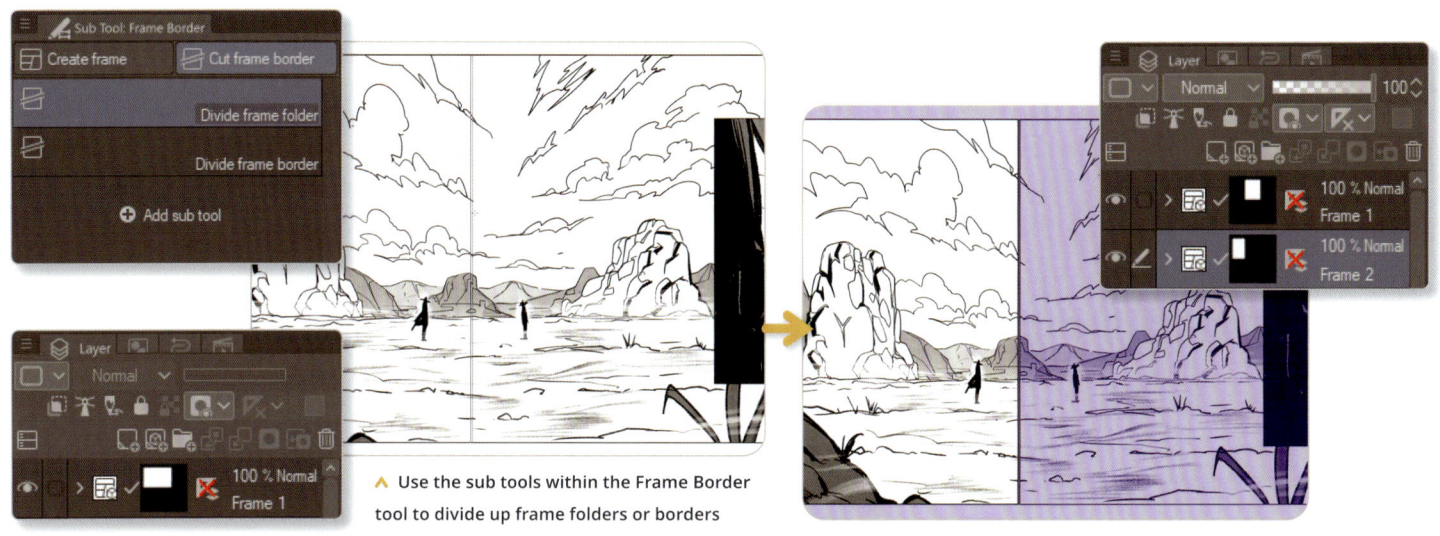

⌃ Use the sub tools within the Frame Border tool to divide up frame folders or borders

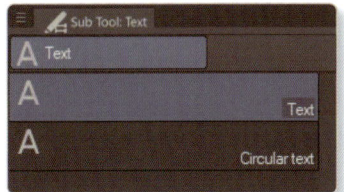

TEXT TOOL

Select the Text tool, found in the Tool palette, to add text to an image. The sub tool palette offers two types of text: Standard and Circular. The size configuration, source, direction, and style can all be modified in the Tool Property palette.

The Circular Text sub tool contains functions to edit the margin and text orientation.

To add new fonts, select the drop-down font menu, scroll to the bottom of the list, and click the Add Font From Files button.

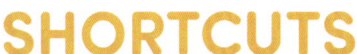

SHORTCUTS

• Press the T key to activate the tool.

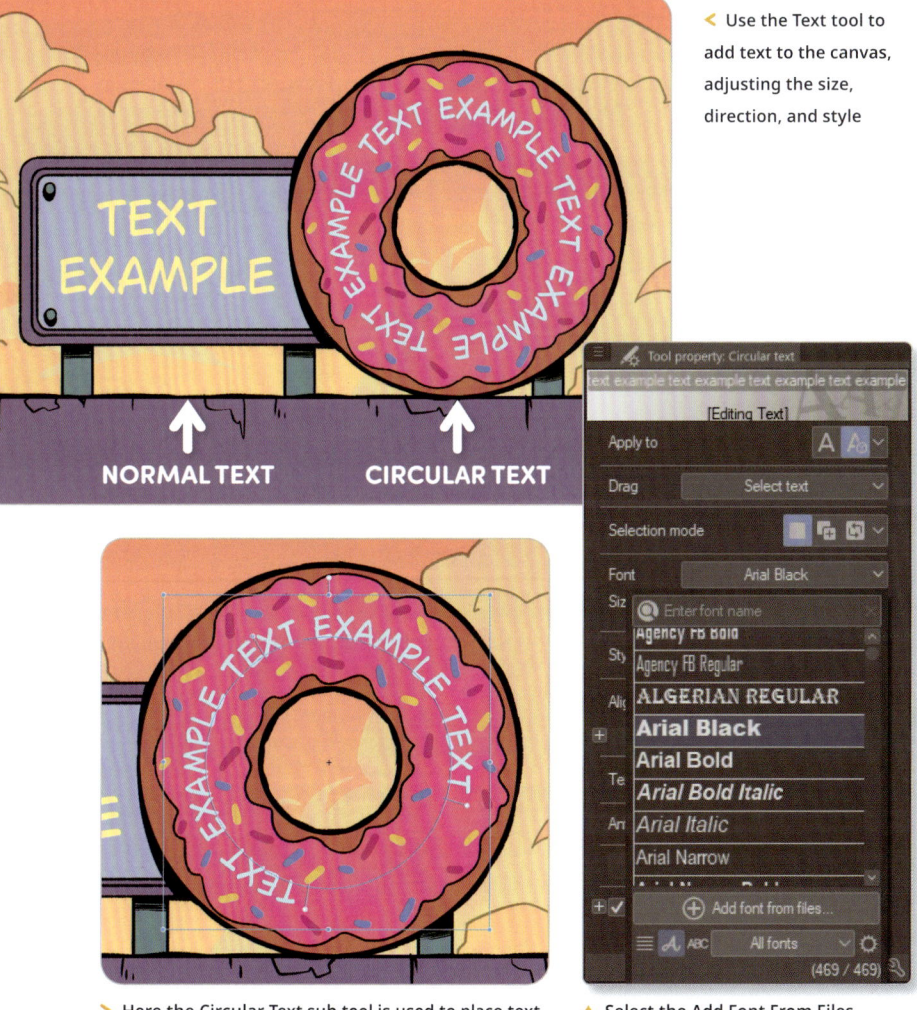

‹ Use the Text tool to add text to the canvas, adjusting the size, direction, and style

NORMAL TEXT **CIRCULAR TEXT**

› Here the Circular Text sub tool is used to place text on the doughnut

⌃ Select the Add Font From Files button to import a font

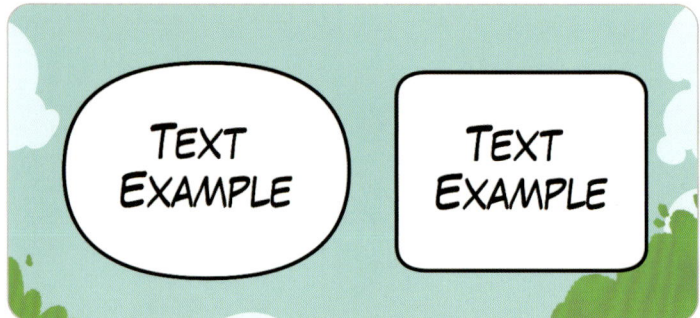

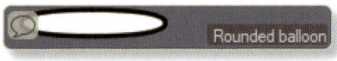

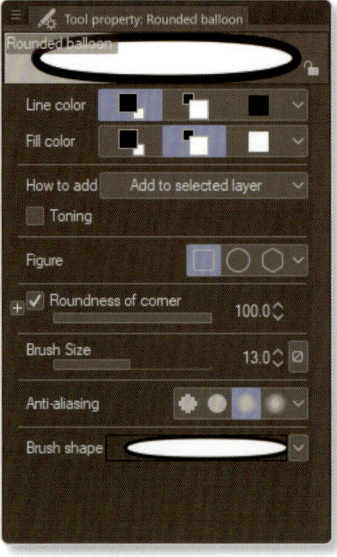

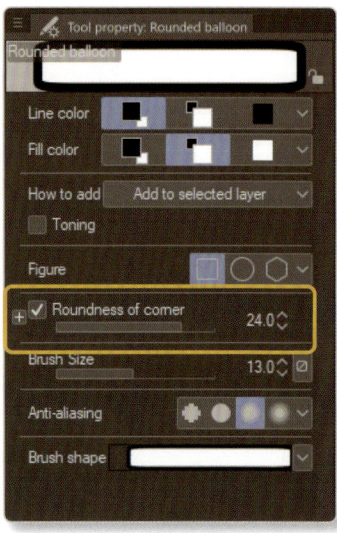

BALLOON TOOL

Comics, manga, and webtoons communicate story through imagery, as well as through written text displayed in text balloons. The Text Balloon tool is located beneath the Text tool in the Tool palette. The first category offers a list of sub tools that create balloons of various shapes, as well as options for the tails of the balloons.

ROUNDED BALLOON

This sub tool creates rounded balloons. The Tool Property palette provides the option to decrease or increase the roundness of the corners.

∧ Reducing the number will make the corners less rounded, whereas increasing the number will make them more rounded

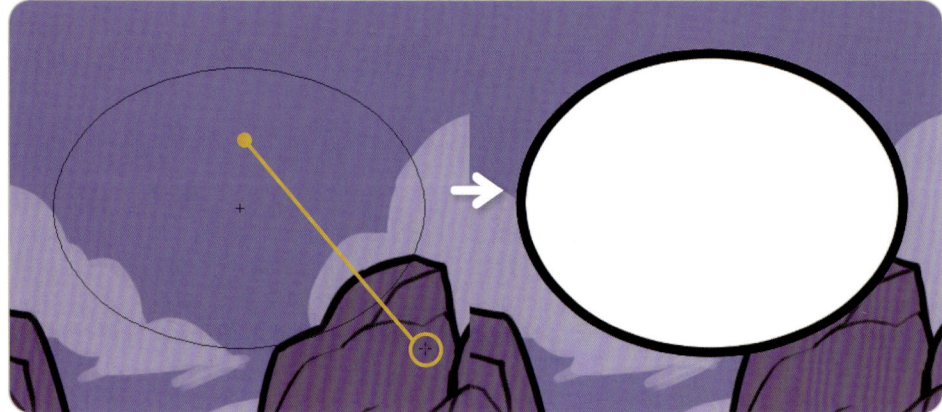

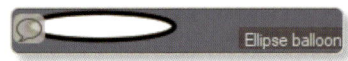

ELLIPSE BALLOON

The Ellipse Balloon sub tool lets you create elliptical balloons. Simply click and drag the cursor to establish the size and shape, then release click to finish.

< Click and drag to set the shape and size of an Ellipse Balloon, which you can then fill with text

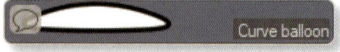

CURVE BALLOON

You can also create a text balloon using continuous curves. To create the form, click on a starting point, then click to create another point, and so on. Use the cursor to shape the curvature, then complete the balloon by clicking back on the starting point.

> The Curve Balloon tool can be used to create text balloons with interesting and unique shapes

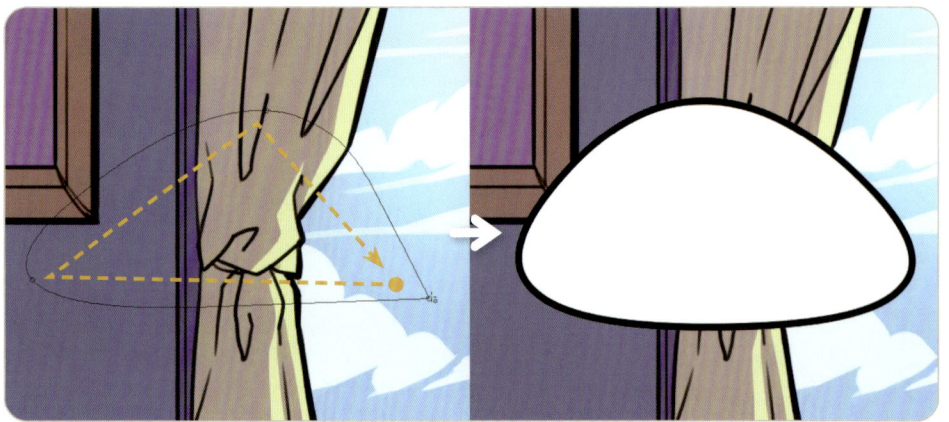

RECTANGLE BALLOON

Use this sub tool to create rectangular text balloons.

> A rectangle text balloon is a classic shape often used in comics, manga, and webtoons

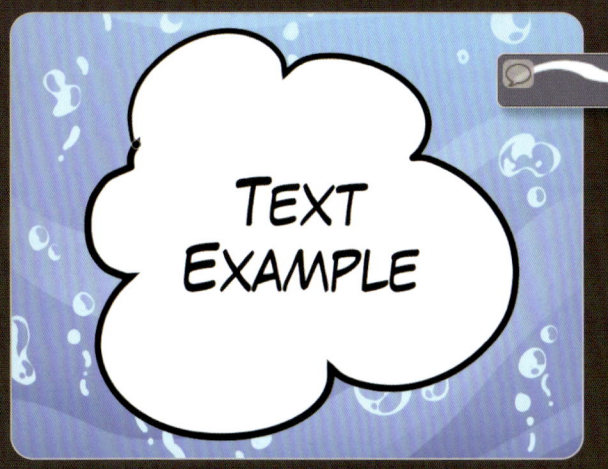

BALLOON PEN

This sub tool lets you create text balloons by drawing them freehand with a pencil.

< Select Balloon Pen to draw a text balloon freehand

BALLOON TAIL & THOUGHT BALLOON TAIL

This sub tool can be used to create the tails that accompany text balloons. You can adjust the width of spacing and how to bond them in the Tool Property palette.

> Use the Object sub tool to modify the size and shape of the tails

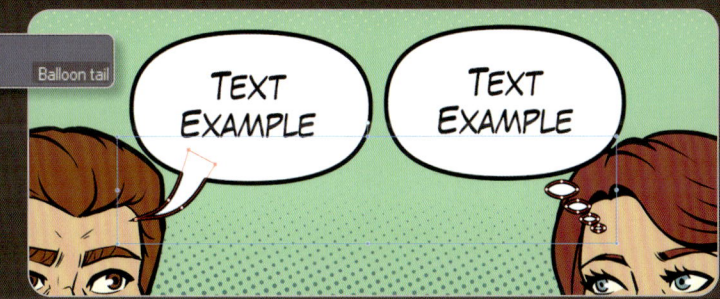

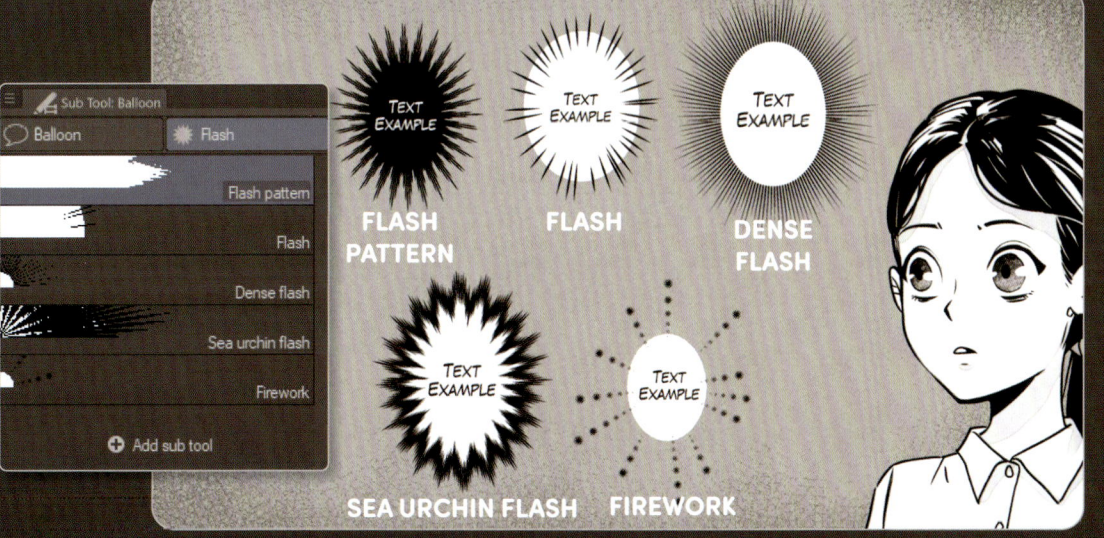

FLASH BALLOONS

The next category in the Balloon tool is Flash Balloons, which come in various styles. These are typically used to communicate loud sounds, dynamic narration, and punchy dialogue.

< The different styles of Flash Balloons include: Flash Pattern, Flash, Dense Flash, Sea urchin Flash, and Firework

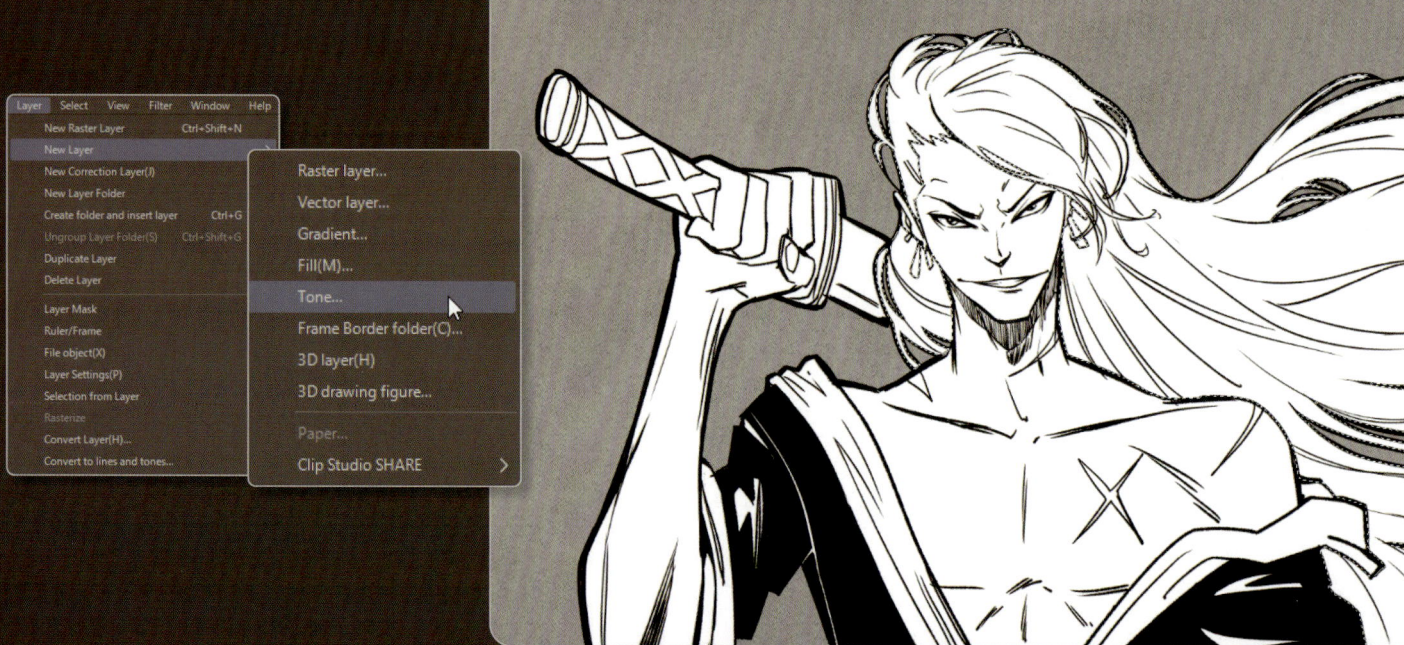

∧ Use the Lasso sub tool to make a freehand selection of the area you wish to fill with screentone

TONE LAYERS

These layers add textures and shades to large areas of an image. This is commonly known as 'screentone'.

Before creating a new Tone layer, you need to first set the area using the Selection Area tool. If you create the new layer without first setting the area, this will fill the whole canvas with screentone. After selecting the area, select Layer > New Layer > Tone. A window will appear with options to adjust the frequency, density, type, and angle of the tone pattern. Once you're happy, click OK to create the Tone layer.

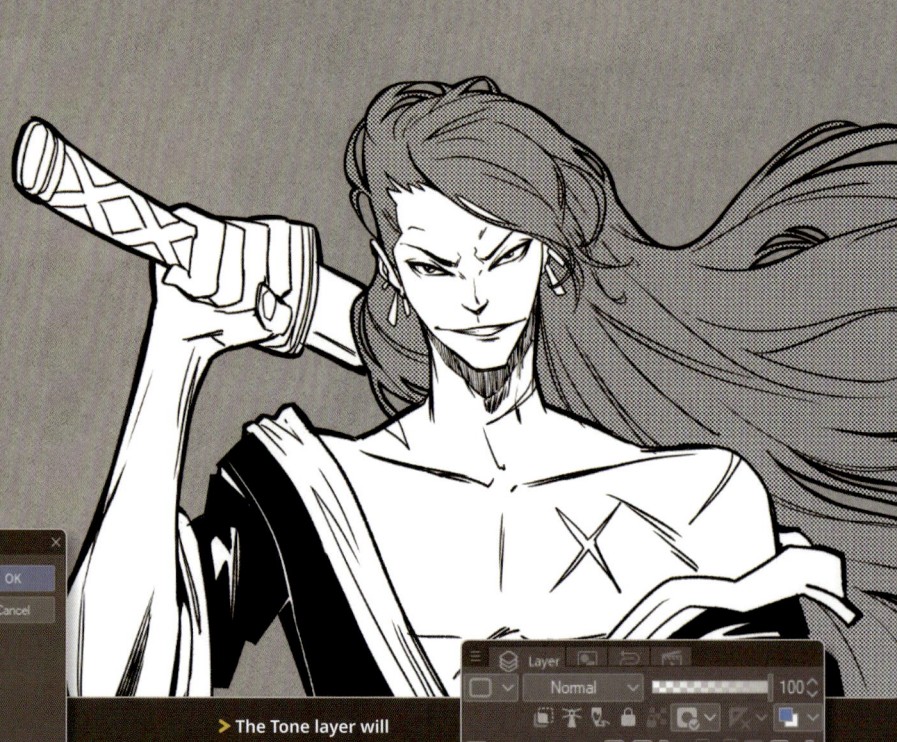

> The Tone layer will then be shown in the Layer palette

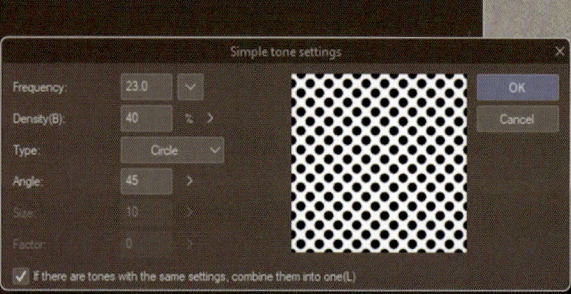

∧ This menu allows you to adjust the Tone layer settings

> The screentone effect can be activated and deactivated on a layer

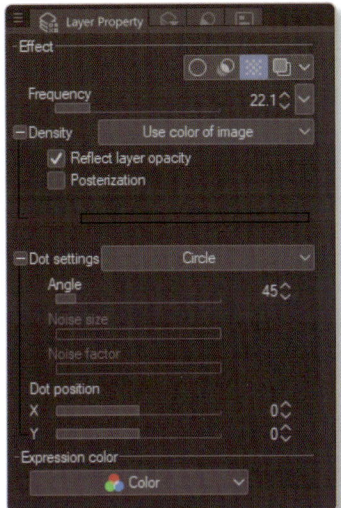

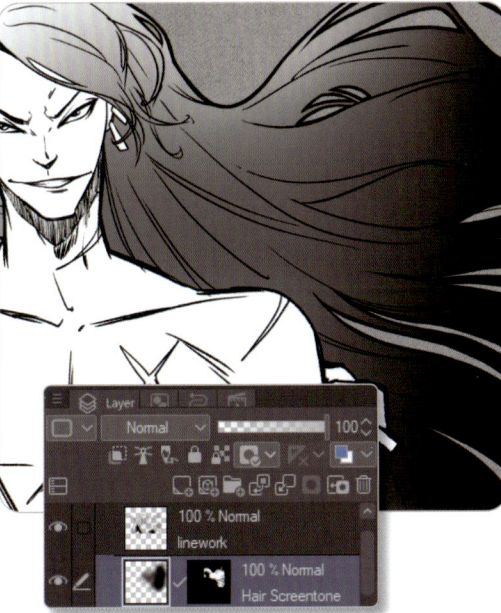

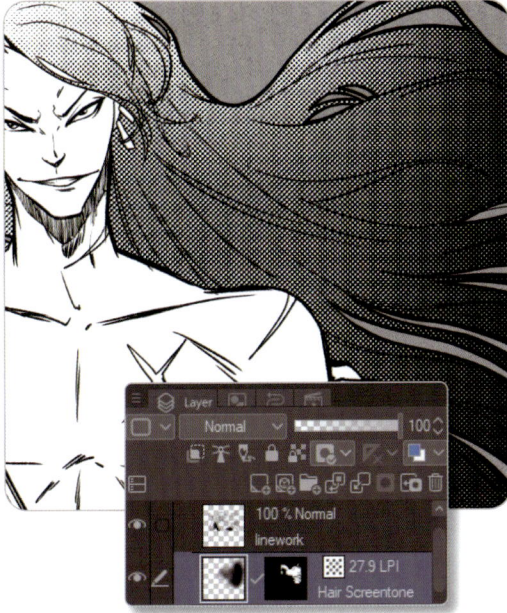

Another way to create Tone layers is to activate the Tone Effect in the Layer Property palette on a normal layer you've already created. After the layer becomes a Tone layer, the settings found in the Layer Property palette can be edited. This includes settings for density, type of tone, position, and more.

You can also activate the screentone effect on an empty layer. This can be used to paint or add elements to the layer and convert them into screentones in real time. To modify the colour of the screentone, you must also activate the Layer Colour effect. It can only be applied in one colour.

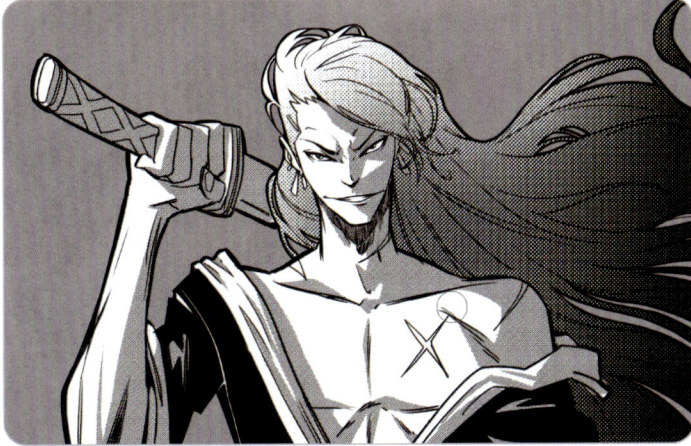

< You can use brushes or gradients to achieve different screentones

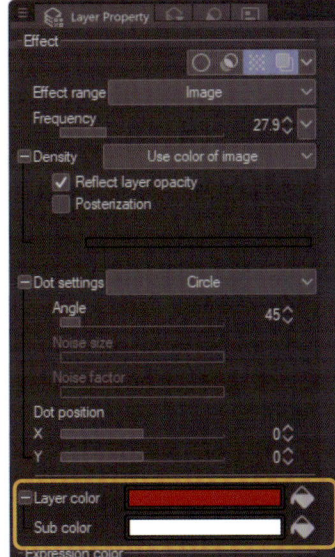

> Choose the colour of the screentone

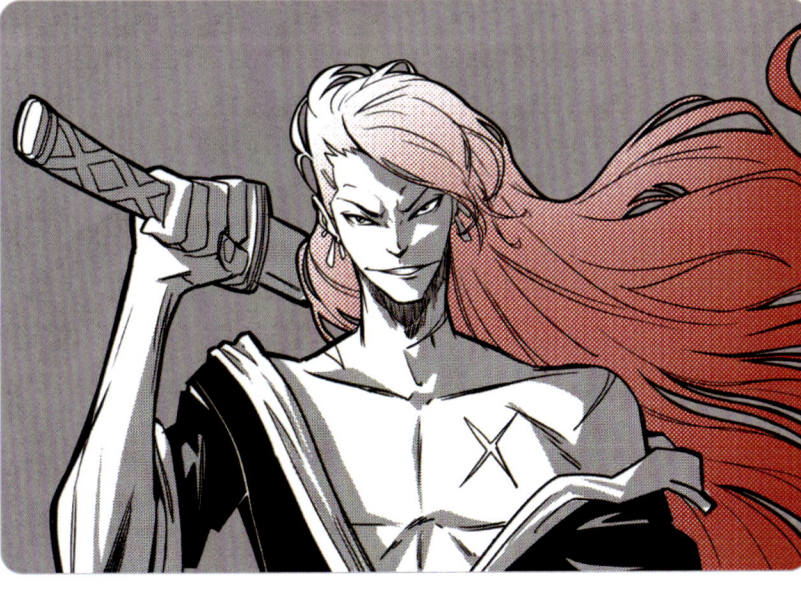

Drawing guides

LEARN HOW TO

- Use the drawing guide tools.
- Recognize and use the Ruler bar, Ruler tool, and Perspective ruler.

Previous chapters have explored how you can use various Figure tools to create drawings with an exact stroke, and then adapt them as needed. Another way to achieve more precise drawings is by using ruler tools and drawing guides. These are more flexible to use and will help you to create drawings with more exact details.

Not only can these guides be used to create digital paintings of architecture, inorganic objects, and symmetrical drawings, they can also be applied to creating characters and scenes with a more realistic perspective.

RULER BAR

In addition to providing canvas measurement numbers, the Ruler bar allows you to create horizontal and vertical guides. You can then draw lines from these guides, as if you were using a traditional ruler. To create a guide, simply click and drag from either of the vertical or horizontal rulers onto the canvas. Releasing the click will create the guide and a new Rule layer will also be created. To delete a guide, simply drag it back onto one of the Ruler bars.

If the Ruler bar isn't visible, you can activate it from the View menu or using the keyboard shortcut **Ctrl+R**.

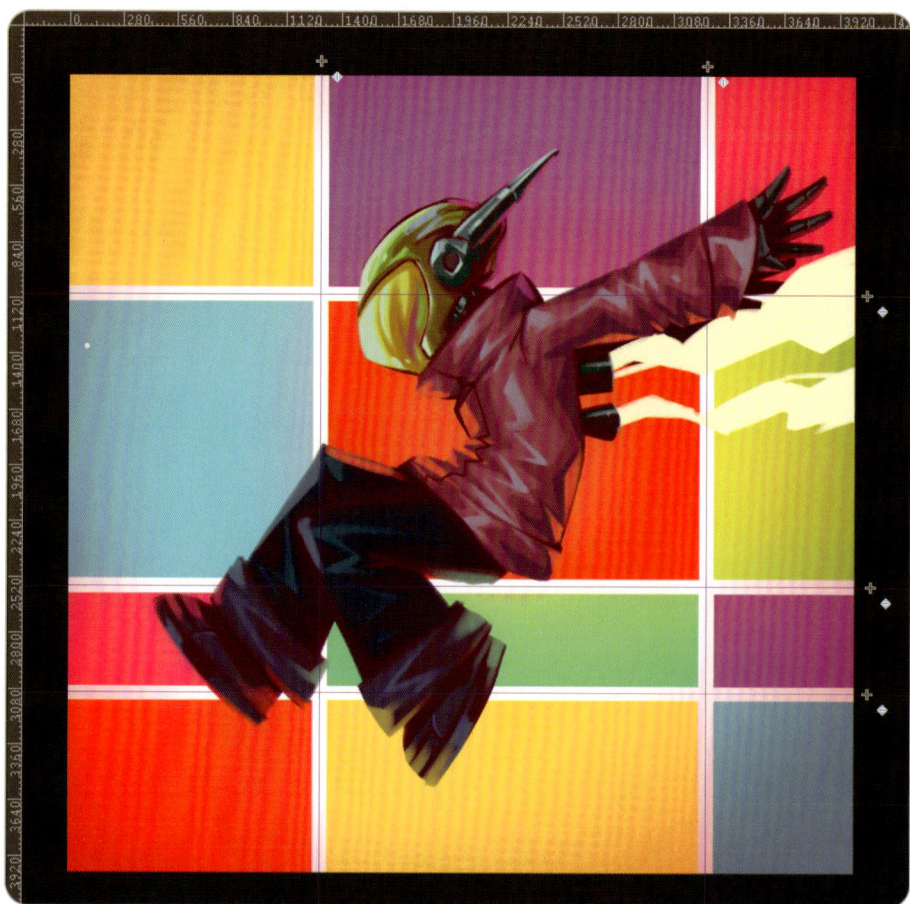

HORIZONTAL

VERTICAL

∧ Guides are used to help create the multicoloured mosaic background

RULER TOOL

The Ruler tool has a list of sub tools that are useful for creating various types of rulers to draw straight or curved lines more accurately. It can be found in the Tool palette on the left-hand side of the interface.

The following pages provide a brief definition of each sub tool on the list.

> **The Ruler tool is represented by the small triangle icon on the Tool palette**

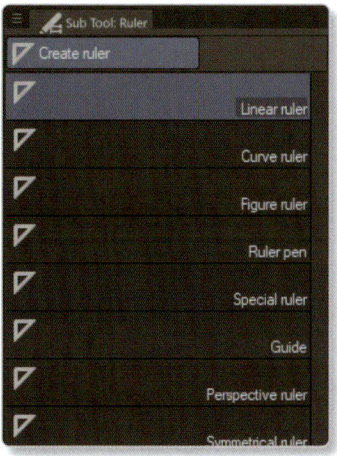

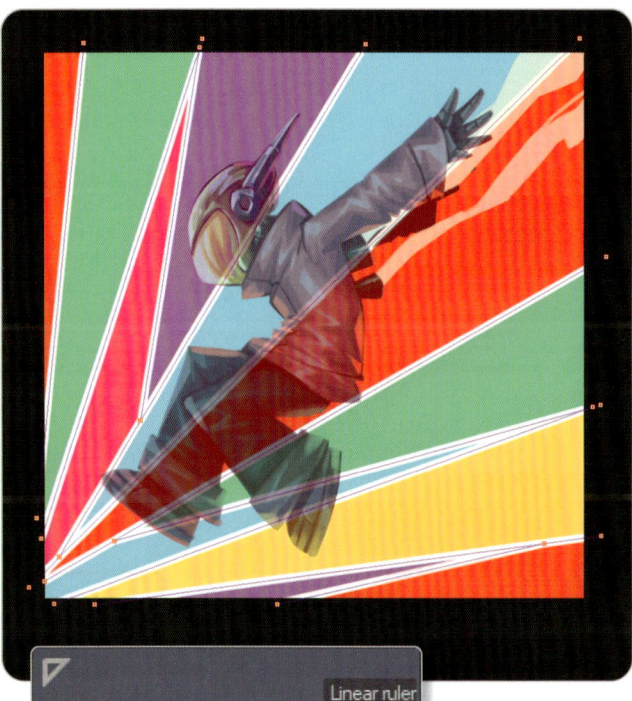

LINEAR RULER

This sub tool creates straight-line rulers at different angles on the canvas. If you hold down the **Shift** key while dragging, you can create a straight-line ruler at preset angles.

^ **Here the Linear Ruler is used to create straight lines at various angles, creating a dynamic background**

CURVE RULER

This option creates curved rulers on the canvas. Start by clicking where the origin of the ruler will be on the canvas, then click on another area to set it, and use the cursor to define the curvature of the ruler.

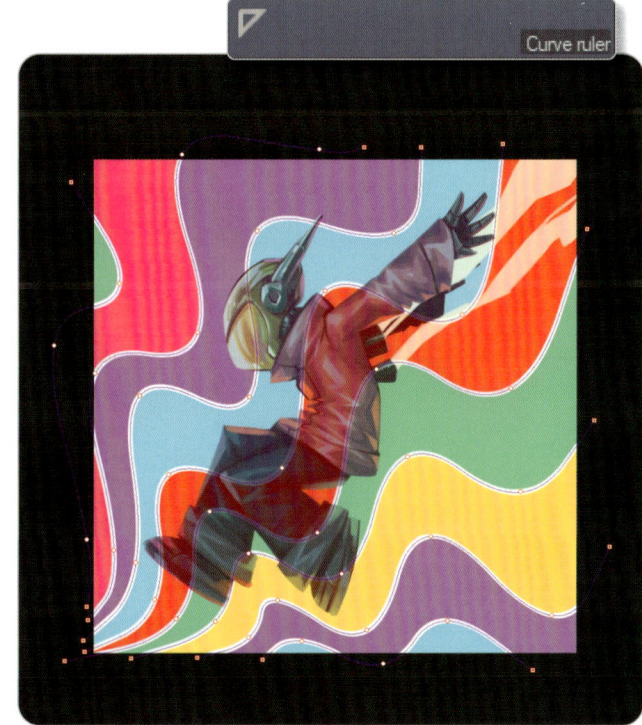

^ **The Curve Ruler is used here to create dynamic rulers with custom curvature**

FIGURE RULER

This sub tool creates rulers with geometric shapes, including ellipses, rectangles, and polygons with more corners. To create a Figure Ruler, click and drag to define the size of the ruler, then use the cursor to define the orientation, and finally click to complete the creation of the ruler. If you hold down the **Shift** key, you can make size of the ruler more accurate.

> **Here the Figure Ruler is used to create rulers of various shapes**

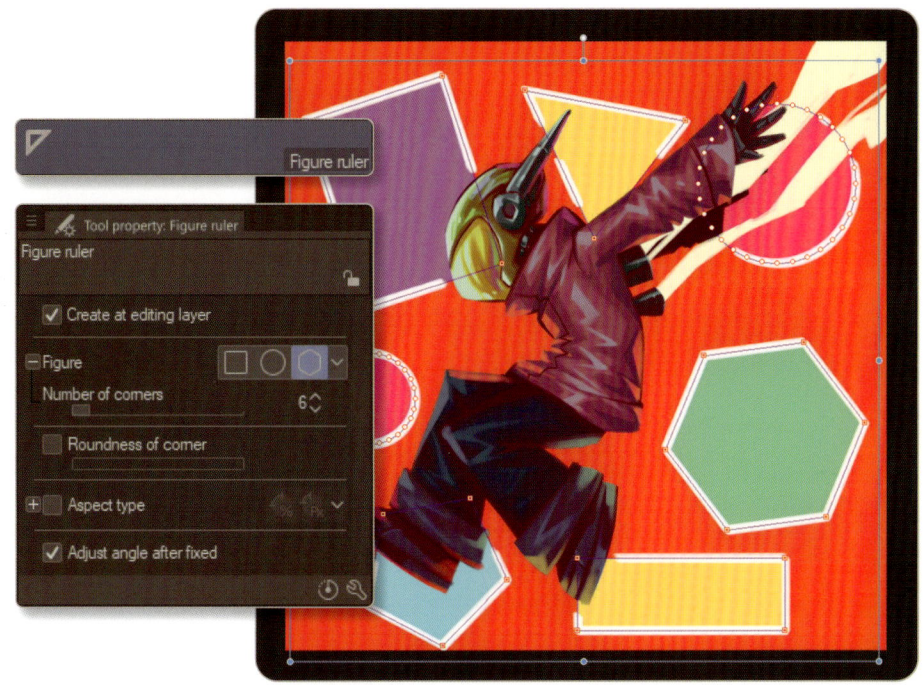

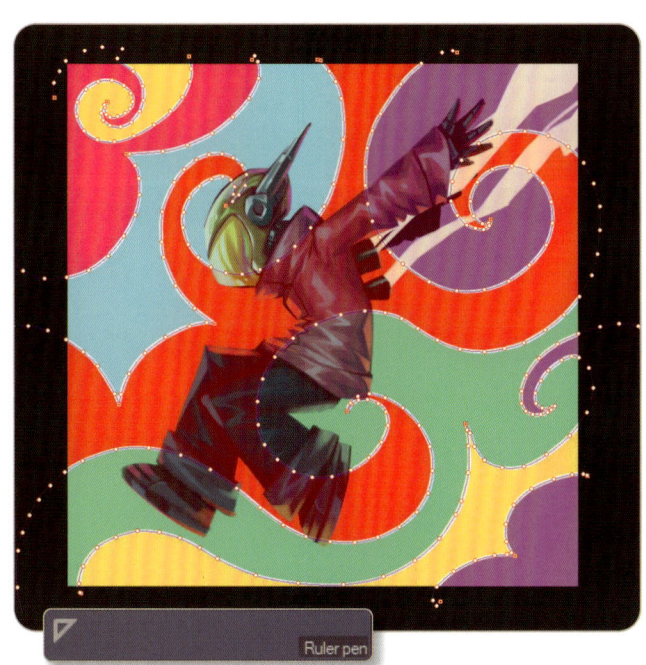

∧ Select the Ruler Pen to draw rulers freehand

RULER PEN

This option enables you to draw rulers manually on the canvas. These can be any shape you like!

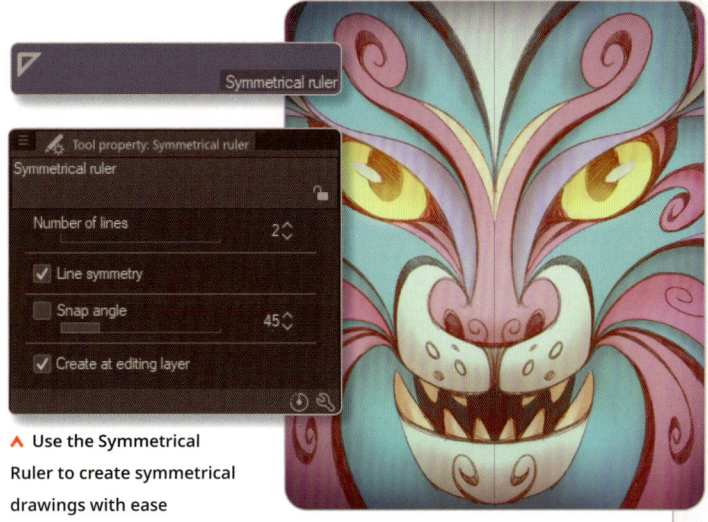

∧ Use the Symmetrical Ruler to create symmetrical drawings with ease

SYMMETRICAL RULER

This sub tool makes the creation of symmetrical drawings quick and easy. To create a Symmetrical Ruler, click on the canvas and hold to define the angle. If you just click, the ruler will automatically be created at a right angle. To define the ruler at a preset angle, hold down the **Shift** key while creating it. You can modify the number of lines that will be drawn simultaneously in the Tool Property palette.

SPECIAL RULER

This allows you to create various types of lines, such as parallel or radial. To choose the type of line, select the Special Ruler option in the Tool Property palette.

∨ Here the Special Ruler sub tool is used to create a series of eye-catching background designs

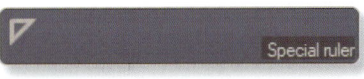

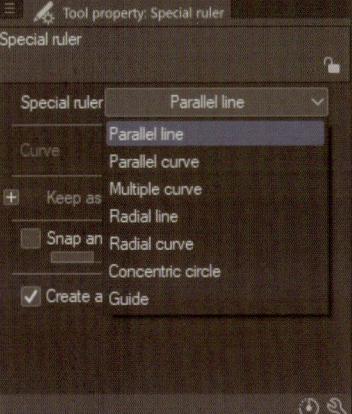

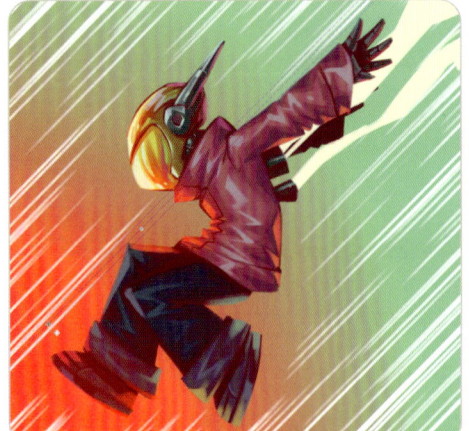

PARALLEL LINE

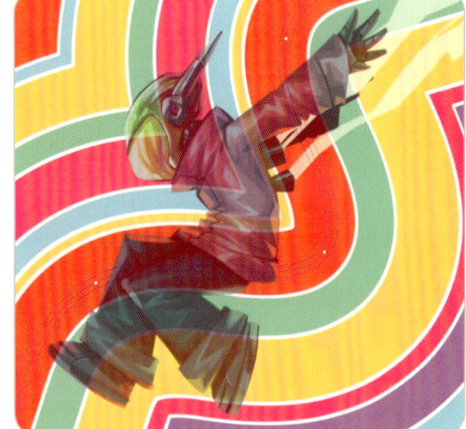

PARALLEL CURVE

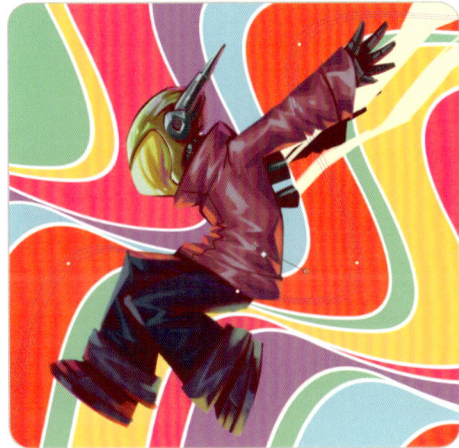

MULTIPLE CURVE

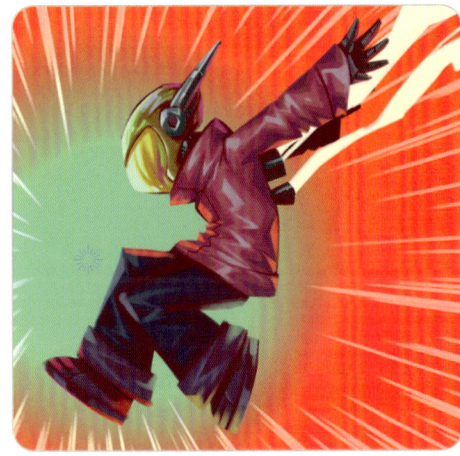

RADIAL LINE

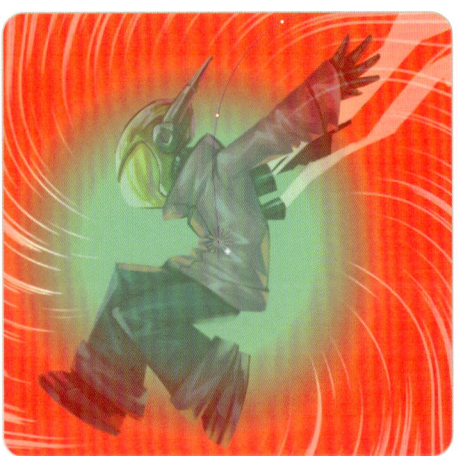

RADIAL CURVE

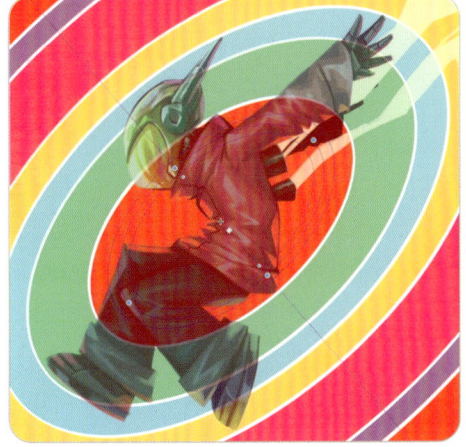

CONCENTRIC CIRCLE

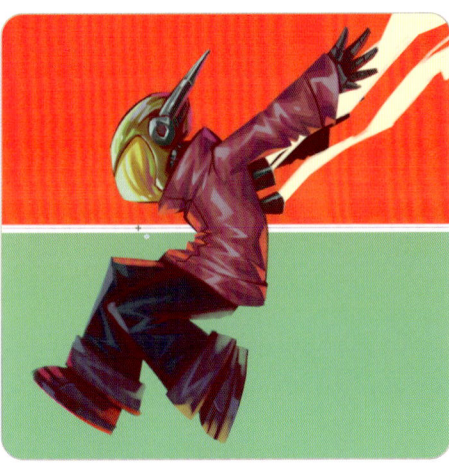

GUIDE

GUIDE

This sub tool allows you to create rulers for drawing horizontal or vertical lines, similar to the Ruler bar and Special Ruler.

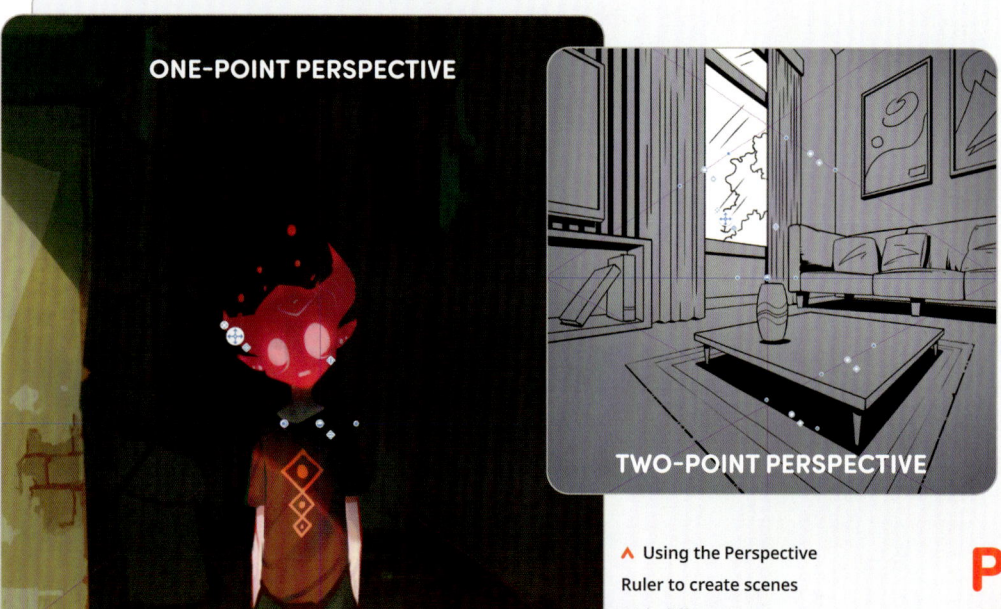

ONE-POINT PERSPECTIVE

TWO-POINT PERSPECTIVE

THREE-POINT PERSPECTIVE

∧ Using the Perspective
Ruler to create scenes
with different types
of perspective

PERSPECTIVE RULER

Perspective Rulers can be used when drawing scenes freehand, as well as for those that have a more precise point-of-view. On creating a ruler, the drawing will adjust to the established perspective. However, you may not be able to draw elements such as ovals or organic shapes unless you deactivate the Perspective Ruler on the layer you wish to draw them on.

Create a Perspective Ruler by selecting **Layer > Ruler/Frame > Create Perspective Ruler**. A window will appear where you can set the type of perspective. The Perspective Ruler will then appear on the canvas.

You can also create a Perspective Ruler by using the Ruler tool and Perspective Ruler sub tool. Click and hold the cursor to orient the first perspective line, then release and click again to orient the second line, which will intersect with the first line. The intersection of these two lines will create a vanishing point. To create another vanishing point, simply repeat the process.

Use the Operation tool's Object sub tool to edit a Perspective ruler that has already been created. You can temporarily activate this sub tool by holding down the **Ctrl** key.

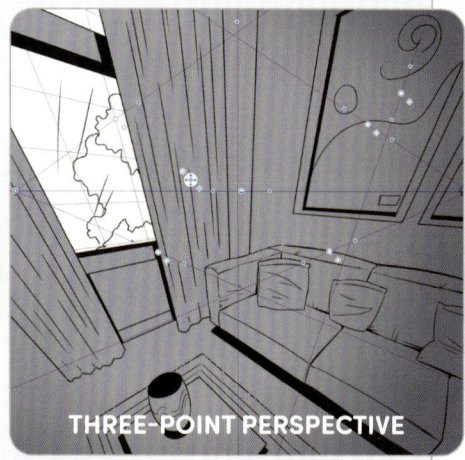

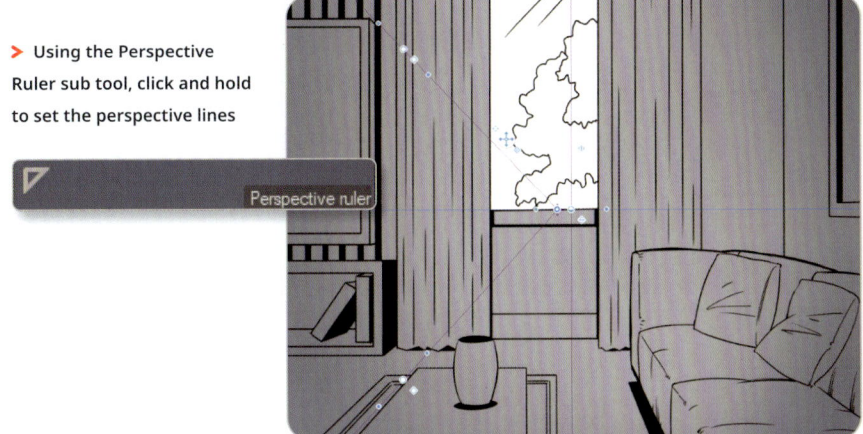

> Using the Perspective Ruler sub tool, click and hold to set the perspective lines

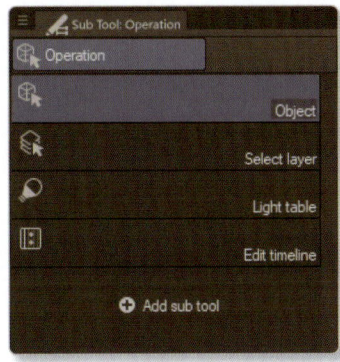

∧ To edit a Perspective ruler, select the Object sub tool of the Operation tool

FISHEYE PERSPECTIVE RULER

This distinctive Perspective Ruler allows you to draw objects or scenes as if you were viewing them through a fisheye lens. Select Layer > Ruler/Frame > Create Perspective Ruler, then in the window that appears, activate the Fisheye Perspective option and select OK.

Another way to create fisheye perspective is with the Perspective Ruler sub tool within the Ruler tool. Simply activate the Add Fisheye option in the Tool Property palette.

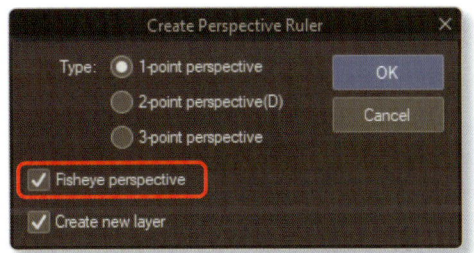

< Activating the Fisheye Perspective option

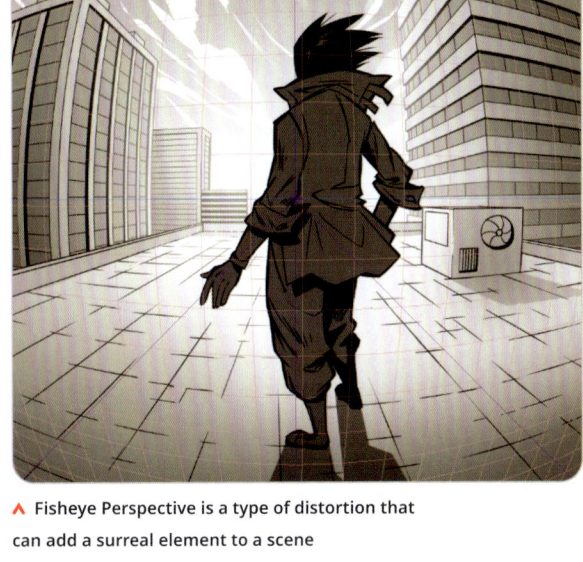

∧ Fisheye Perspective is a type of distortion that can add a surreal element to a scene

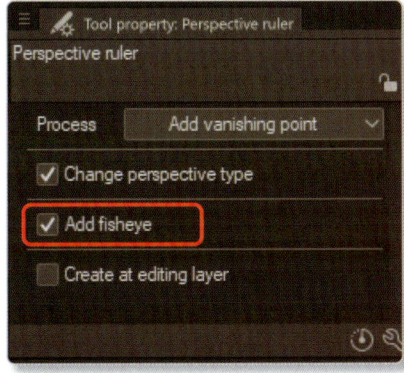

> Select the Add Fisheye option in the Tool Property palette

PERSPECTIVE

Practise sketching with a roughly defined perspective, without the use of rulers, to increase your skill and develop a better sense of space. When you then apply Perspective Rulers to a drawing, over time you will find there won't be so many corrections to make.

Tonal correction

When an artwork is almost finished or you're midway through the process, you may start to question whether the lighting, contrast, or choice of colours is correct. You might decide that the entire piece, or a particular layer, requires a small adjustment or balance. The Tonal Correction tool can be used to make these types of value adjustments to your work.

There are two ways to apply Tonal Correction. The first way is by directly applying Tonal Correction to your layer, and the second way is by creating a Tonal Correction layer that affects the layers below.

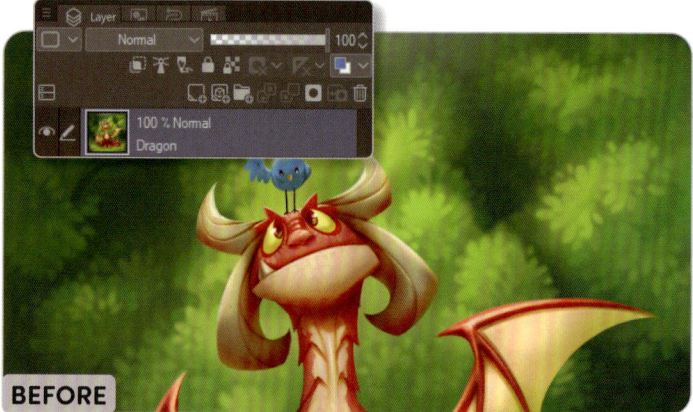

APPLYING TONAL CORRECTION

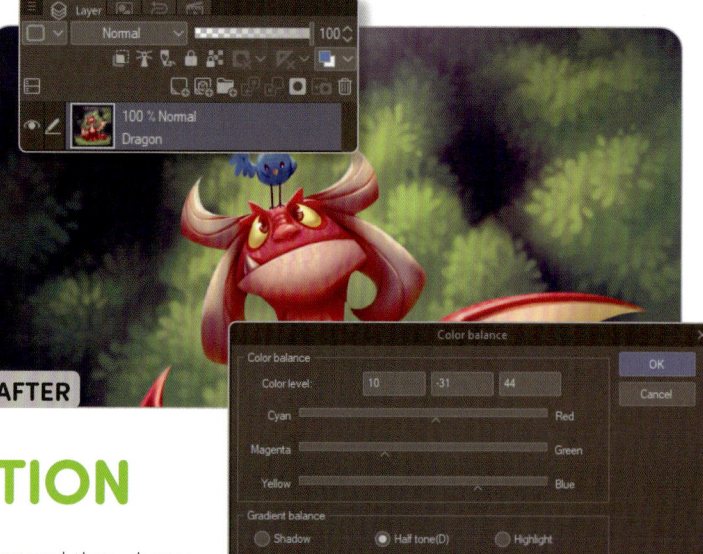

▲ The Colour Balance Tonal Correction is applied to this image

To create a Tonal Correction directly on a layer, select **Edit > Tonal Correction** and then choose the Tonal Correction effect you want to apply. The types of effect available are covered over the next few pages. You can also apply Tonal Correction to a selection area. After applying the effect, it cannot be modified later.

TONAL CORRECTION LAYER

To create a Tonal Correction layer, select **Layer > New Correction Layer** and then choose the type of correction you want the layer to have. These types of layers are useful because you can modify the correction after creating it, except for the Reverse Gradient correction. To modify a correction, double-click on the thumbnail of the Correction layer you created.

As it's a layer, you can also apply various settings such as Mask, Clip to the Layer Below, and Blending Modes.

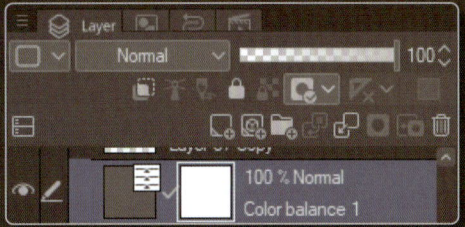

^ Masks can be used to modify specific areas of a layer

⌄ The Hue/Saturation/Luminosity Tonal Correction is applied here – the mask helps to control the skin's Tonal Correction and prevent it from applying to other parts of the image

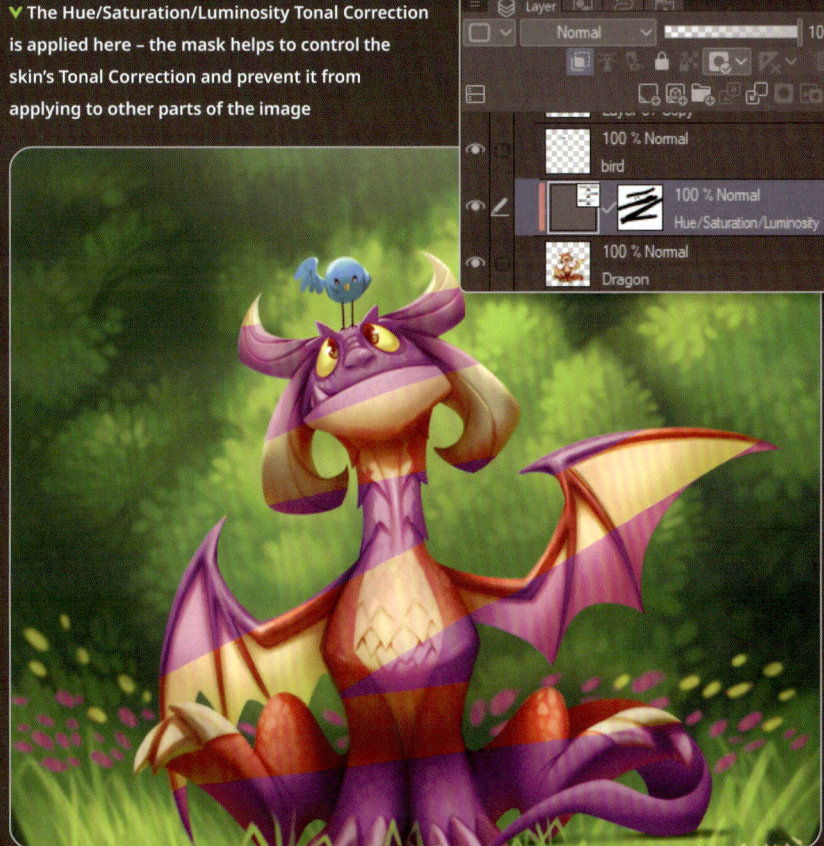

TONAL CORRECTION EFFECTS

We will now look at the Tonal Correction effects available in Clip Studio paint and how they change an image.

BRIGHTNESS/CONTRAST

This effect adjusts the brightness and contrast of a layer or selection. It also has an auto-adjust option that automatically applies brightness and contrast, but this is only available when you apply the correction from the Tonal Correction submenu of the Edit menu.

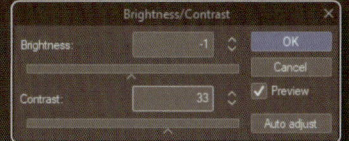

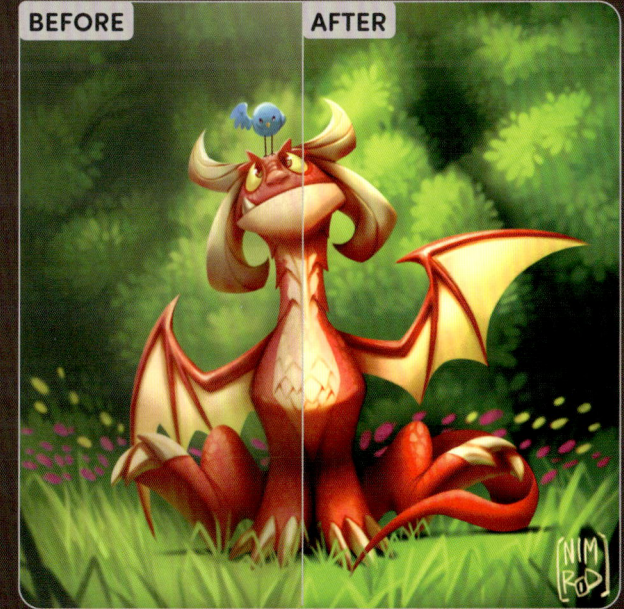

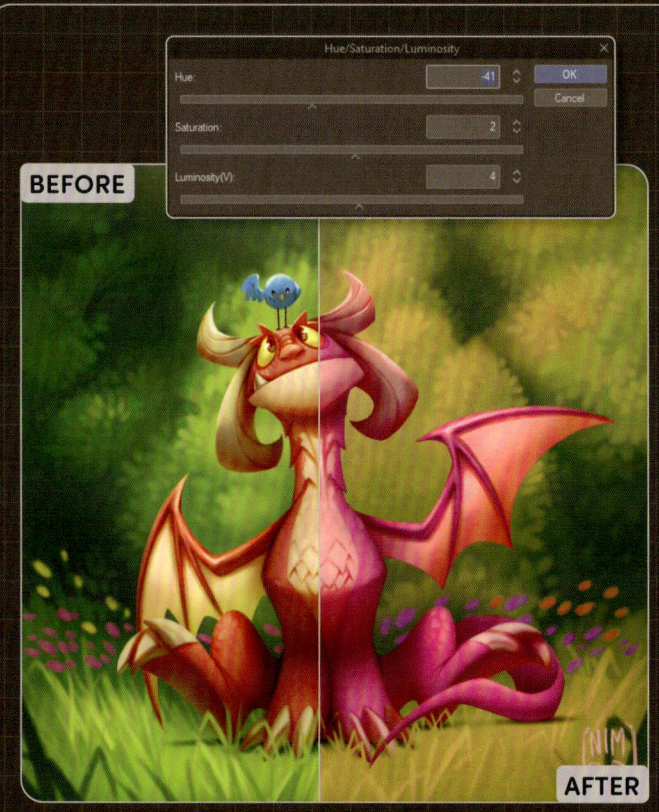

HUE/SATURATION/LUMINOSITY

Based on the HSV colour model, this effect modifies the hue, saturation, and luminosity of a layer or selection.

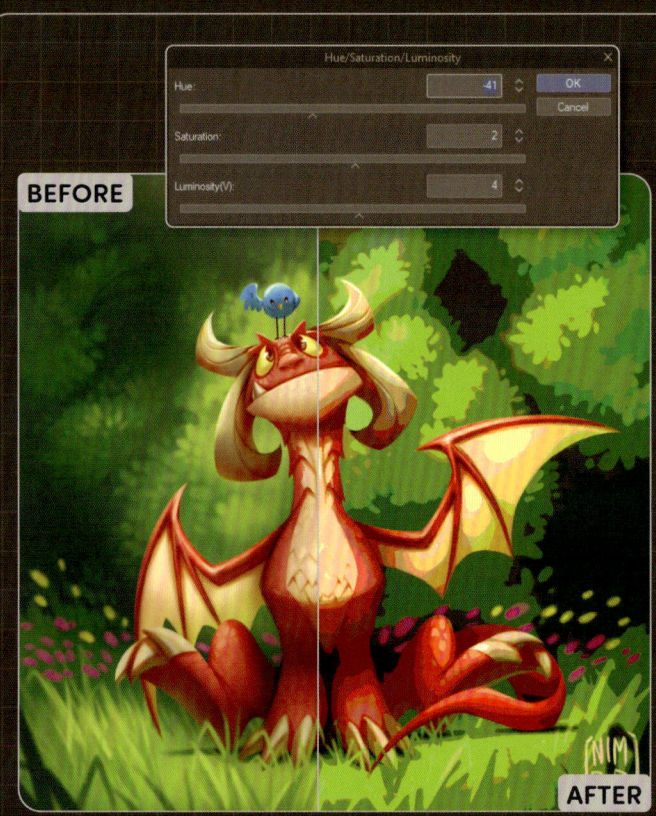

POSTERIZATION

This effect reduces the colour levels of the image. If the levels are lower, the image will have a coarser effect.

REVERSE GRADIENT

This effect inverts the colours of the image. There are no editing options if you create this effect as a Correction layer.

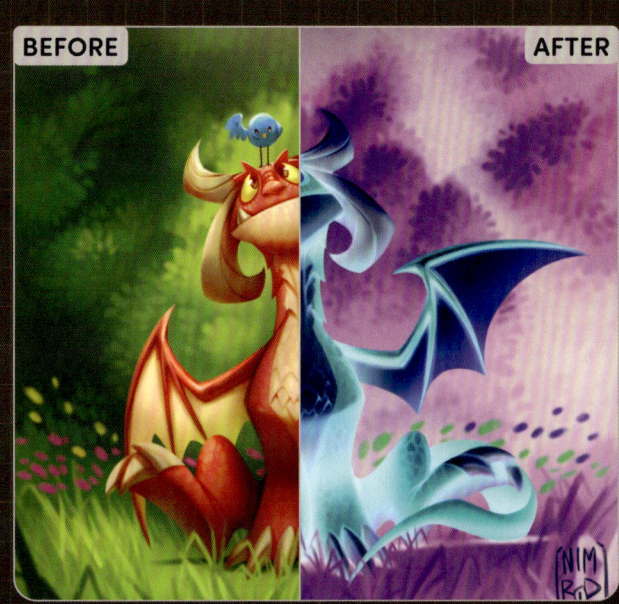

LEVEL CORRECTION

This effect adjusts the dark, mid, and light tones of the image. These tones can be modified using the three buttons located along the edge of the histogram. The button on the left-hand side controls the dark tones, the middle button controls the midtones, and the button on the right-hand side controls the light tones. For example, if you slide the dark tones button to the right, there will be higher contrast in the image due to the increase in dark tones. Conversely, the light tones increase if you slide the light tones button to the left.

The Output controls allow you to reduce the dark and light tones. If you slide the dark tones control to the left, the dark tones of the image will decrease. The image will appear brighter and softened.

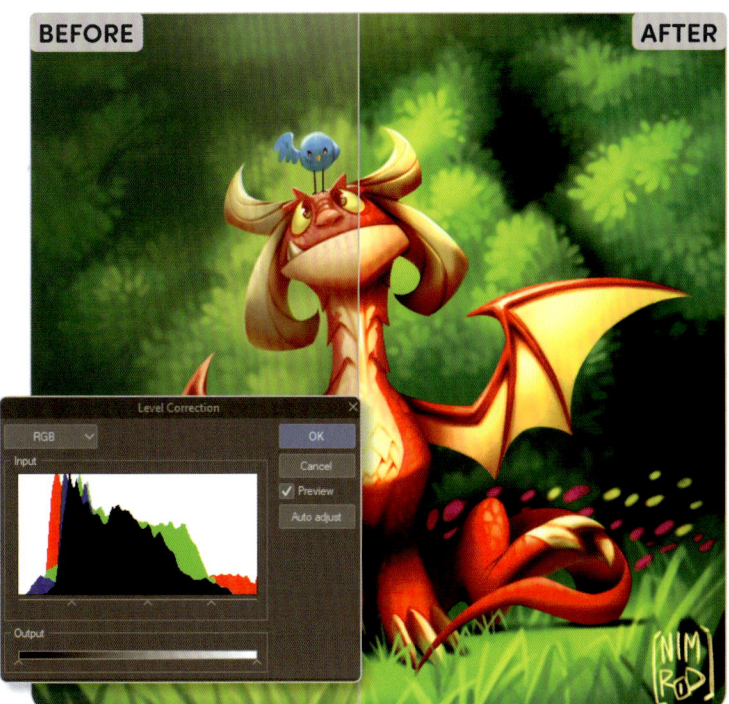

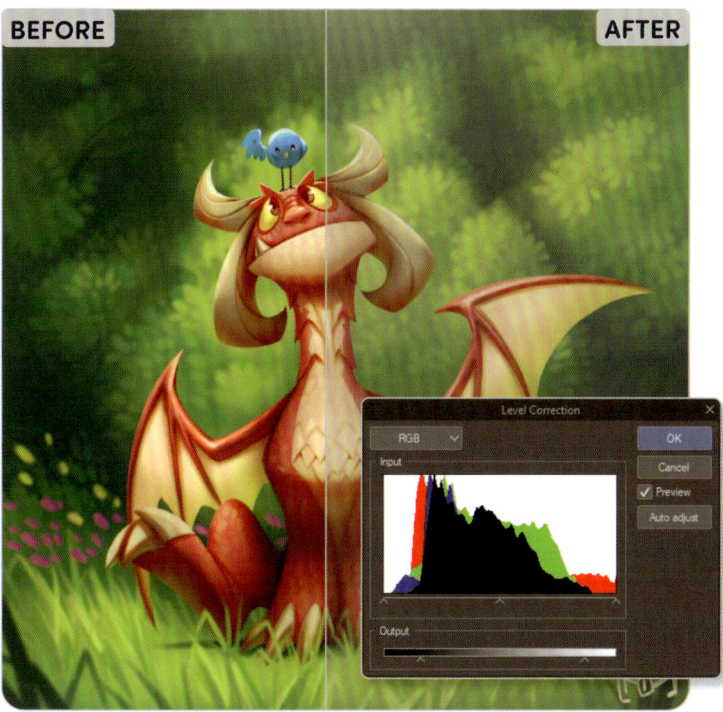

⌃ **Level Correction with the output adjusted**

TONE CURVE

This effect allows you to adjust the brightness of the image using a graph, where the horizontal axis represents the input values (original brightness of the image) and the vertical axis represents the output values (brightness of the image after applying the adjustment). To modify the tone, you need to create a control point on the graph. To create a control point, click on the graph and hold down the mouse button to drag the point to the desired position. You can adjust the brightness of each RGB channel individually or all three at once. This option also serves to increase or decrease the amount of red, green, or blue in your artwork.

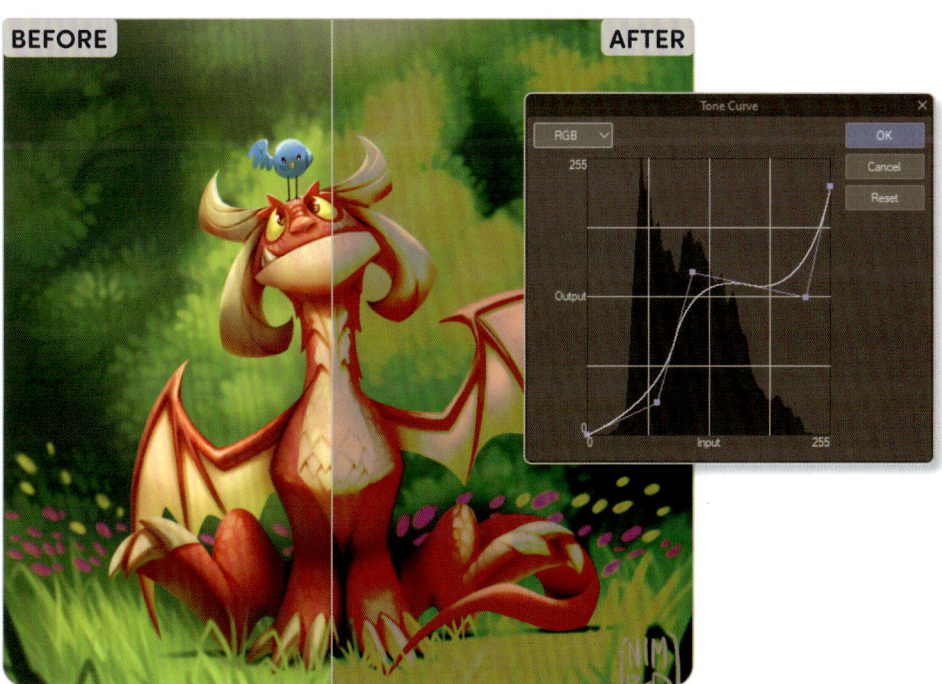

COLOUR BALANCE

This effect allows you to adjust the colour blend of an image by modifying the RGB colour balance. Each tonal range (Shadow, Half tone, Highlight) has an independent colour balance that can be edited. It also has a Keep Brightness option to preserve the brightness values of the image, if activated.

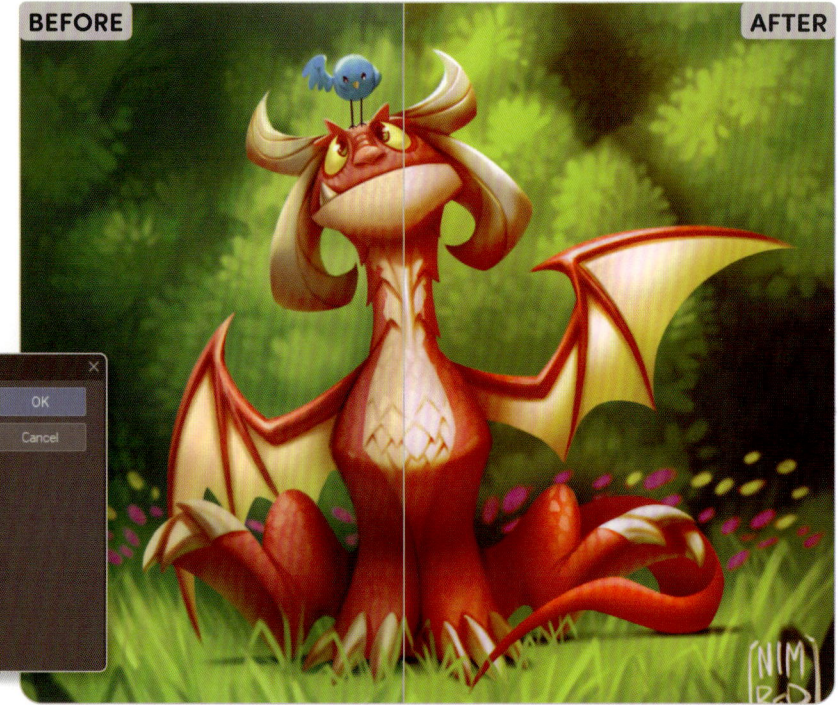

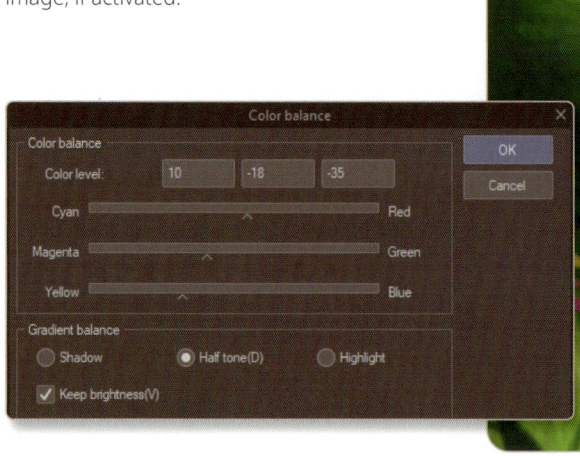

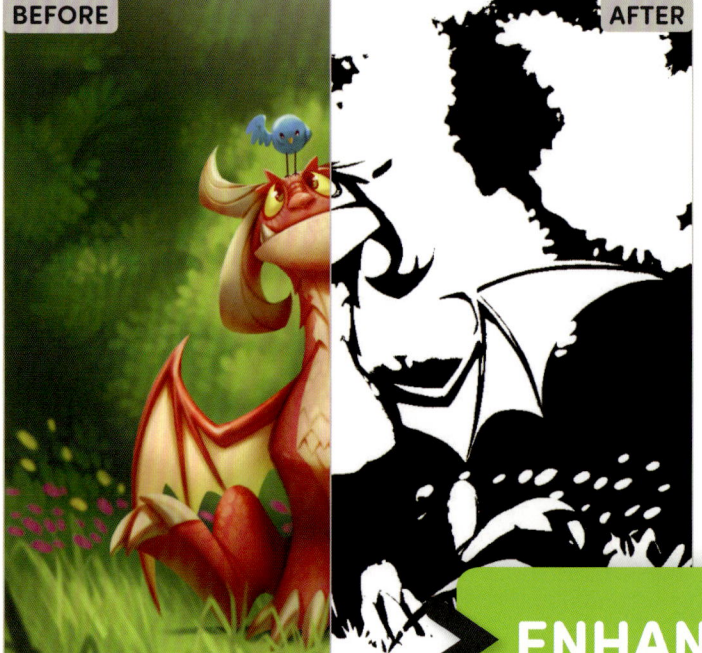

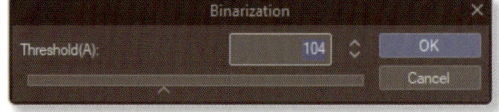

BINARIZATION

This effect allows you to convert a layer or selection to black and white. The darker colours of the image will be considered black, and the lighter colours will be considered white. When you create this effect from the Tonal Correction submenu of the Edit menu, there is an extra setting called Leave Transparency. This setting maintains the transparency of the original image.

ENHANCE YOUR ARTWORK

Applying Tonal Corrections can further correct and enrich your artwork during the creation process or after finishing it. You can apply Tonal Corrections at the beginning of the process to create quick colour palettes. You can also use them after completing an artwork to correct values and colours.

GRADIENT MAP

This effect replaces the colours of the original image with colours from a chosen gradient. These colours are replaced according to the darkness and brightness values of the original image. In the settings window, you can choose a pre-established gradient, modify it, or create a new one. You can also change the colours, define the quantity, and edit the proportions of colours that the gradient will have.

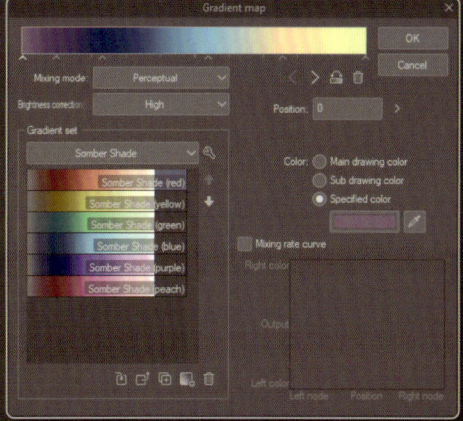

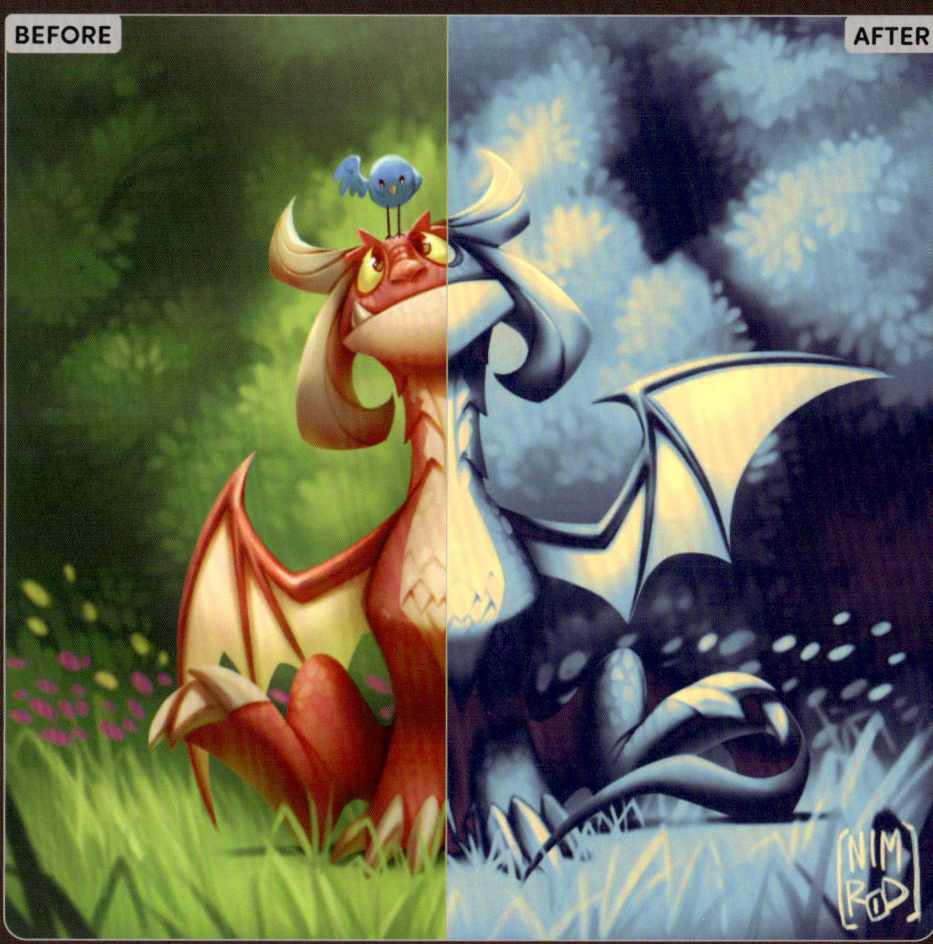

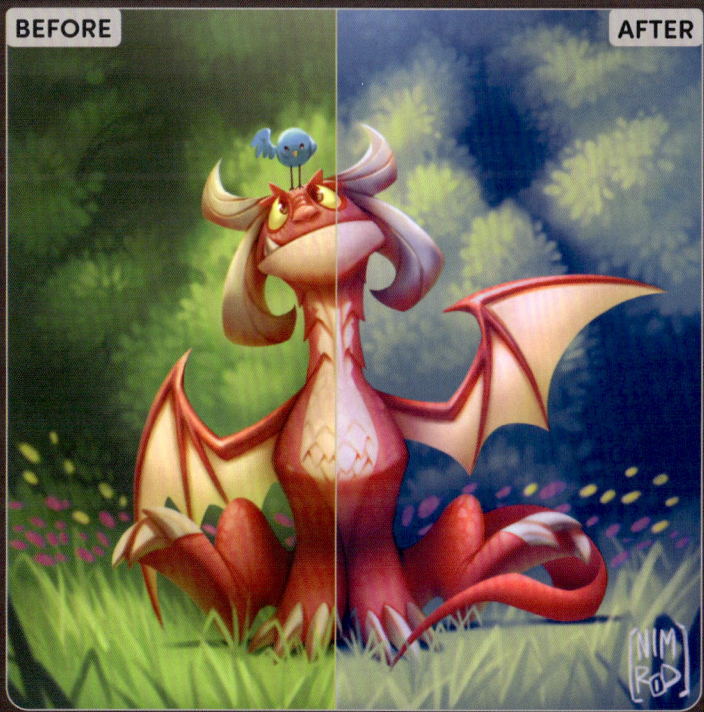

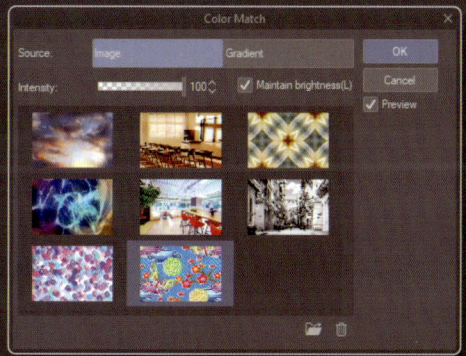

COLOUR MATCH

This effect allows you to adjust the colours of your artwork to the colours of a pre-established or imported image or gradient. This effect is only available when you create it from the Tonal Correction submenu of the Edit menu.

Filters

Filters are used to apply special effects to a Raster layer or selection. These effects can generate visual changes to your artwork that are difficult to replicate manually. The list of available filters can be found in the Filter menu. A submenu will appear, from which you can select the filter you want to apply to your layer or selection.

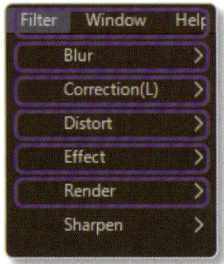

Filter	Window	Help
Blur		>
Correction(L)		>
Distort		>
Effect		>
Render		>
Sharpen		>

∧ **The Filter menu offers six types of filter**

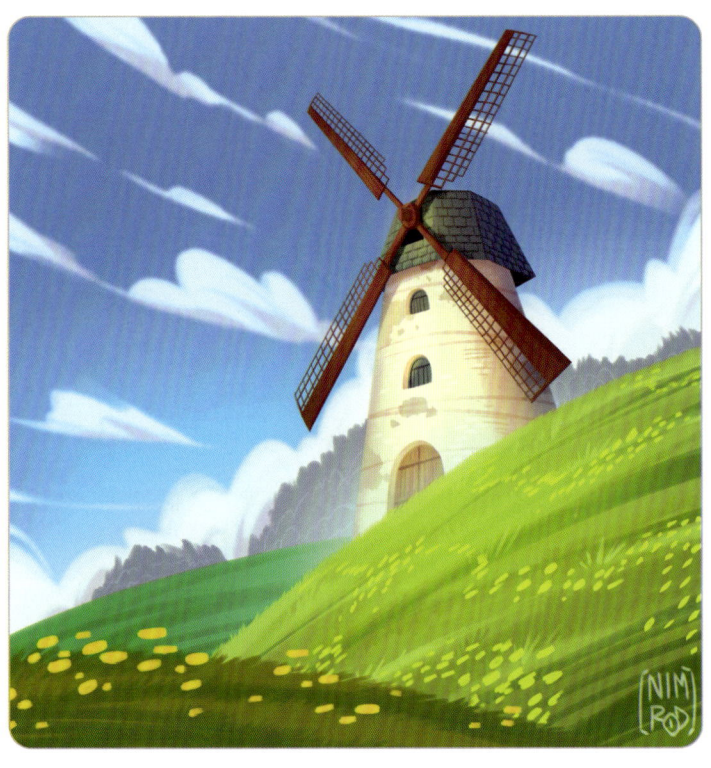

BLUR

These options apply different types of blur effects to the Raster layer or selection.

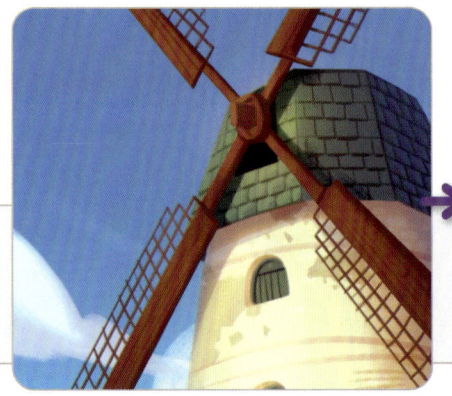 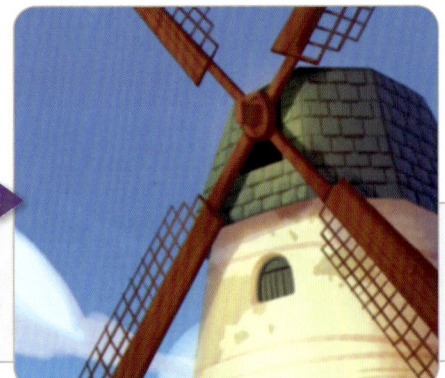

BLUR/BLUR (STRONG)
These apply a basic blur, either normal or strong in effect.

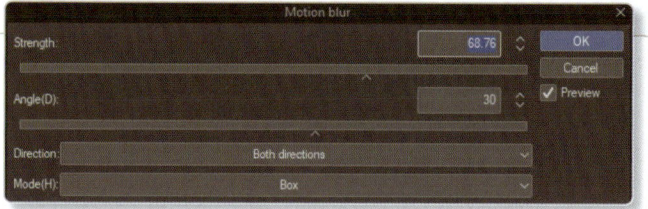

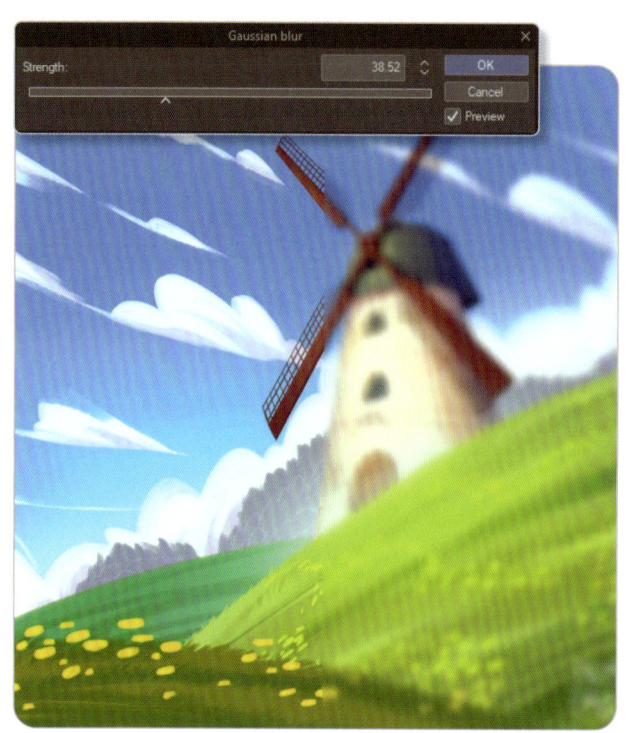

GAUSSIAN BLUR

Gaussian Blur evenly smooths out noise and detail. A window will appear that has a sliding scale, which allows you to control the level of blur. This filter can be used to soften areas that you don't want to draw so much attention to, therefore highlighting the area with more focus. Applying Gaussian Blur to the sky makes the windmill stand out and creates depth.

MOTION BLUR

Motion Blur applies a directional blur effect to the layer or selection, creating an illusion of movement in the image. This effect creates an illusion of speed on the selected layer.

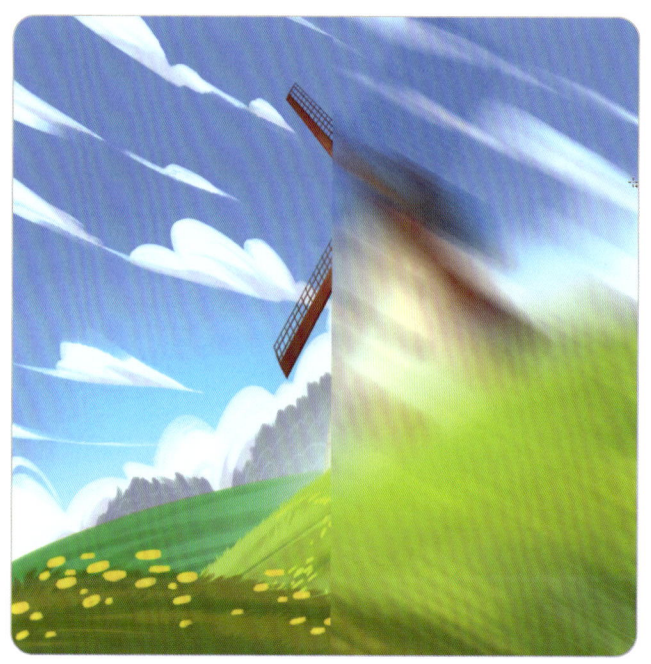

RADIAL BLUR

Radial blur creates a radial blur effect. As well as being able to configure the strength, direction, and mode in the window that appears, you can choose where to place the centre of the effect using the small red cross that appears on the canvas.

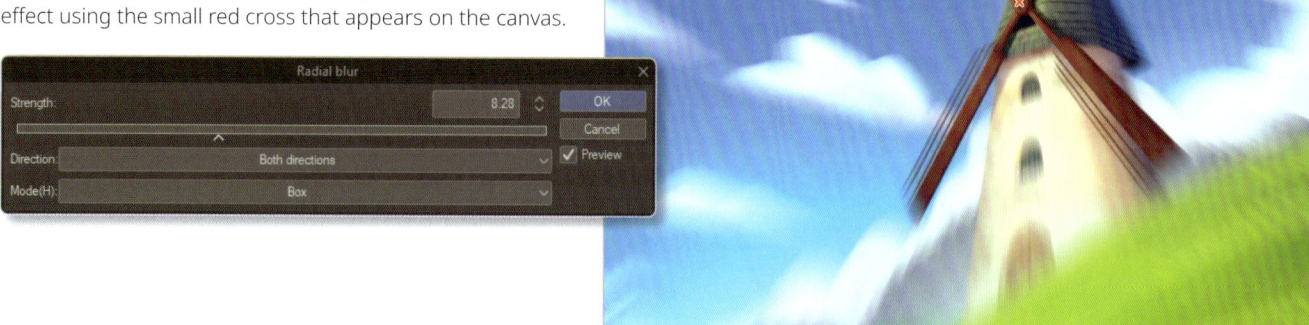

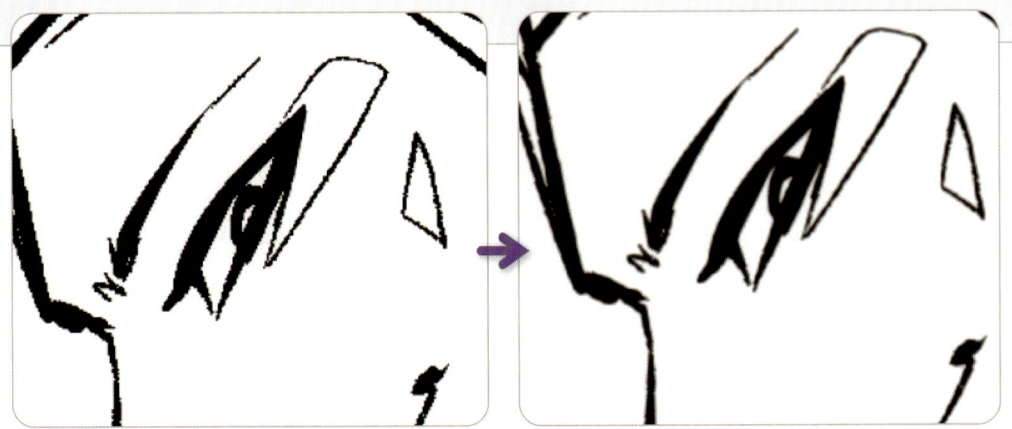

SMOOTHING

This filter smooths the edges of any lines that have irregular outlines. There are no additional settings.

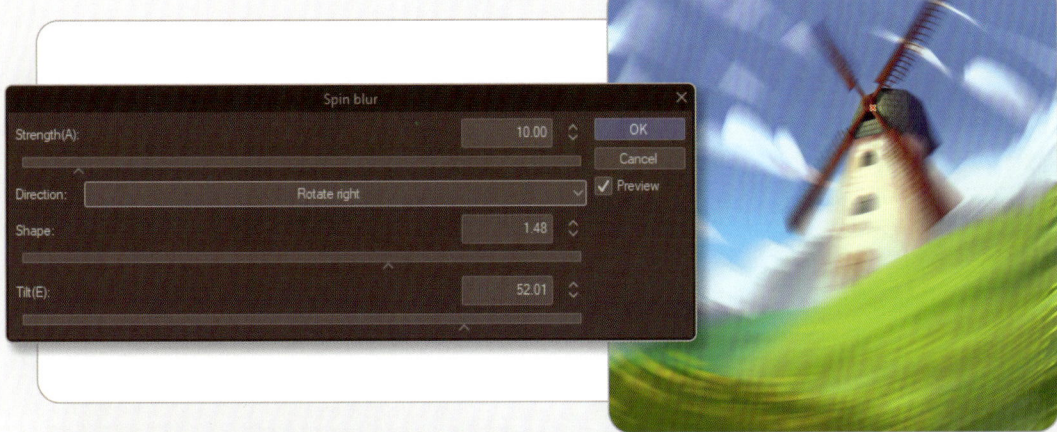

SPIN BLUR

This filter circularly blurs the image. You can adjust the settings – such as the strength, direction, shape, and tilt – in the window that appears.

CORRECTION

These filters can correct drawing lines.

ADJUST LINE WIDTH

This filter thickens the lines of the selected drawing. For this filter to work, the drawing must have a transparent background. You can select the At Least 1 Pixel option if the line adjustment type is set to Narrow and you want to leave one pixel of the central line.

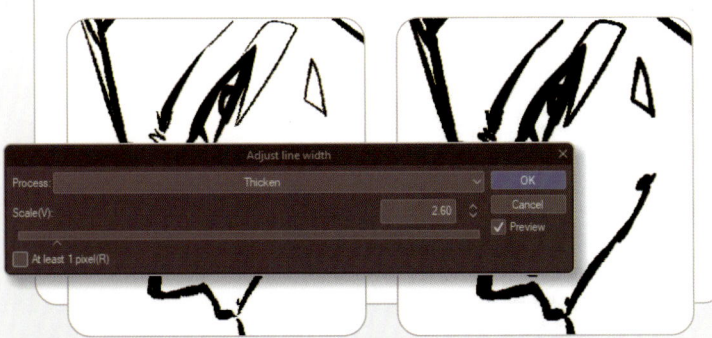

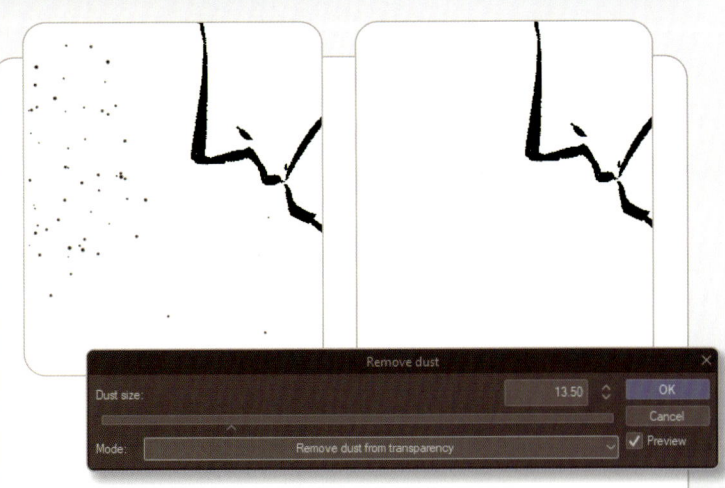

REMOVE DUST

This filter automatically removes dust and small particles that often occur on scanned drawings. For it to work effectively, it's recommended that the drawing has a white or transparent background. If the strokes are thin lines, the filter may erase some fragments of these.

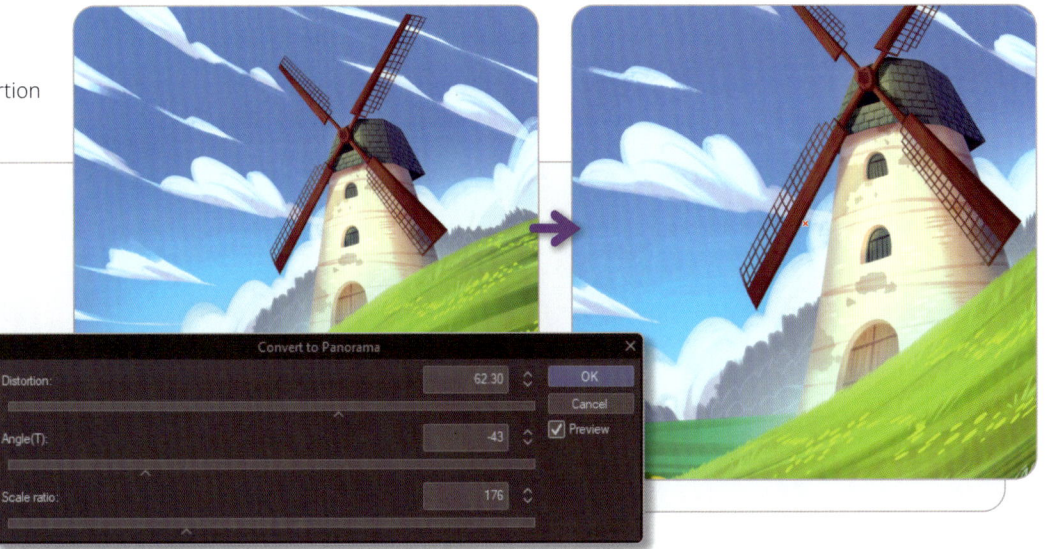

DISTORT

These filters apply different forms of distortion and deformation to images.

CONVERT TO PANORAMA

This filter distorts the image as if it were viewed through a panoramic lens. Use the sliders to adjust the Distortion, Angle, and Scale Ratio.

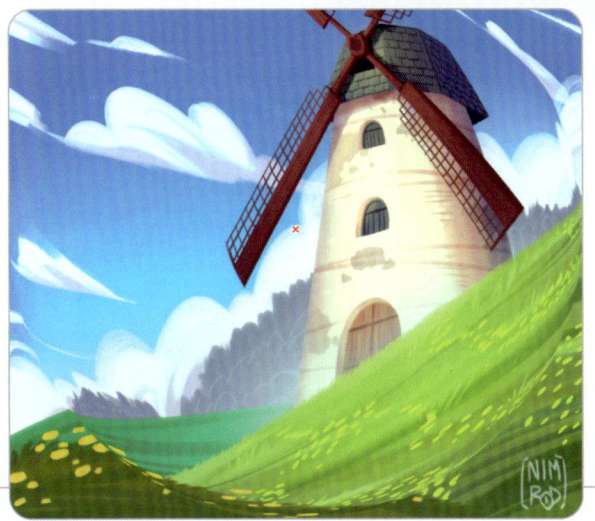

CURVED SURFACE

This filter deforms the image into a spherical or cylindrical shape. The window provides options to alter the Strength, Angle, Radius, and Shape, plus more.

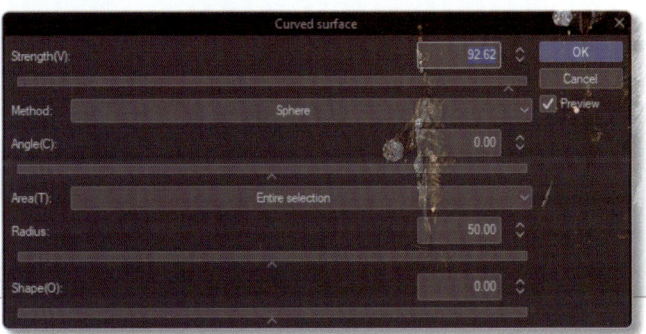

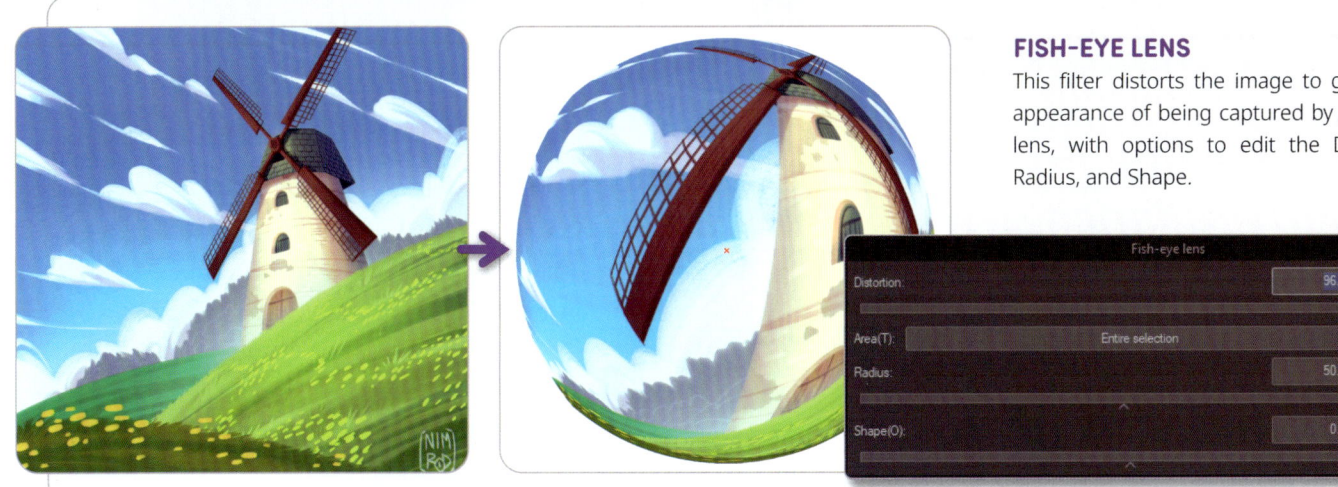

FISH-EYE LENS

This filter distorts the image to give it the appearance of being captured by a fish-eye lens, with options to edit the Distortion, Radius, and Shape.

GEOMETRIC DISTORTION

This filter warps the image by curving it inward or outward, with options to adjust the level of Distortion and Scale Ratio.

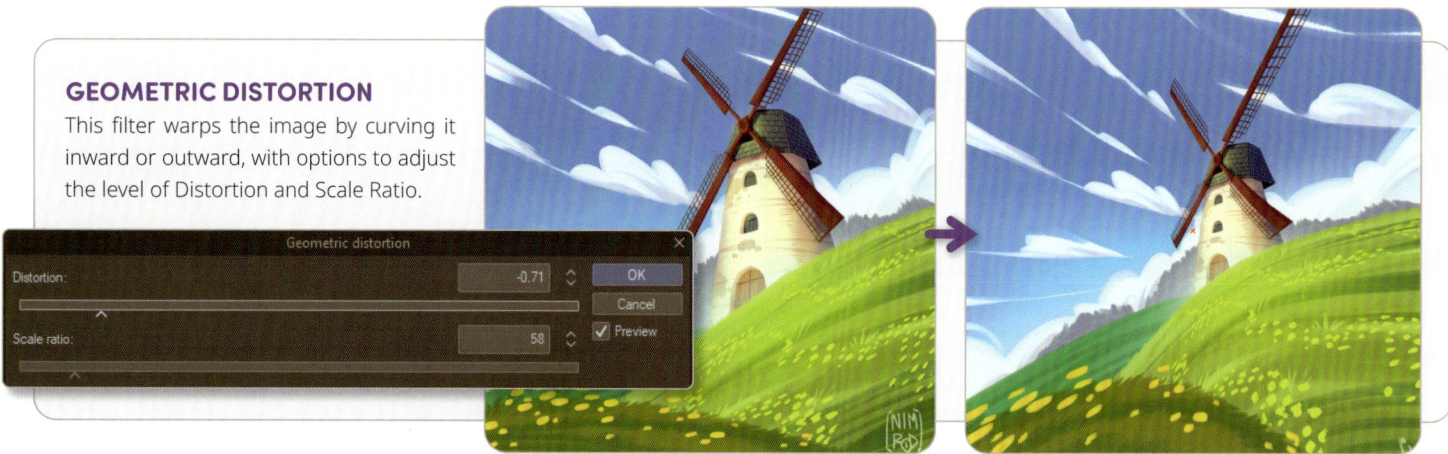

PINCH

This filter pinches the central point of the effect to distort the selected image. Edit the Strength, Radius, and Shape to alter the distortion.

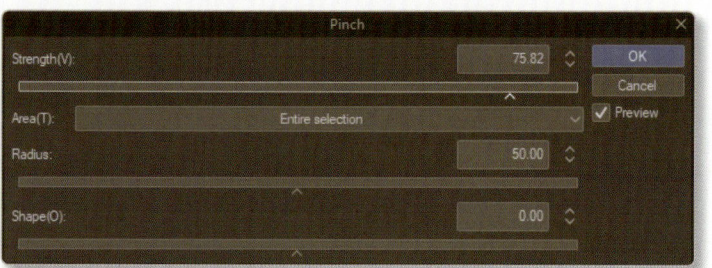

CONNECTION

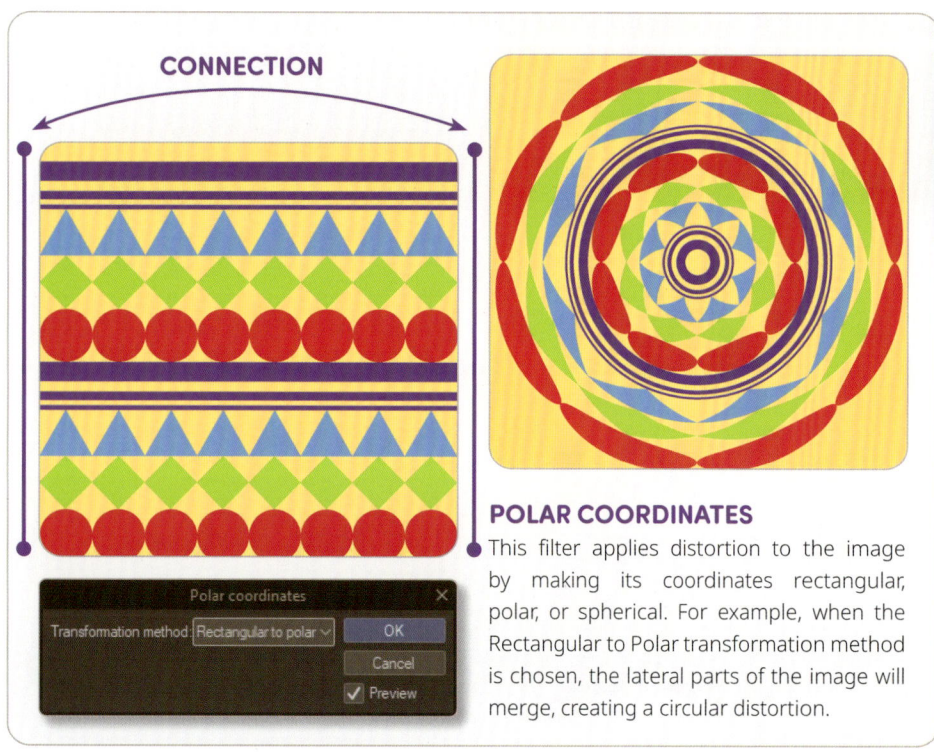

POLAR COORDINATES

This filter applies distortion to the image by making its coordinates rectangular, polar, or spherical. For example, when the Rectangular to Polar transformation method is chosen, the lateral parts of the image will merge, creating a circular distortion.

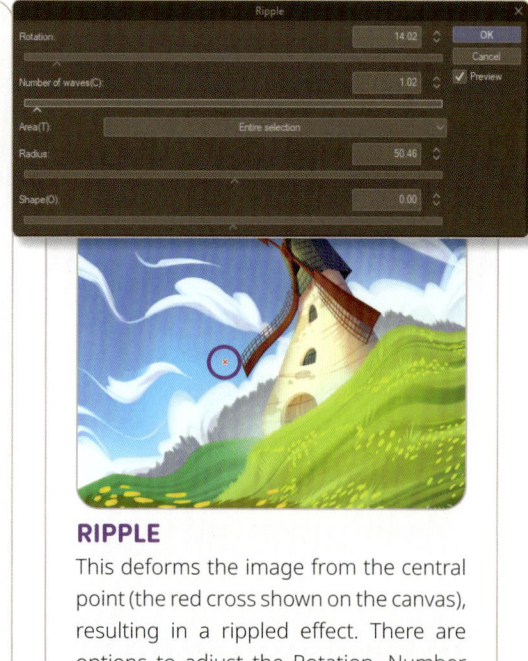

RIPPLE

This deforms the image from the central point (the red cross shown on the canvas), resulting in a rippled effect. There are options to adjust the Rotation, Number of Waves, Area, Radius, and Shape.

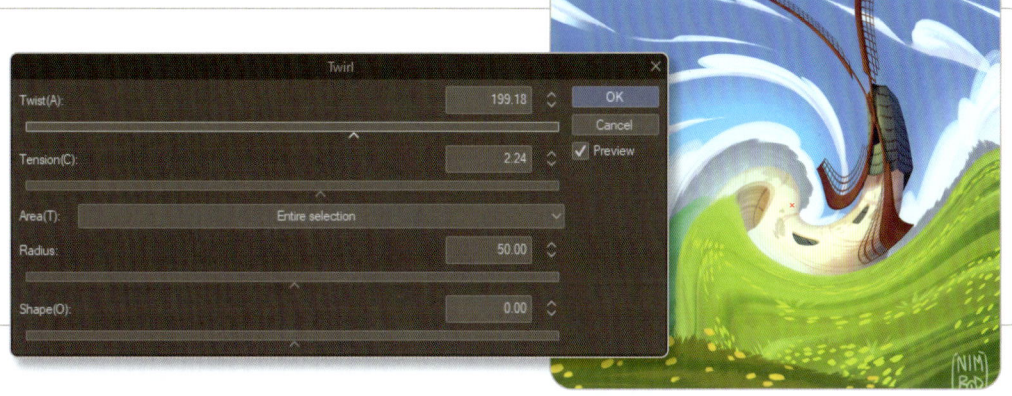

TWIRL

Twirl warps the image with a swirling effect originating from the central point. There are options to adjust the Twist, Tension, Area, Radius, and Shape.

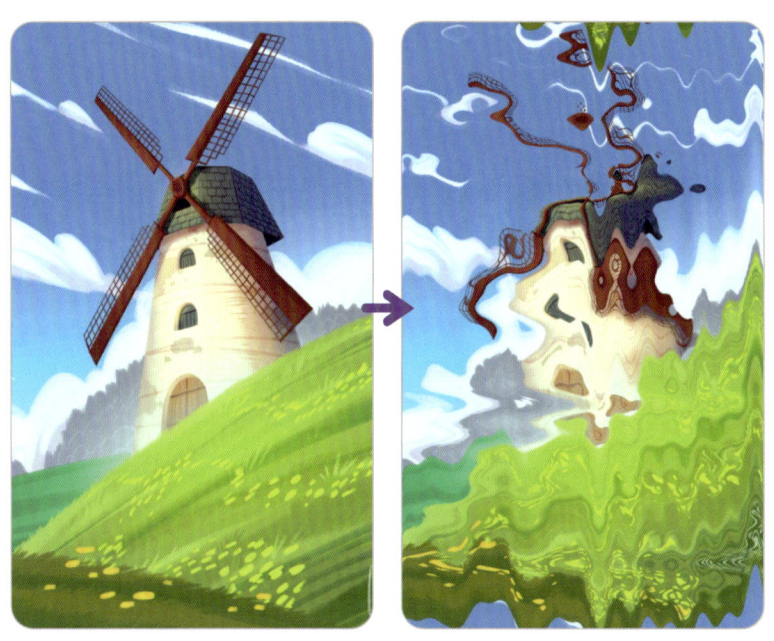

WAVE

This filter creates a water wave effect on the selected image. In the window that appears there are options to configure the Number of Waves, Wavelength, Amplitude, Ratio, and how Undefined Areas should be filled. There is also a Regenerate function if you want to create a wave effect with random values.

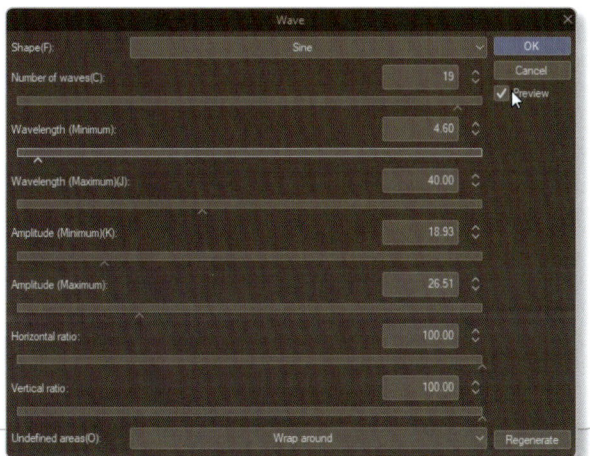

ZIGZAG

This filter creates a pattern of parallel waves in the image. The window provides options to change the Angle, Wave Height, and Number of Waves.

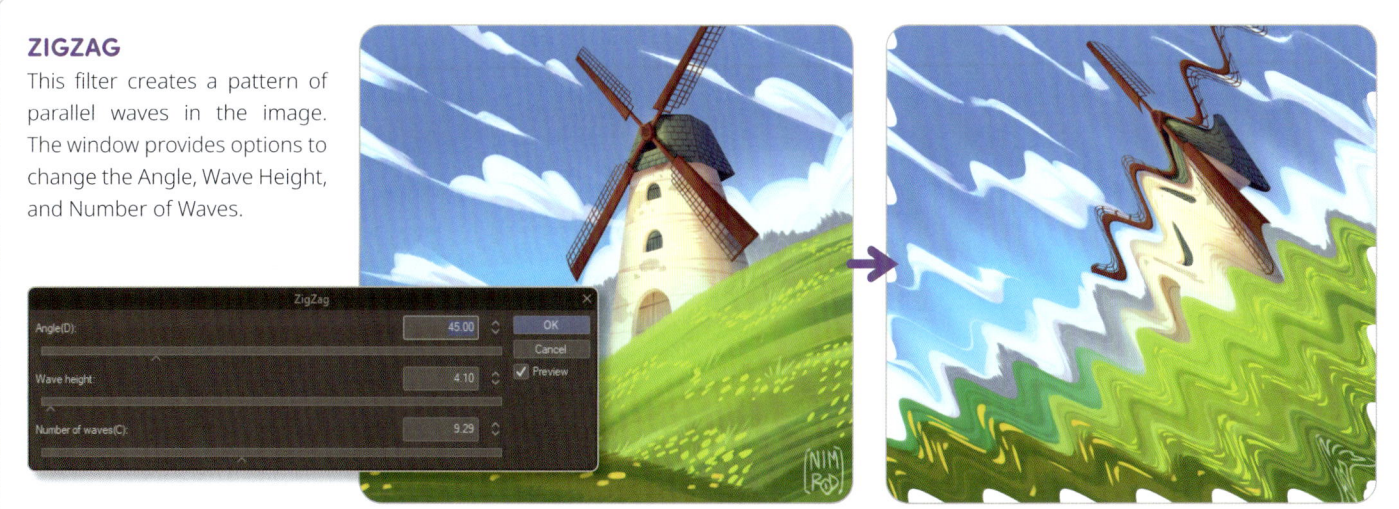

NOISE

This filter creates a photographic noise effect on the selected image. It can be used when a painting is finished to give it a more rendered style, with options to alter the Noise Strength and Colour Mode.

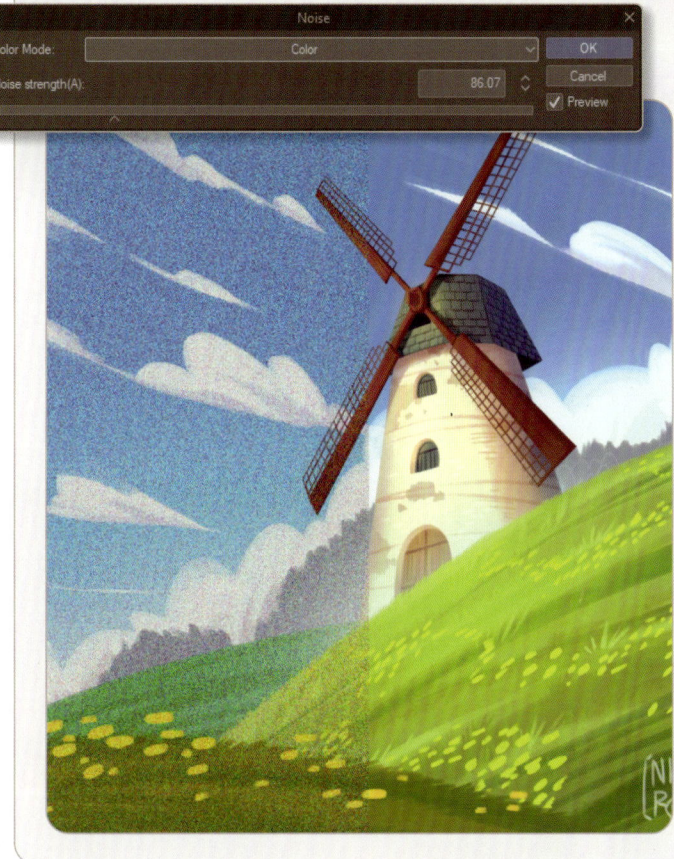

NORMAL MAP

This converts the image into a normal map to provide lighting and relief details to a 3D model after export. You can alter the Strength and Orientation in the window.

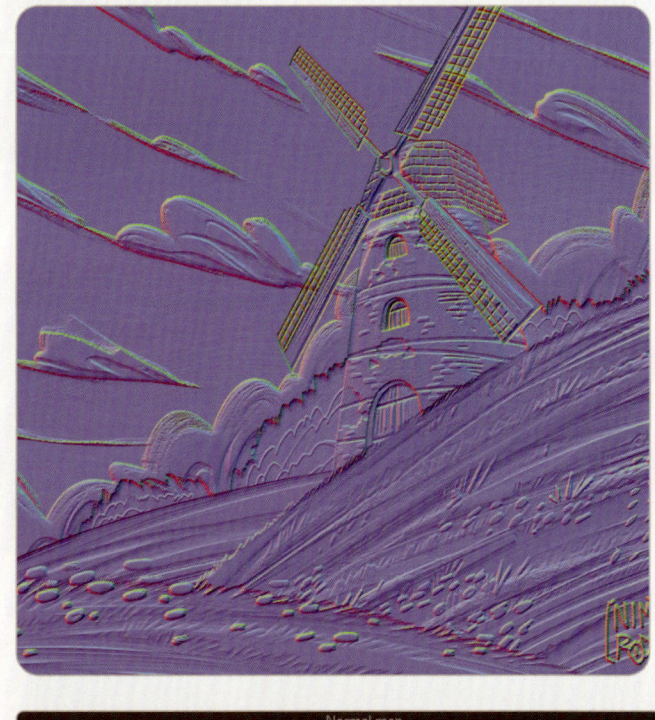

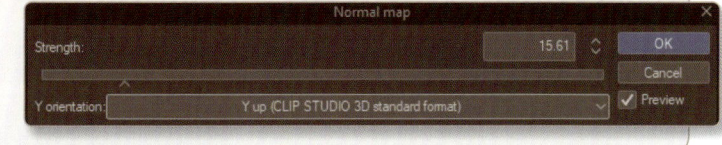

PENCIL DRAWING

This filter adds an artistic pencil-drawing effect to your image. The window contains options to set whether you want the result to be in colour or greyscale, as well as Hatching options.

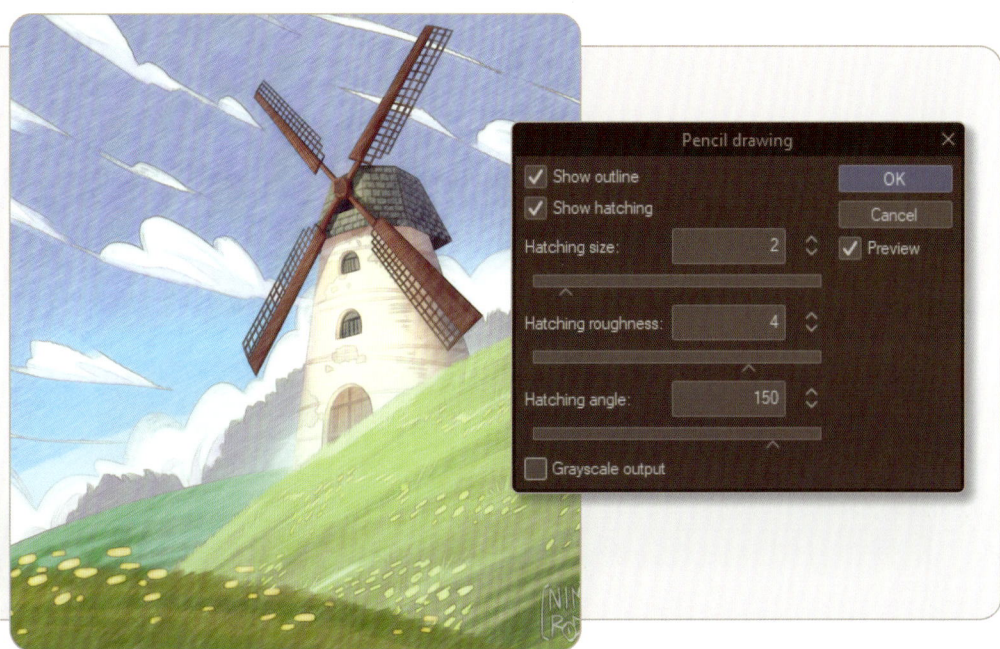

EFFECT

These filters are used to create special effects on the selected image.

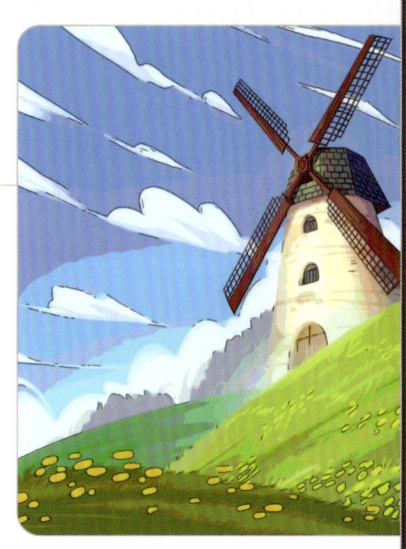
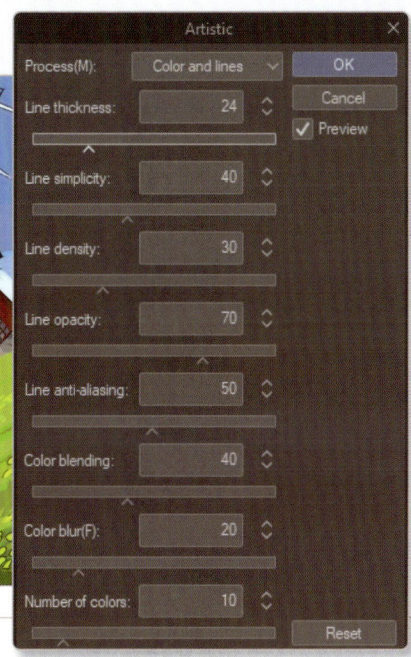

ARTISTIC

This filter applies a painterly effect to the image. (The effect will be more noticeable if it's a photo.) The image's colour palette will be simplified and drawing lines will be added. The window provides options to set whether you want the filter to apply colour and lines, only colour, or only lines.

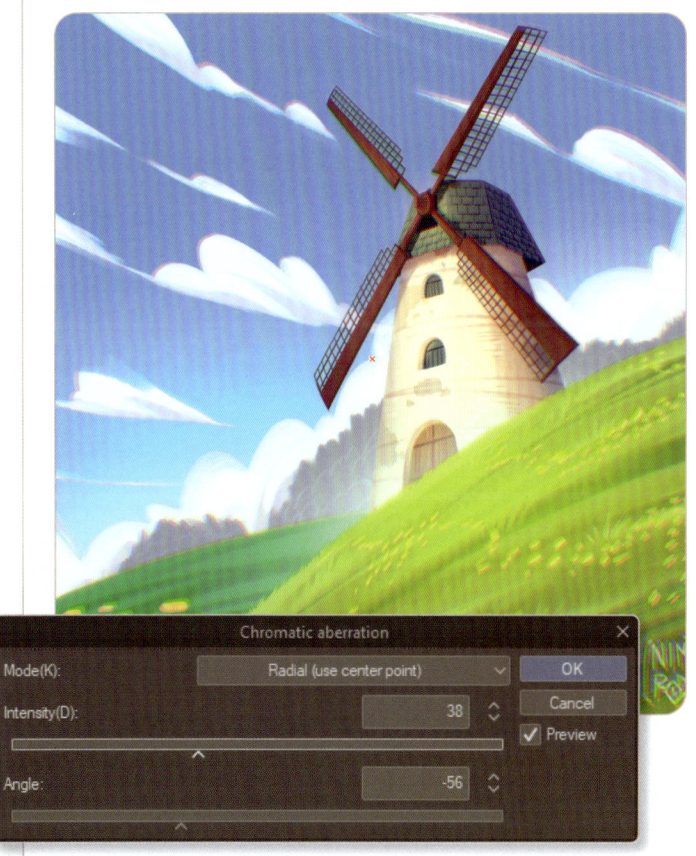

CHROMATIC ABERRATION

This filter creates a colour distortion effect, similar to that found in photographs. You can adjust the Mode, Intensity, and Angle.

MOSAIC

Mosaic creates a pixelation effect on the selected image. You can change the size of the pixels/tiles.

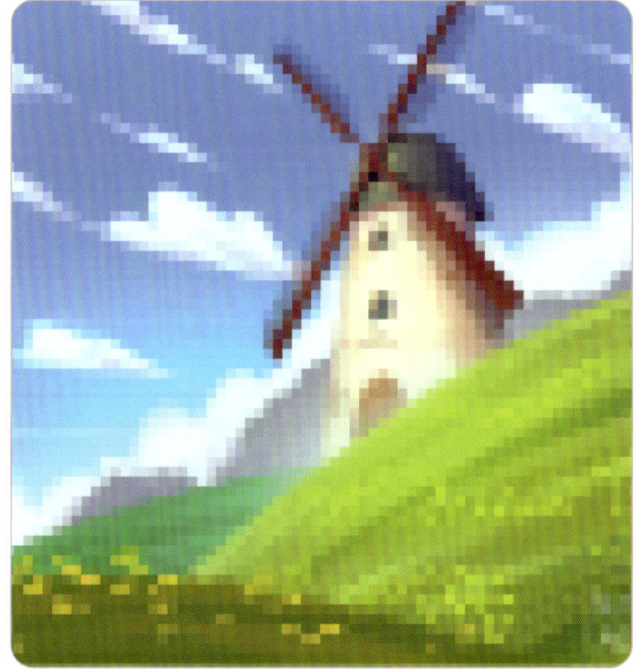
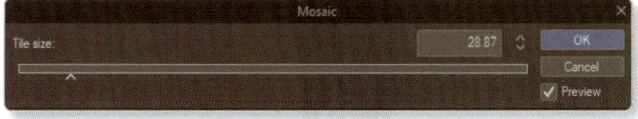

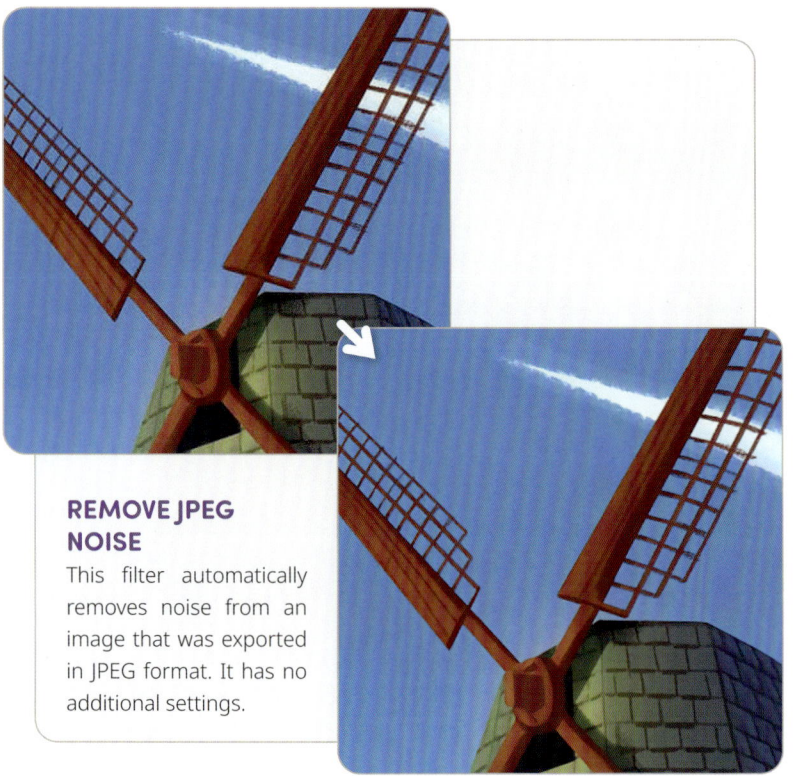

REMOVE JPEG NOISE

This filter automatically removes noise from an image that was exported in JPEG format. It has no additional settings.

RETRO FILM

This adds lighting effects, colour, noise, and chromatic aberration to your image, giving it a retro style. The red cross on the canvas enables you to control the direction of the chromatic aberration. You can also alter the Effect, Intensity, and Noise Strength.

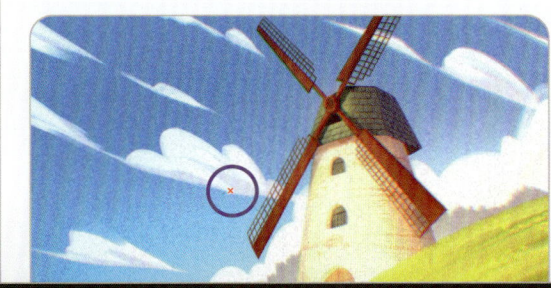

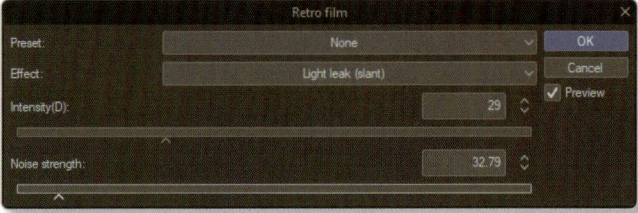

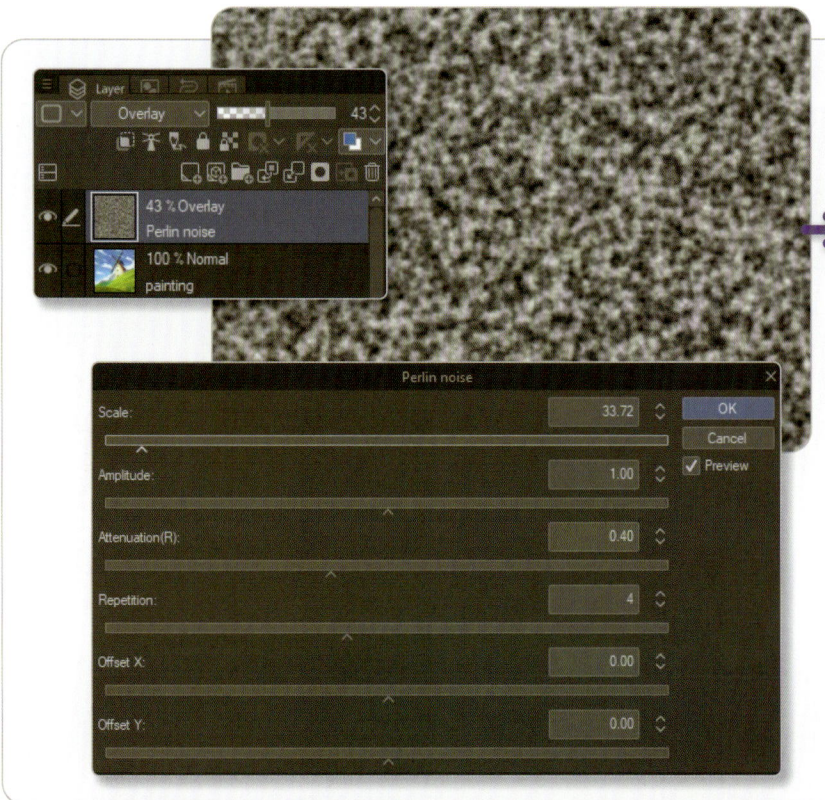

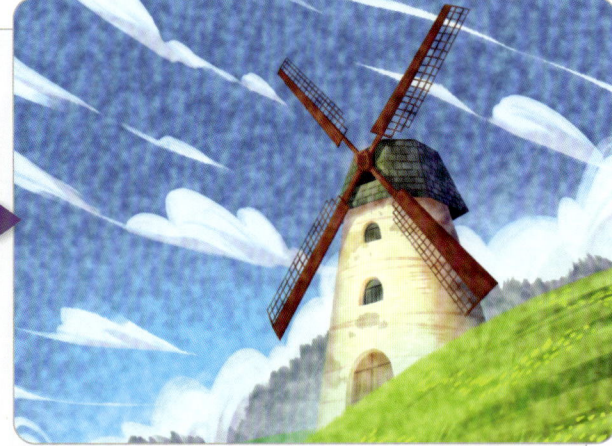

RENDER

PERLIN NOISE

This filter converts the image into a noise pattern resembling grey clouds. The content of the layer doesn't influence the final result of the effect. One way to use the layer with the noise is by changing its blending mode and opacity so that it becomes a texture for the lower layers. There are options to adjust the Scale, Amplitude, Repetition, and more.

SHARPEN

SHARPEN/SHARPEN MORE

These filters automatically sharpen the selected image, adding more contrast and refinement to the strokes. Both filters differ in the level of intensity they apply. They do not have additional settings.

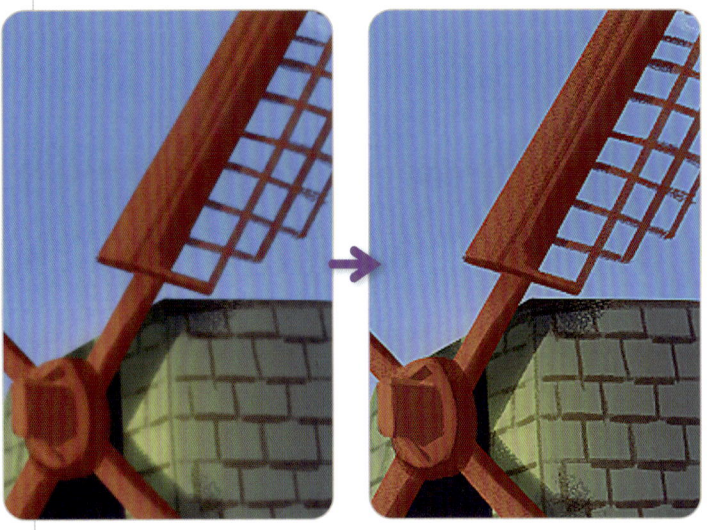

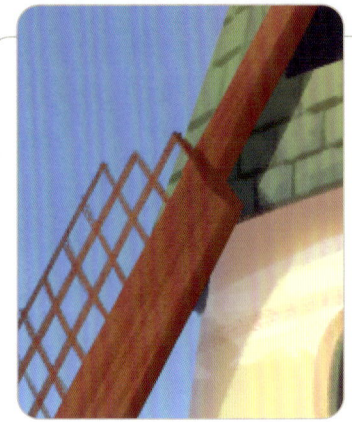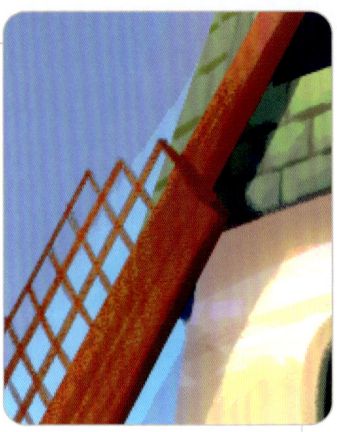

UNSHARP MASK

This filter enhances and refines the contrast of the image at the edges of colours, resulting in sharpness. The Radius, Strength, and Threshold can be adjusted in the window.

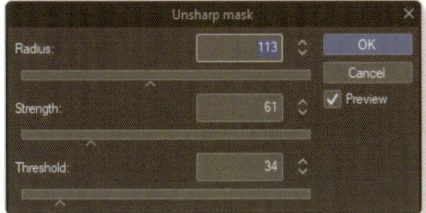

COMBINE FILTERS

Try using more than one filter in the final stages of your painting to enrich your artwork further. For example, the use of Sharpen, Chromatic Aberration, Blur, and Noise filters can give a painting a more photographic look.

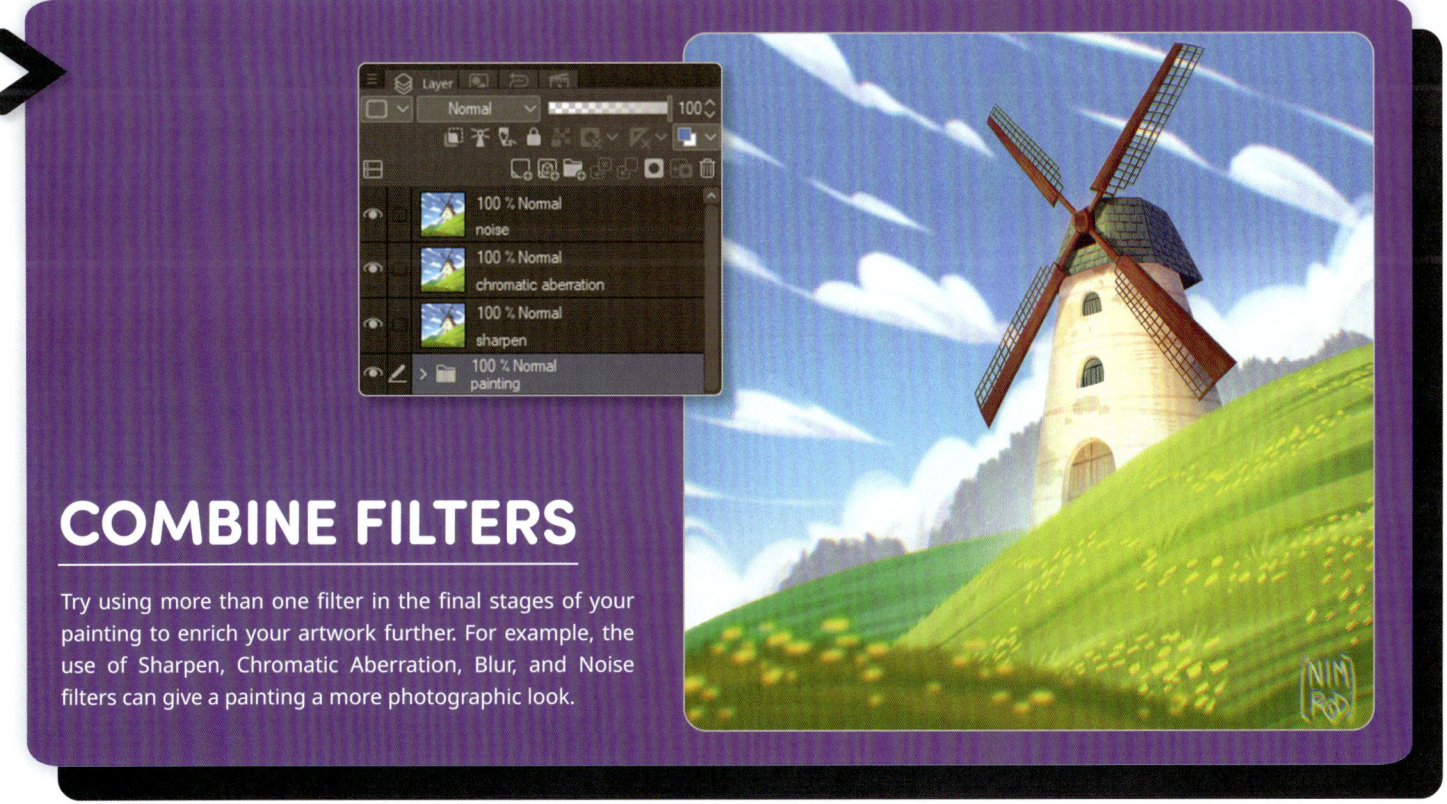

Materials

Clip Studio Paint offers a diverse variety of default materials. Organized into folders in the Material palette, they can be applied to your artwork and can even facilitate your workflow. The materials include brushes, images, 3D models, textures, workspace settings, action sets, automatic actions, animations, comic materials, and templates. The types of materials found in the Material palette are:

1. COLOUR PATTERN
These materials are textures and colour patterns of abstract, natural, and effect types.

2. MONOCHROMATIC PATTERNS
These are black-and-white textures and patterns organized into categories including Basic, Gradient, Cross-Hatching, and Background. They have a transparent background, which helps to blend them with other layers.

3. MANGA MATERIAL
This category contains pre-made elements that are useful for developing comics or manga. You can find framing templates, balloons, line effects, sound effects, and signs.

4. IMAGE MATERIAL
This category includes photos, illustrations, and sampled brushes. The latter are applied as brush tips and can be configured in the Sub Tool Detail of the selected brush sub tool.

5. 3D
This contains 3D materials typically used for references, drawing guides, or perspective grids. These materials can be posed in different ways, depending on the type of model they are.

6. DOWNLOAD
Materials downloaded from the material library Clip Studio Assets are automatically added here.

> **The Material palette contains folders of different types of materials**

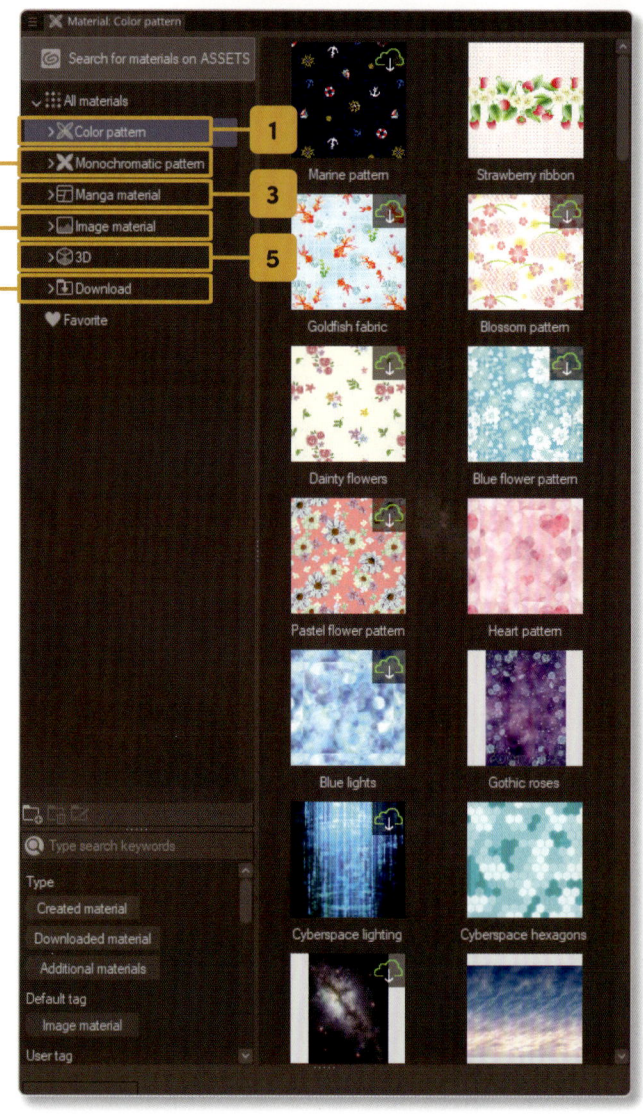

MATERIAL PALETTE

The Material palette is located on the right-hand side of the interface, between the canvas window and the Navigation, Layer Property, and Layer palettes, and below the Quick Access button. Each button corresponds to a category of materials, but within each you can select any category when you press any button. To customize the interface, drag the side of the Materials palette to expand its size. You can also click and hold the top of the palette to drag and detach it if you want to move the palette to another part of the screen.

The Materials palette can also be activated by selecting **Window > Material**, and then selecting the material folder you want to view.

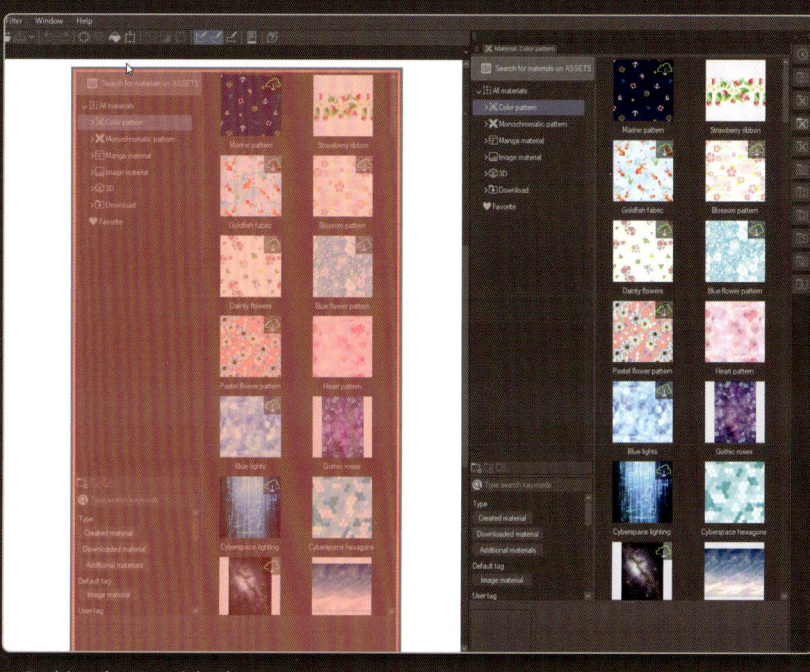

∧ **Position the Material palette wherever you like on the interface**

1. MATERIAL LIST

This section displays the available materials, according to the category you choose. Each material will have a name and preview.

2. SEARCH FOR MATERIALS ON ASSETS

This opens the Clip Studio launcher, which allows you to search for new materials that don't come as default. When you download new materials, they're stored in the Downloads category of the Material palette.

3. MATERIAL FOLDER

This area contains the various Material palettes organized by folder and subfolders.

4. FOLDER COMMAND BAR

This Command bar has buttons to create, delete, or rename folders. You can only delete and rename folders that you have created. Default folders cannot be modified or deleted.

5. MATERIAL FILTERS

In this area, you can filter materials by tags according to the selected folder.

6. MATERIAL INFORMATION

This section provides detailed information about the selected material.

7. COMMAND BAR

This Command bar allows you to organize materials, paste the selected material onto the canvas, add it to the Favourites folder, or delete it.

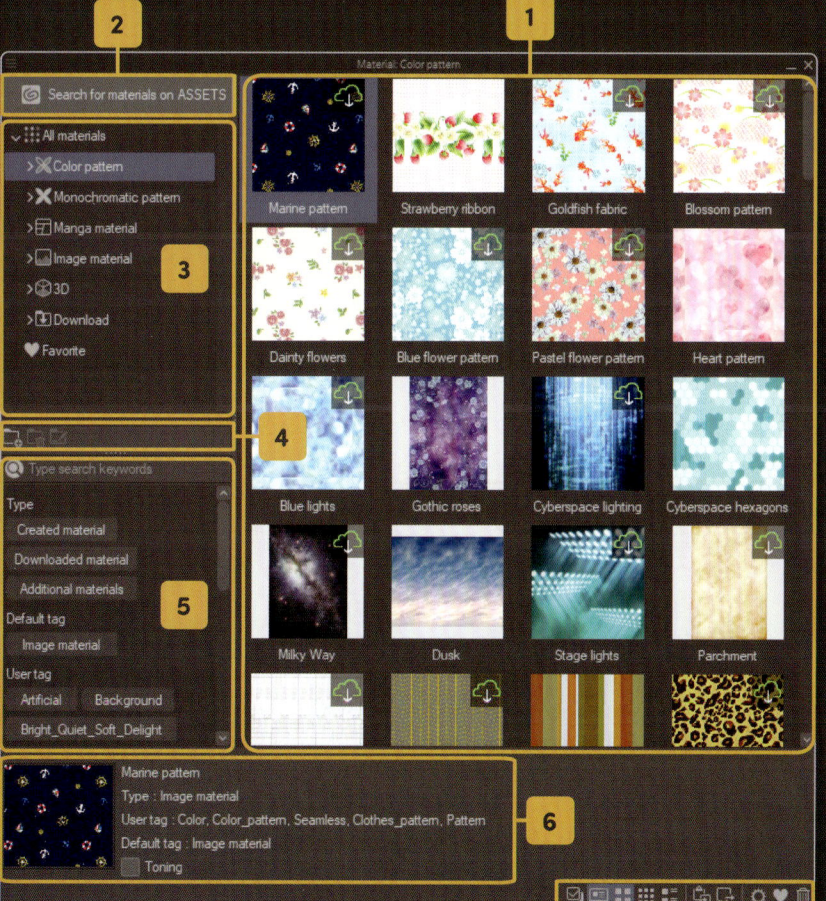

∧ **Material palette**

APPLYING MATERIALS

There are several ways to apply a material, depending on the type of material you want to use.

> For images, textures, patterns, or 3D materials, simply drag the material you want to use onto the canvas. Another way to add it to the canvas is by selecting the Paste Select Material to Canvas icon located in the Command bar of the Material palette

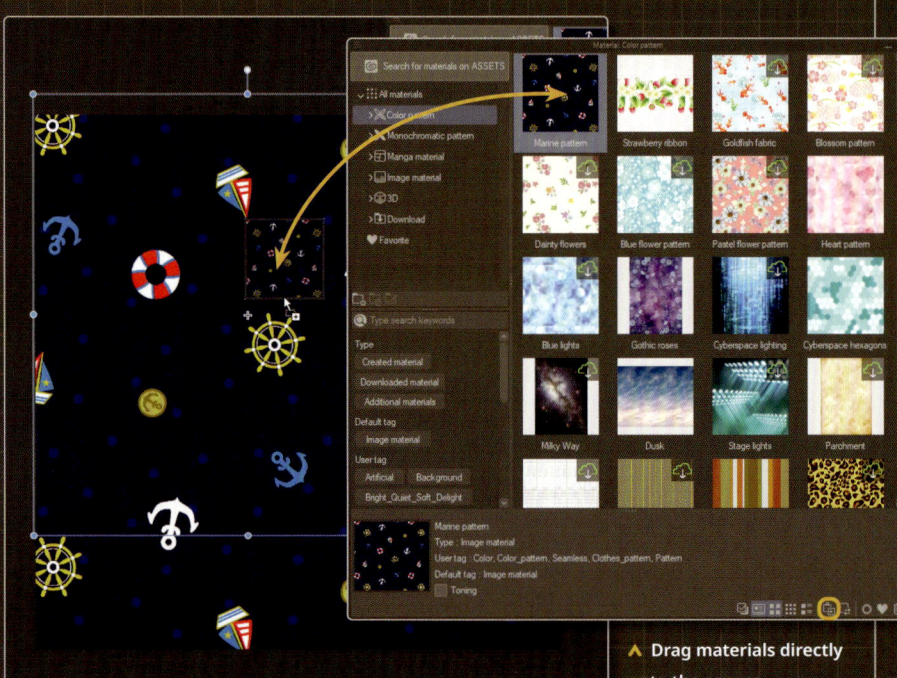

∧ Drag materials directly onto the canvas

∨ For colour-set materials, import them by selecting Add Colour Set on the Colour Set palette

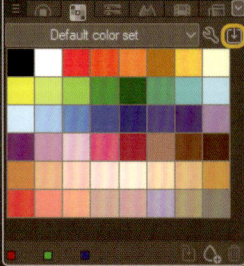

> Add gradient materials from the Gradient Sub Tool palette. To add a gradient set, select Add Gradient Set in the Edit Gradient window

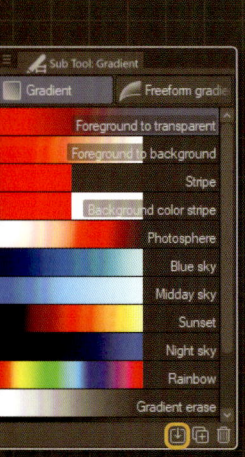

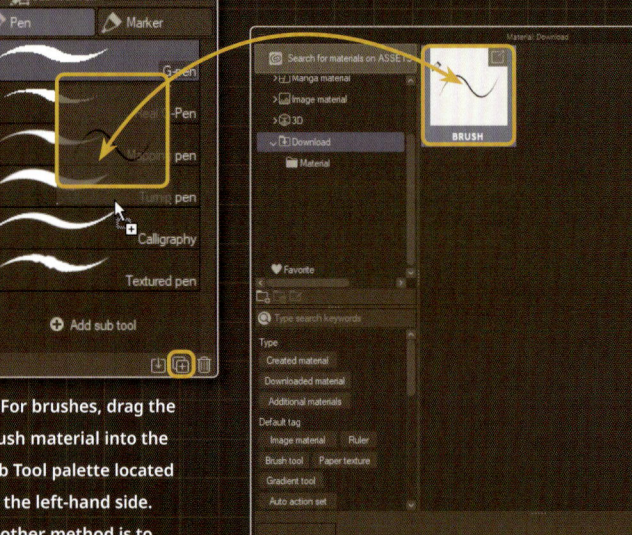

∧ For brushes, drag the brush material into the Sub Tool palette located on the left-hand side. Another method is to select Add Sub Tool from the Sub Tool palette

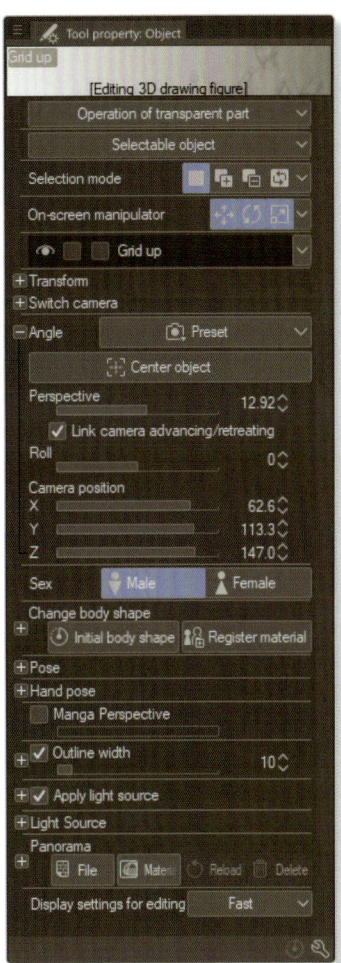

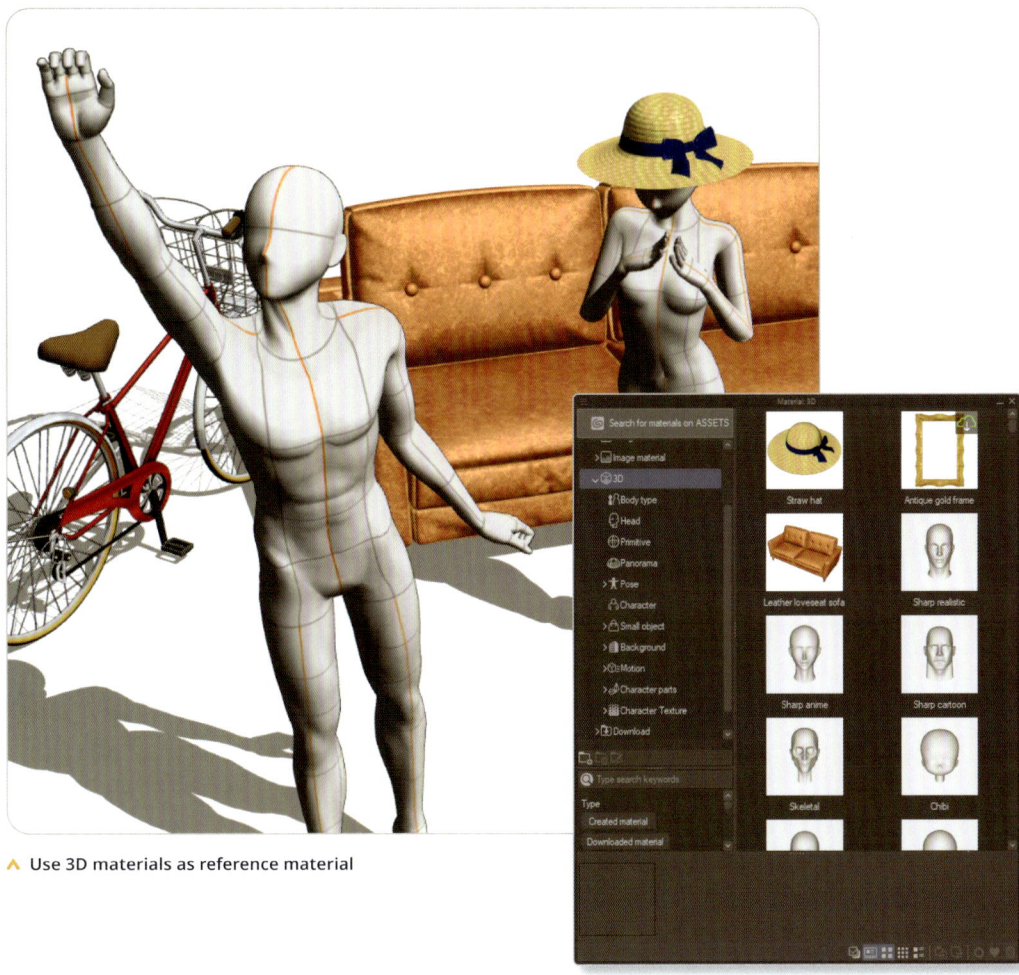

∧ Use 3D materials as reference material

3D MATERIALS

3D materials are elements that can facilitate the drawing process when used as references. There are various types of 3D materials available, such as objects, models – including head models and hand models – pose materials, and more. Each type of material has settings that can be adjusted in the Tool Property palette, as well as from the Sub Tool Detail window.

Two of the most commonly used materials are Body Type and Pose. These have a human form and allow you to adjust the pose as required.

< Body Type models have highly detailed joints that you can alter

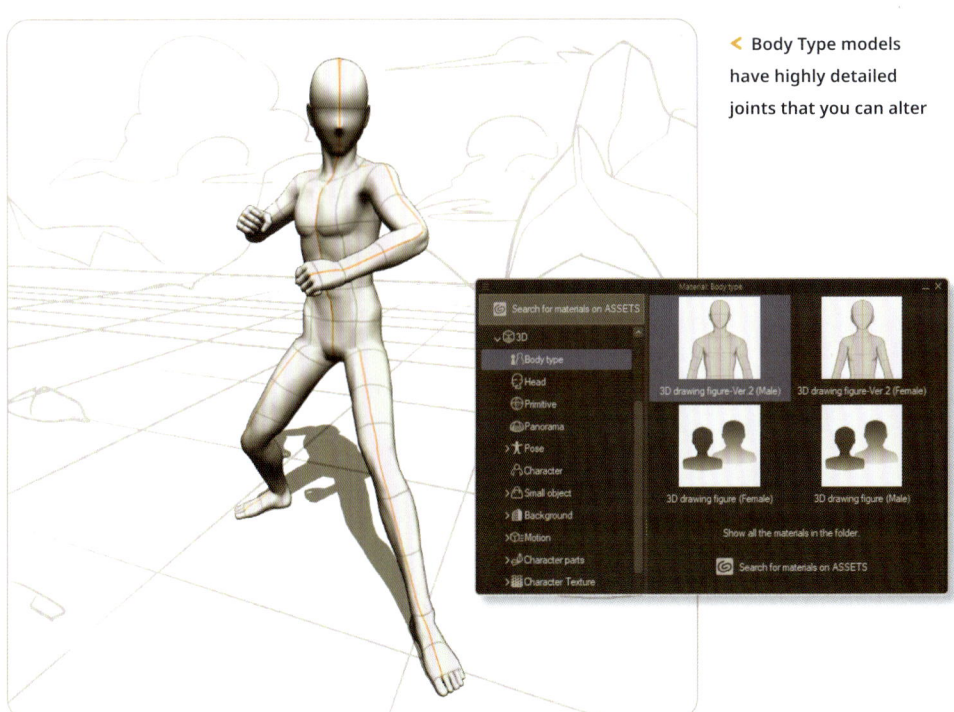

When a 3D model is selected, the Movement Manipulator will appear. This allows you to control the displacement, size, rotation, and focal point of the 3D material. The Movement Manipulator Functions control bar directs the camera and position of the 3D material. The Object Launcher control bar has some settings that you can find in the Tool Property palette and the Sub Tool Detail dialog box.

Click on a joint or body part of the pose model to select it. You can then move it by clicking and dragging. It's also possible to rotate the selected part using the rotation controls that activate when you make the selection.

Add more 3D elements to the scene by dragging the object from the Material palette onto the canvas with the original selected 3D material.

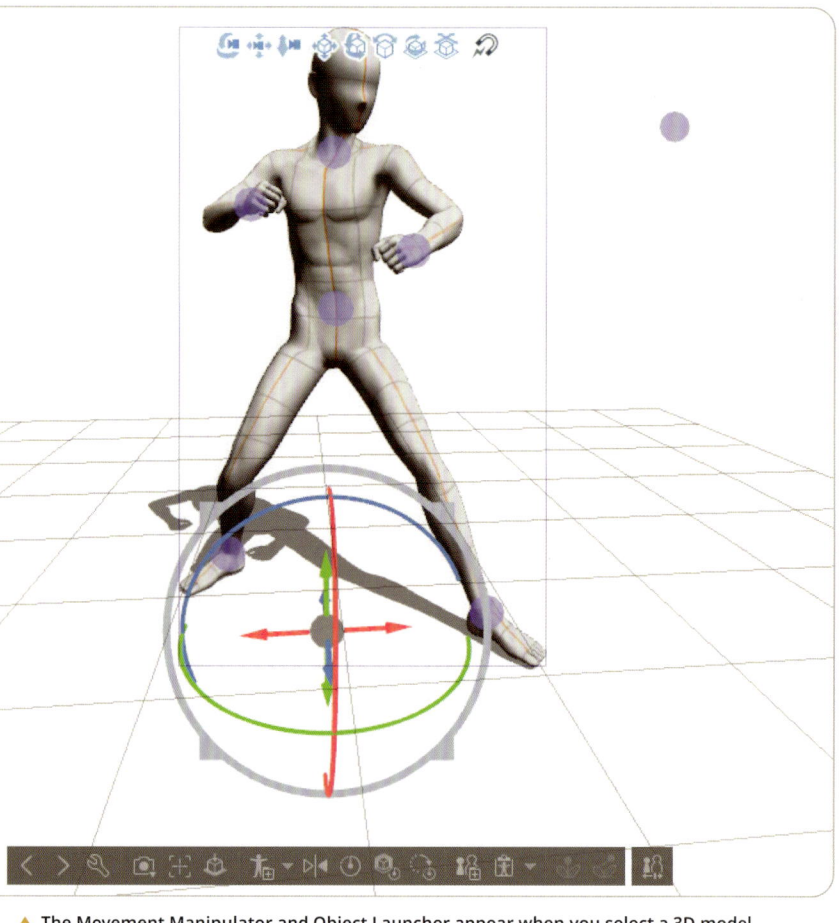

∧ The Movement Manipulator and Object Launcher appear when you select a 3D model

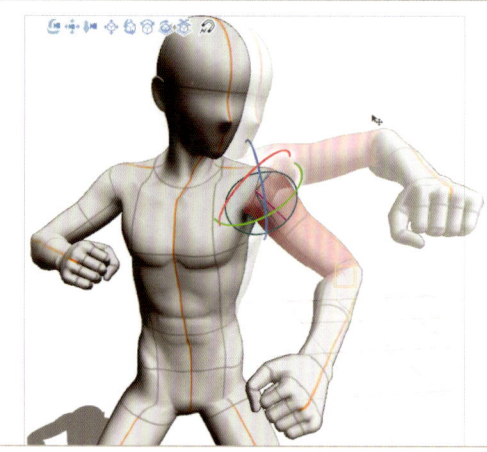

∧ Click on a joint or body part to move or rotate it

REFERENCE ONLY

3D materials are great for using as reference material when planning composition or perspective, but avoid relying on them as an exact guide for the pose or drawing you're aiming to create. While visually the materials may seem to appear as a perfect and realistic pose, they may often lack dynamism or can look odd when you disable the 3D layer and only see the drawing. 3D models are a good starting point, but avoid making them part of your drawing style.

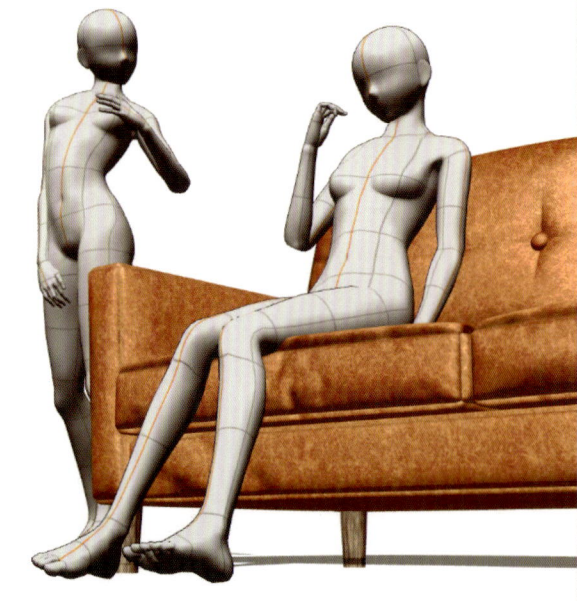

∧ Drag additional elements onto the canvas from the Material palette

CLIP STUDIO ASSETS

Clip Studio Assets lets you download new materials for free or for a fee. Before downloading them, you must first log in with your Clip Studio account.

Access Clip Studio Assets by opening the Clip Studio launcher application. Select the Menu icon in the top right of the window, then click Search for Materials. This will automatically take you to the Assets page.

In the search bar, specify the type of material you're looking for. You can filter the results by tags.

Once you have downloaded the material, it will be stored in the Download folder of your Material palette.

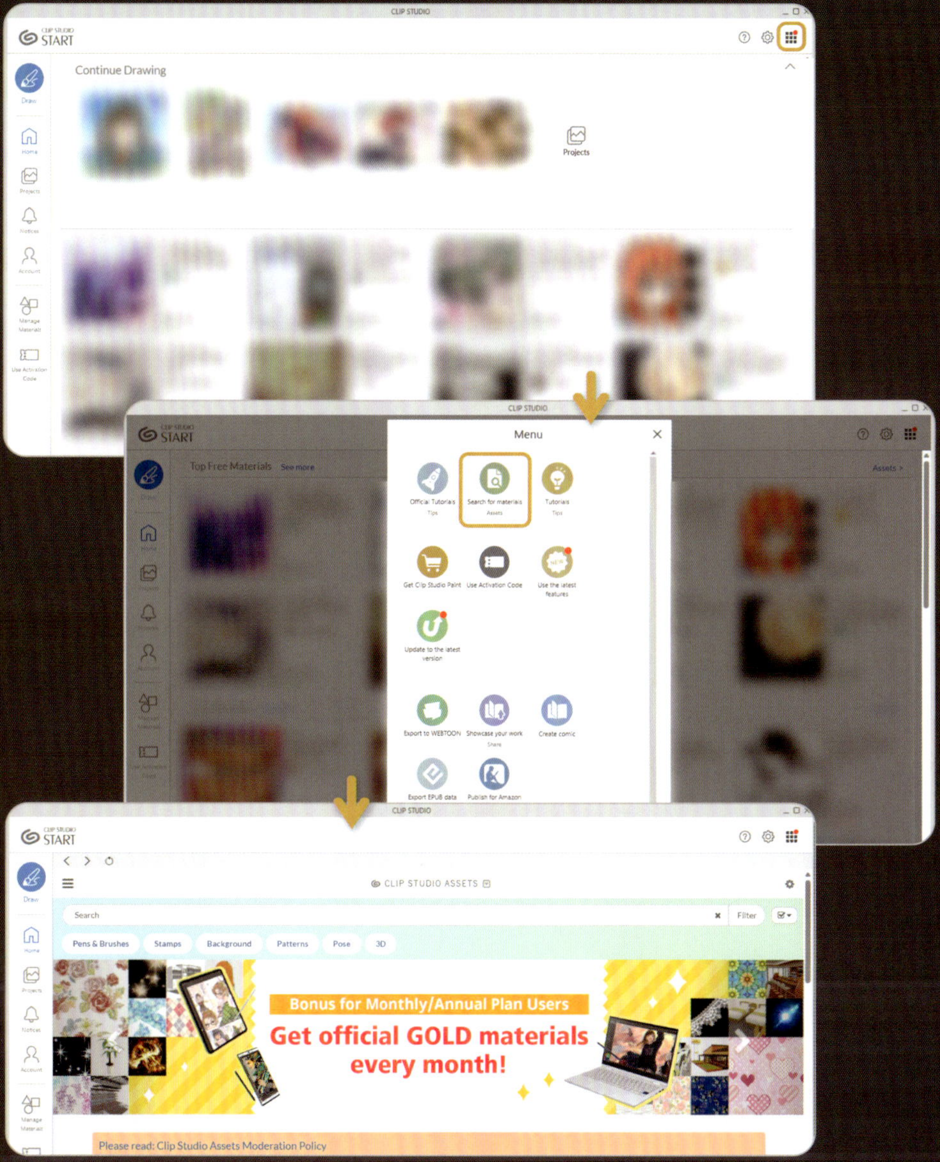

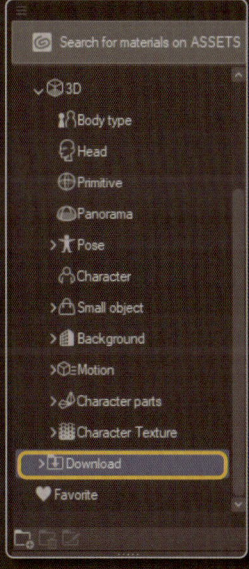

∧ The downloaded materials will appear in your Download folder

∧ Search for materials on Clip Studio Assets

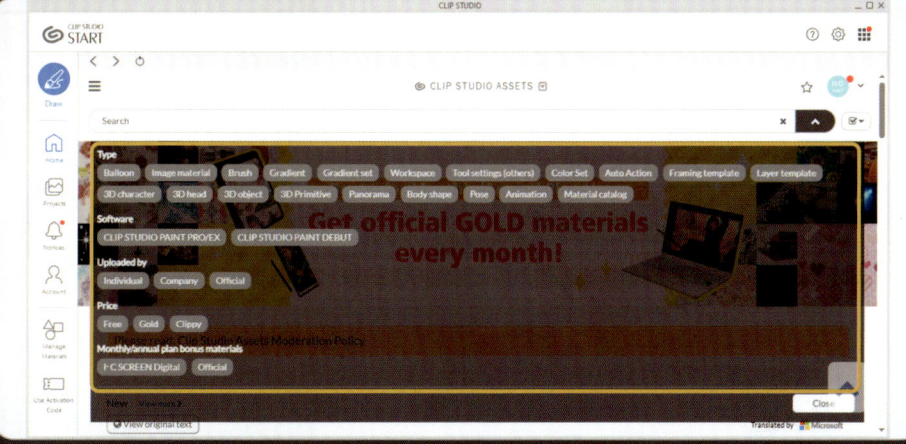

∧ Browse the different types of assets and select filters to refine your search

109

Using Clip Studio Paint on a tablet

LEARN HOW TO

- Use the basic gestures.
- Use the Quick Access palette.
- Use the edge keyboard.

USER INTERFACE

Clip Studio Paint can be used on numerous devices, including a tablet. The tablet software offers two versions of the user interface:

- **STUDIO MODE**
  This version is very similar to the PC interface, except for some palettes that are kept hidden. All of the settings, tools, and palettes offered in the PC version can also be found in this mode.

- **SIMPLE MODE**
  This version has a simpler and more user-friendly interface, but with certain configuration limitations, such as the types of layers you can edit.

To switch from Studio mode to Simple mode, select the Clip Studio Paint menu icon in the top-left corner of the screen. To switch to Studio mode while in Simple mode, select the Settings menu in the top-right corner.

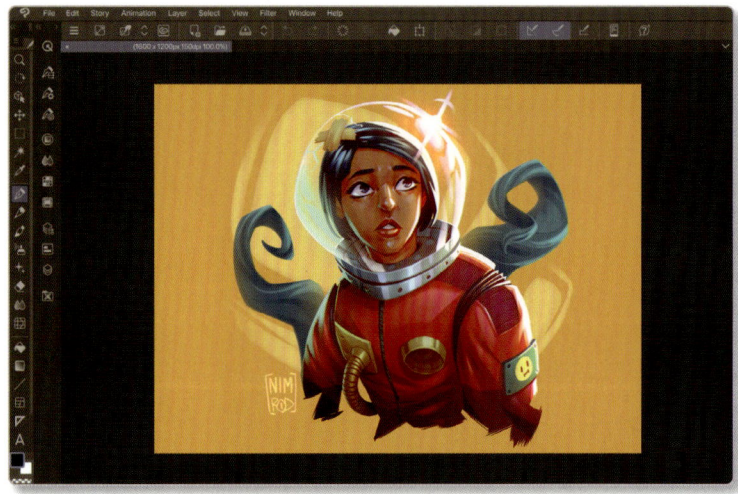

∧ Studio mode contains all of the tools, settings, and palettes available on a PC

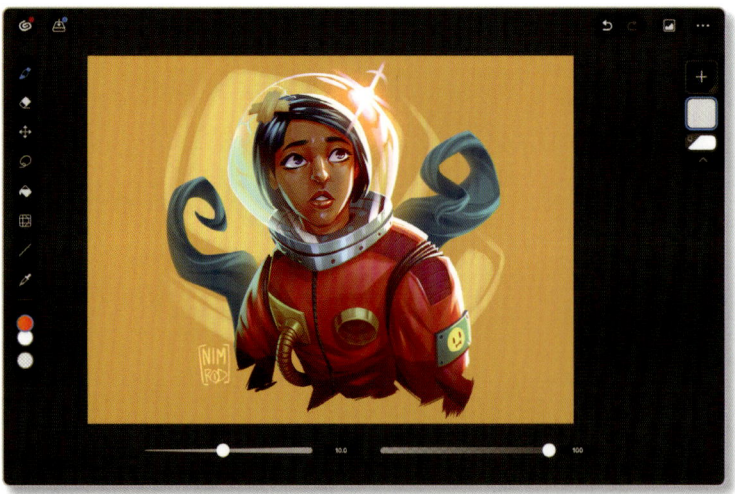

∧ Simple mode has a more user-friendly interface

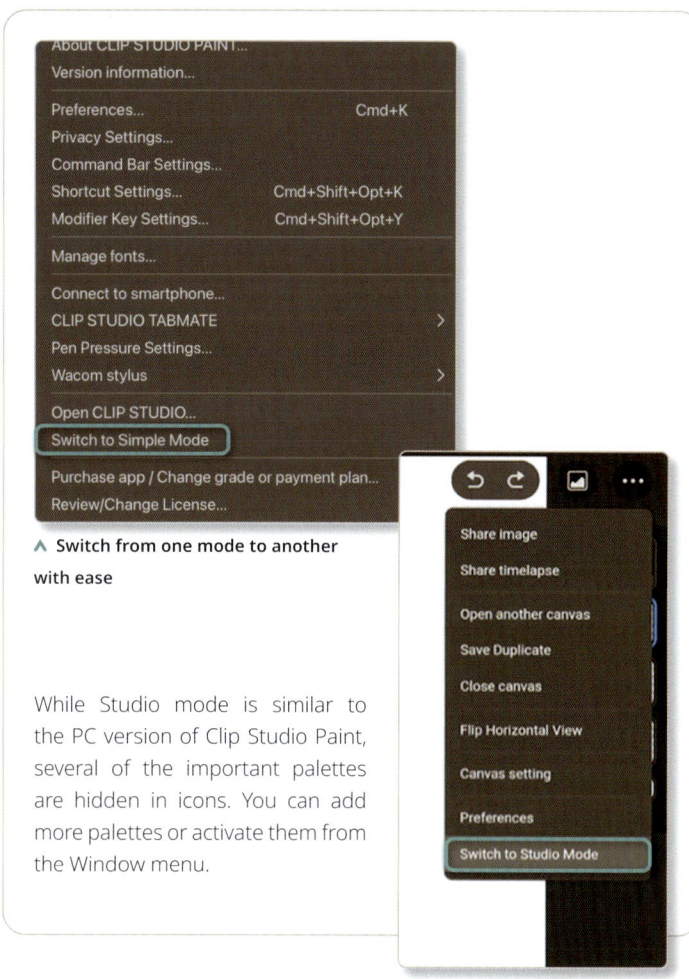

▲ Switch from one mode to another with ease

While Studio mode is similar to the PC version of Clip Studio Paint, several of the important palettes are hidden in icons. You can add more palettes or activate them from the Window menu.

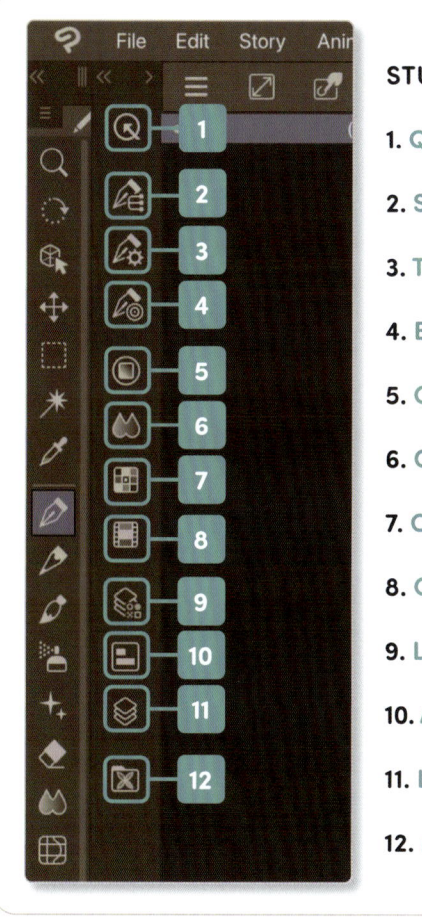

STUDIO MODE

1. **QUICK ACCESS**

2. **SUB TOOL**

3. **TOOL PROPERTY**

4. **BRUSH SIZE**

5. **COLOUR WHEEL**

6. **COLOUR MIXING**

7. **COLOUR SET**

8. **COLOUR HISTORY**

9. **LAYER PROPERTY**

10. **ALIGN/DISTRIBUTE**

11. **LAYER**

12. **MATERIAL**

GESTURES

Touch gestures are an integral part of using Clip Studio Paint on a tablet and are a great complement to a pen/stylus. They work based on the way your fingers touch or glide over the screen.

CSP has a useful setting for adjusting which tools can be used with fingers and pen, located in the Command bar.

USE DIFFERENT TOOLS WITH FINGERS AND PEN

When this option is disabled, you can use your fingers to use any tool that is selected. When this option is enabled your finger will function as the sub tool, or whatever tool you have added in the Modifier Key Settings window.

SET TOOL TO USE WITH FINGER

This option opens the Modifier Key Settings window where you can configure the set of tools to be used with fingers when Use Different Tools With Fingers and Pen is activated.

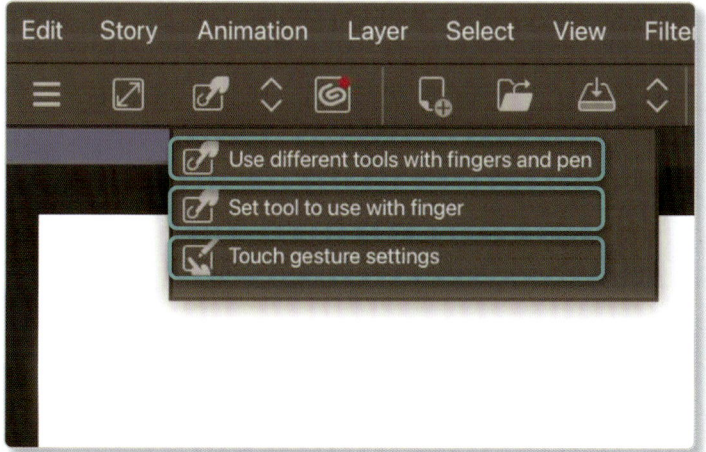

▲ Click the icon in the Command bar to edit touch gesture settings

TOUCH GESTURE SETTINGS

This setting allows you to edit the touch gesture operations.

BASIC GESTURES

1. UNDO

Tap the canvas with two fingers to undo your last action.

2. REDO

To redo your last action, tap the canvas with three fingers.

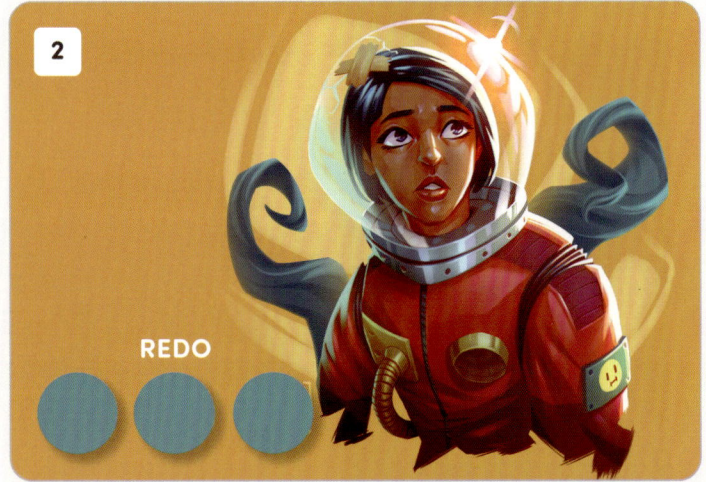

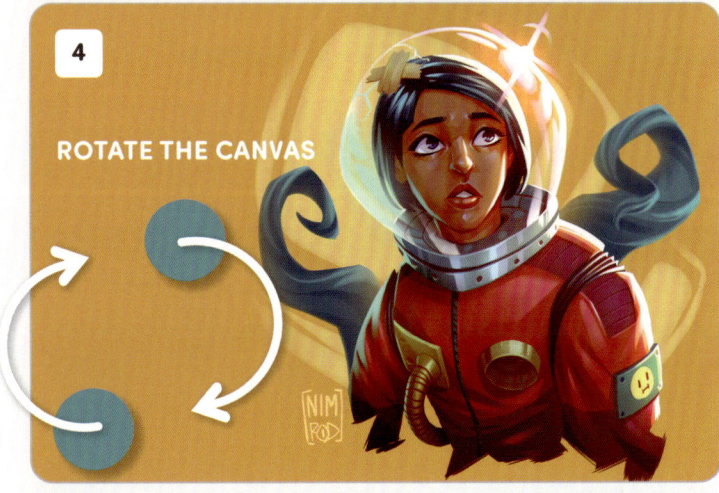

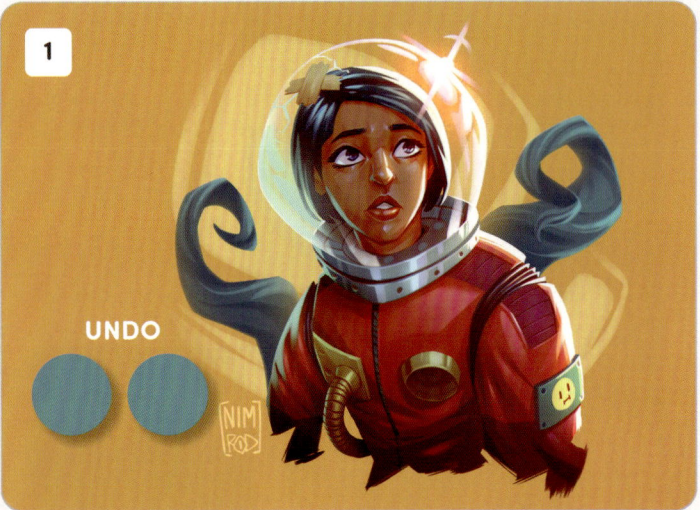

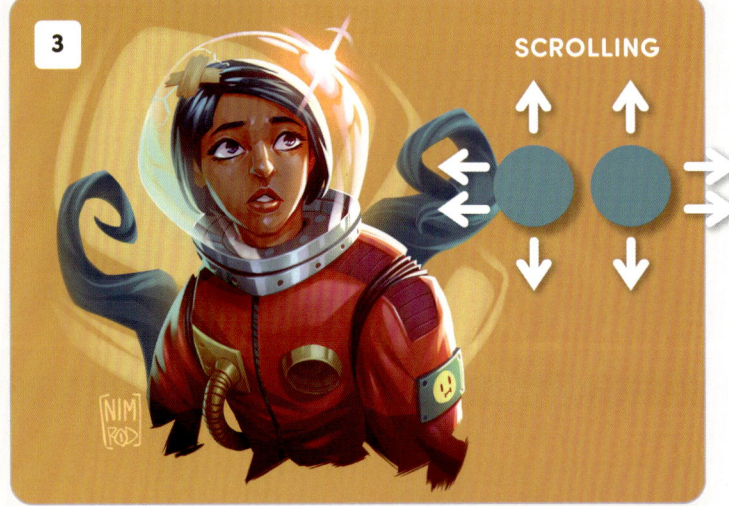

3. SCROLLING

Drag two fingers across the screen to move the canvas in various directions, such as up, down, left, and right.

4. ROTATE THE CANVAS

Rotate the canvas by rotating two fingers on the screen.

5. ZOOM IN & ZOOM OUT

To zoom in on the canvas, drag your fingers away from each another on the screen. To zoom out of the canvas, drag your fingers towards each other on the screen in a pinching motion.

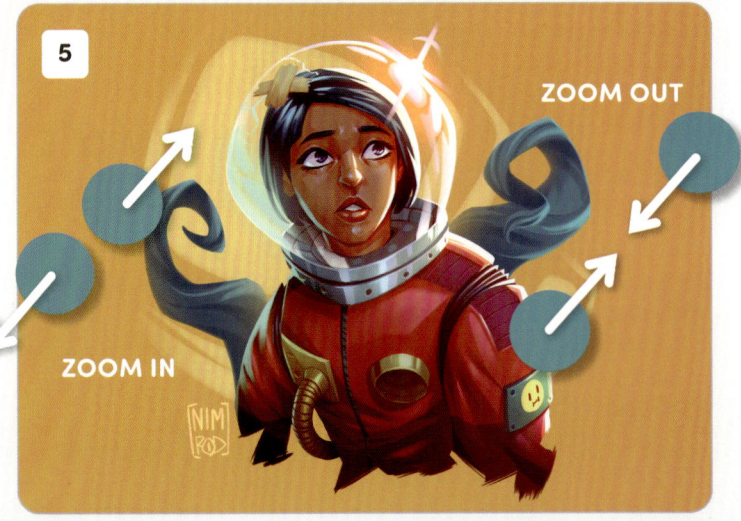

QUICK ACCESS PALETTE

While the Quick Access palette is available in the PC version of Clip Studio Paint, it's incredibly useful on devices such as tablets. It offers quick access to various tools and functions, avoiding using key combinations or activating hidden palettes. You can customize the Quick Access palette with new functions or tools using the Quick Access settings located at the bottom of the palette.

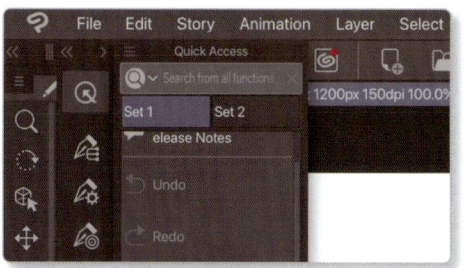

◄ The Quick Access palette allows you to quickly find a variety of tools

EDGE KEYBOARD

The edge keyboard is an exclusive feature to the tablet version of Clip Studio Paint. It emulates basic keyboard keys, while also allowing you to add hotkeys or automatic actions.

To activate the edge keyboard, swipe from either side of the screen towards the centre. To deactivate it, swipe your finger from the edge keyboard away from the screen.

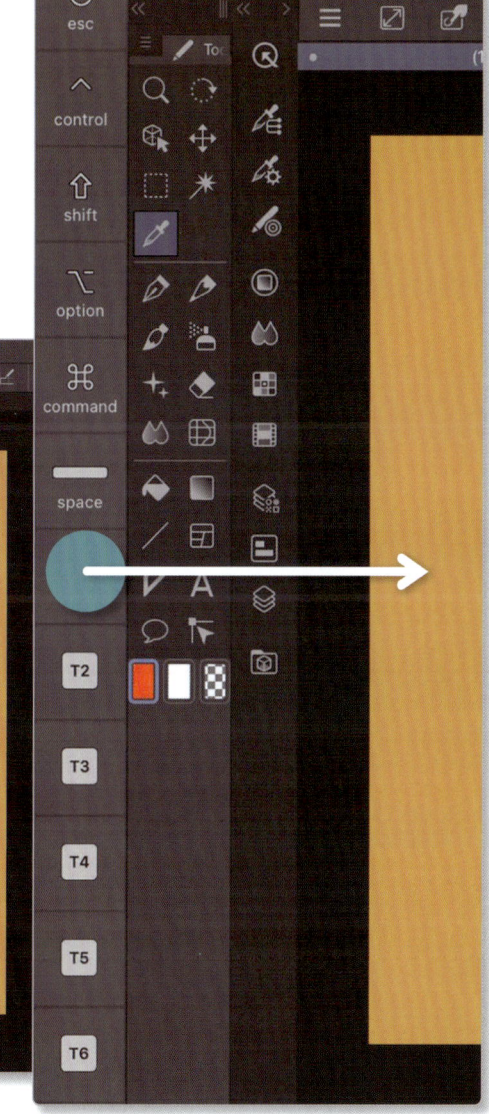

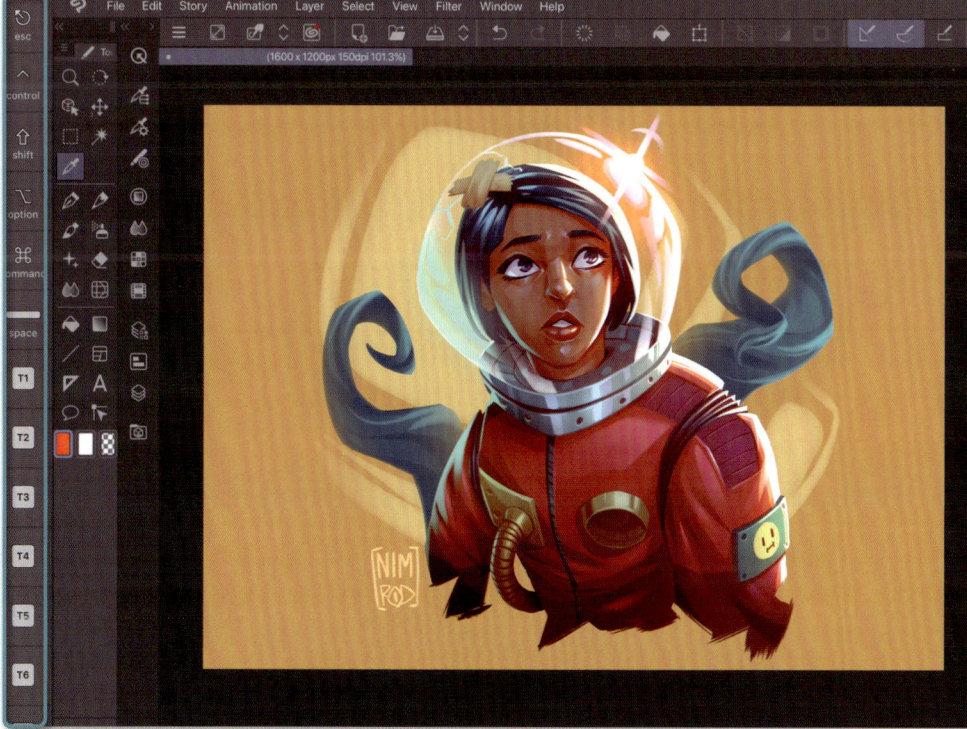

⌃ The edge keyboard

⌃ Swipe to activate the edge keyboard

Tutorials

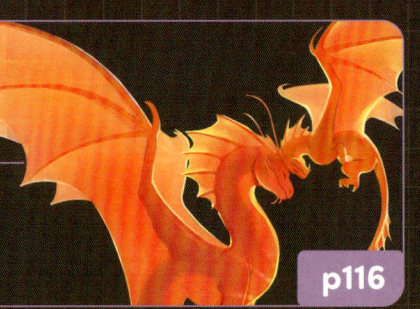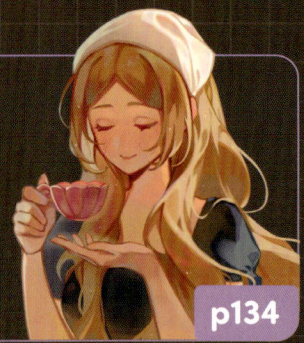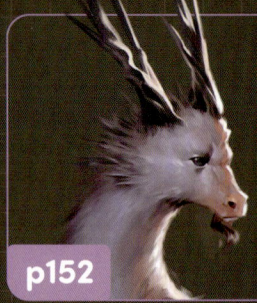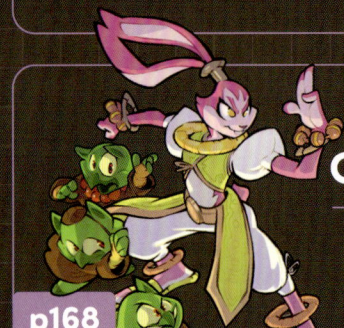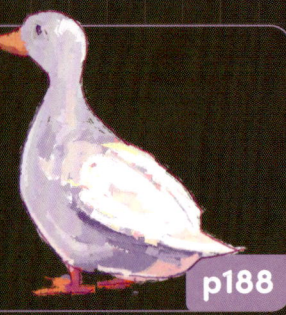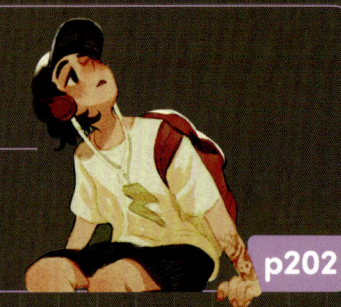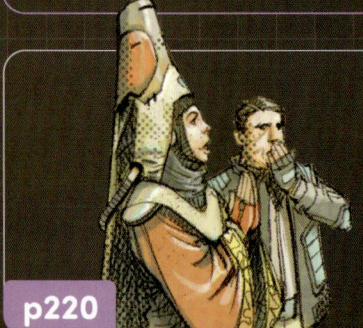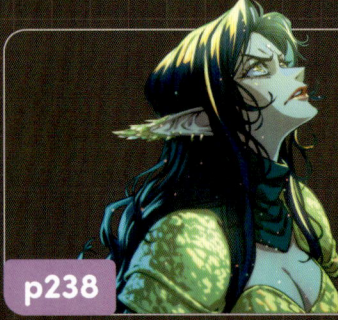

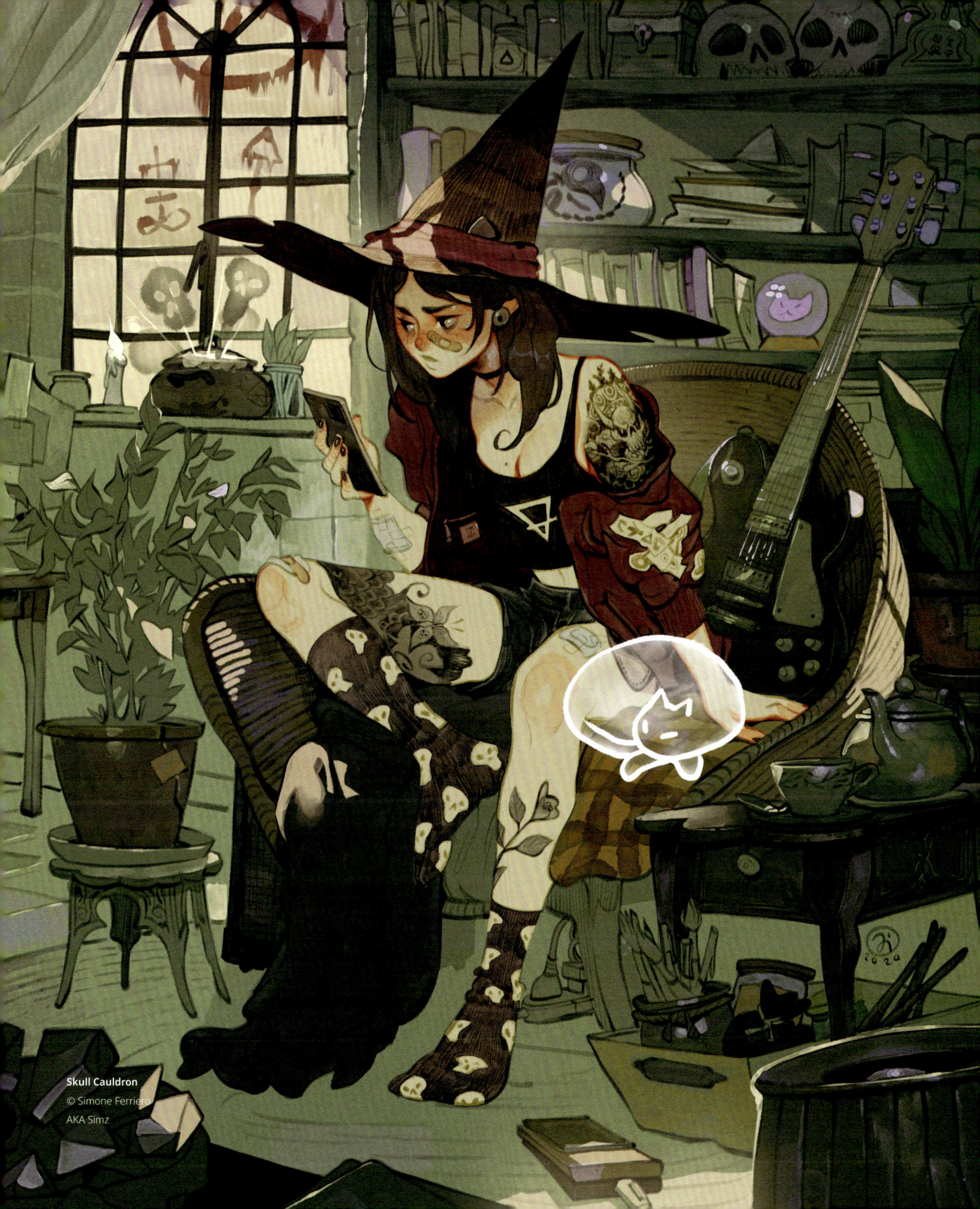

Skull Cauldron
© Simone Ferriero
AKA Simz

Fantasy scene

BY DEVIN ELLE KURTZ

In this tutorial, I will guide you through my process of creating a fantasy city scene in one-point perspective using Clip Studio Paint. I love to paint dragons and other fantasy creatures in modern settings! I enjoy the surprising and delightful contrast that occurs when combining our modern world and the creatures usually found in high-fantasy settings. For this tutorial, I'll be painting a father dragon and his (aeroplane-sized!) son.

Over the course of the tutorial, you will learn how to set up and use a 1-Point Perspective Ruler. This is a skill set that you can carry into all kinds of cityscape and building drawings, once you understand the tools. Next, you'll learn to use the Rectangle Selection tool together with the Perspective Ruler, which will enable you to block in buildings following perfect perspective. Next, the tutorial will cover how to design window brushes and use the Free Transform tool to put them into perspective. Finally, you'll learn how to use blending modes and colour balance Correction layers. Painting a city scene may seem daunting at first, but as you'll soon discover, buildings are really just boxes – the easiest shape to draw in perspective! With a little bit of help from handy digital tools, this process is achievable for artists of any level.

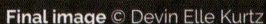

Final image © Devin Elle Kurtz

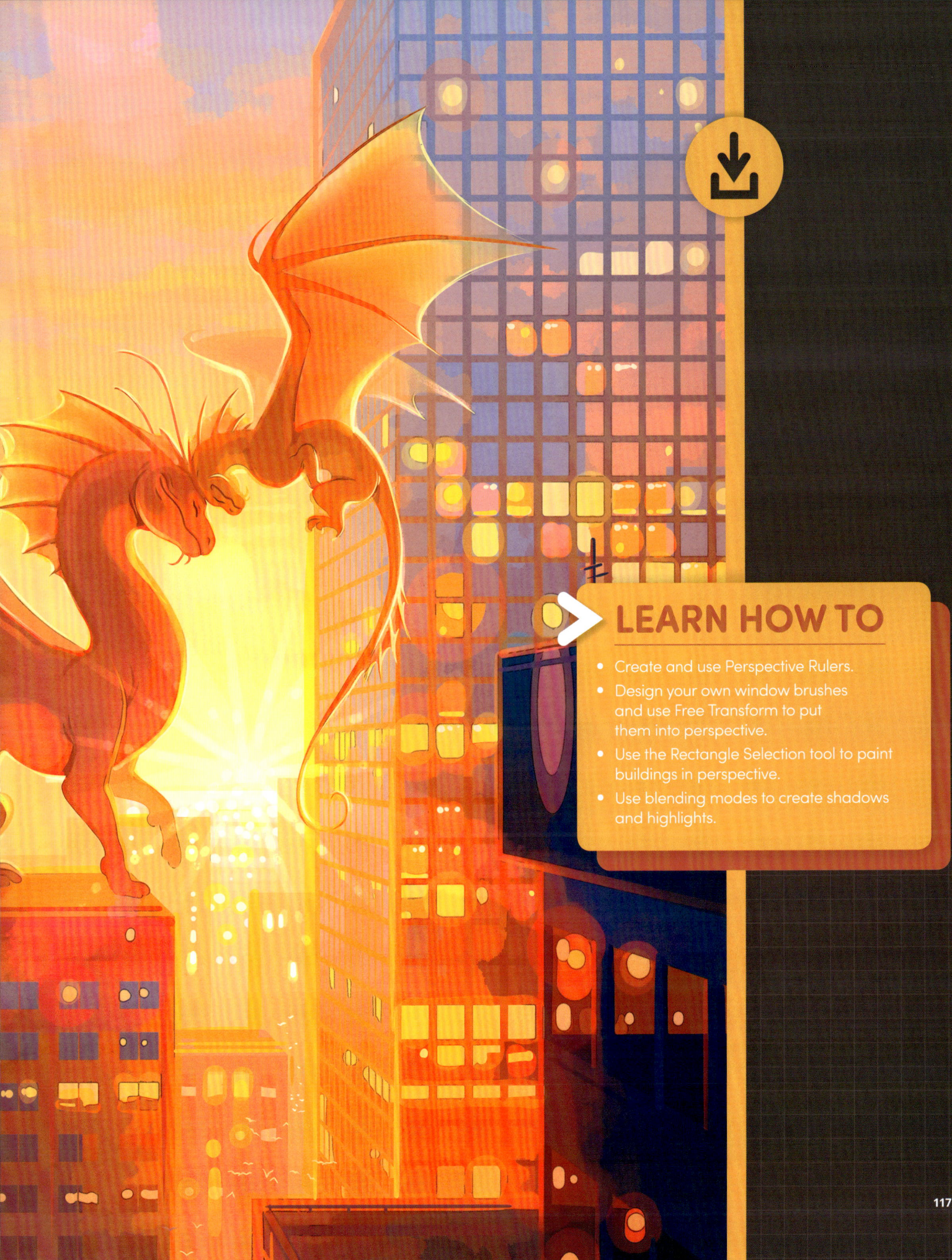

LEARN HOW TO

- Create and use Perspective Rulers.
- Design your own window brushes and use Free Transform to put them into perspective.
- Use the Rectangle Selection tool to paint buildings in perspective.
- Use blending modes to create shadows and highlights.

01 Create a new canvas that is 270 mm tall × 230 mm wide, and with a resolution of at least 300 dpi. Next, create a **1-Point Perspective Ruler**. One-point perspective is a drawing technique that simplifies an image into perfectly vertical lines, perfectly horizontal lines, and lines that converge to a single perspective point. Select **Layer > Ruler/Frame > Create Perspective Ruler** from the top bar. In the pop-up window, select 1-Point Perspective, check the Create New Layer box, and click OK. You will now see an assortment of lines on your screen. The horizontal blue line represents the horizon line and the diagonal lines forming an upside down V represent the perspective lines that will converge to your single perspective point.

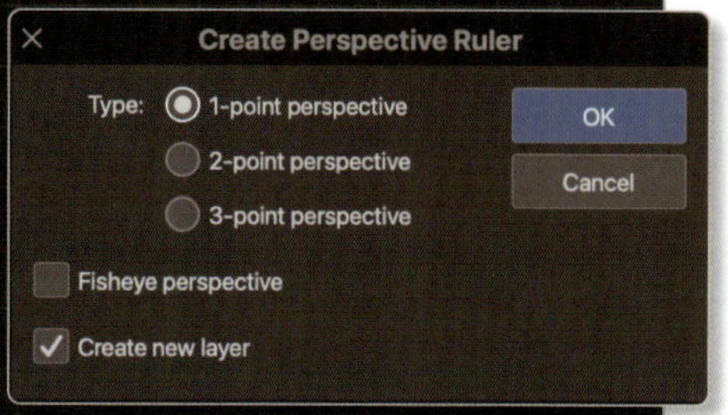

> ∧ The Perspective Ruler pop-up menu, with the correct options selected

> < The Perspective Ruler will appear on your canvas

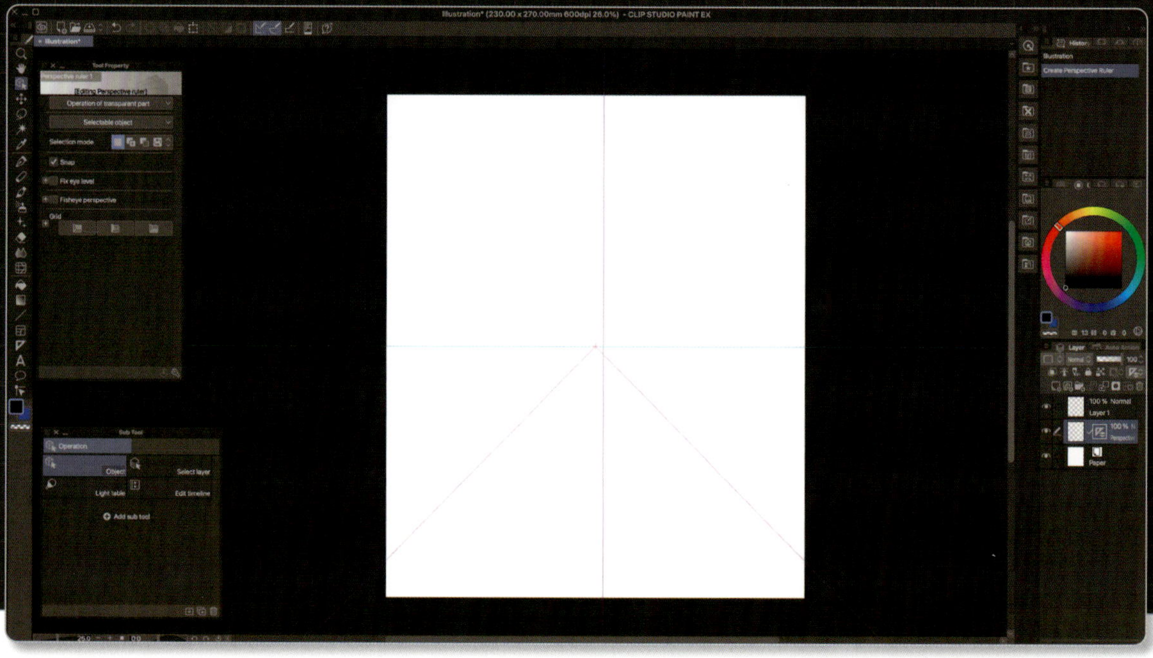

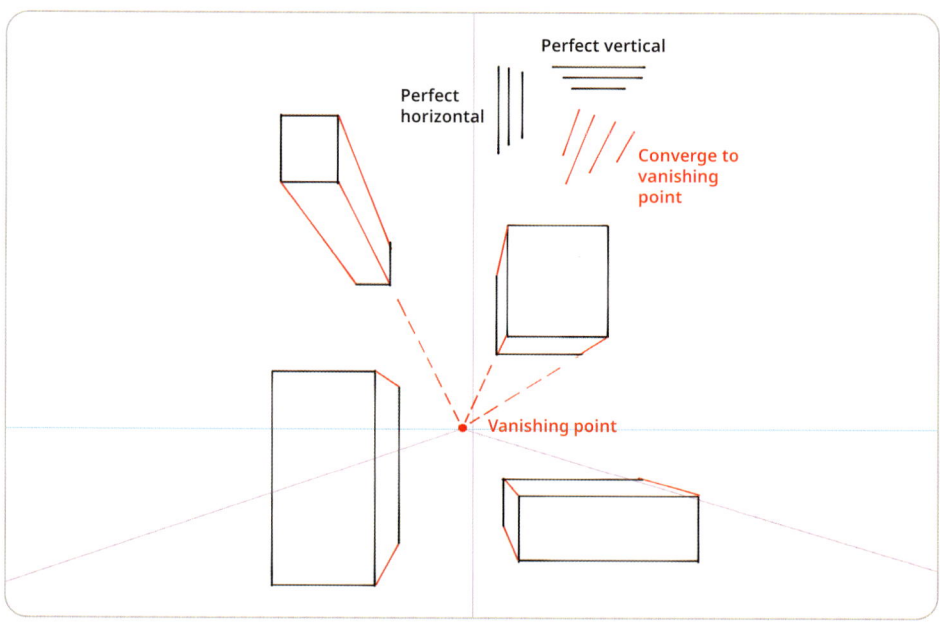

02 If you click and drag the perspective point in any direction, the converging lines will move with it. Drag the point around the canvas to get a feel for the tool, then place it three-fifths of the way down from the top, in the centre. Next, select the **Marker** tool and experiment with drawing lines. You can draw three kinds of lines while this ruler is active: perfectly vertical lines, perfectly horizontal lines, and diagonal lines (highlighted with red) that converge to the perspective point. Try drawing some boxes like the ones in the image.

> < The Perspective Ruler with the horizon line dragged three-fifths of the way down the canvas, with a few demo boxes drawn

Perfect vertical

Perfect horizontal

Converge to vanishing point

Vanishing point

03 The next step is to compose a cityscape out of simple rectangles. Begin by arranging flat 2D rectangles on your canvas. Don't worry about the lines that converge to the vanishing point yet. Focus solely on creating a visually pleasing composition of rectangular shapes. Try to vary the size of the rectangles to ensure there are small, medium, and large shapes present. Layer the rectangles so that surprising shapes are created in the negative space. Avoid adding too many rectangles that are exactly the same size and shape, as too much repetition will distract from believability.

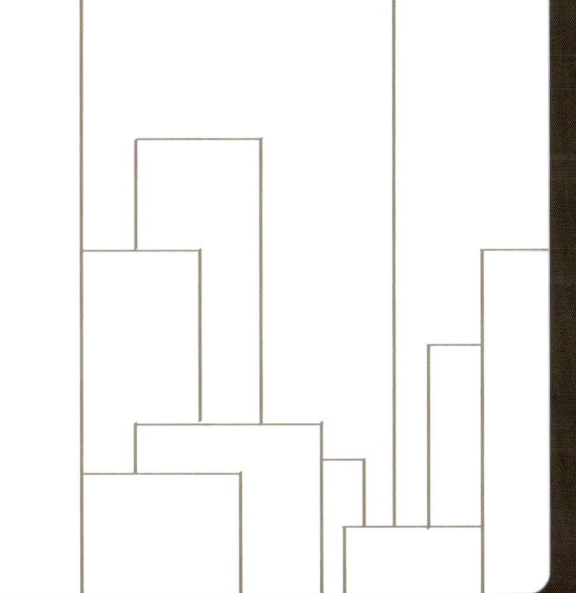

< An arrangement of flat 2D rectangular shapes; all of the lines are perfectly vertical or horizontal, parallel to the top, bottom, or sides of the canvas

> The cityscape begins to take shape, with red highlighting the lines that converge to the vanishing point

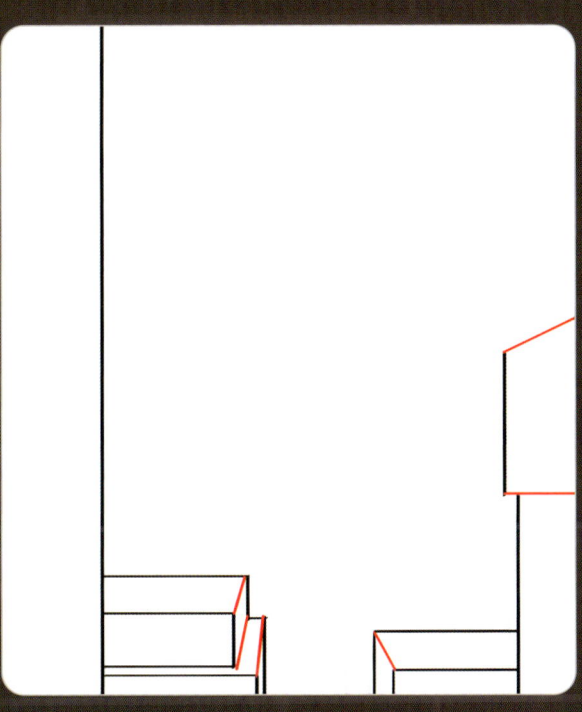

04 Collect reference images of a city of your choice to use as loose guidance for the next step. Next, using all three types of line (vertical, horizontal, and convergent) begin to transform your flat, 2D rectangles into 3D box shapes. Examine your reference images to find visually interesting boxy building shapes, such as layered buildings, or buildings with overhangs. Then, use the **Perspective Ruler** to replicate those building shapes on your canvas. Avoid curved buildings or buildings that aren't four-sided, unless you already know how to draw them – this tutorial won't cover those skills.

05 Finish drawing your cityscape using the skills covered in step 04. If you find it difficult to stack theoretical boxes in the imaginary space of your canvas, try doing it in real life first. Grab some box-shaped household and food items, such as cereal boxes, sticks of butter, and tissue boxes. If you have toy blocks in your house, they will be an even better building stand-in! Use whichever box-shaped objects you have at your disposal to design a cityscape with real 3D items. Take a photo, and then do your best to replicate it using the **Perspective Ruler**.

> The finished cityscape drawing

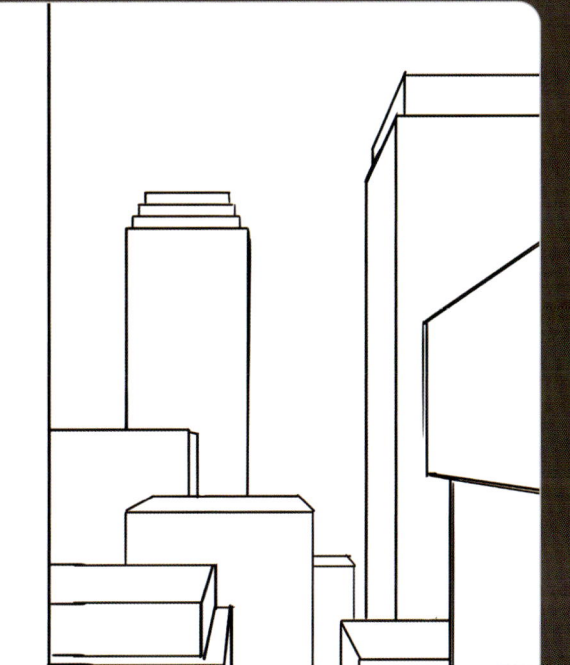

06 Temporarily turn off your Perspective Ruler. You can do this either by hiding the visibility of the layer that your ruler is on, or by right-clicking on the Perspective Ruler icon in the layer panel and unchecking Show Ruler. Now it's time to draw your dragons! Start by looking at photos of real reptiles, such as sailfin lizards, to inspire your design. Next, find a flat building for your dragons to perch on. Place your dragons right in front of the perspective point. When you place a subject on or around the perspective point, the converging lines of the surrounding buildings will all point towards your subject, strengthening it as the focal point of the painting. If no good perch exists in this area, flex your new perspective skills and raise or lower an existing building to create a perch.

> Adding the dragons to the drawing; arrows demonstrate how the converging lines all point towards the dragons

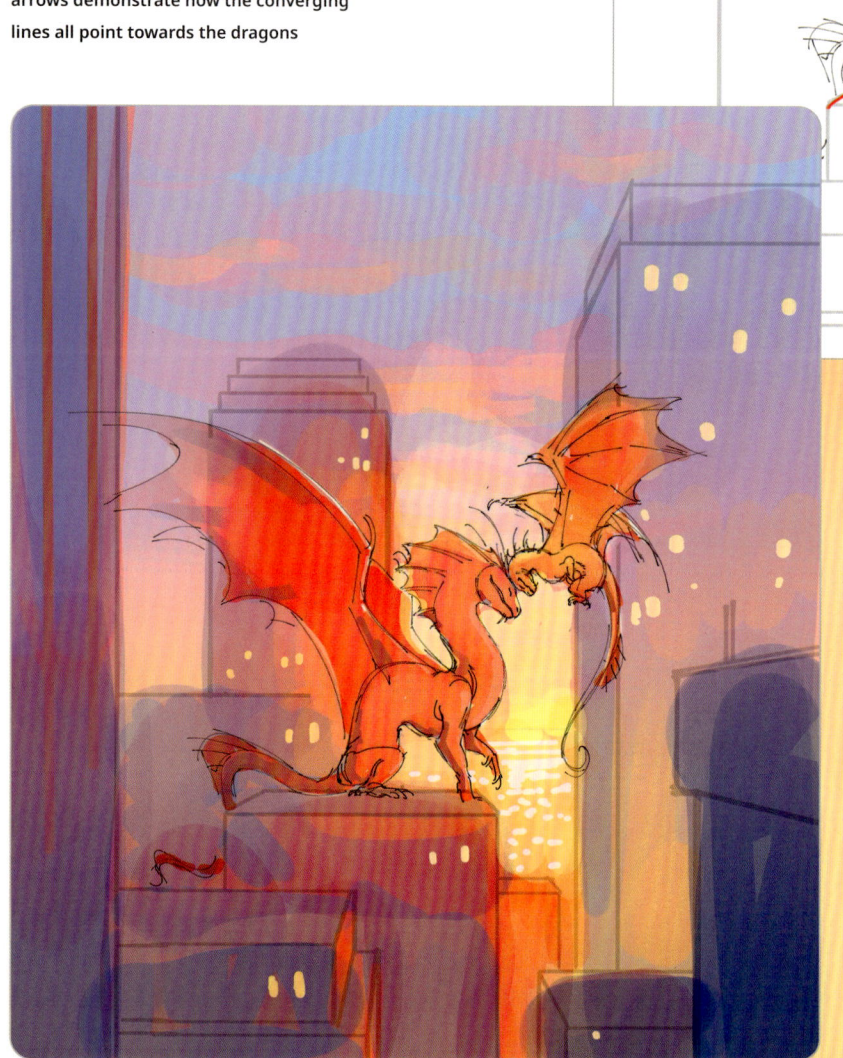

07 Next, use a limited colour palette of three or four colours to block in a loose colour key for your painting. Cities can become overly busy, to the point of distraction, very easily. To combat this, begin with a very simple colour palette and only deviate from that palette for small details as you paint.

< The colour key for this painting uses blues, purples, peaches, and yellows

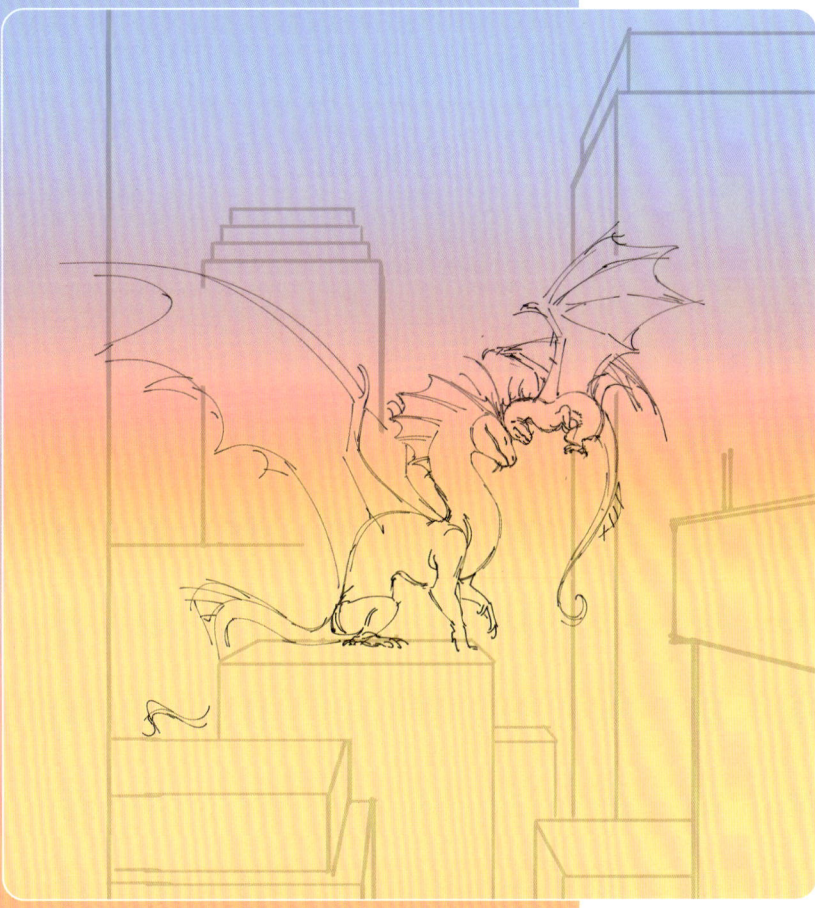

08 Save a copy of your colour key as a separate file, and then delete all of the layers associated with the colour key from your main file. Open the colour key file, and drag it into its own window to the left of your main file by hovering on the left side of the screen until a red bar pops up. Now, you can reference it easily. In the main file, set your sketch to the Multiply blending mode. Use the Eyedropper tool to colour-pick the sky colours from your colour key, and use the Gradient tool to create a sky gradient.

< Use the Gradient tool to create a smooth sky gradient under the sketch layer

09 Turn your Perspective Ruler on again. In the next few steps, you will use the Rectangle Selection tool and Gradient tool to block in your buildings.

Get a feeling for how the Rectangle Selection tool operates in conjunction with the Perspective Ruler. Drag out a selection, then hover and move your cursor back and forth to switch the selection between a flat rectangle or a rectangle that converges to the perspective point. Tap the tablet pen to finalize a selection. Once you feel ready, begin blocking in buildings by selecting a plane and then filling it with gradients of colour to match your colour key.

> Block in the first building using the Rectangle Selection tool and the Gradient tool

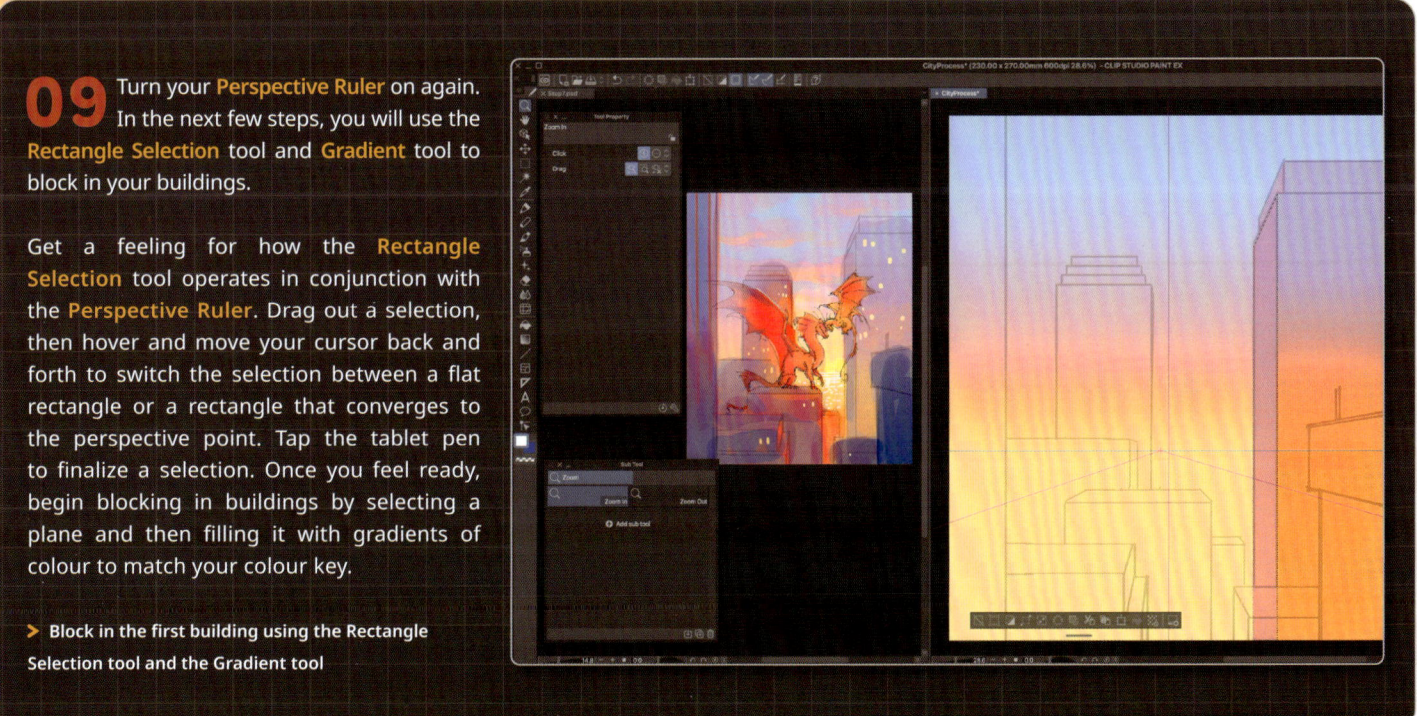

10 Continue using the Rectangle Selection tool together with the Gradient tool to block in the buildings. Work from back to front in space, and create a new layer for every plane of each building. This will allow you to lock transparent pixels on the layers to easily make colour or texture adjustments later on. Collect each building into its own group, and then label the building groups 1, 2, 3, and so on.

> **The city is half blocked-in at this stage**

> ## ARTIST'S TIP

With what you've learnt so far, you already have the majority of the knowledge you need to draw and fill in a city using one-point perspective. To further improve these skills, try replicating a reference photo of a city using the tools you have learnt about in the previous steps. Reference images will add new information to your mental library, which you can then draw from later when painting from your imagination!

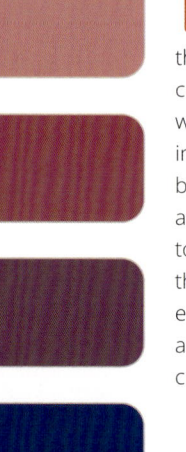

11 Use richer, darker colours for the buildings in the foreground to separate them in space. Imagine that a beam of sunlight is cutting through the central buildings. That light warmly illuminates the lower half of the buildings in the middle ground (the middle section of buildings between the background buildings and foreground buildings). This warmth will help to focus attention towards the middle, where the dragons will be. Dividing sections of your environments into foreground, middle-ground, and background planes can be helpful for creating the illusion of distance.

< **The city is now fully blocked-in with base colours**

12 Now you're going to make a window brush. Select **Material palette > All Materials**, then right click and create a new folder titled 'Window Brushes'. Next, create a new canvas for brush-tip creation. Set the width to 1,000 pixels and the height to 300 pixels. Set the **Basic Expression Colour** to Grey. Using pure black, create a small black rectangle on the left side. Now, using the **Move** tool, with the rectangle selected, hold **Option+Command+Shift on Mac** (or **Ctrl+Alt+Shift** on PC) simultaneously to copy and drag the rectangle over to the right. Repeat this process until you have a full line of windows across the canvas. Try grouping windows into sets of two or four to create a more unique window line.

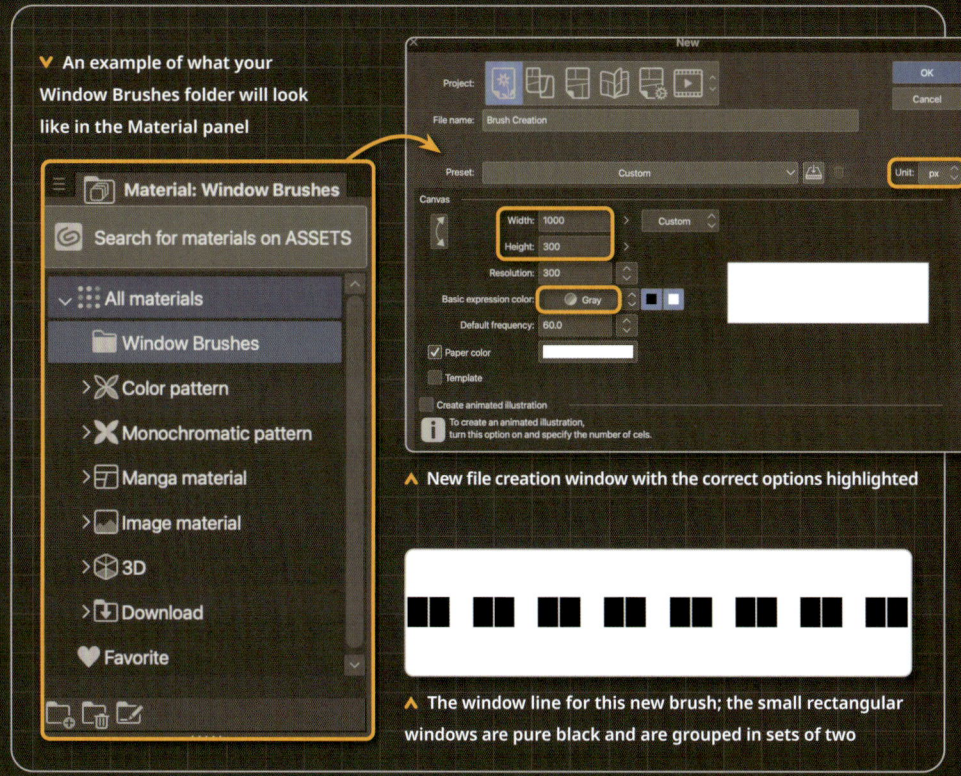

ᵛ An example of what your Window Brushes folder will look like in the Material panel

ᴧ New file creation window with the correct options highlighted

ᴧ The window line for this new brush; the small rectangular windows are pure black and are grouped in sets of two

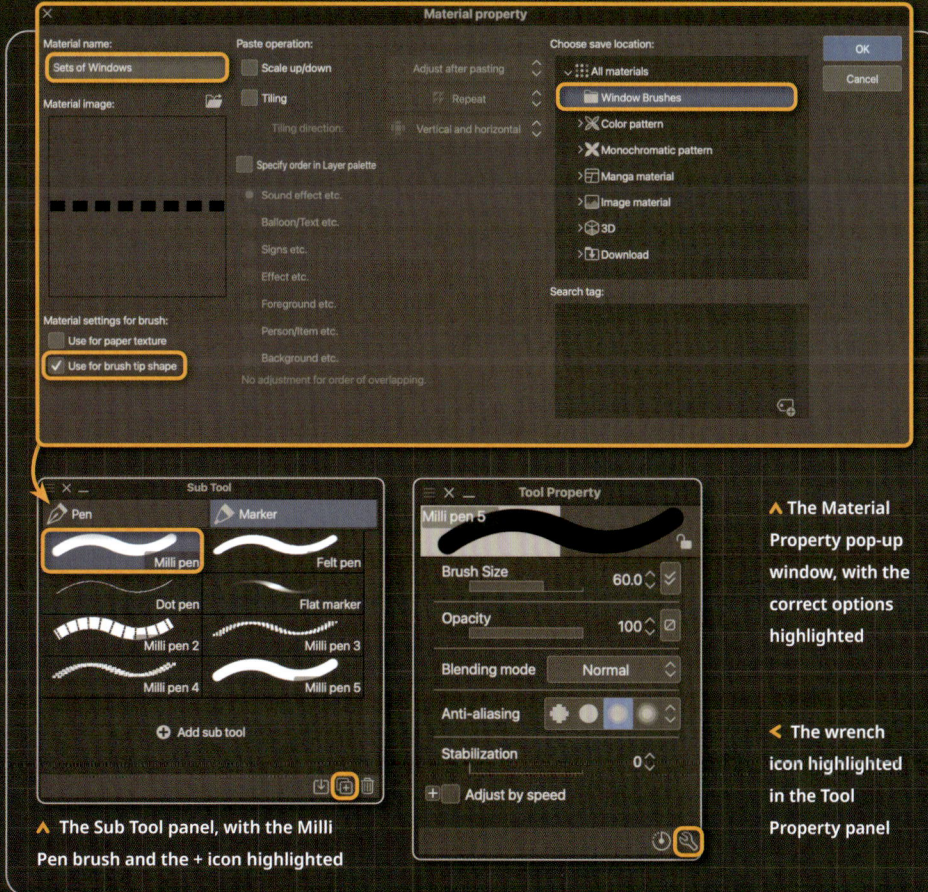

ᴧ The Material Property pop-up window, with the correct options highlighted

◄ The wrench icon highlighted in the Tool Property panel

ᴧ The Sub Tool panel, with the Milli Pen brush and the + icon highlighted

13 In the top bar, select **Edit > Register Material > Image**. On the bottom left of the pop-up window, check the **Use For Brush Tip Shape** box. In the **Choose Save Location** area, save the brush in your Window Brushes folder. Name your brush on the upper-left box, and then click OK. You will now see the brush tip in your **Material** panel. Select any basic brush in your **Sub Tool** panel and click the + in the bottom right to copy it. With the copied brush selected, click the wrench icon on the bottom right of the **Tool Property** panel to open the brush in the **Sub Tool Detail** panel.

14 In the Sub Tool Detail panel that pops up, click on Brush Tip, then to the right of Tip Shape, click on Material, followed by Click Here and Add Tip Shape. In the Select Brush Tip Shape panel that opens, click on Created Material on the left side of the panel, and then find and click on the windows you just created. Click on Stroke and at the top of that panel set Gap to Fixed, then slide the bar to adjust the gap between each line of windows. Make sure Correct velocity Input is checked. Next, click Save All Settings as Default on the bottom right. You now have a window tower brush! Test it out by drawing a brushstroke down your canvas.

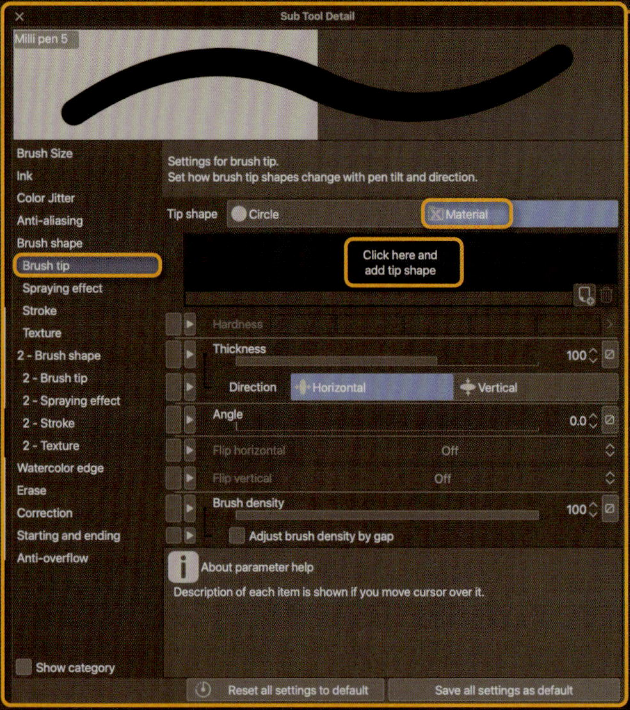

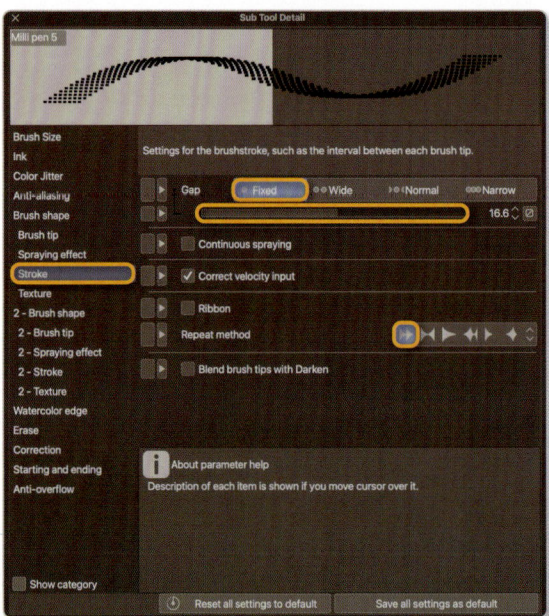

∧ The Select Brush Tip Shape pop-up window, with the location and brush tip highlighted

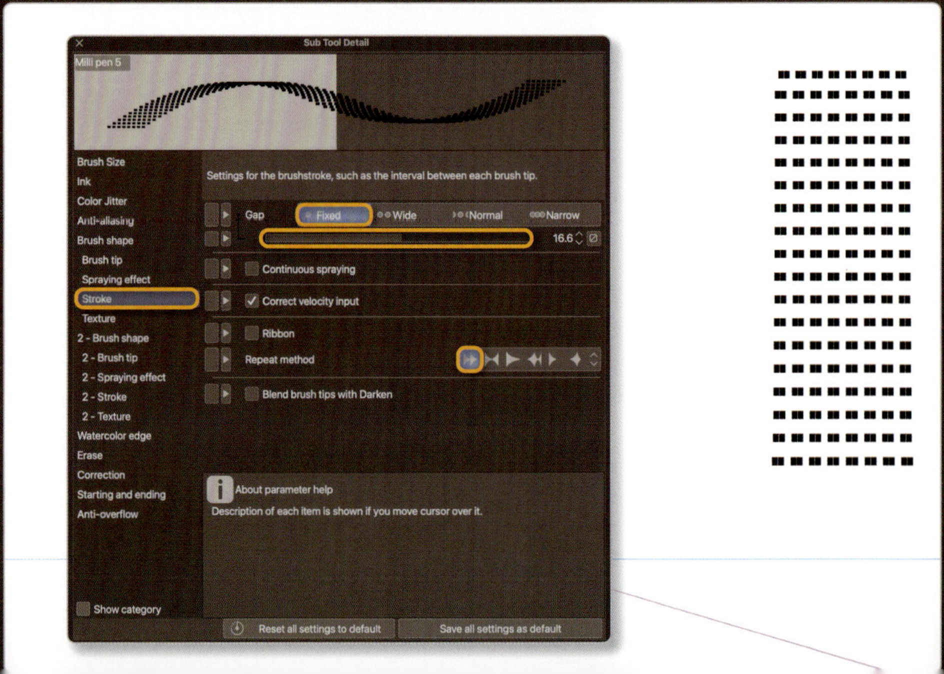

∧ The Sub Tool Detail panel on the Brush Tip page, with the options highlighted

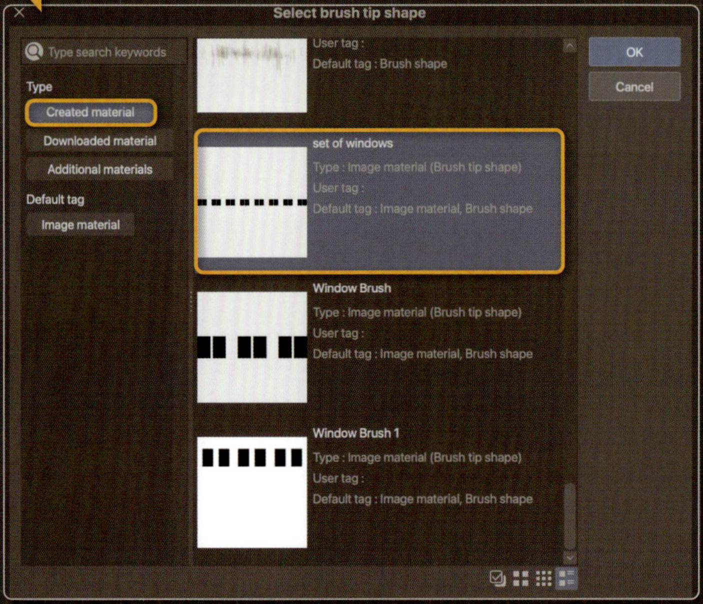

> The Sub Tool Detail panel on the Stroke page, with the options to click on and adjust highlighted

15 On a new layer, draw a line of windows down the front of one of your buildings with the brush you created in the previous step. In one-point perspective, the side of the building that faces the viewer is perfectly parallel to the canvas, so you don't need to transform your windows at all – they are already in perspective! Many buildings have a gap between the top of the window tower and the roof. Leaving a space there can add a touch of realism and believability to your buildings.

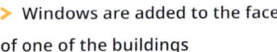

> **Windows are added to the face of one of the buildings**

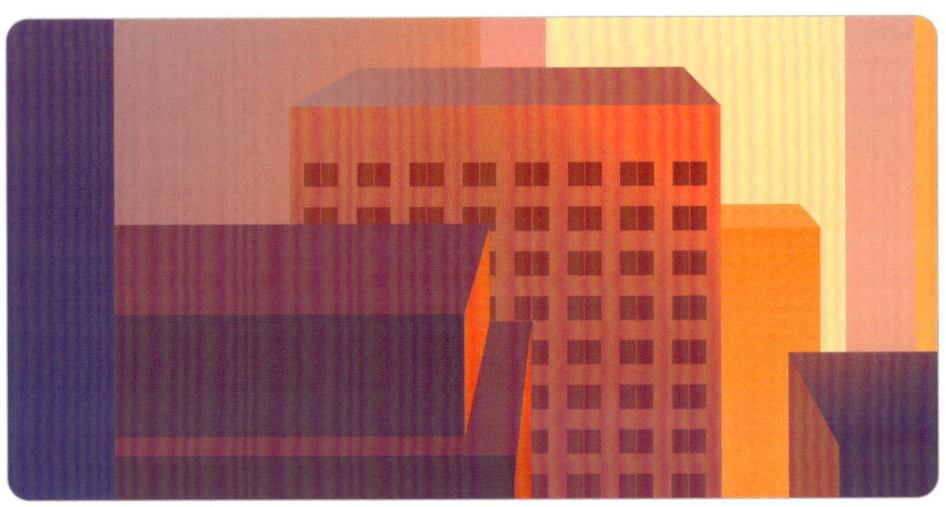

16 Continue to add windows to your buildings. If you're feeling confident, you can make a new window brush for each building. Repeat steps 12 and 13, then duplicate your previous window brush and simply swap in the new brush tip you created.

To put windows into perspective on the side of a building, draw temporary guidelines for where the top and bottom of the window tower should fall in perspective using your Perspective Ruler. Next, copy and paste the window tower from the front of the building to a new layer. Finally, use Free Transform on your duplicated window tower and drag the corners of the bound box in line with the guides you've drawn.

< **The city scene with the windows on the big building transformed into perspective**

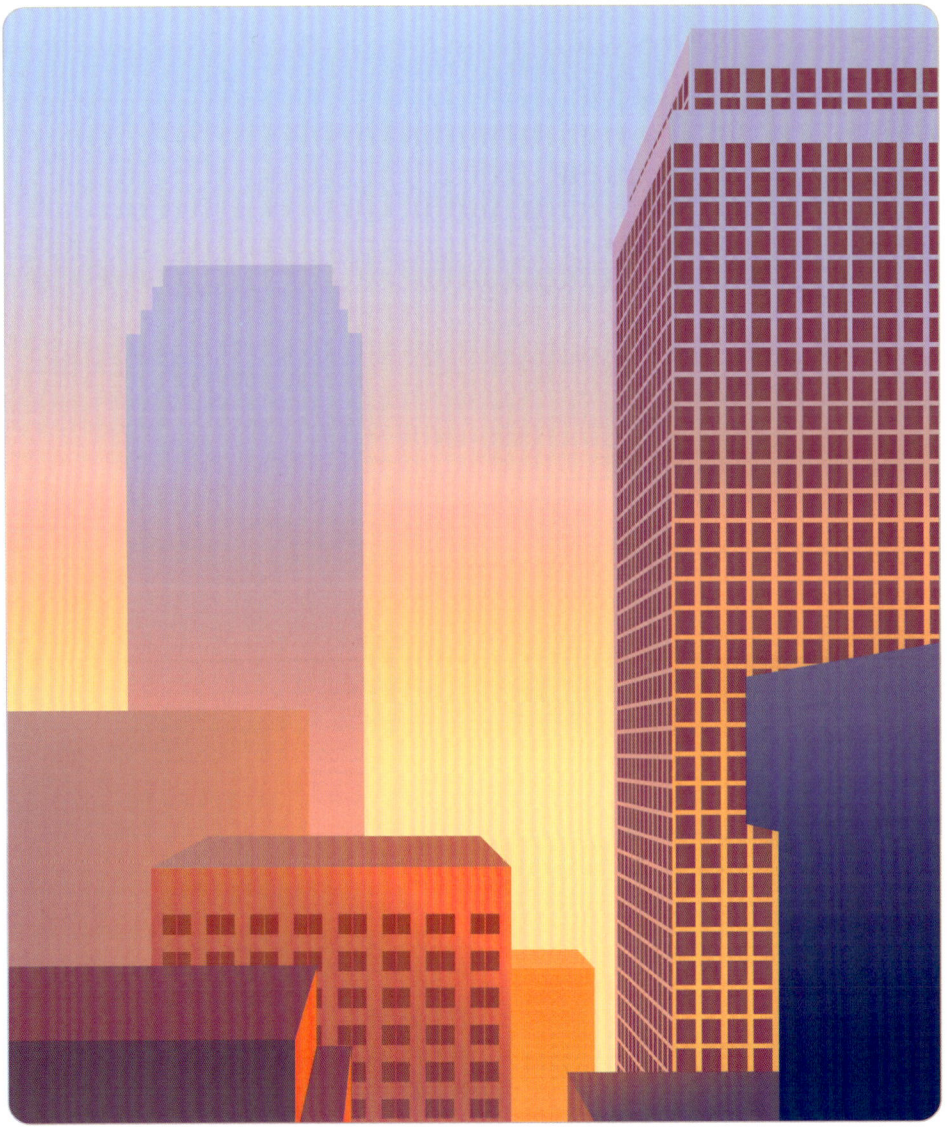

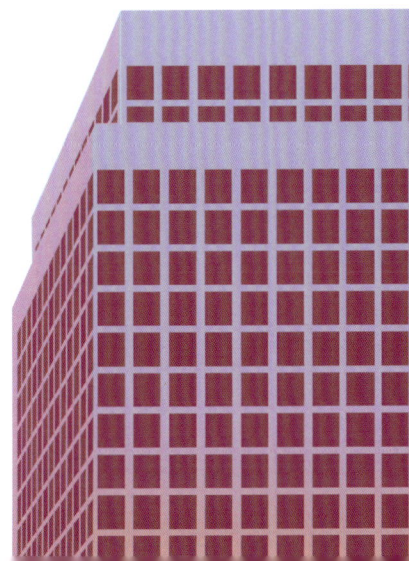

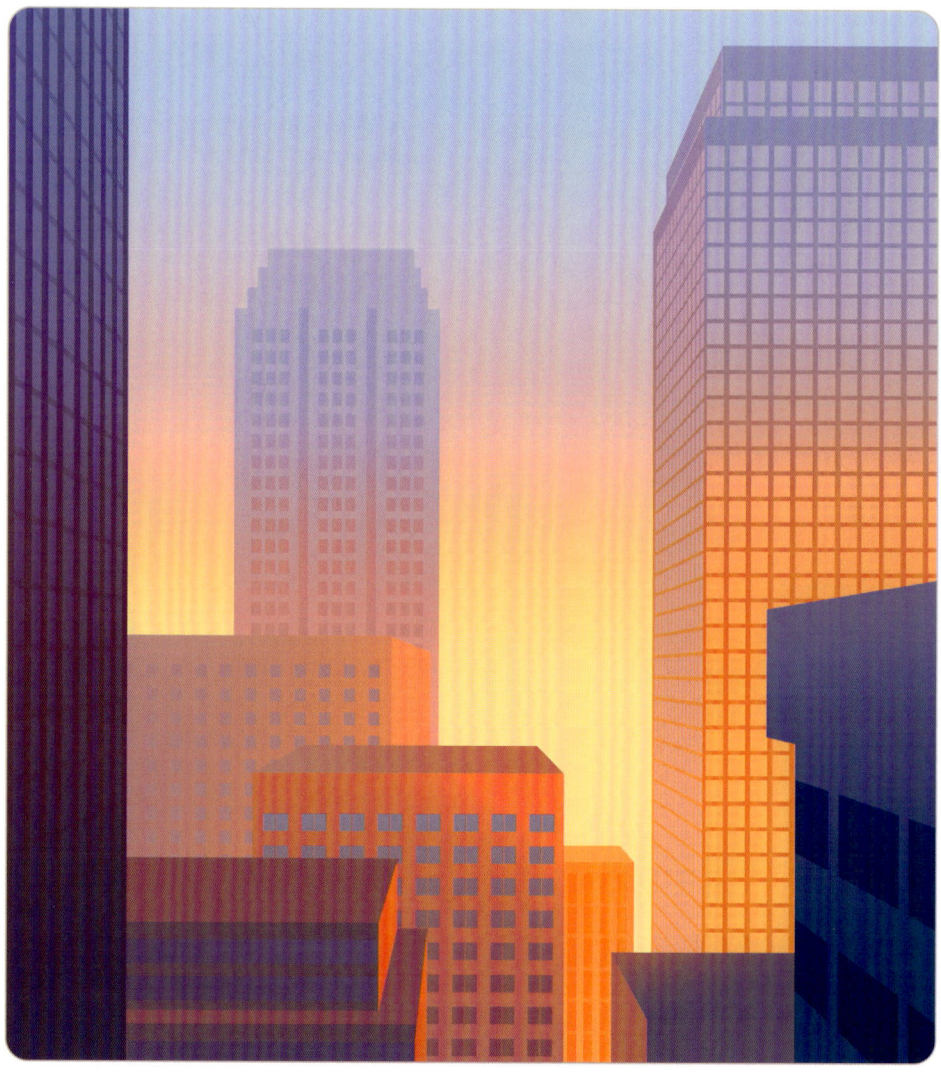

17 Finish adding windows to each of your buildings. Make sure each window tower is on its own layer, so you can lock it to change the colour and add detail. Next, experiment with the colour of your tower windows, creating gradients within them to reflect the colour of the surrounding skies, or sky colours from off-screen.

To put windows in perspective on a building that hits the top and bottom of the canvas, you will need to draw guidelines within the pixels of the canvas, and then match up horizontal lines from within the tower windows to the guides, rather than relying on the top and bottom of the tower.

< The tower windows are now completed

18 Lock your tower window layers, then use the Rectangular Selection tool to add the reflections of other buildings into the windows. This will add depth and realism to your scene, and will make your windows look glass-like and reflective. Look at reference photos of real cityscapes to get a feeling for how tower windows reflect their surroundings in the real world.

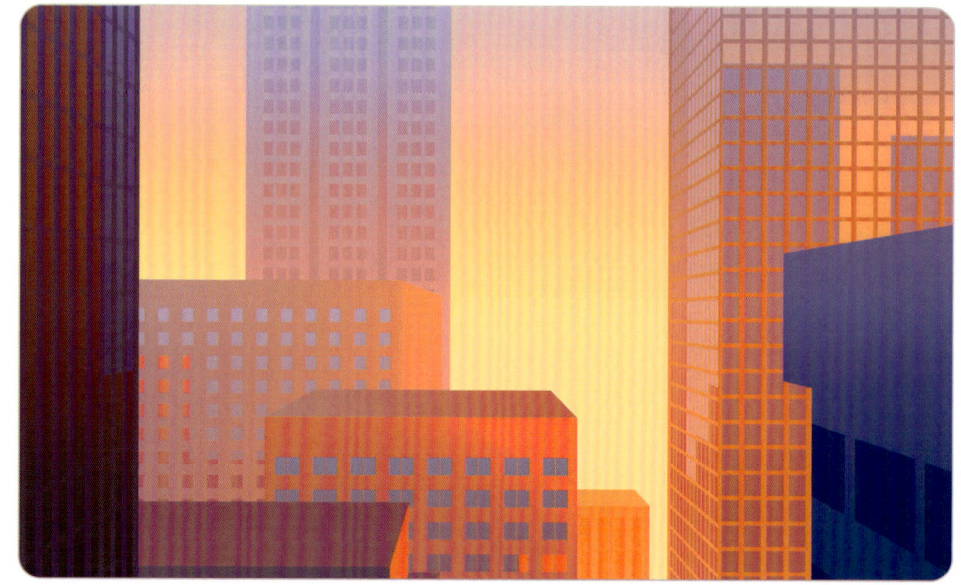

> Lock the tower window layers and add reflections of other buildings using the Rectangle Selection tool

19 Continuing on your locked window tower layers, turn on some lights inside of the buildings. Take a look at reference photos of real cities in the evening to determine the patterns of window lights. These patterns are not completely random – often an entire floor will still be working late into the evening, or perhaps an entire department on a few adjacent floors. Your cityscape will feel more believable if you include little details like this in your environment design.

> Illuminate some windows on the locked window tower layers

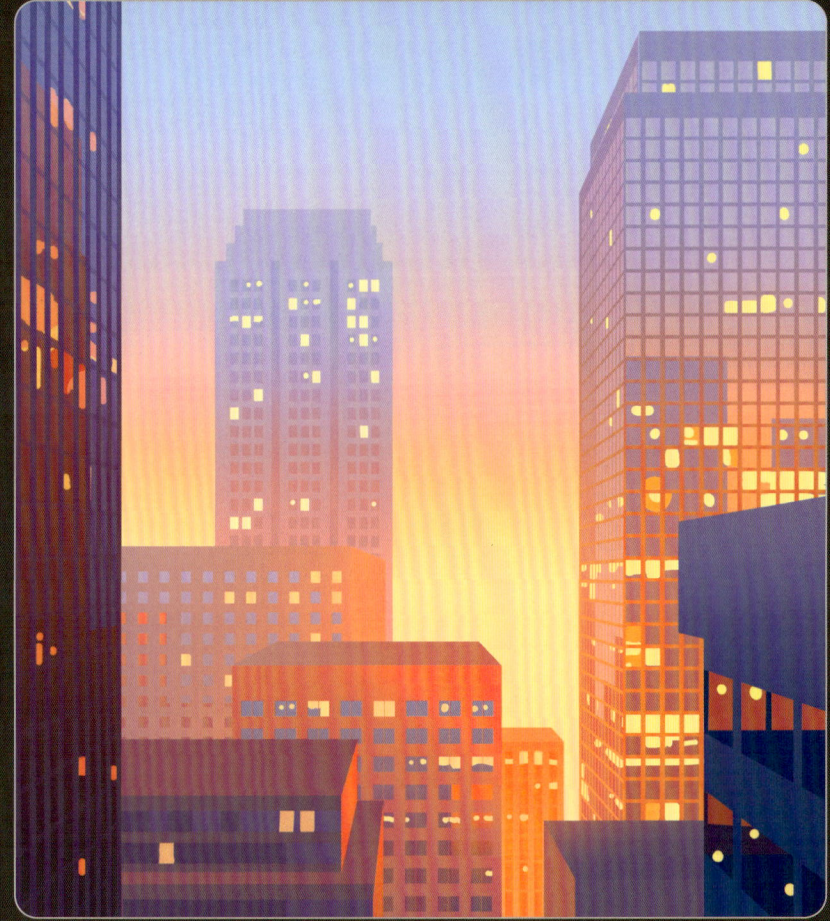

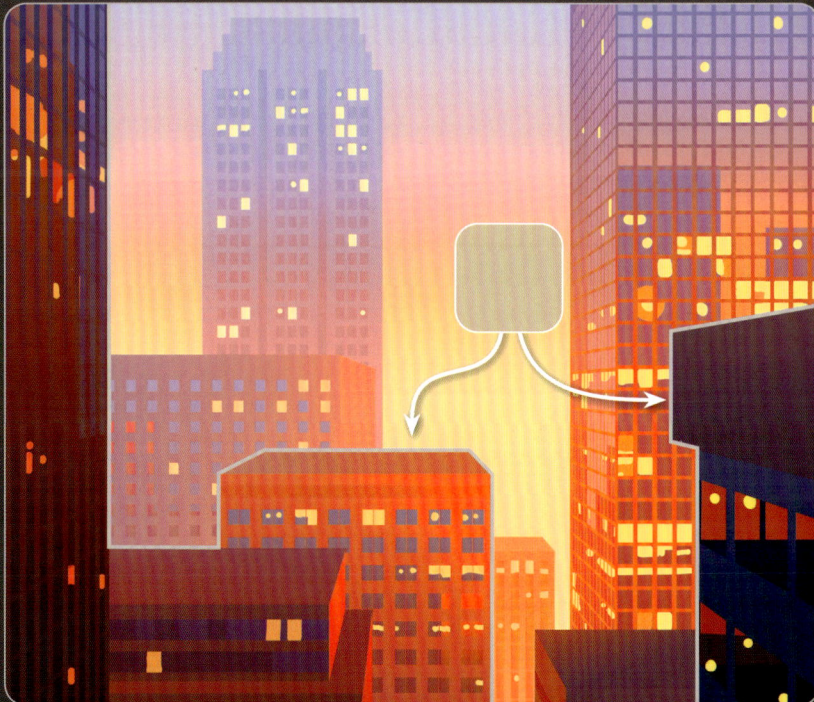

20 Clip new layers, all set to Colour Burn, to the top of each of your foreground and middle-ground building groups. Next, selecting dull beige and grey colours, use the Gradient tool to bring more contrast and drama to the scene. Drag gradients up from the bottom of the buildings with the Gradient tool set to Foreground to Transparent. Experiment until you're happy with the added depth and drama.

< The colour used on the Colour Burn layers is noted with arrows

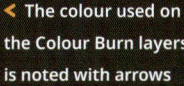

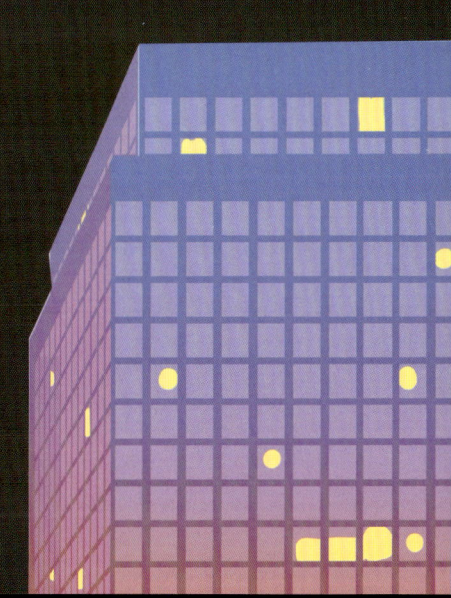

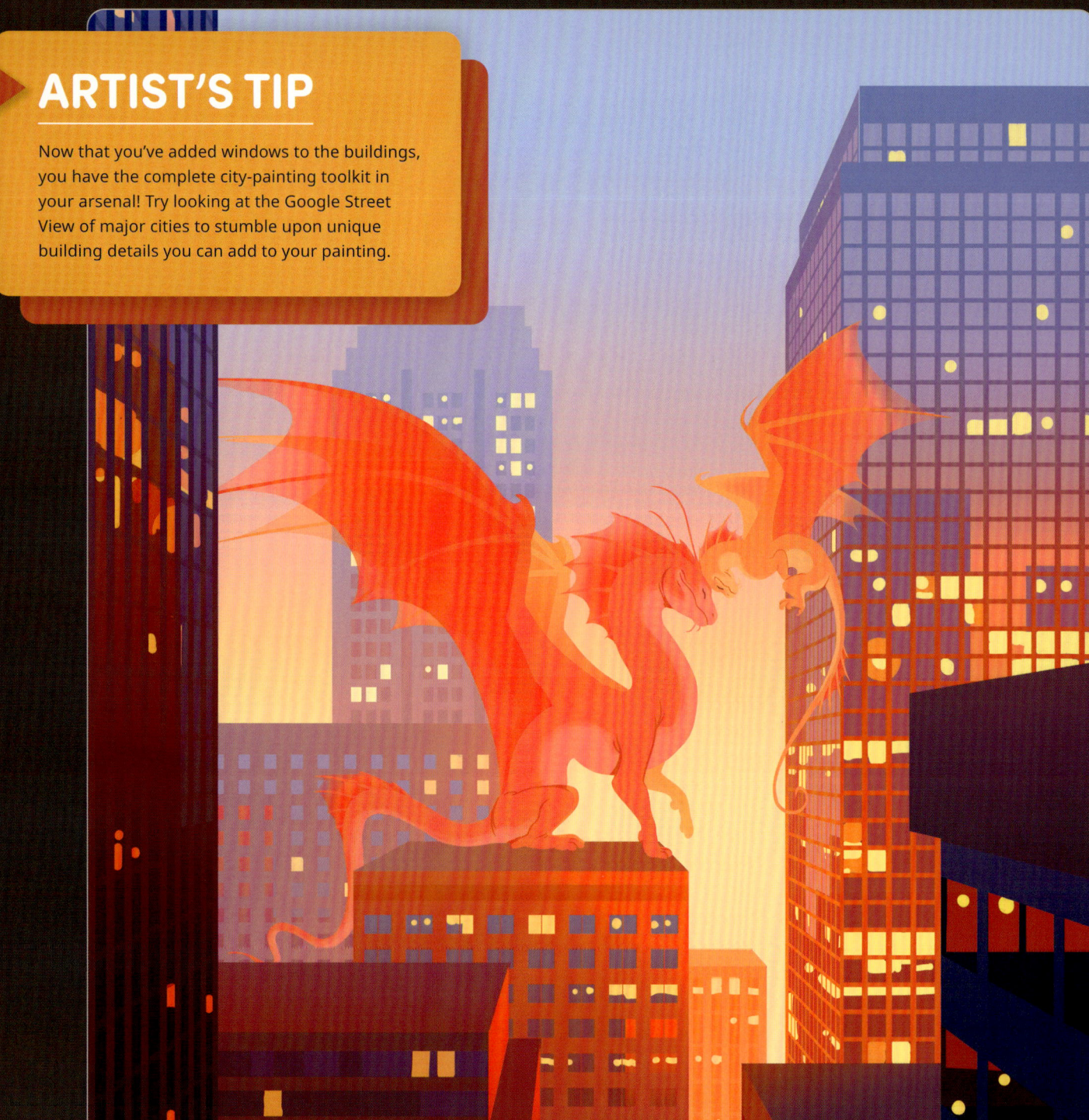

ARTIST'S TIP

Now that you've added windows to the buildings, you have the complete city-painting toolkit in your arsenal! Try looking at the Google Street View of major cities to stumble upon unique building details you can add to your painting.

21 In a new layer group, block in the dragons. Use a hard round brush to draw around the periphery of each dragon's silhouette, then use the **Magic Wand** tool to select and fill in each dragon. The **Close Gap** setting will allow you to select all

the way into the lines you drew, avoiding the pixelated gap that is present in other programs. Paint semi-translucent areas – such as the wings and fins – on a new layer, then use a big, soft eraser to erase out the edges slightly.

∧ Block in the dragons; the larger dragon is pink and the younger dragon is orange

22 Paint the sky using watercolour brushes. The Rough Wash brush works very well for large, fluffy clouds. The Wet Wash brush is another great choice. Painting clouds is all about creating a mix of soft and hard edges. Use the softer blending brushes to lose some of the hard edges. This will make the clouds appear more realistic. The clouds should be fluffiest at the top of the canvas, and then become narrower as they get closer to the horizon. After painting the sky, add the ocean in the distance, and the simple rectangular silhouettes of the distant buildings.

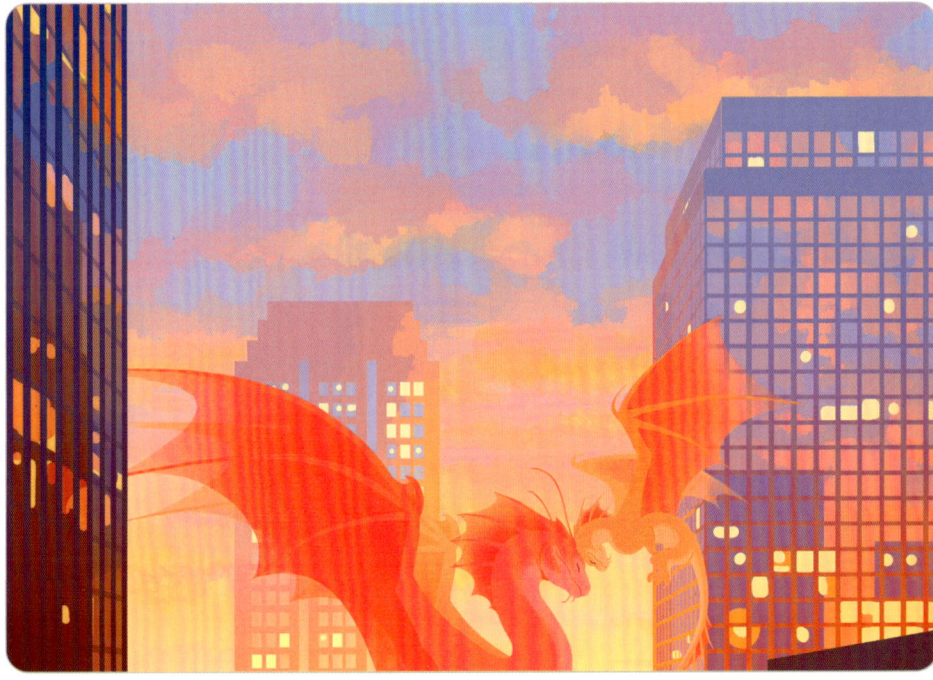

> Paint the sky using a variety of Clip Studio Paint's watercolour brushes

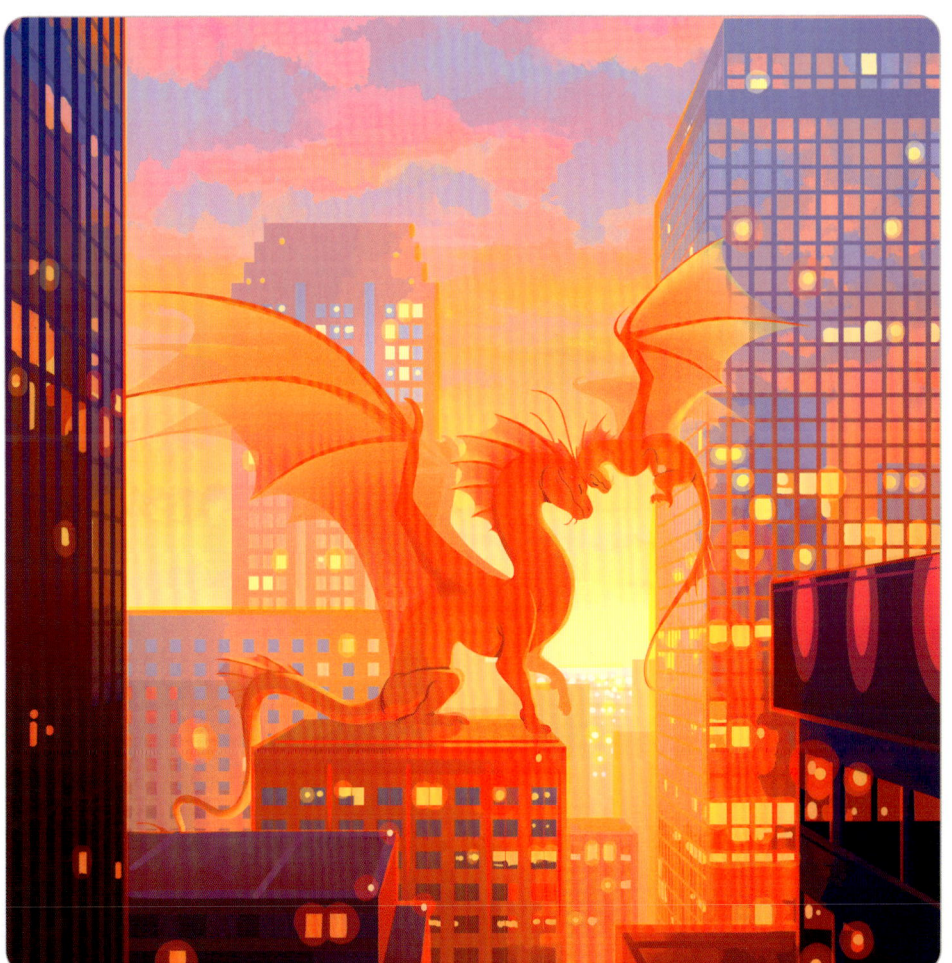

23 Now is the time to experiment, be free, and finalize your painting. Don't feel constrained by your initial colour key – you can make changes or different choices at any point in the process. To backlight the dragons, use a Multiply layer to darken them into shadow. Next, create an Overlay layer to paint yellow highlights that give the wings and bodies dimension. Now that all of the building blocks are in place, add shading and detail across your painting, including small touches such as extra glow around the lit windows. Look at the scene as a whole and try to improve the entire image.

< Add shadows and highlights to the dragons, as well as to some areas of the environment

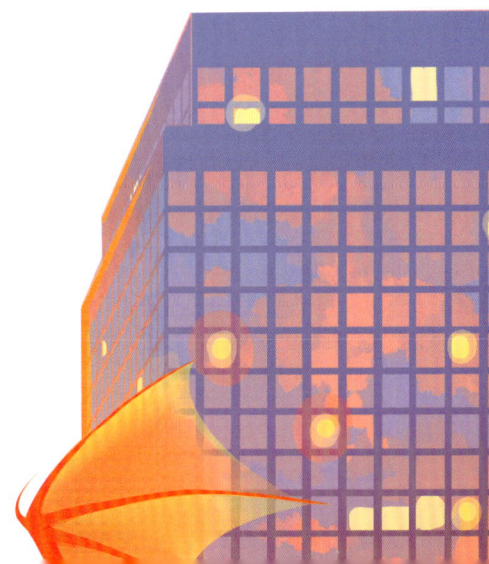

24 Use a Symmetrical Ruler, a sub tool of the Ruler tool, to add a sun flare to the painting. Try using eight to ten lines to start with – you can adjust the number of lines at the top of the image. Next, on a layer set to the Glow (Add) blending mode, draw the blooming sunlight. Finish off your painting by using colour balance Correction layers to make final adjustments to the colour of the image. You can find Correction layers under Layer > New Correction Layer. Colour Balance allows you to adjust the hues in your shadows, midtones, and highlights by moving the sliders towards and away from each colour on the three slider bars. Once you're satisfied with the look of your painting, you can save and export it.

> Add sun flare to the scene and make final colour corrections to the overall image

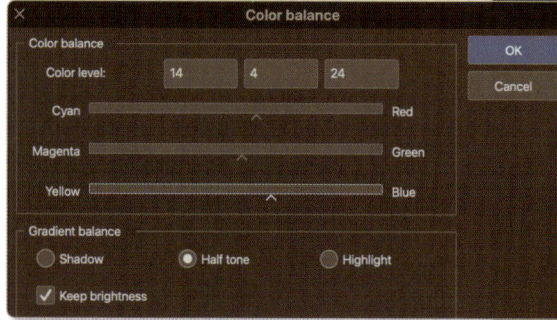

Conclusion

You've completed your first city painting using one-point perspective in Clip Studio Paint – take a moment to congratulate yourself! Drawing and painting in perspective can be a daunting task, and simply being willing to try is a huge achievement. If you found the steps overwhelming on your first try, keep practising. This is a process that gets easier with repetition. Whether you had a difficult or easy experience, you can walk away from this tutorial feeling confident in your ability to try out challenging things with digital art.

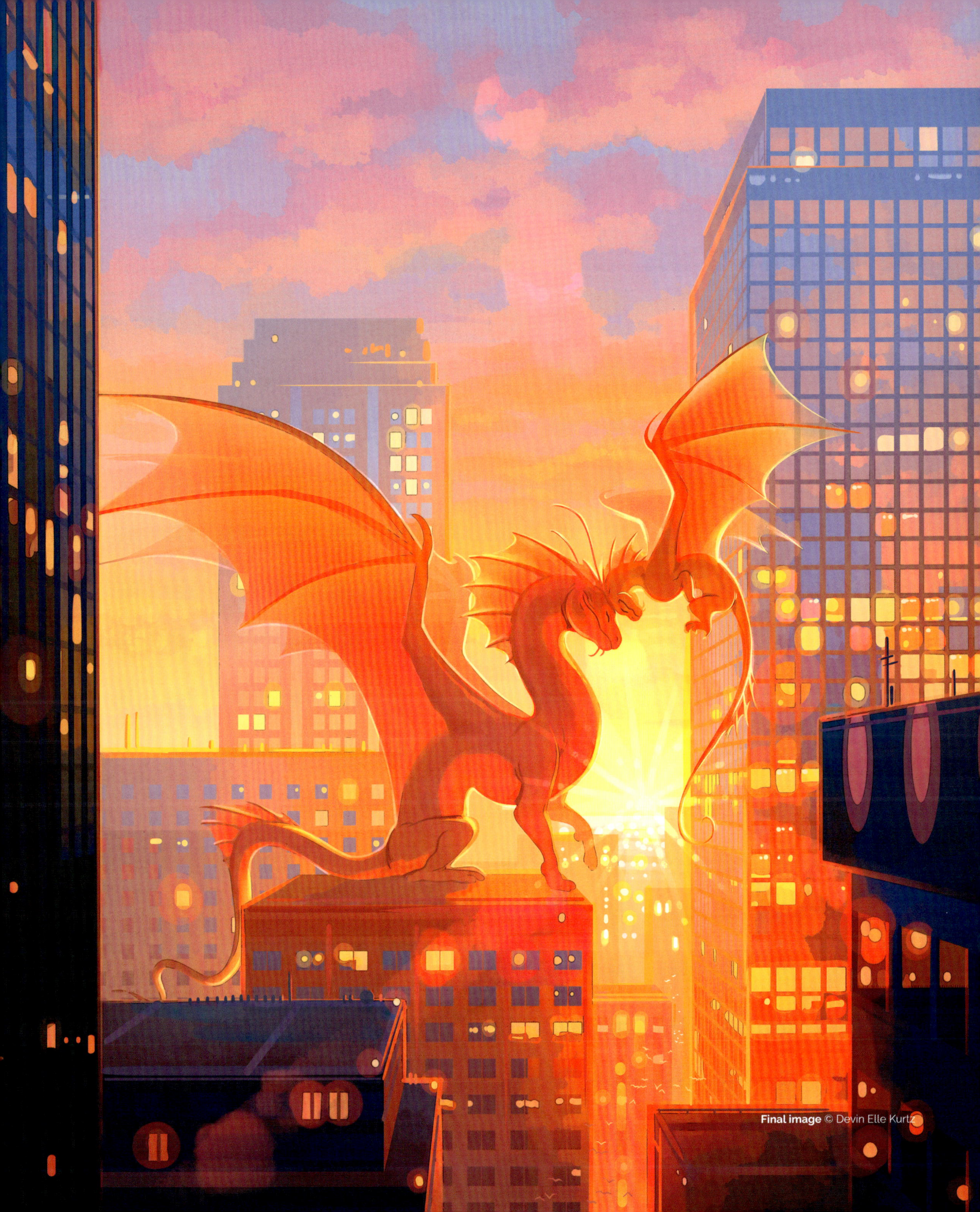

Final image © Devin Elle Kurtz

Gallery

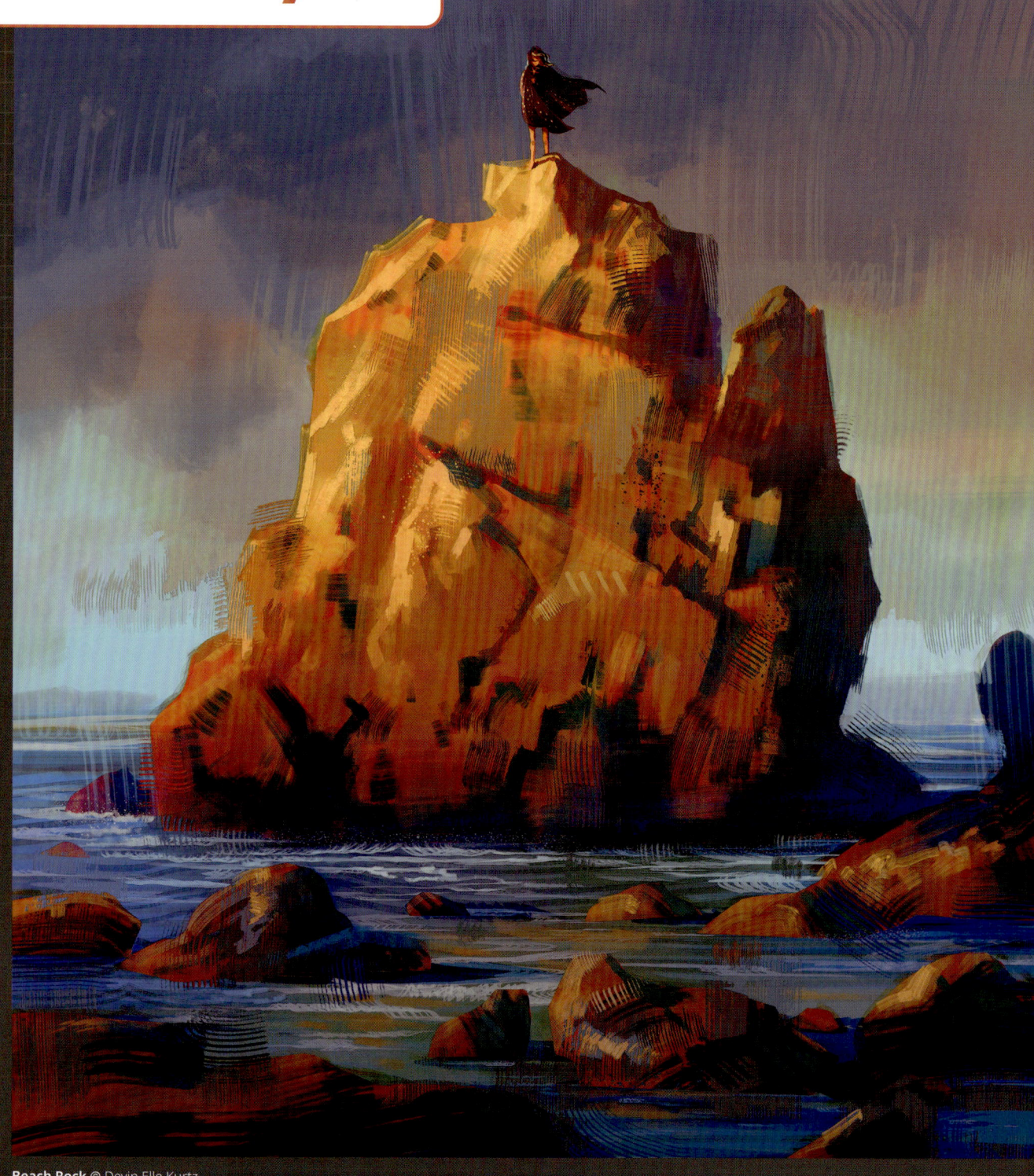

Beach Rock © Devin Elle Kurtz

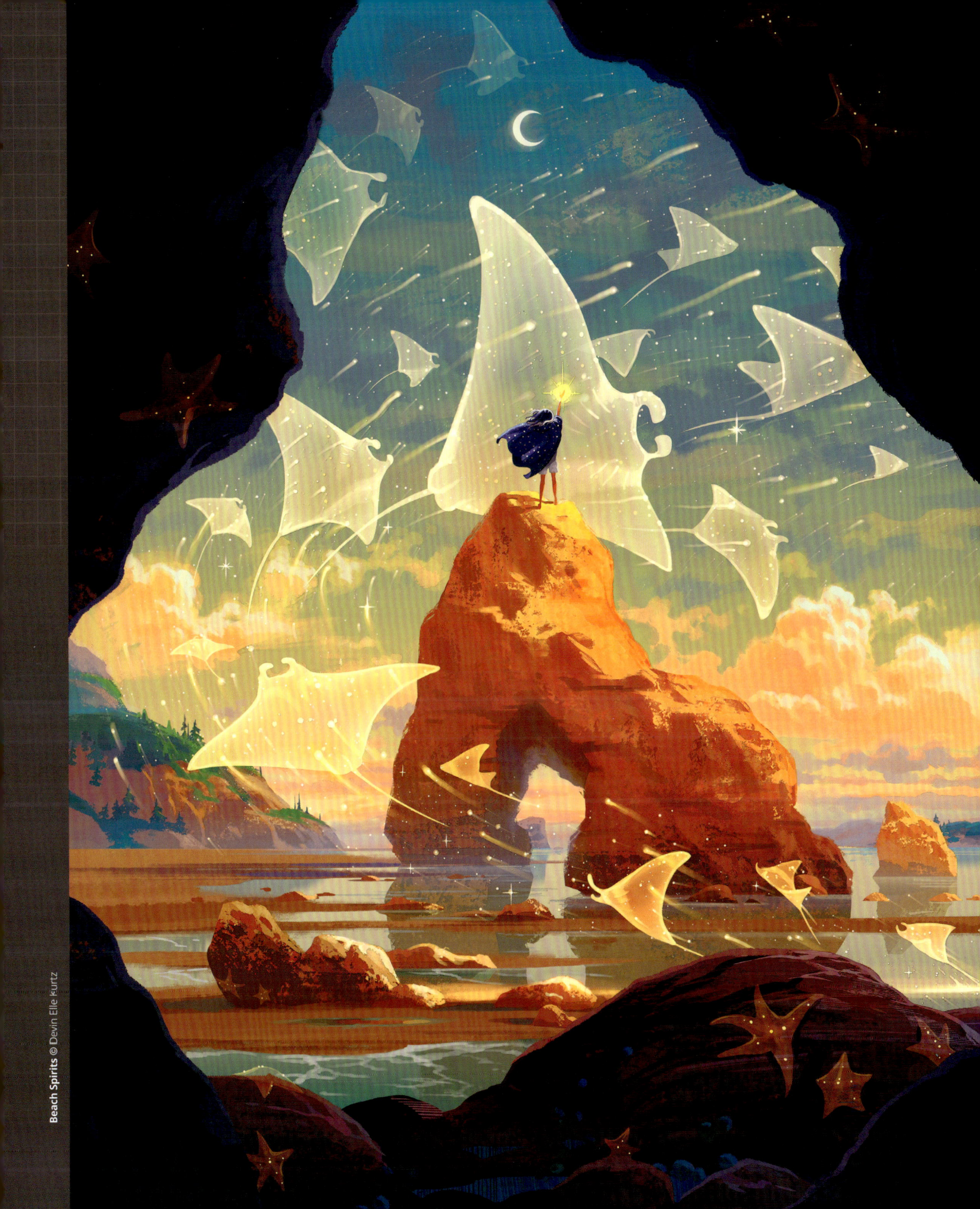

Manga character design

BY MMUMECHII

This tutorial will walk you through the steps involved in using Clip Studio Paint to paint a captivating scene featuring a cute manga character. It's not uncommon to feel overwhelmed and a little lost when painting detailed backgrounds, so I will share broken-down, simplified steps for you to follow. Plus, there will be tips and tricks for how to create scenes in which viewers can feel the emotion and mood of the piece. Let's get started!

Final image © Mmumechii

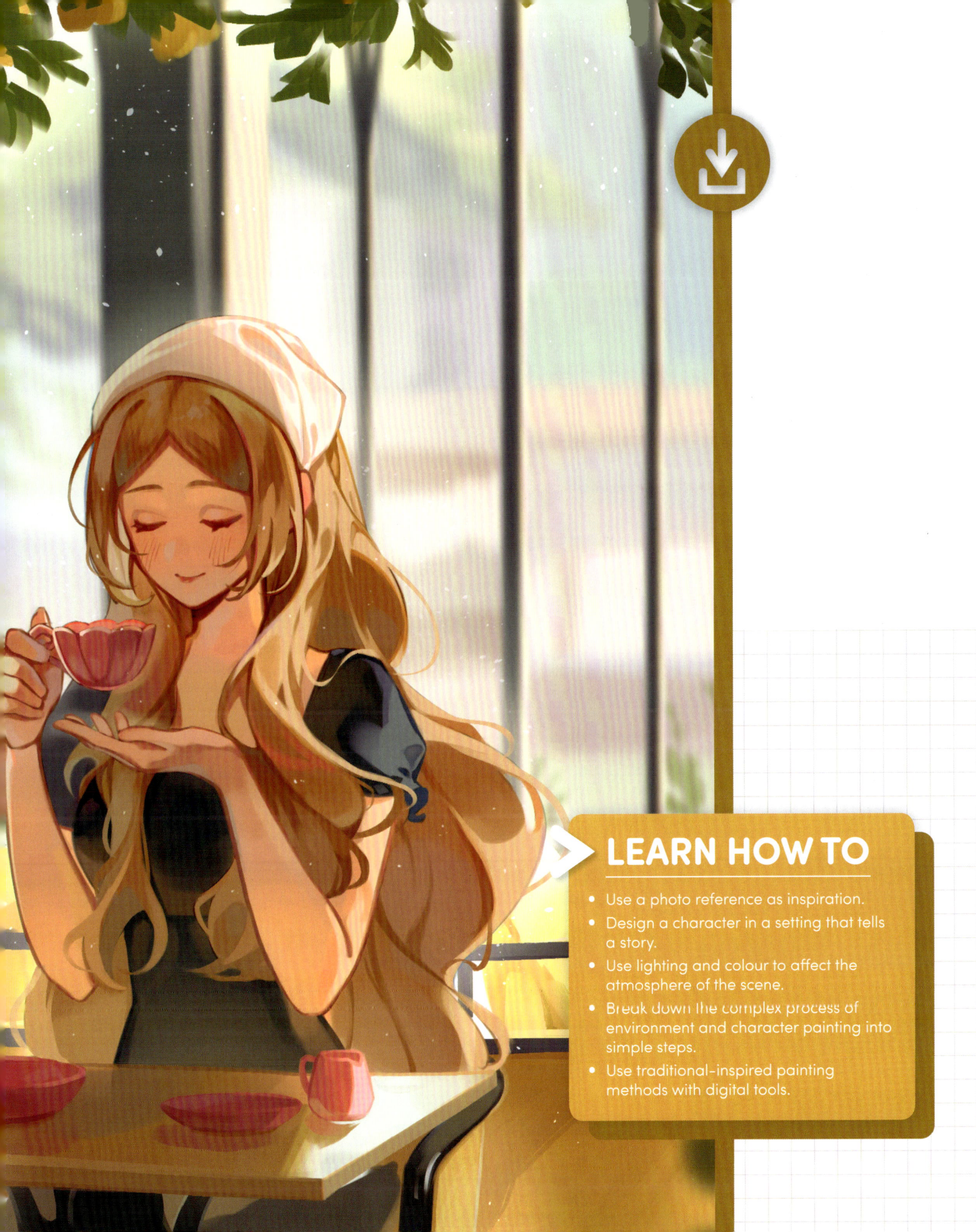

LEARN HOW TO

- Use a photo reference as inspiration.
- Design a character in a setting that tells a story.
- Use lighting and colour to affect the atmosphere of the scene.
- Break down the complex process of environment and character painting into simple steps.
- Use traditional-inspired painting methods with digital tools.

01 Begin by selecting a reference photo (or photos) to inspire and inform your illustration. You will still be able to add your own imaginative elements and style to the final painting, but having one or two references to refer to will help you to create accurate lighting and perspective. Insert the reference photo into Clip Studio Paint by selecting **Window > Sub-View** and click on the image icon to import a photo. This will position your photo in the Sub-view window, allowing you to draw while also looking at the reference on the same screen.

> **When choosing a reference image, keep in mind the lighting and composition**

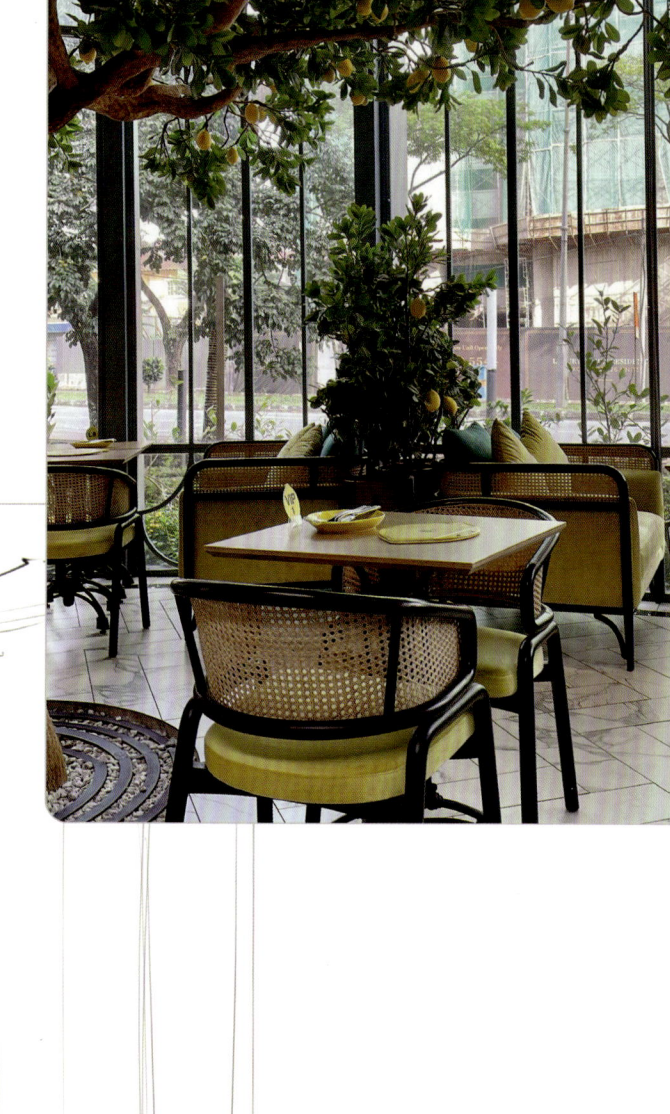

02 Create a rough sketch using a **Pencil Brush** from the Sub-Tool window. You can adjust the stabilization of the brush in the Tool Property palette, which will make your lines smoother. Avoid focusing on the small details or accuracy just yet. Instead, think about the atmosphere you want to capture. You don't need to zoom into the canvas at this early stage. Your attention should be concentrated on the overall drawing. A good artwork should be clear and easy to read, even from far away, so pay attention to the big shapes and whole composition.

< **Draw rough foliage around the outside of the canvas – this will frame the scene nicely, as well as introduce some natural elements**

03 Refine your sketch, starting with the background. Observe the perspective of your reference photo and try to replicate it on the canvas. Draw in some guidelines to help understand the angles and perspective. Here the chair and table are broken down into a simple cuboid, then lightly sketched in, guided by the construction lines of the 3D shapes. You can also use the **Rectangle Line** tool for added precision, found by selecting the square icon on the toolbar. This tool will allow you to draw parallel lines easily.

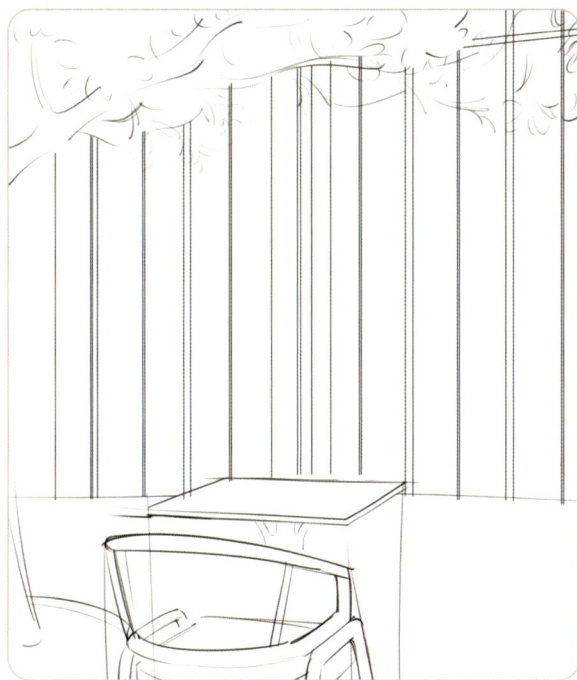

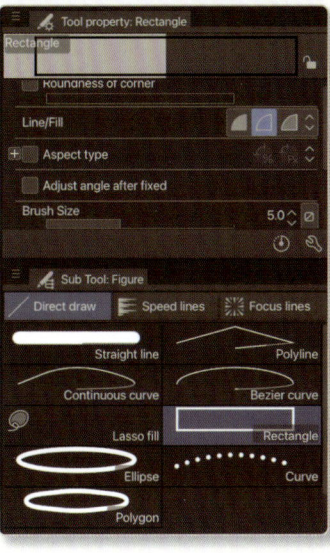

< The parallel lines and cuboid block strike a nice contrast beside the organic leaf shapes

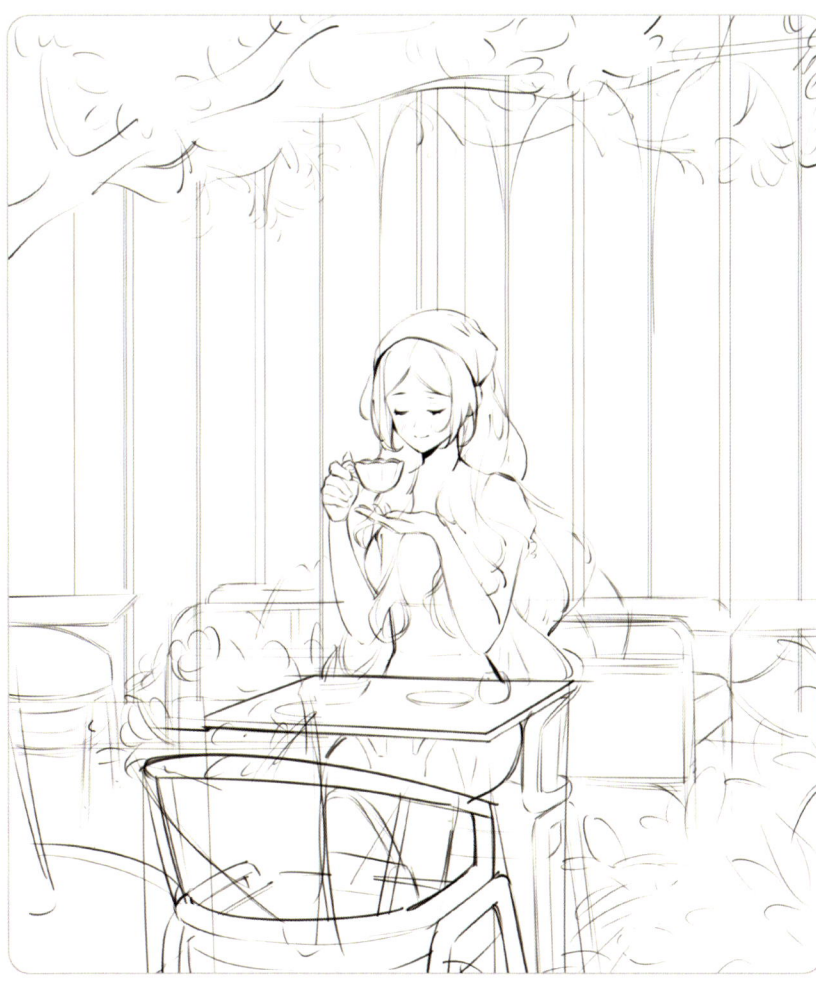

04 Now sketch in your character and the surrounding details. Finding a reference for the character's pose will help you to capture the anatomy and proportions accurately, while also keeping in mind the overall perspective of the scene as a whole. To double-check the anatomy is correct, flip your canvas using the **Flip Horizontal** icon on the Navigator window. Use the **Liquify** tool to adjust your sketch if it appears slightly off. Next, introduce expressive, flowing lines – such as in the hair – to make the character more visually interesting to look at. Sketch in some small props to help tell the story. A delicate teacup, cute bandana scarf, and bouquet of flowers work together to create the elegant and feminine personality of this manga girl.

< Draw in the character, plus props to help tell her story

ARTIST'S TIP

Sketching is often the most difficult stage of the process. If you're struggling, take a step back, zoom out, and block in the overall shapes and lines. It's easy to get caught up in the smaller elements of the drawing, only to zoom out to realize there's a huge mistake or the overall drawing looks strange. Working from big to small – from macro to micro – will help to prevent this error. Once you're satisfied with the simplified sketch, you can zoom in and add those finer details.

05 Use a **Round Brush** to begin blocking in colour on a new layer under the sketch. A useful technique for shading is to use lighter, desaturated colours on objects that are further away from the viewer. This helps to replicate real-life scenes, while also immediately drawing the viewer's attention to the character in the foreground. When following this rule, the elements in the far background will add detail to the scene without distracting the viewer's eye from the focal point. Here a cold blue tint is added to the background, which will contrast with the warm foreground to be added later.

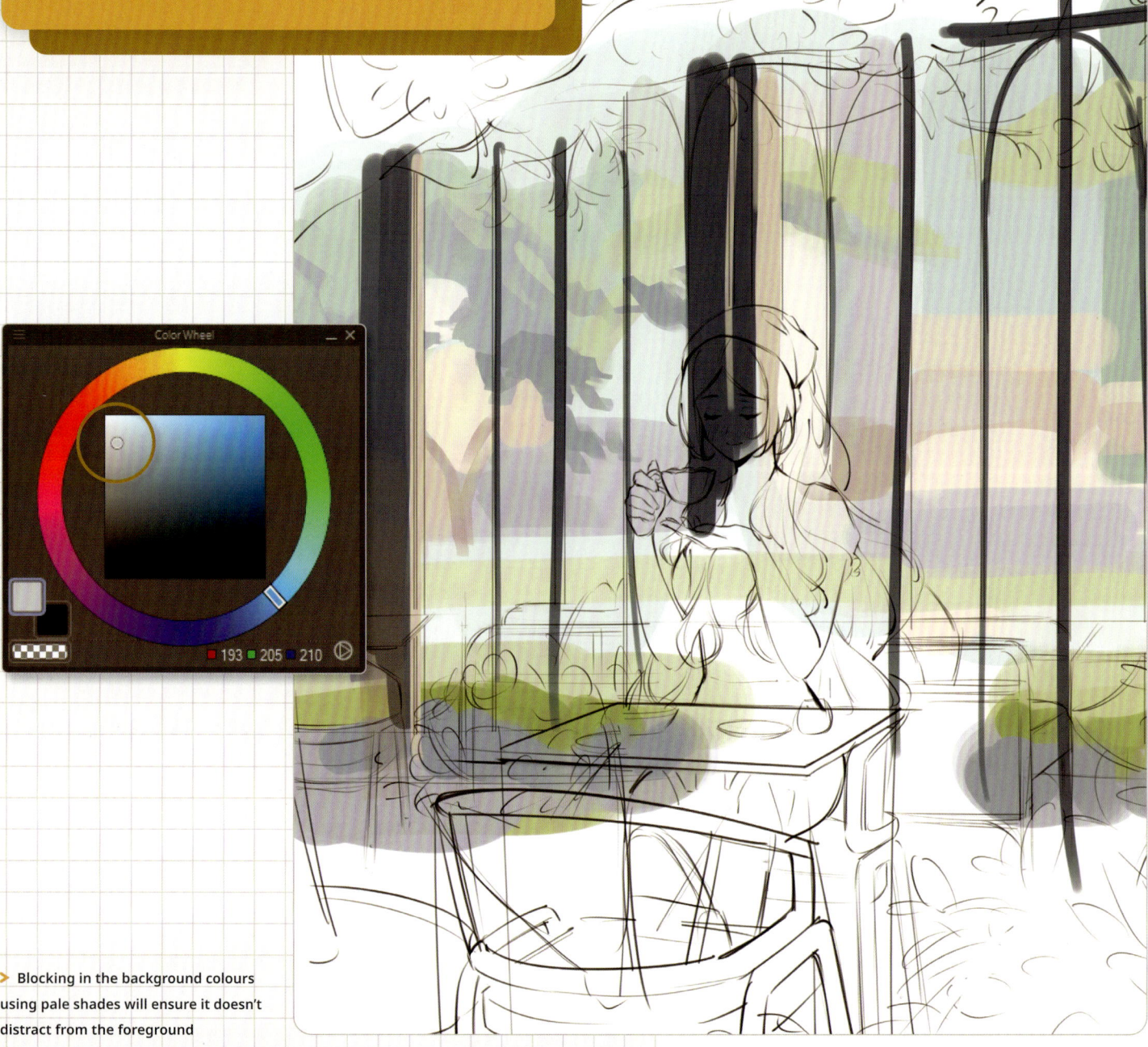

> Blocking in the background colours using pale shades will ensure it doesn't distract from the foreground

06 As you move on to the foreground, keep the various elements separate by painting each new object on a new layer. This will make them easier to adjust, if needed. Keep the light direction in mind at all times. Here the light source is behind the character, so her front side is cast in shadow. Use the **Overlay** and **Multiply** blend modes to adjust the colours and tones. To make a blend mode layer, create a new layer on top of the layer you want to adjust. Right-click on it in the Layer panel and select **Layer Settings > Clip to Layer Below**, so it will only affect the painted areas of the layer beneath. From the dropdown menu at the top of the Layer panel, set to Normal by default, experiment with different modes and see how they affect the colours below when you paint.

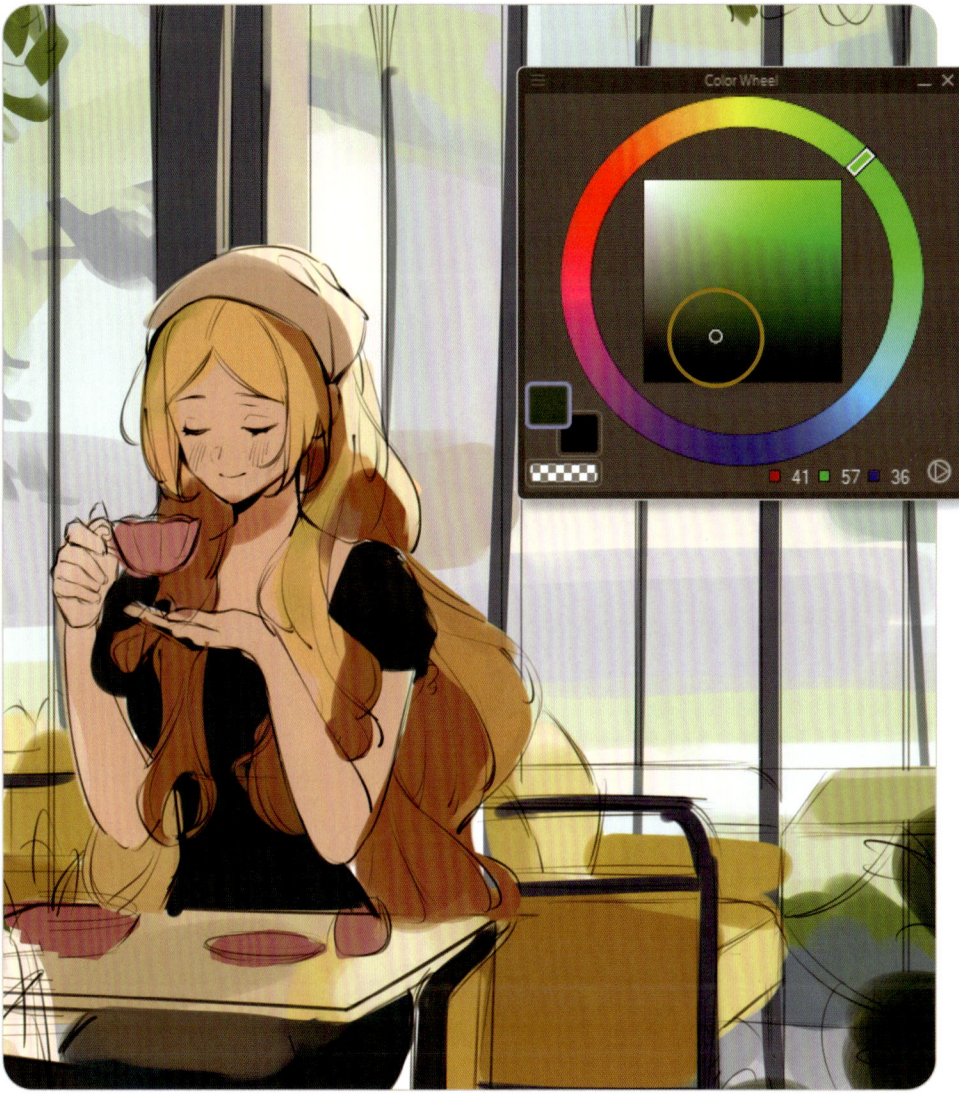

> Begin to introduce more saturated colours, found closer to the bottom-right corner of the colour picker

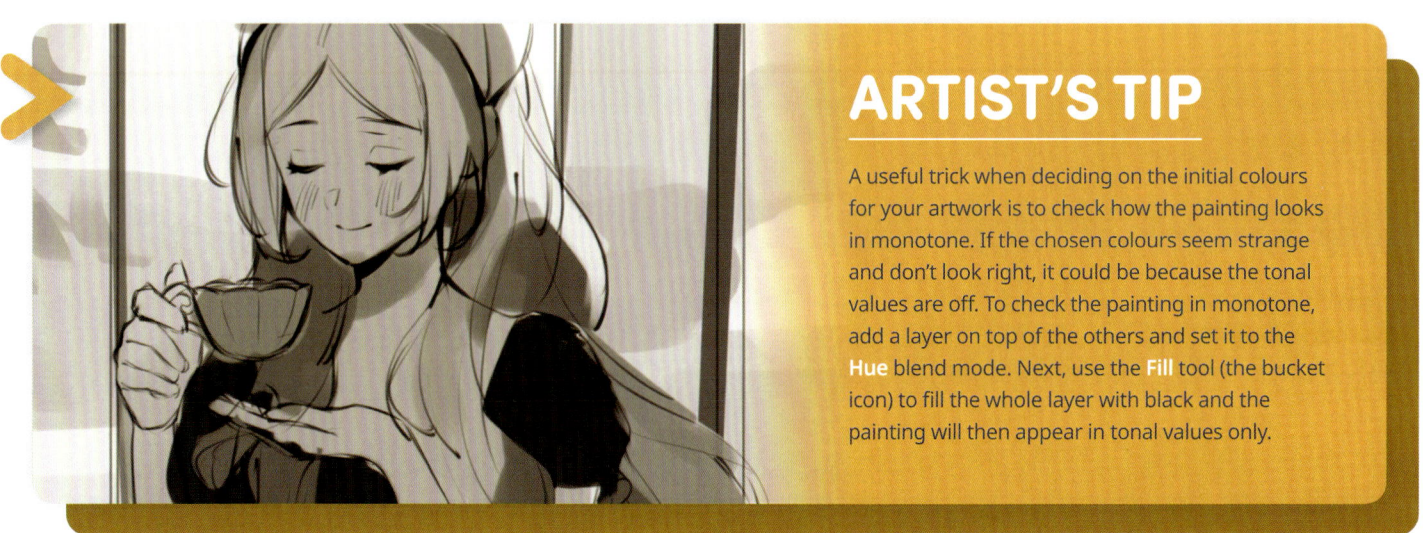

ARTIST'S TIP

A useful trick when deciding on the initial colours for your artwork is to check how the painting looks in monotone. If the chosen colours seem strange and don't look right, it could be because the tonal values are off. To check the painting in monotone, add a layer on top of the others and set it to the **Hue** blend mode. Next, use the **Fill** tool (the bucket icon) to fill the whole layer with black and the painting will then appear in tonal values only.

07 Once you're happy with the base colours, add in some more details as well as a few gradients. Using an **Airbrush** on an **Add (Glow)** blending layer, gently paint some light behind the various objects to mimic light bouncing off of them. Notice how a small amount of yellow glow is applied behind the character and foliage here. Next, paint in some ambient light, such as the yellow glow of the chair cushion on the girl's dress and green from the tree leaves on her skin. Ambient light is soft light that is reflected off the surroundings and will make your painting appear more realistic and harmonized.

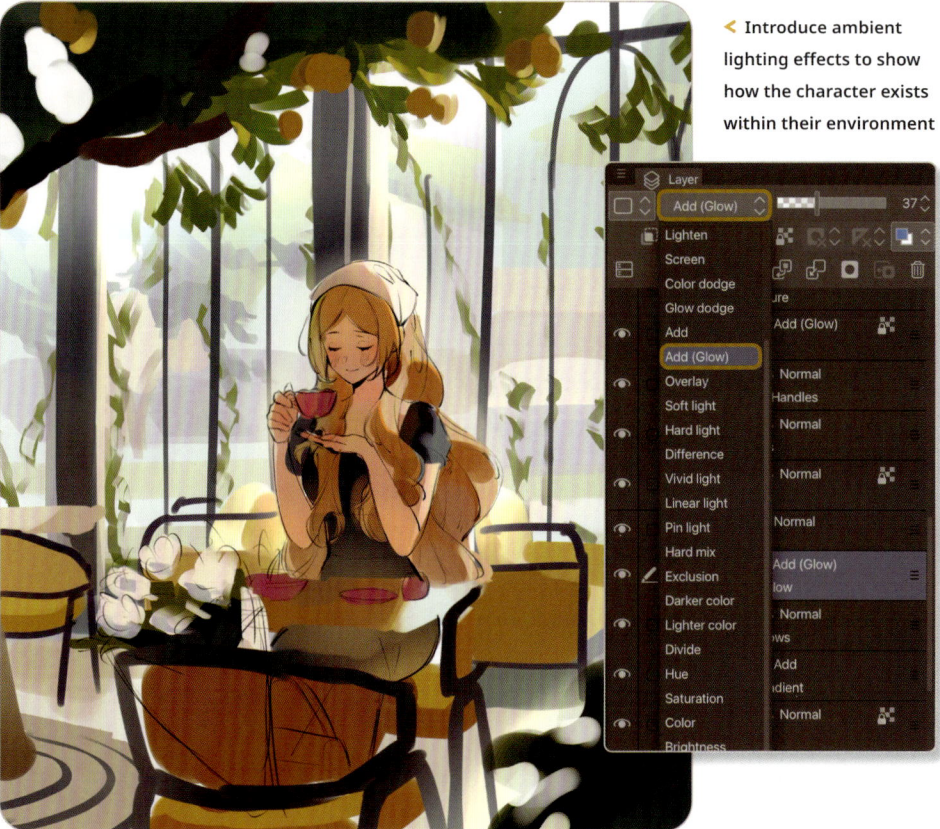

‹ Introduce ambient lighting effects to show how the character exists within their environment

08 The next step is to begin rendering the character. Clean up your rough colours, filling in any gaps you may have missed and erasing any parts that go over the outline. You can also add more gradients using the airbrush, such as blending some of the skin colour into the hair area around the face to give the skin a slight glow.

‹ Tidy up the messy parts and fill in any gaps

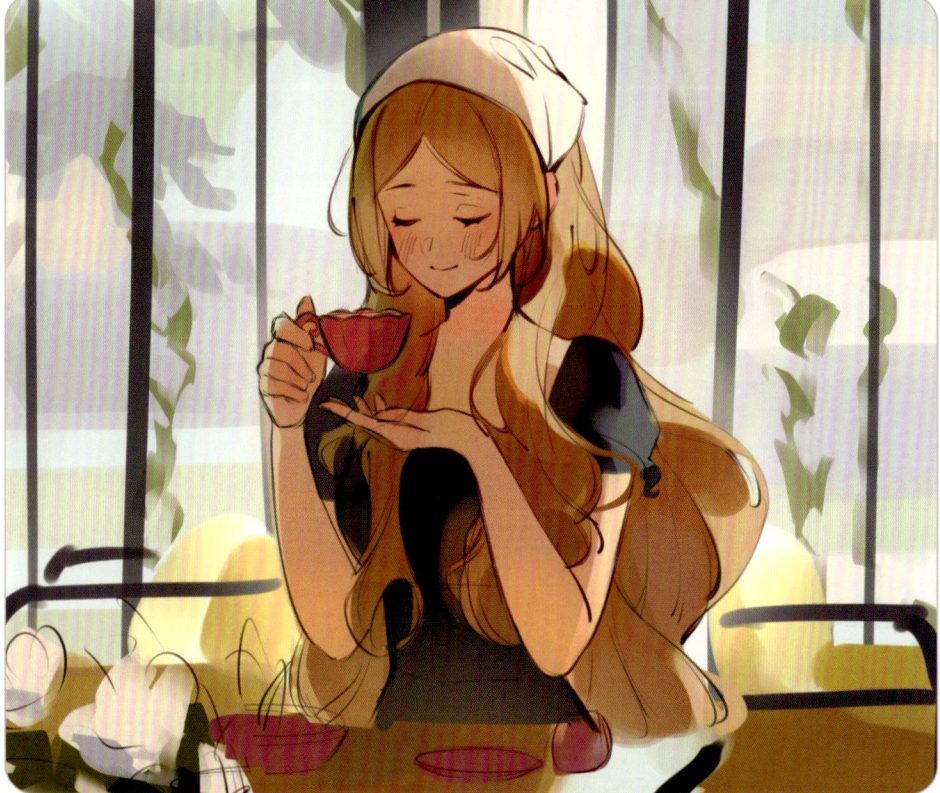

09 Now it's time to render the character's face. Select **Lock Transparent Pixels** on the skin layer and eyedrop the skin colour, then make it slightly redder and lightly airbrush it onto the cheeks as blush. Use the same colour to paint in the bottom of the character's lips, which will help to accentuate her soft smile. Next, select a subtle grey colour to paint in the nose shadow and eyelid highlights. You can also select **Lock Transparent Pixels** on the outline layer and airbrush in some dark red for the face outline area. Try to make sure the skin colour appears more vibrant the closer it is to the light.

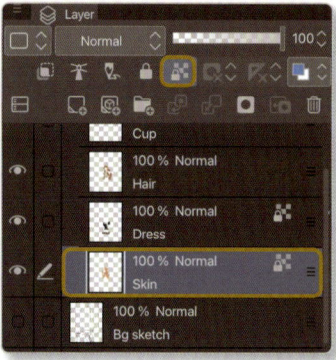

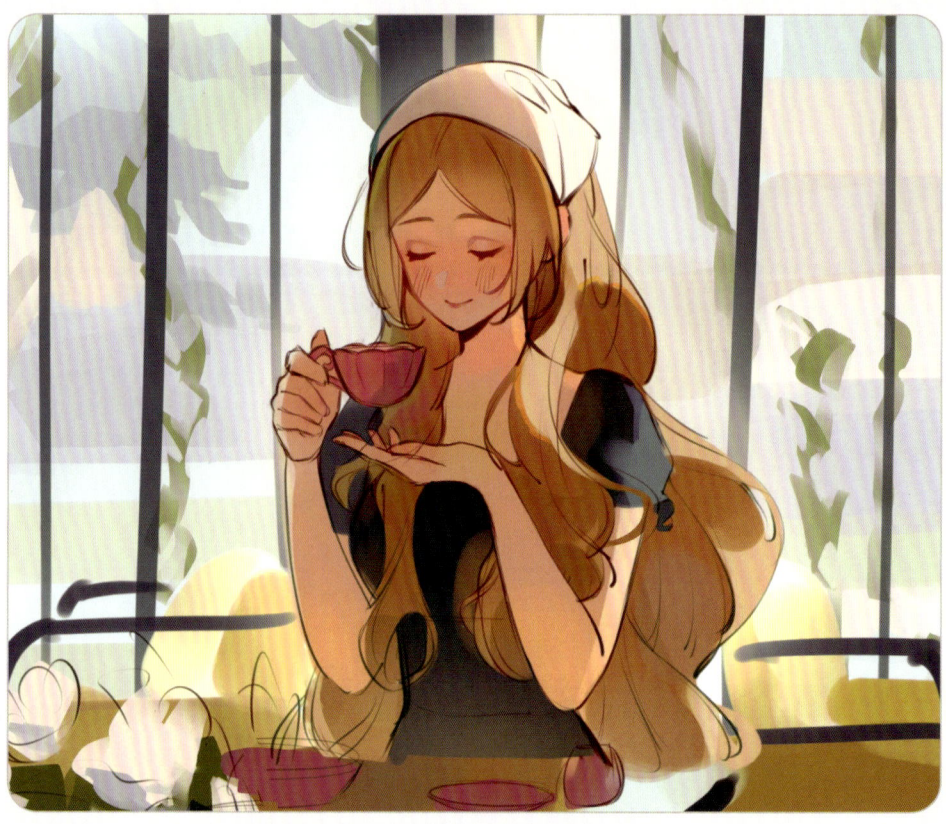

∧ Paint in airbrush shading on the character's face, including shadows and blush

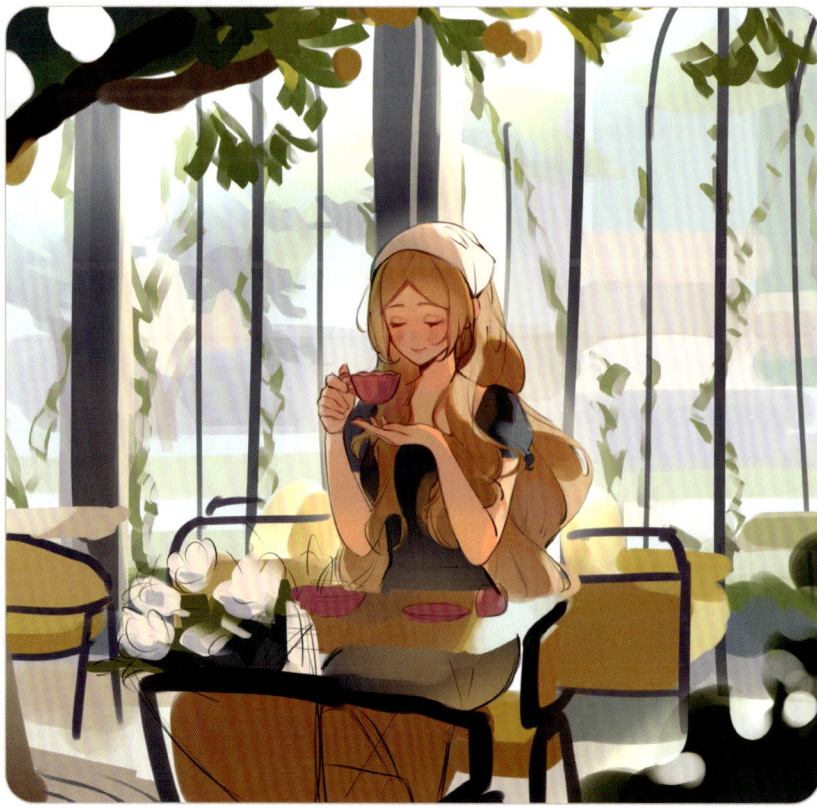

10 Create a new layer on top of the line-art layer and paint stronger shadows in areas most hidden from the light source. For example, on the character's neck and where strands of her hair cast a shadow over her face. Painting over the line art can give the image a more natural and interesting look, as it provides more detail and makes the shading more realistic. Next, clean up any messy sketch lines. You can also paint some lighter brushstrokes into the eyelashes to give them more detail.

< Add some bolder shadows to areas on her face where the light doesn't reach

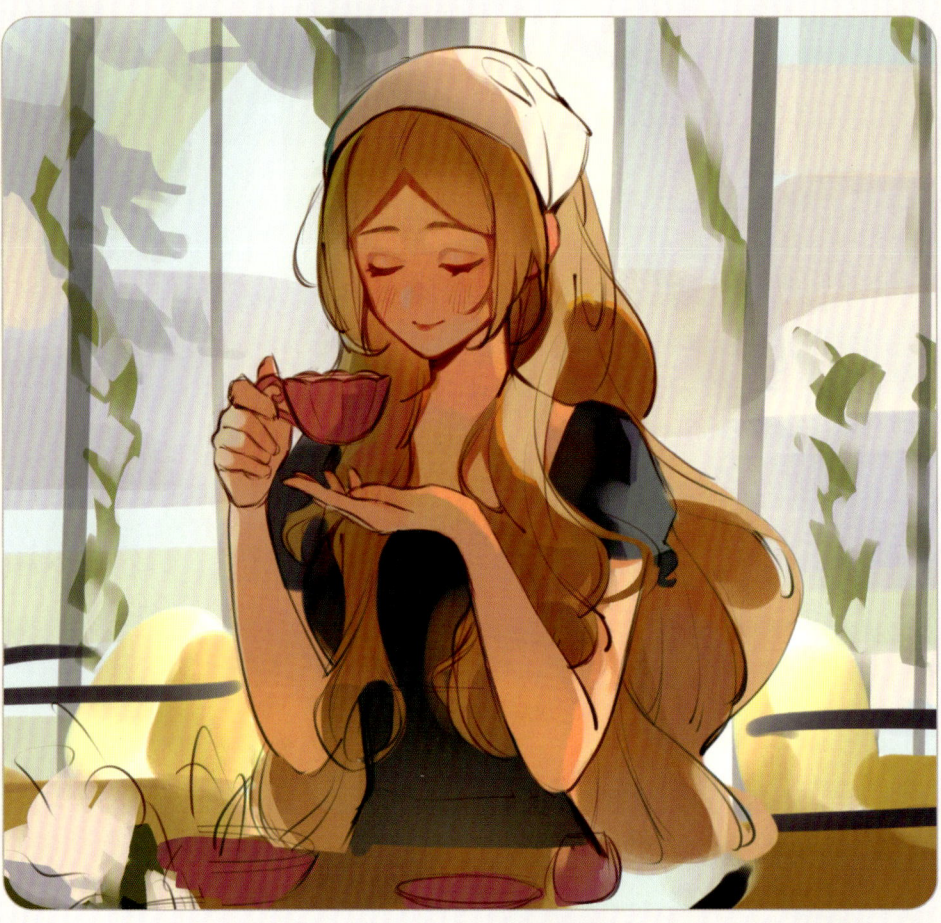

11 Once you're happy with the character's face, you can begin to paint the rest of the skin, keeping the light source in mind the whole time. Take a photo in the same pose with similar lighting to use as reference – this will help you to capture it in a realistic and believable way. Shading skin is simple when you do it step by step. Start by laying out the parts of the skin that are in light and shadow. Focus on the light for this step. The light hits the top part of the girl's fingers on her right hand from behind.

< **Keep the light source in mind as you paint the character's skin**

12 Following your reference photo, paint in some more shadow details. Try to understand the character's anatomy and how you can convey the shape of their features. Add a slightly darker colour at the intersection of where the light meets the shadow, and then blend it out on the darker side.

> **Use a smooth paint blending brush to paint in shadow details**

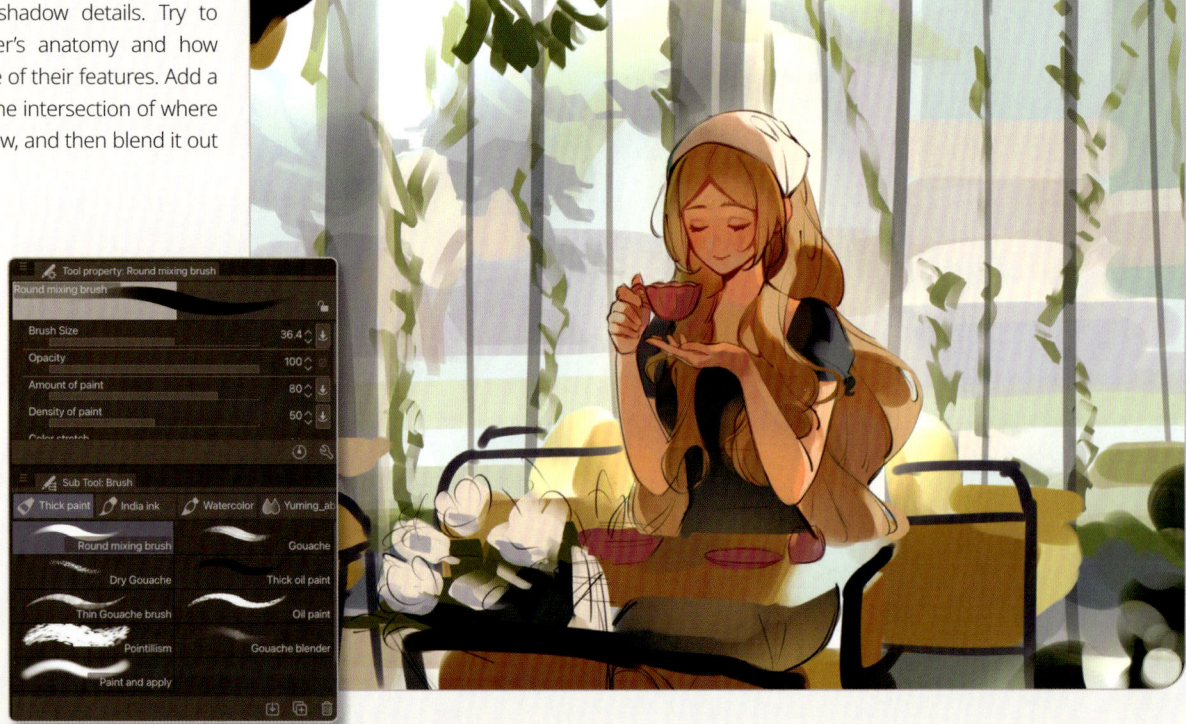

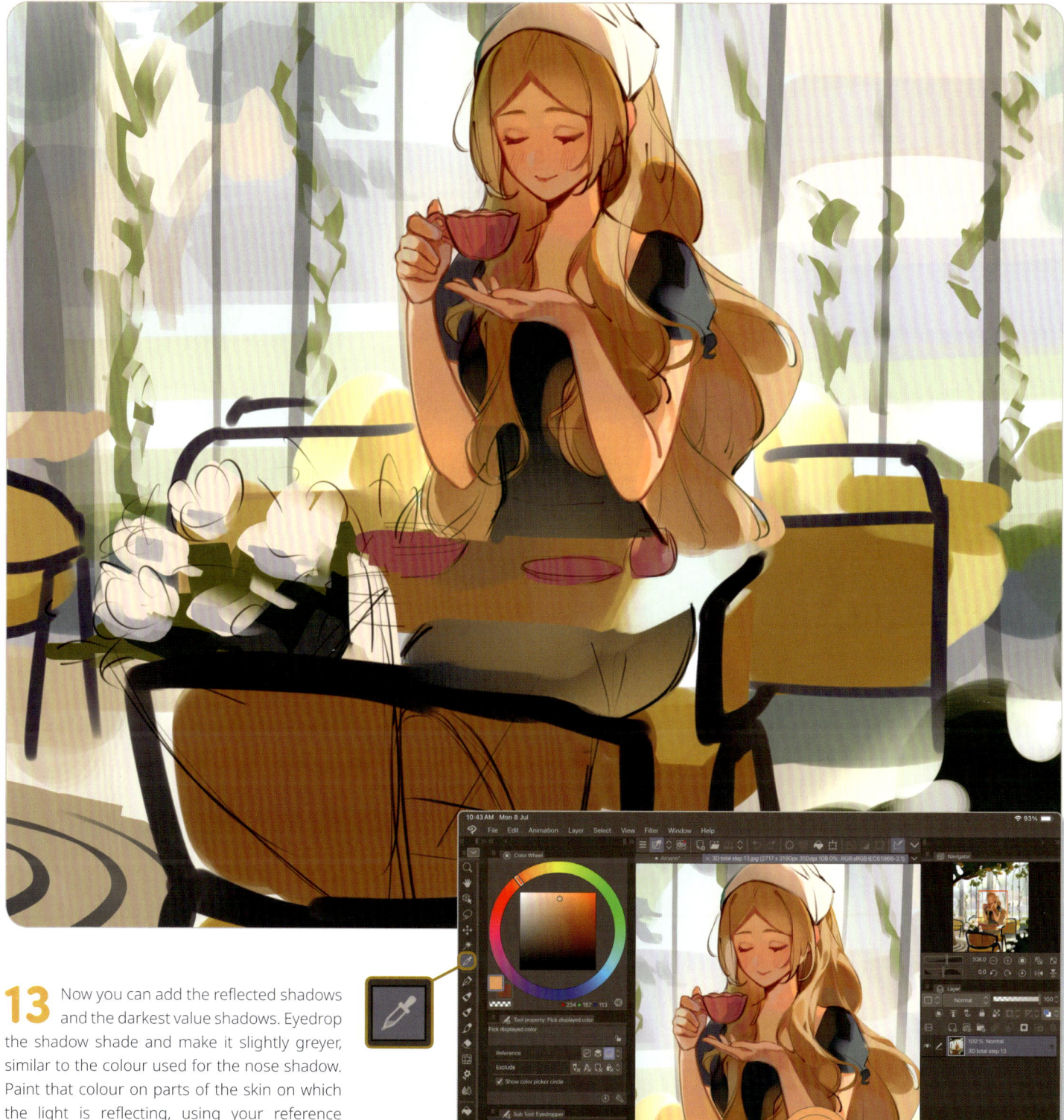

13 Now you can add the reflected shadows and the darkest value shadows. Eyedrop the shadow shade and make it slightly greyer, similar to the colour used for the nose shadow. Paint that colour on parts of the skin on which the light is reflecting, using your reference photo as a guide. You can also add ambient light from other objects. For example, where the hair touches her arm, paint in a slight yellow hue to emphasize their proximity. Lastly, select a dark brown and go over the outline to help blend it with the skin.

∧ Shade the skin, painting in reflected shadows and darkest value shadows

14 The next step is to render the character's hair. Start with the light section, adding smaller details and shadows. The flow of the hair is very important. Map out the direction of each section of hair strands on a new layer. At the same time, don't be afraid to be expressive and confident in your brushstrokes. Painting hair is where you can let your creativity flow freely! A clever trick for blending shadow and light is to paint a bright, saturated colour between them. Bright orange is used here.

> To create wavy hair, paint the hair flow in various shapes of the letter 'D'

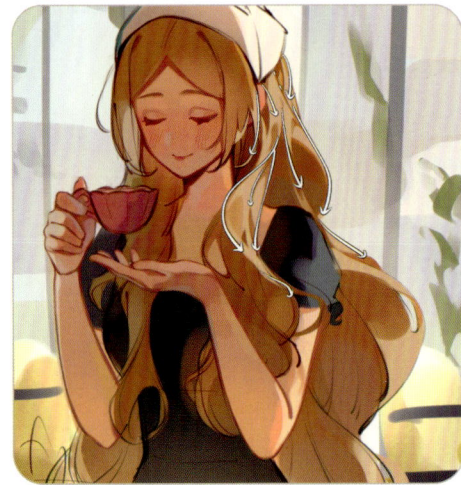
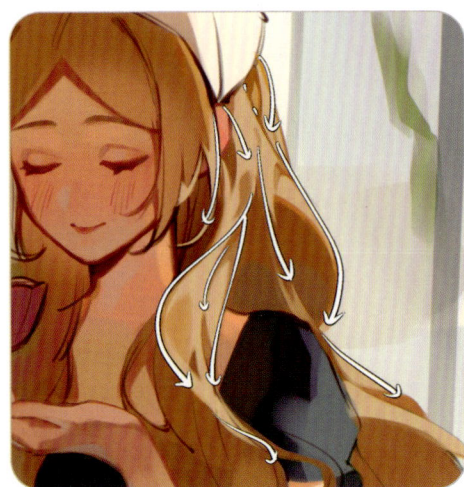

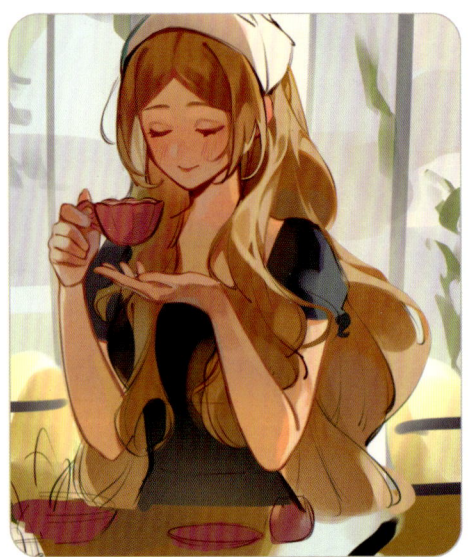

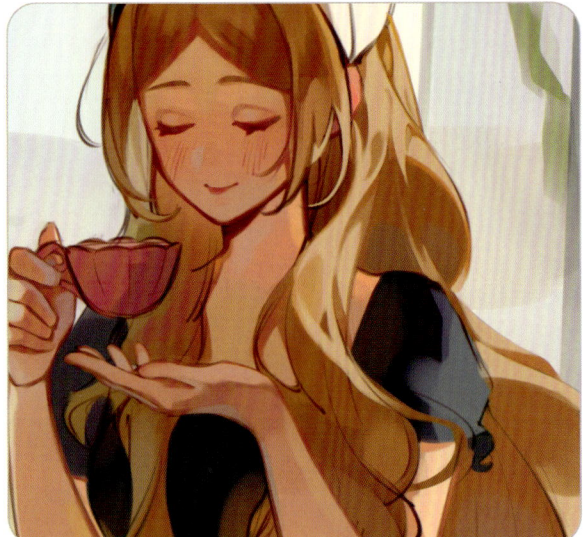

15 Next, focus on the shaded section of the hair. Using a dark shade to paint rough vertical strokes, while following the hair direction, can help to make the hair look shiny. What's more, using a balance of sharp and blended lines can make the hair look more realistic. Use a flat brush to achieve the thin shape.

< Paint in reflections, making use of complementary colours for added vibrance

16 Now it's time to introduce some reflections using the same technique. Paint a little blue where her frontmost hair tucks behind her ears. This will contrast nicely with the blonde as yellow and blue are complementary colours, sitting opposite each other on the colour wheel. Using complementary colours will make the colours stand out more and give the painting more vibrance. Don't forget to add in ambient light from other objects, such as the pinkish hues from her skin and the cup. Next, finish the hair by cleaning up the brushstrokes and adding in the darkest colours and outlines. Painting in a dark brown outline will help the hair to stand out.

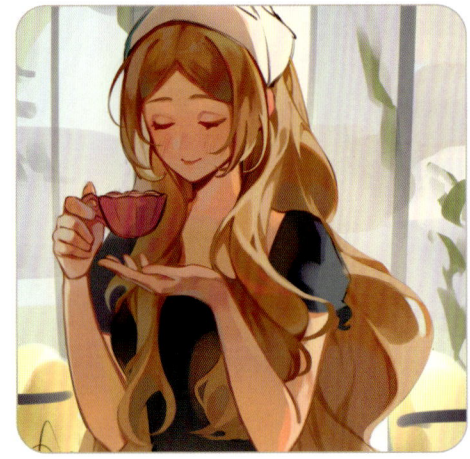

∧ Add some shine and realism to the hair

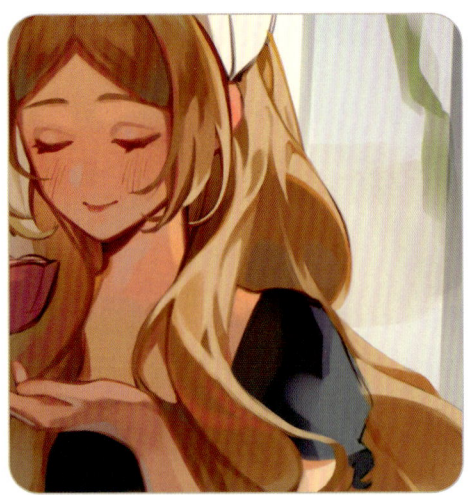

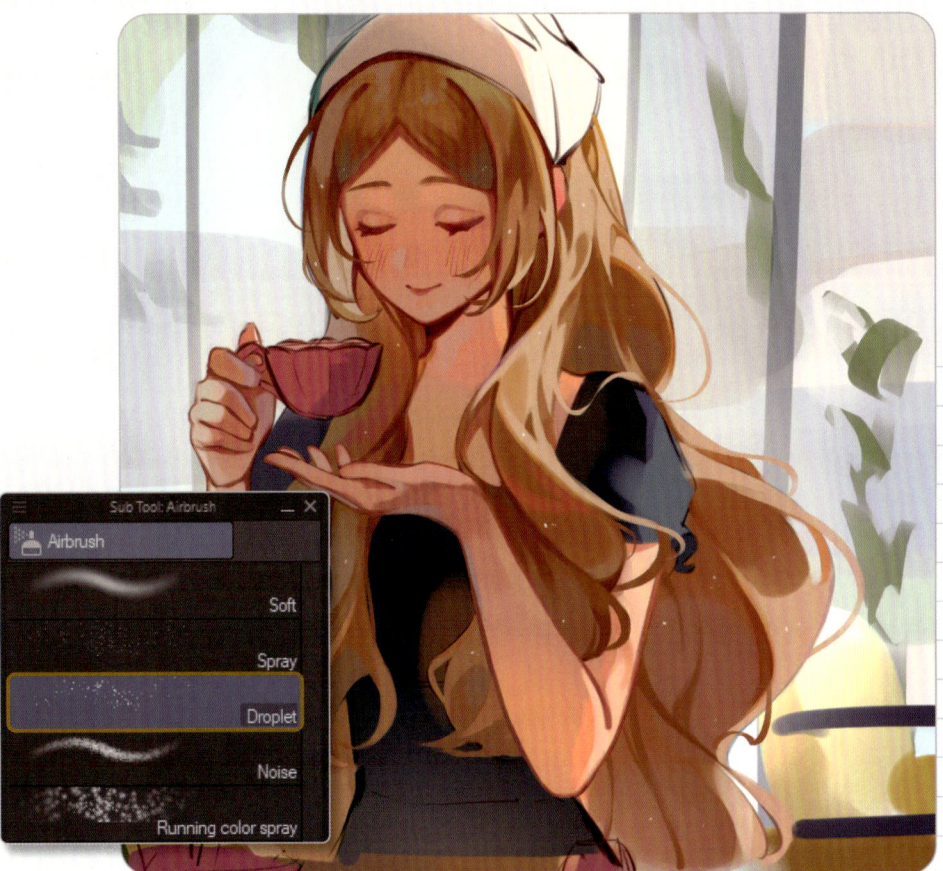

17 Finalize the hair by adding a few aesthetic details. Create a new layer and select **Tool > Airbrush > Droplet**, then paint white sparkle dots around the hair. Next, use various colours to paint some extra-vibrant outlines on the hair. Purple and blue are used here. Even though these are barely noticeable, they still add vibrance and uniqueness to the character. Lastly, use the **Flat brush** to paint in a few stray hair strands for added detail.

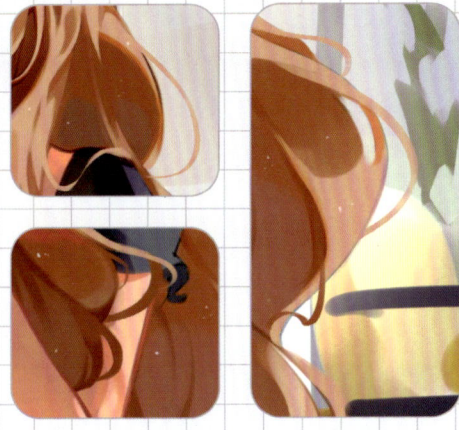

⌄ **Add the finishing touches to the hair, including coloured outlines and a few stray strands**

18 Paint in the accessories using the same techniques used for the hair and skin. The accessories don't need to be too detailed, as you want the character to be the main focus of the image. Again, keep the direction of the light source in mind. When there are multiple objects, it can sometimes become difficult to distinguish what's happening in the painting when you zoom out. To prevent this, use a white **Airbrush** on a **Glow** layer to lightly add some glow behind objects in the foreground. Here this effect has been added behind the cup and hand.

> **Paint a soft glow behind the hand and cup to prevent them from blending into the hair**

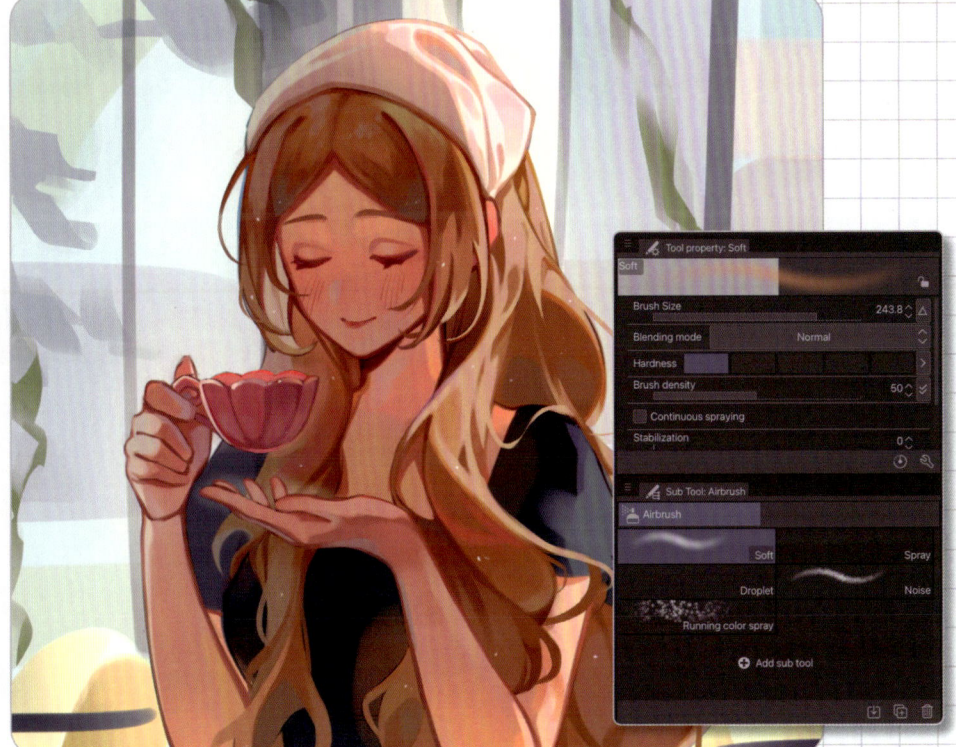

145

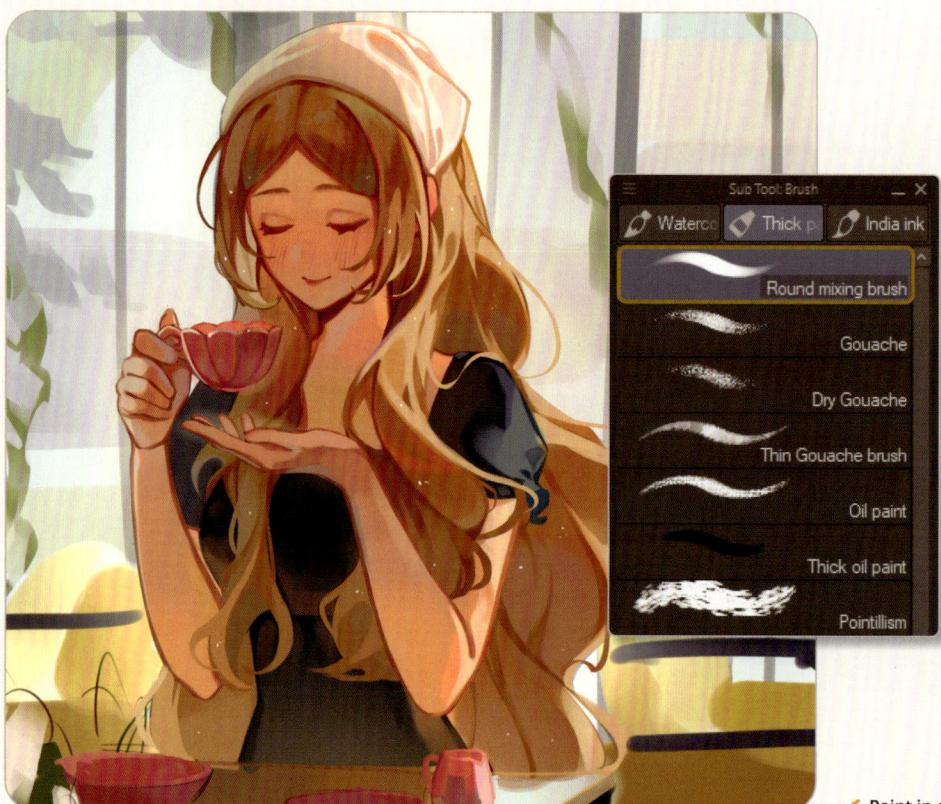

19 The next step is to render the clothes. Most of the character's clothing is cast in shadow, so there won't be lots of small details to paint. Sometimes less is more! Using a regular round brush, such as the **Round Mixing Brush**, paint in the lighter areas of the fabric. As you did with the hair, keep in mind the character's direction, shape, and anatomy. Contrast big, rough brushstrokes with finer, more controlled ones. A triangle shape is useful for conveying clothing that's creased or at an angle, like in the ruffles on her sleeves. To make it appear more realistic, add in some bounced light reflecting from the hair and table.

‹ **Paint in details on areas of the clothing not in shadow**

20 When painting shiny objects, the same technique can be used to create a realistic effect every time. Start by using an **Airbrush** to lightly show the direction of the light. Shade the dark and light areas and blend them with a transition colour (a mix of the two shades with a slightly stronger hue). Next, add a reflection colour from nearby objects. For example, paint the yellow seat cushions as a reflection on the black metal chair bars, and the reflection of the pink dishes on the smooth brown tabletop. Lastly, add in the highlights and outline.

› **Paint the reflection of nearby objects onto any smooth or shiny surfaces**

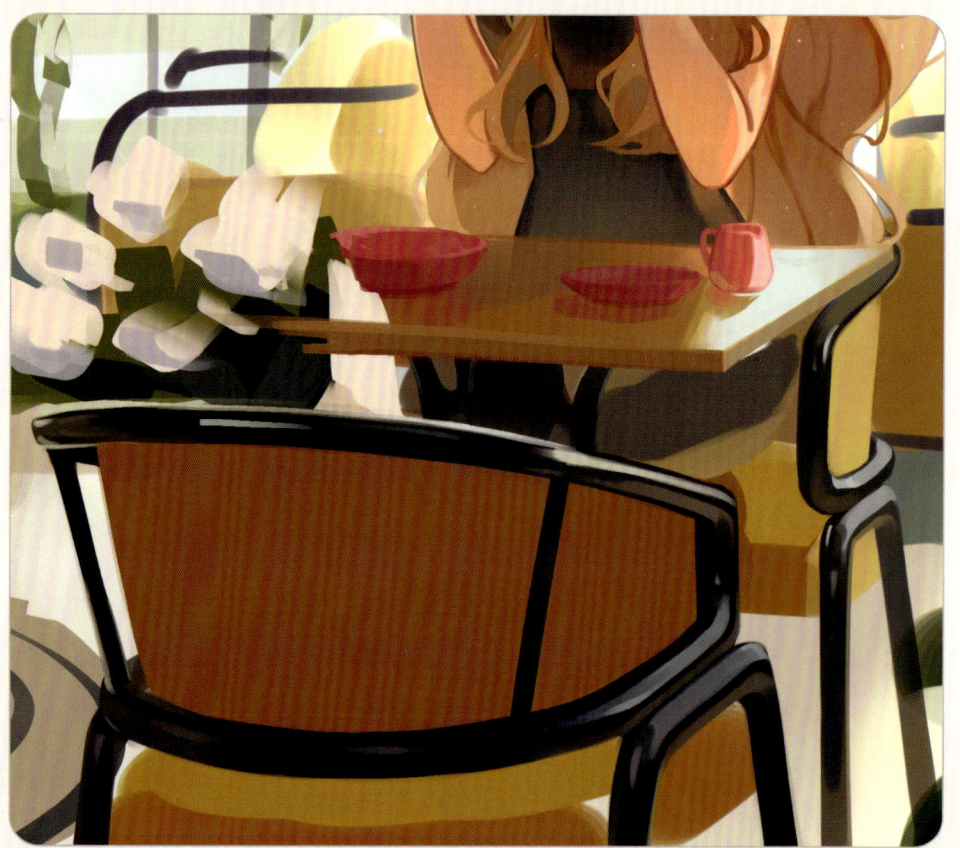

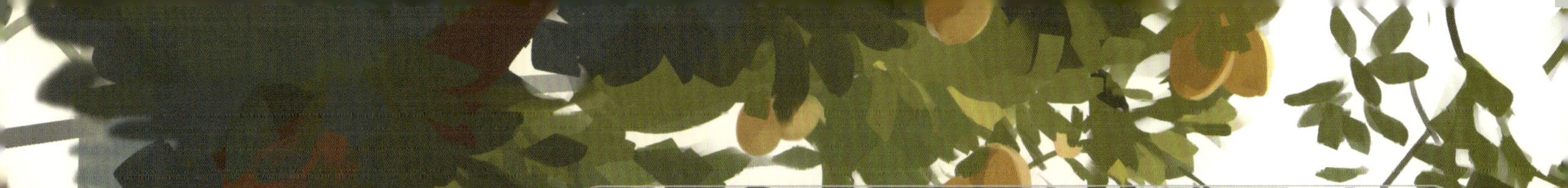

21 For the remainder of the background, use a **Thick Paint Flat Brush** to provide some traditional painting texture. For this lighting situation, objects closer to the viewer need a darker tone and deeper colour, whereas objects further away should be light and desaturated to create a sense of distance. The scenery outside the cafe will be blurred in the final step, so you don't need to pay too much attention to it here.

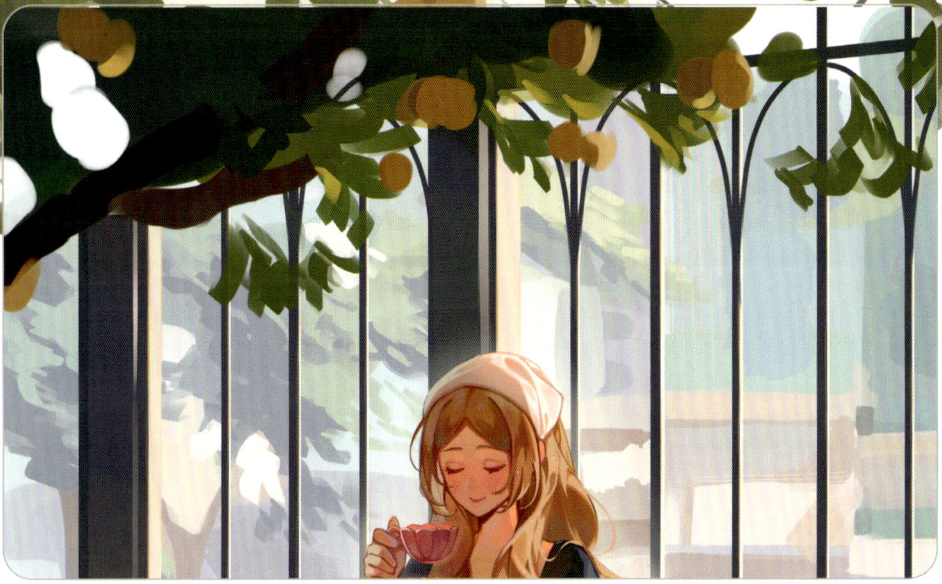

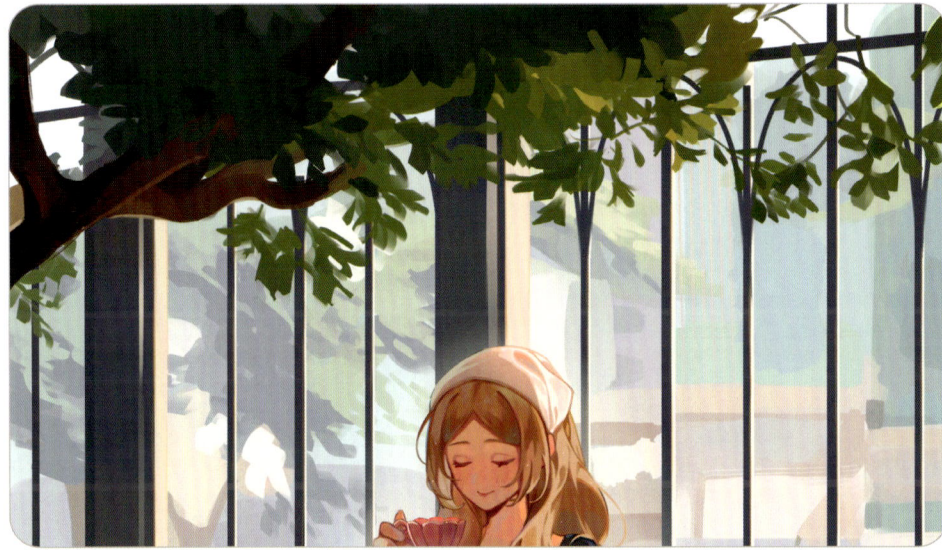

∧ Give the objects further away a light, desaturated colour, and those in the foreground a deeper, darker colour

22 Paint the leaves with a simple **Flat Brush**, using the addition and subtraction technique. This is where you add in brushstrokes, then use a transparent brush to erase some parts to create a more realistic and natural look. To make your brush transparent, simply select the clear colour box on the colour wheel window. The colours of the leaves should range from a warm green in the light, to deep green, and then to a blueish green in the shadow. Colour and tone shifts like these mimic real light environments, so make sure to change your hue as well when choosing a shading colour.

< Create dimension by using the addition and subtraction technique to alternate between broad and thin leaves

23 Add lemons to the tree above to contribute to the yellow theme of the illustration. Next, paint the flowers using a strong pink to match the tableware and help attract the viewers' eyes to the central focal point: the character. To paint rose-like flowers, layer horizontal brushstrokes to create the illusion of overlapping petals.

> Add lemons to reinforce the yellow colour palette, plus vivid pink flowers to draw attention to the centre of the painting

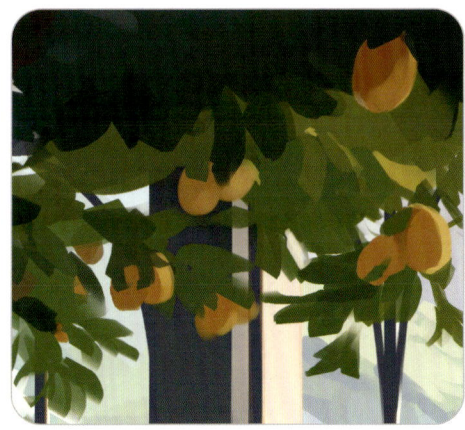

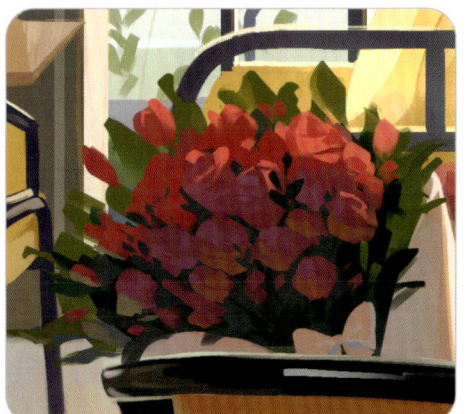

24 Blurring some areas of the illustration will create a sense of depth. Add a slight blur to objects in the far background (everything outside of the window), and the objects closest to the viewer (the leaves at the front that frame the image). To blur a layer, start by making sure it's not set to Lock Transparent Pixels, then select **Filter > Blur > Gaussian Blur** and select your preferred level of blur. For a finishing touch, paint in a glow behind the foliage by duplicating the layer, setting the layer to the **Add (Glow)** blend mode, filling it in white, and then applying a **Gaussian Blur** to the layer. You can also add some sparkles or floating specks of dust using a **Spray Droplet Brush**.

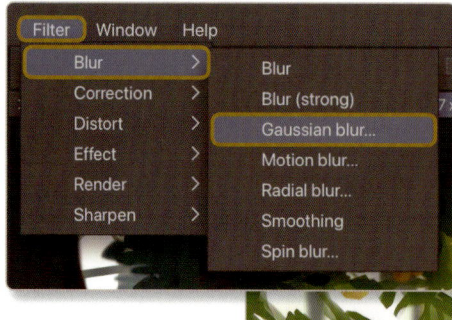

> Blur elements in the far background and extreme foreground to direct focus to the character

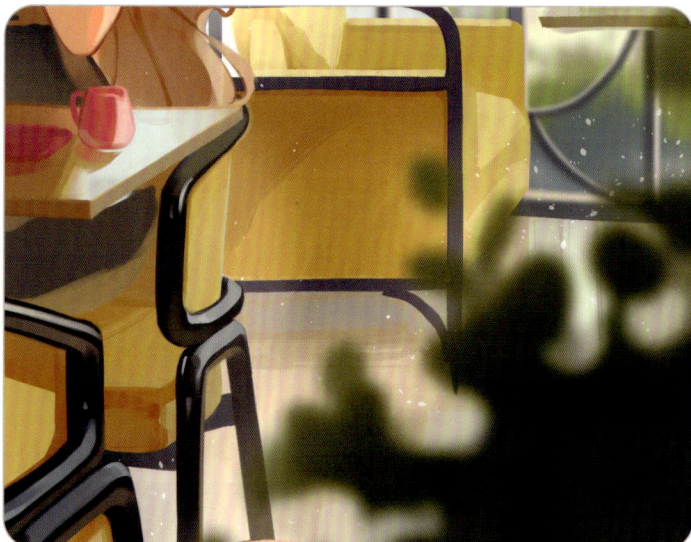

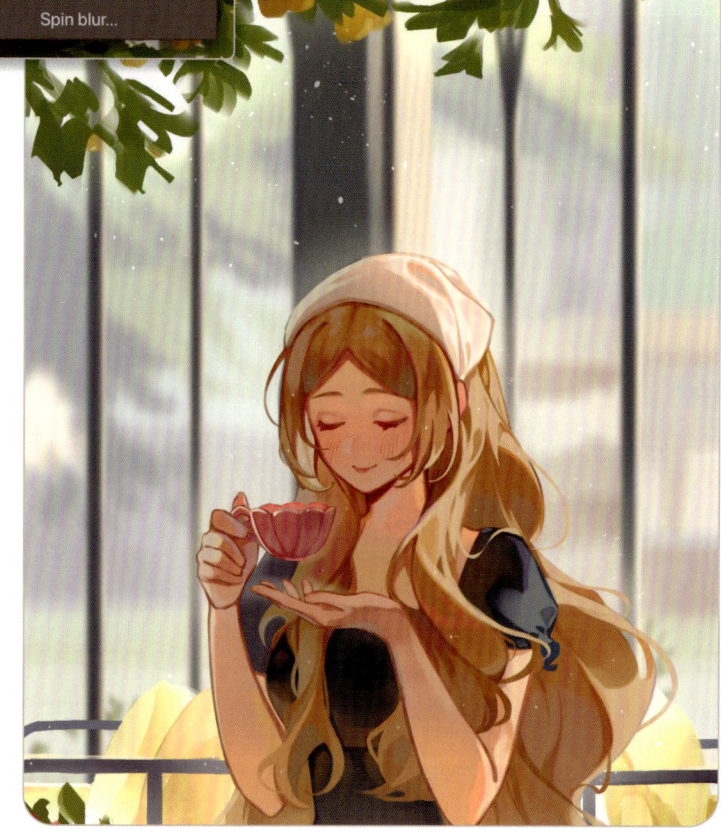

Conclusion

The final illustration is a gentle and warm piece intended to brighten the mood of the viewer. The story behind it is simply a manga girl enjoying her tea in the ambience of a lemon-themed glasshouse cafe. The illustration strikes an interesting mix of two contrasting elements: leafy natural foliage and the manmade furniture and architecture. You can now use the tools and techniques you have learnt in this tutorial to create your own character-focused scenes in Clip Studio Paint. Take photos of people and environments to use as references, then see how you can use Clip Studio Paint's colour and lighting tools to influence the mood and emotion of the scene.

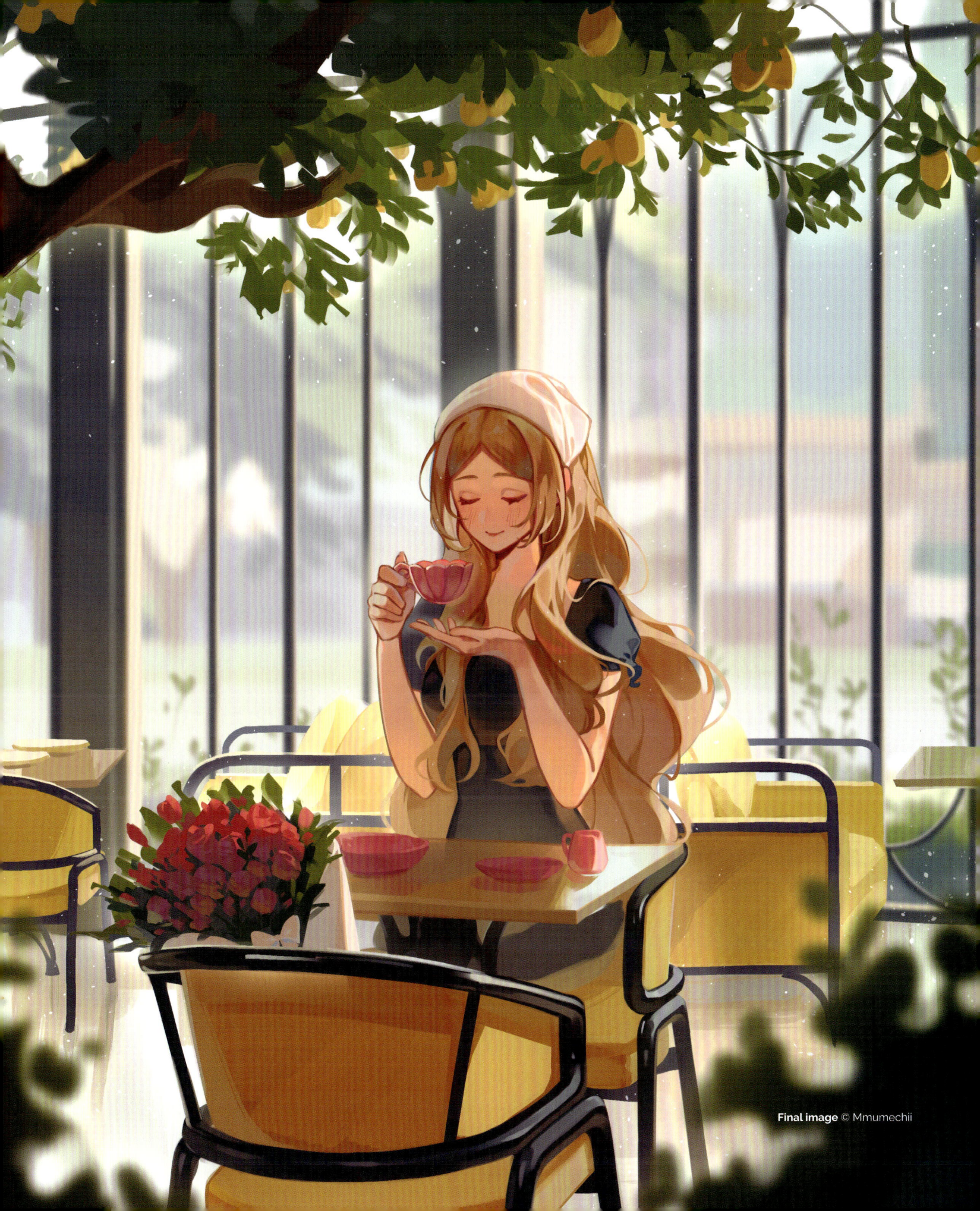

Final image © Mmumechii

Gallery

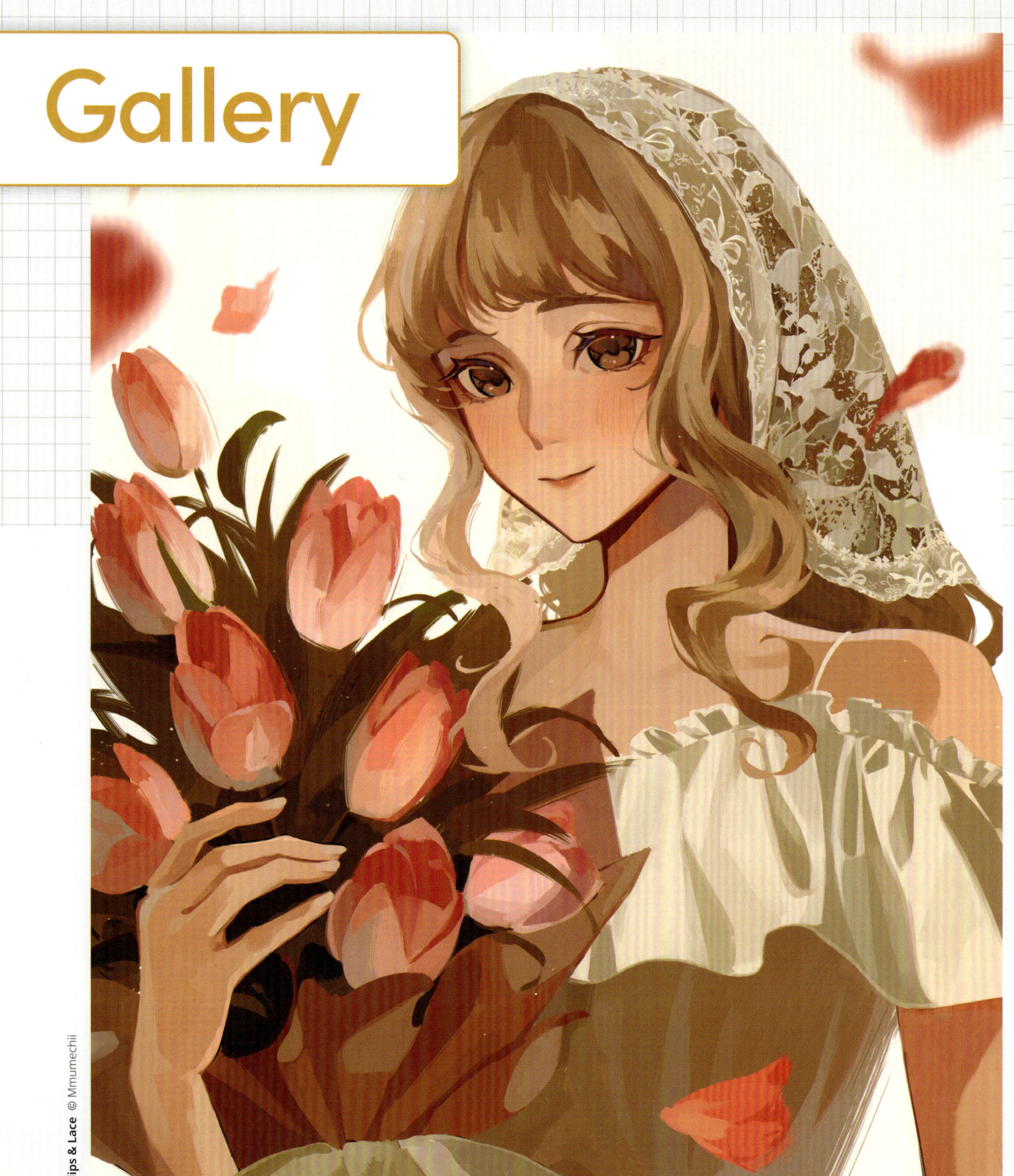

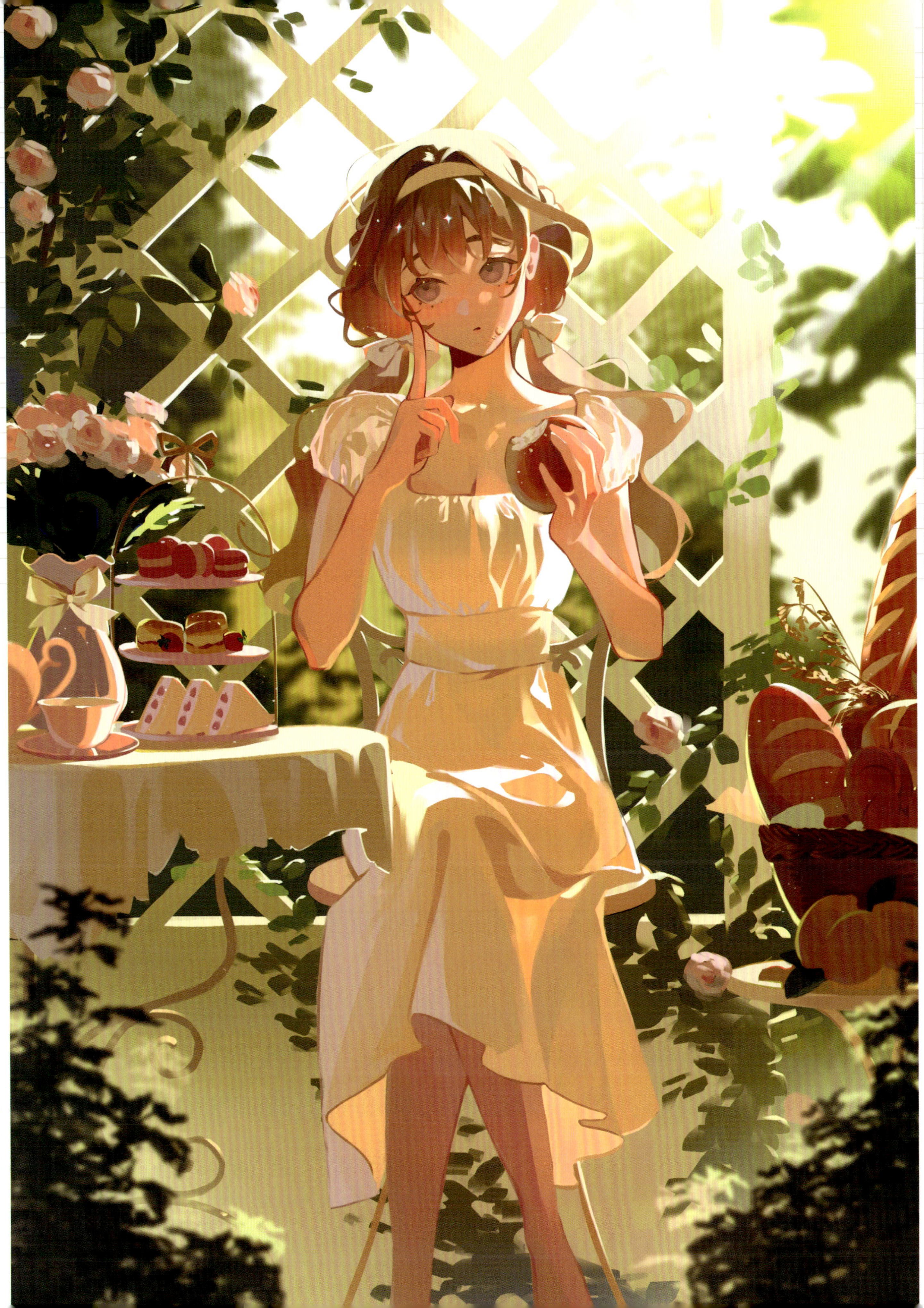

Noelle's Garden Tea Party © Mmumechii

Fantasy creature

BY STEFAN KOSTIC

Painting from life and trying to recreate a subject as faithfully as possible is one of the best ways to improve in your artistic practice. At other times, it's a great feeling to be able to let your imagination run wild and create something new and fantastical. Maybe something never seen before! My tutorial will focus on just that. I will take you on a journey of imagining a new type of a fantastical creature, while addressing the technical aspects of illustrating as well.

Step by step, you can follow the process of creating a fantastical creature from start to finish. Beginning with an exploratory thumbnail and rough sketch, the tutorial will progress through to line art and colour block-ins, and then on to a fully coloured and rendered illustration... all from your imagination! I will also share the various useful tools Clip Studio Paint has to offer. By the end of the tutorial, you will have learnt how to progress through a simple creature-design workflow, including how to transition between steps, how and where to apply texture, what brushes to use, how to use blending modes, and a host of other tips and tricks.

Working through this tutorial will serve as a strong foundation as you move on to tackle future digital paintings in Clip Studio Paint. With approaches and techniques that are applicable no matter the subject, it will take you a step further into the magical world of digital painting.

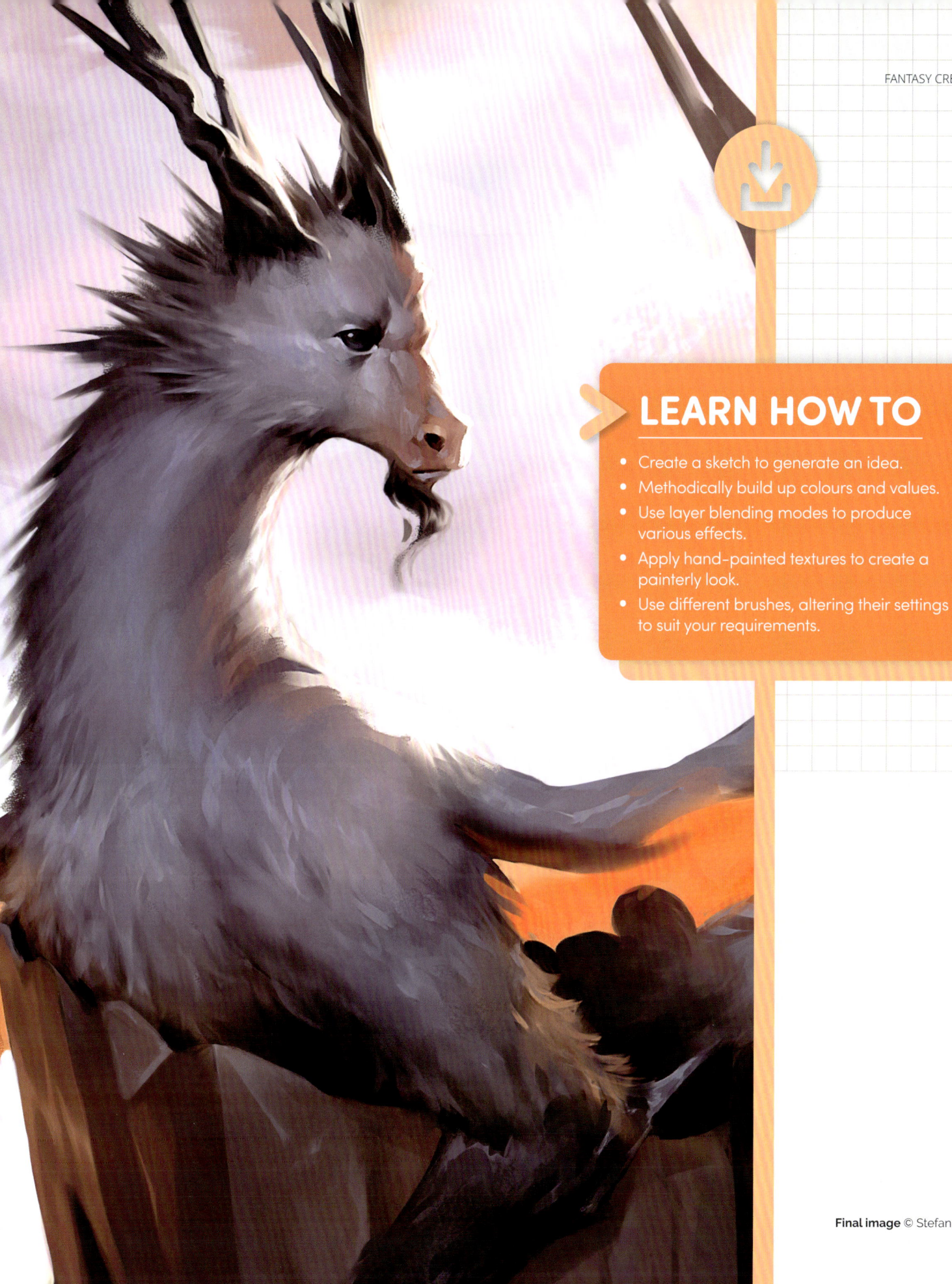

LEARN HOW TO

- Create a sketch to generate an idea.
- Methodically build up colours and values.
- Use layer blending modes to produce various effects.
- Apply hand-painted textures to create a painterly look.
- Use different brushes, altering their settings to suit your requirements.

Final image © Stefan Kostic

01 Everything starts with an idea. Take some time to think about the type of fantasy creature and the composition of the scene you wish to create. Next, choose a brush for sketching, such as the Pencil Brush, selected by pressing **P** twice. Alternatively, if you want cleaner and less grainy lines, try the Round Mixing Brush. This is selected by pressing **B** and then selecting the Thick Paint tab. Next, choose a dark grey, almost black, colour and start sketching. Be loose and focus on the big, simple shapes of the creature first. Only indicate where the various elements will go – don't sketch in any detail just yet.

> Sketch rough simple shapes that establish the composition and proportions of the creature

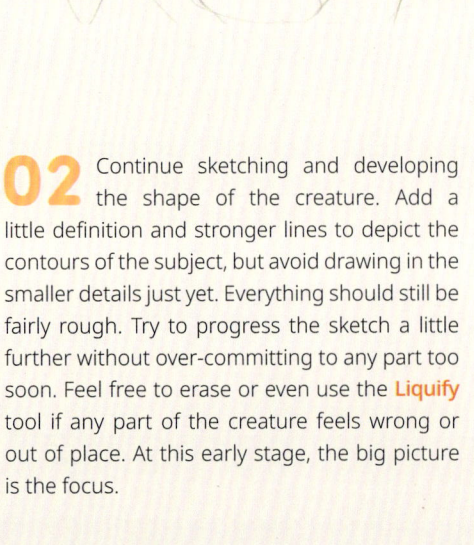

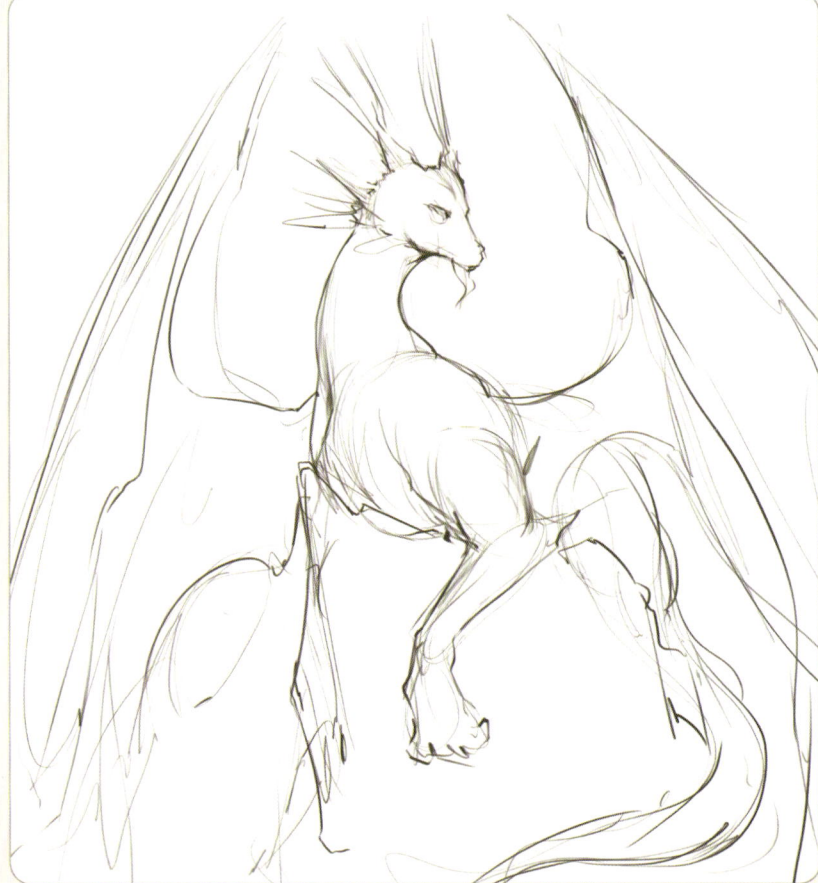

∧ The sketch is still pretty loose, but a clearer idea is slowly emerging

02 Continue sketching and developing the shape of the creature. Add a little definition and stronger lines to depict the contours of the subject, but avoid drawing in the smaller details just yet. Everything should still be fairly rough. Try to progress the sketch a little further without over-committing to any part too soon. Feel free to erase or even use the Liquify tool if any part of the creature feels wrong or out of place. At this early stage, the big picture is the focus.

03 Now that you have a rough sketch, it's time to add some more defined details. First, select the sketch layer you've been drawing on and reduce its opacity to about 40–50%. You can do this by selecting your layer, then sliding the slider above the layers to the left. Second, create a new layer above the sketch and select it by selecting the New Raster Layer icon above the layers. Using the transparent sketch as a guide, you can now trace over your rough sketch with cleaner, stronger lines. Continue to add medium-sized details, such as the horns and beard, plus some anatomy, like muscles on the arm.

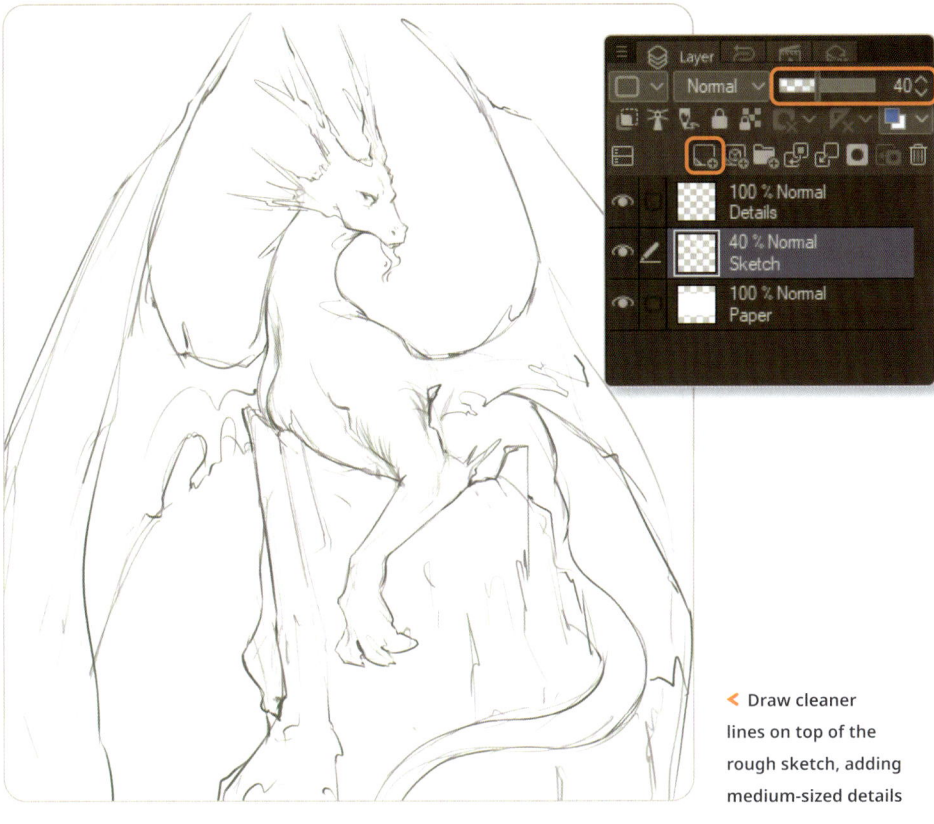

< Draw cleaner lines on top of the rough sketch, adding medium-sized details

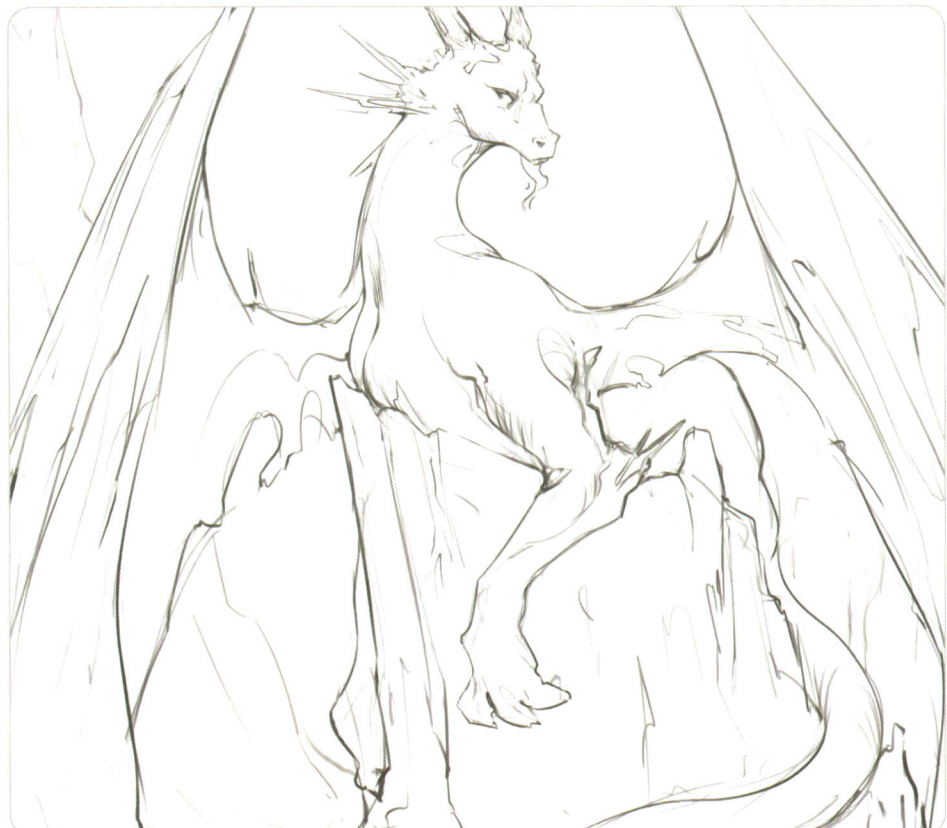

04 With the drawing of the dragon slowly coming to life, it's time to refine and finalize the sketch. Merge the two layers from the previous step by selecting the top layer and pressing Ctrl+E, or selecting Layer > Merge with layer below. Continue drawing on that merged layer. You can now sketch in even the tiniest details. Feel free to zoom in and draw every little element that requires defining. Keep in mind that proportions and the overall composition still need to work, so if something feels off, use the Liquify tool to adjust it, or erase and draw it again. Lastly, create a layer below your sketch on which to draw an indication of the background behind the creature.

< Finish the sketch, adding detail to the creature and roughing in a simple background

05 Now that the sketch is finished, it's time to prepare it for painting. Start by selecting your sketch layer and then clicking on the Lock Transparent Pixels icon located above the icon for creating a new layer. This will allow you to paint directly onto the sketch layer, but only on already established pixels. Pick a warmer, reddish colour for this; warmer lines are much easier to incorporate into the painting than pure black ones, as they are more of a natural shade. This will change the colour of the sketch, but otherwise keep it intact. Change the sketch layer blending mode to Multiply. This will ensure that your sketch is always visible, no matter what is painted underneath. Repeat the same process for the background sketch layer.

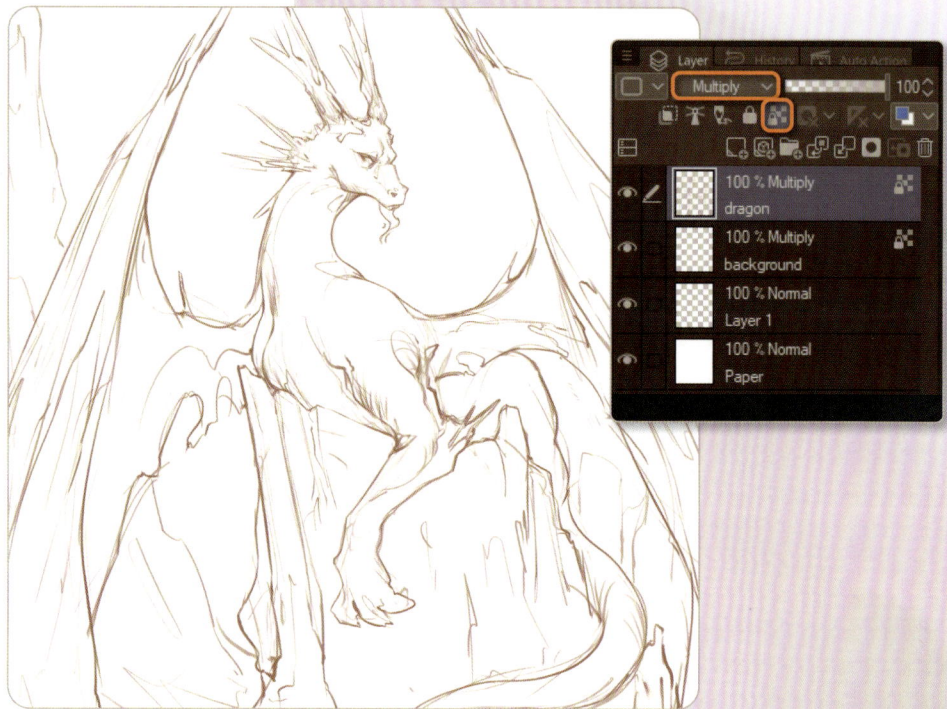

> **Prepare the sketch for painting by introducing a warmer colour**

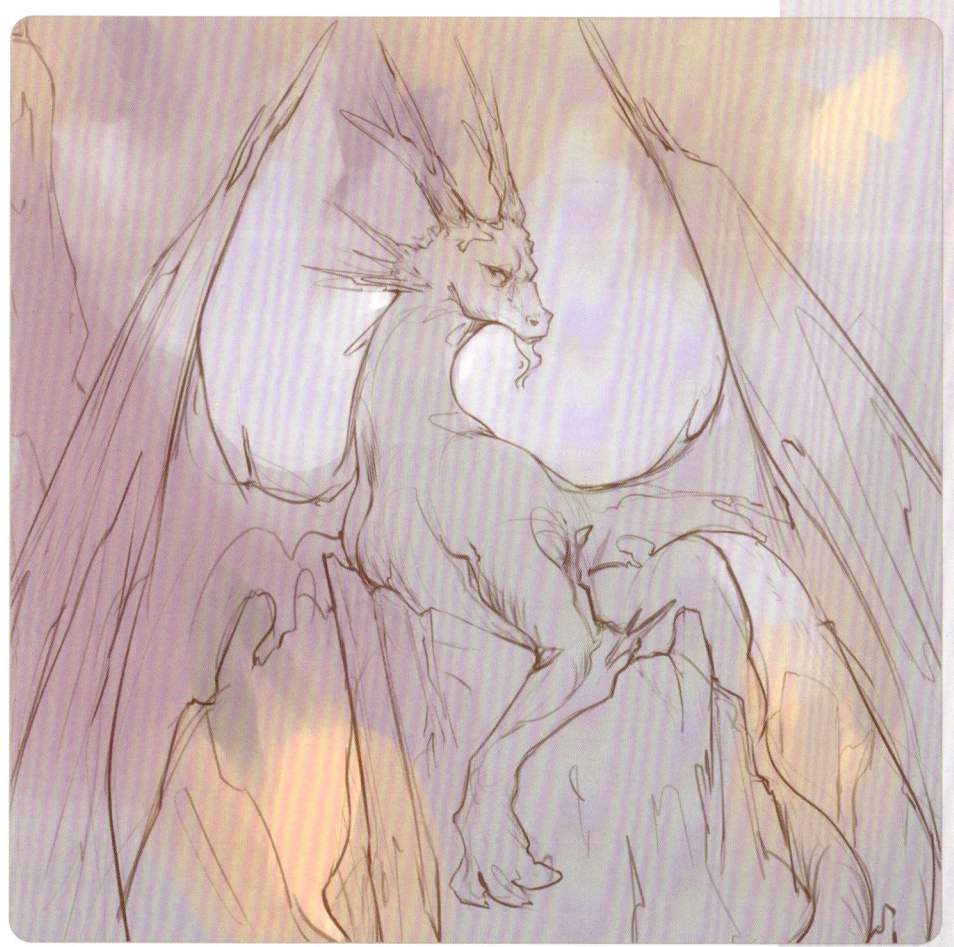

06 Create a new empty layer below all your sketches – this is where you will paint. Select that new empty layer and block in the background with a soft, textured paintbrush. You only need to paint the sky at this stage, as it's the most distant element. Paint an array of colours of varying temperatures, starting with natural sky-blue tones, before slowly introducing yellows, oranges, and even purples for some darker areas.

< **Block in the sky background with a variety of colours**

07 Now it's time to block in other elements, including the mountains, the rock, and the dragon. Start with the mountains in the background by creating a new layer on which to paint them. Select the Lasso sub tool by pressing **M** or by selecting it from the **Tools** bar. The **Lasso** sub tool is used to create clean selections, so that you can easily block in your elements. Use it to draw a selection for the mountains, following the sketch lines. When you have a selection, choose a lighter colour and paint inside that selection. Once you're finished, press **Ctrl+D** or select the **Select > Deselect** menu option to remove your current selection. Repeat the same process for the rock and dragon, each on its own layer.

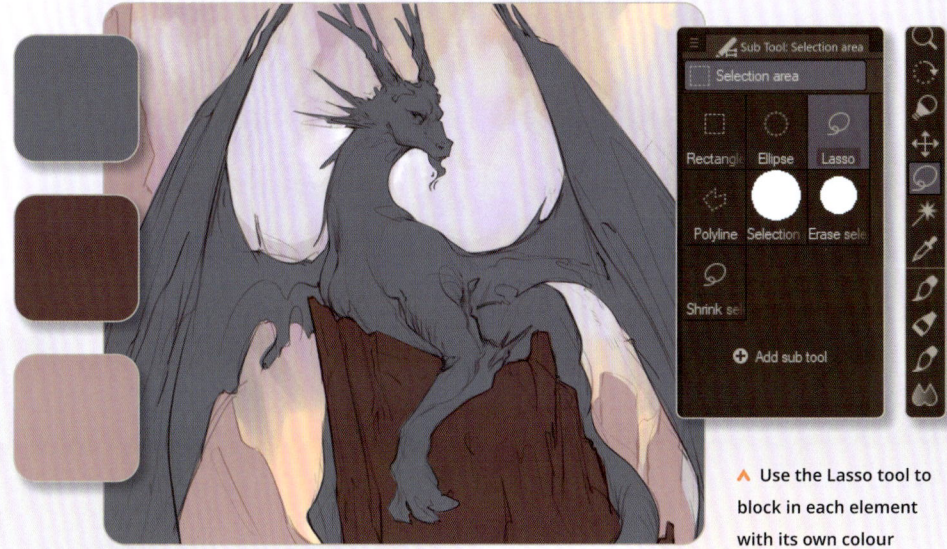

∧ **Use the Lasso tool to block in each element with its own colour**

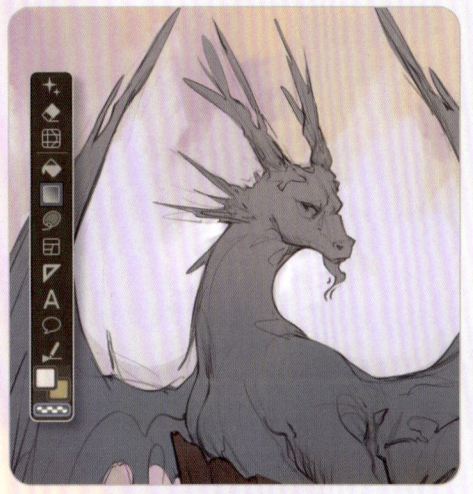

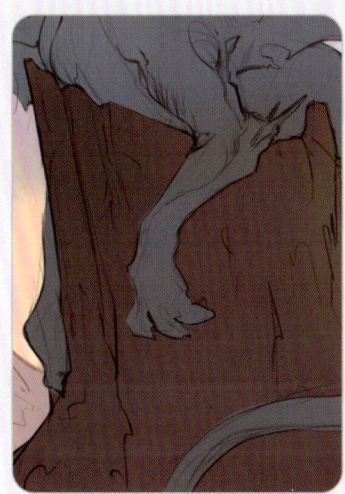

08 Ensuring you have separate, clean shapes for each element will make the painting process much easier, but at this stage they look a little flat and unnatural. To improve this, select the **Gradient** tool by pressing **G** or selecting it from the **Tools** bar. First, select the layer with the dragon shape and lock its transparent pixels. Select a slightly lighter and warmer colour, then paint a gradient from top to bottom. This will indicate that the light source is above. Repeat the process, this time from the bottom up, signalling some reflected light bouncing off the ground. Next, do the same for the rock, except this time paint a lighter gradient coming from the right and a darker one from the left.

< **Paint in gradients for each of the main elements to indicate lighting**

09 Not all parts of the dragon should remain the same colour and value. Some parts should be darker, while other parts need to be lighter. To adjust the colours, select the dragon shape layer, then pick a dark colour and a soft round brush, all while having **Lock Transparent Pixels** activated. Go over areas that should be darker, such as his front feet, horns, tail, wings, eye, and beard. This will create a nice transition, while also drawing focus to certain parts of his body. If you don't have a soft brush, you can easily make one. Simply select the **Round Mixing Brush** and reduce its hardness in the **Tool Property** palette.

> **Paint darker colours on various parts of the dragon's body**

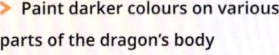

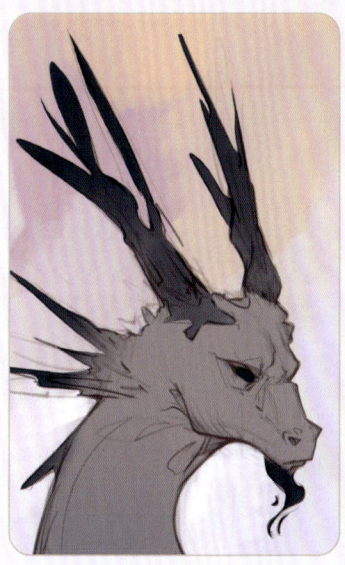

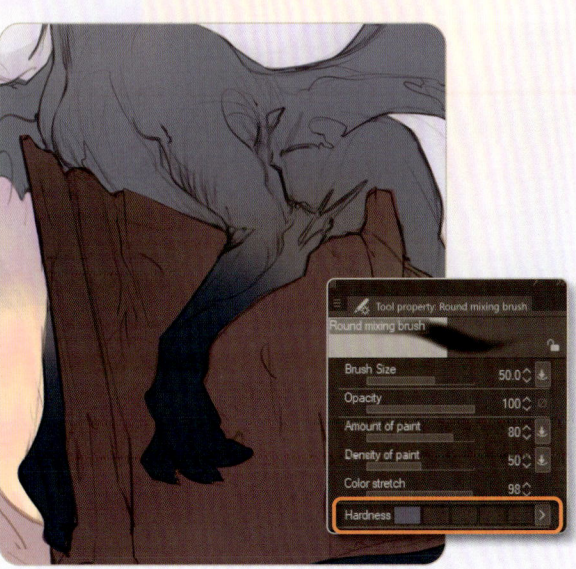

10 In the previous step you painted in the darker, shadowed areas. Now it's time to introduce some strong lighting coming from the side to accentuate the dragon's silhouette and create a visually pleasing contrast. On the same layer as the previous step, select a very light yellowish colour and, still using a soft brush, paint that colour onto the right side of the dragon's form. This simple but effective change will add depth and dimension to the creature.

> ## ARTIST'S TIP
>
> When working on a digital painting for a long time, it can be hard to spot mistakes, as you get used to seeing the image in a certain way. Flipping your canvas horizontally is a crucial tool for mitigating this problem. When you flip the canvas, you trick your brain into seeing an image as if you're viewing it for the first time. This allows you to spot errors you might not have seen previously. To flip your canvas, select View > Rotate/Flip > Flip Horizontal. Getting into the habit of flipping your canvas regularly will enable you to identify and fix mistakes sooner rather than later.

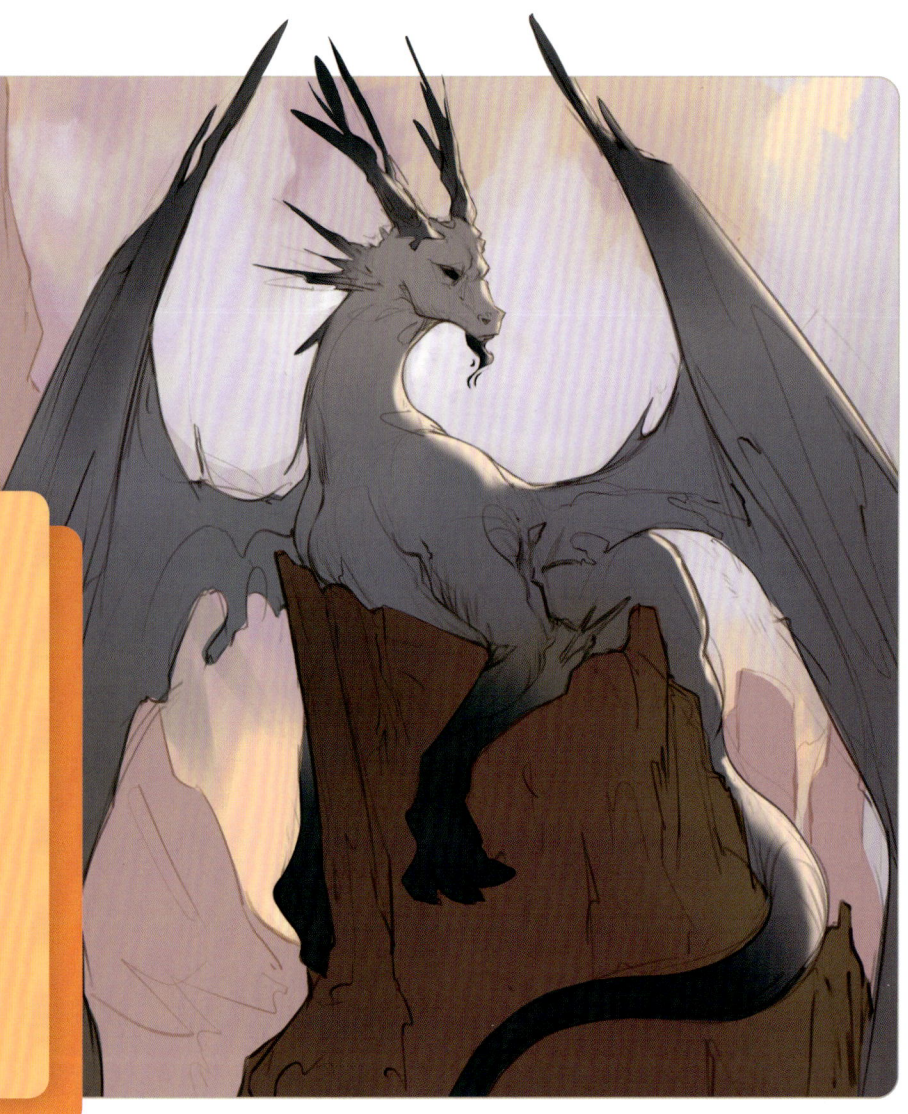

∧ **Paint in lighting coming from the side of the image**

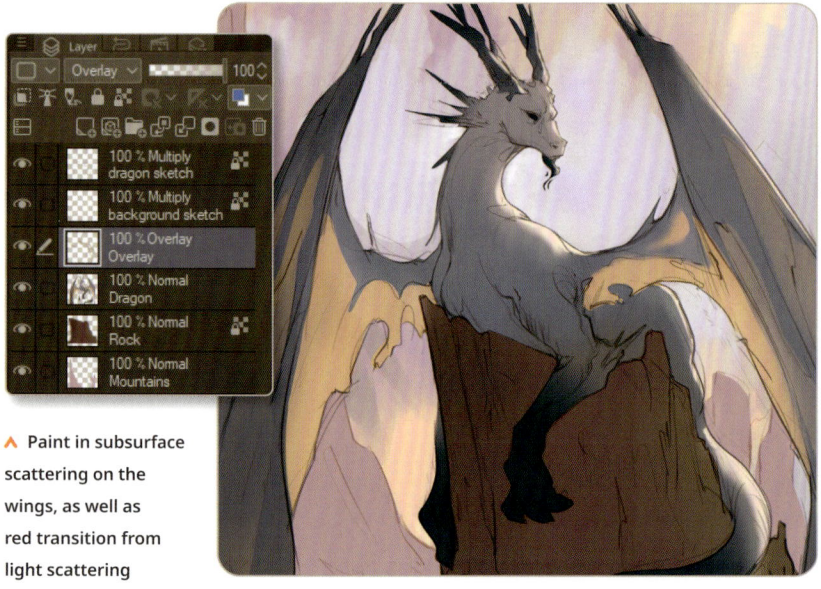

∧ **Paint in subsurface scattering on the wings, as well as red transition from light scattering**

11 At this stage you can introduce some more colour, as well as begin to indicate the various textures in the scene. Dragon wings are generally very thin and made mostly of skin, therefore a lot of light will be visible passing through them. This is called subsurface scattering and including it in your design can bring in a lot of interesting contrast. Remaining on the same layer, choose a light saturated-orange colour to paint the insides of the wings. As the dragon is cooler in colour and largely in shadow, this little change will add an appealing colour interplay. Additionally, since the light scatters when hitting the surface, you want to capture that as well. Create a new layer on top of the dragon shape layer and set it to Overlay blending mode. Using a mid-value orange colour with a low saturation and a soft brush, glaze over the areas hit by light, slowly reaching into the shadows as well. This creates a natural reddish colour transition. After adding this effect, merge the Overlay layer back into the original dragon shape layer.

12 Now it's time to add some depth to the dragon, namely the ambient occlusions in the shadows. Remain on the dragon layer with locked transparent pixels. Select a much darker but also warmer colour and slowly paint the deep pockets of ambient occlusion, specifically the darkest parts on the dragon's form. This shadow type mostly occurs in crevices and areas that are fully covered, meaning very little light can enter. Such areas include below his chin, front legs, wings, and stomach, since he's lying on a rock. Conversely, the top side of the dragon is directly exposed to the sky and overhead light source. This will reflect light onto those top-facing areas, meaning they need to be a slightly lighter shade of blue. Choose such a colour and paint this in.

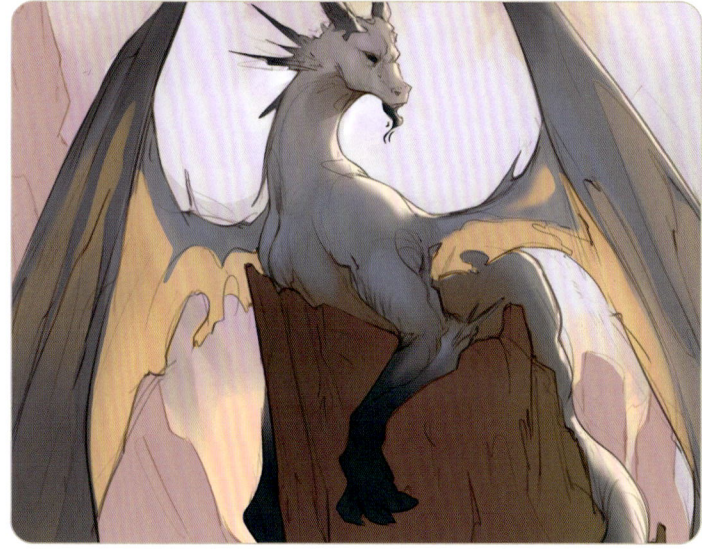

> Paint in the darker, shadowed areas of the dragon, as well as those parts that reflect light from the sky

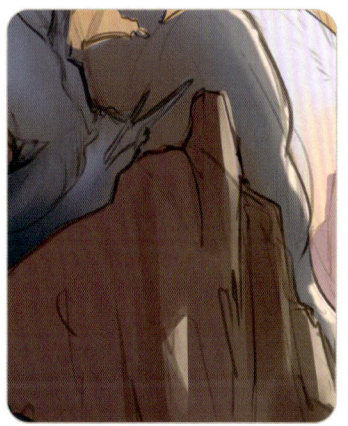
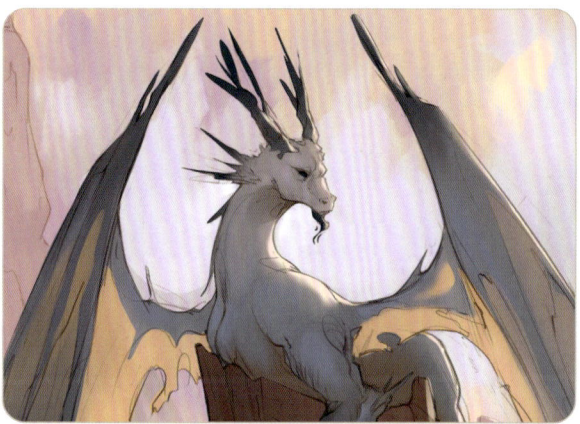

∧ Add light and ambient occlusion to the rock and mountains

13 Now you've added lighting information to the dragon's form, the creature is starting to stand out from its surroundings. Take some time to add the same lighting and ambient occlusion to the central rock, as well as the mountains in the background. Repeating the process from the previous step, paint light onto the rock and mountains – coming from the same direction as it hits the dragon – and then do the same for ambient occlusion. The mountains furthest away should be much lower in contrast due to atmospheric perspective, so keep their lights quite similar in value to their shadows.

14 You have now established a good starting point for rendering. Value, light, and colour information are all present, so you can use this as a base to continue working on the details. Since rendering requires painting numerous parts at the same time, you may wish to merge any layers that describe the same thing. For example, move the dragon shape layer up so it's below its sketch, then select the dragon sketch layer and merge them together. Repeat the same process for the rocks and mountains. This will leave you with four distinct components on which you can work separately: dragon, rock, mountains, and sky.

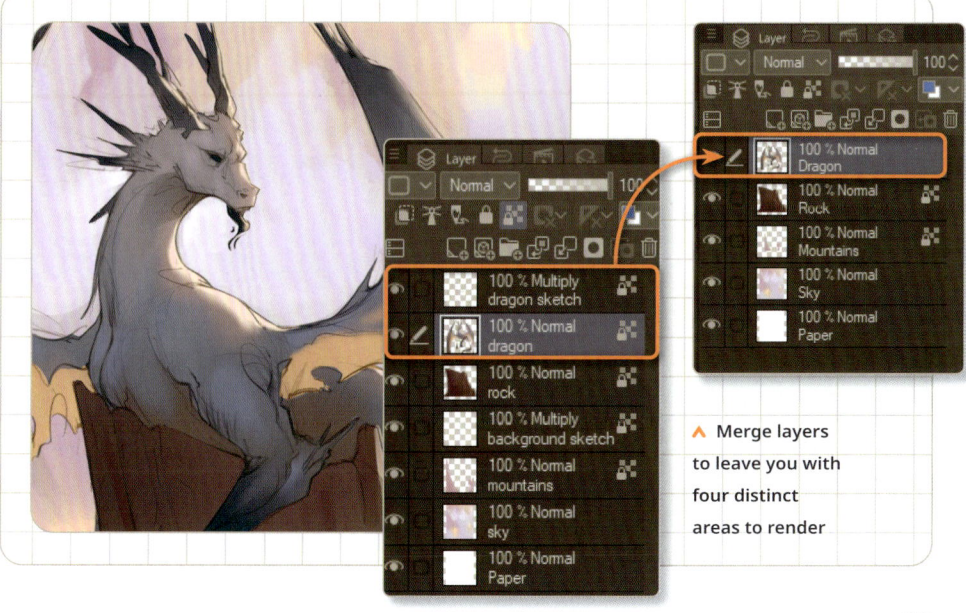

∧ Merge layers to leave you with four distinct areas to render

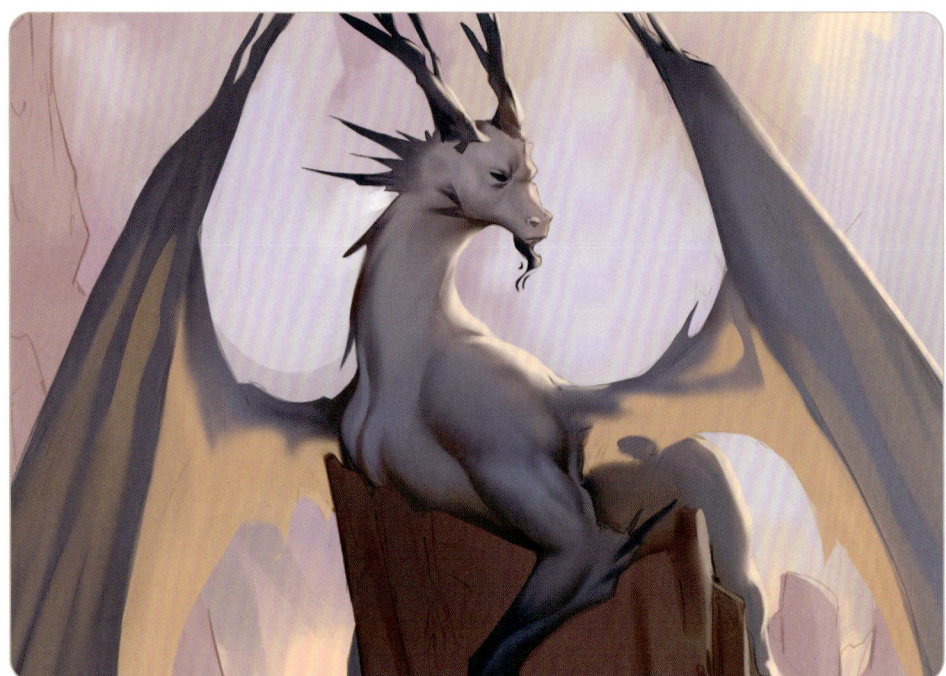

15 Before you start to render, you might want to assimilate the line work into the painting. Currently, the lines are doing much of the heavy lifting, carrying the shapes and enabling overall readability. To remove them, select a soft brush and paint over the lines with a darker colour. In a sense, lines represent areas of form changes, so try to replace each line with either a hard or soft edge that shows the roundness of the dragon's form; for example, the muscles on his front leg. Don't worry if the image looks too smooth at this stage – this is only a base on which to paint all of the interesting textures later.

< Blend the line work into the painting, introducing hard and soft edges to depict the creature's three-dimensional form

16 Now the lines are integrated into the design, you can start to define the soft forms added in the previous step. Still using the soft brush, add further muscle definition, as well as painting in bigger details along the way. Avoid focusing on the smallest details just yet, as this step is still about establishing the core of the painting. Additionally, pay attention to the colours. Introduce warmer areas, such as around the highlights, as well as the face and snout. This will give the dragon more of a natural look.

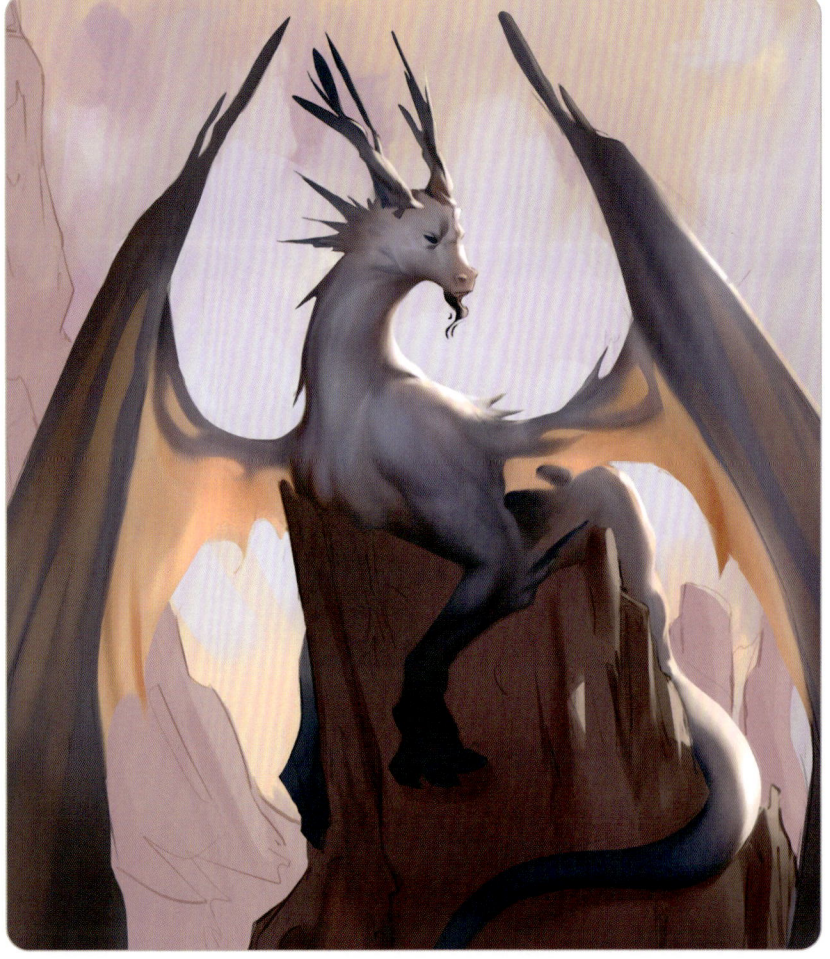

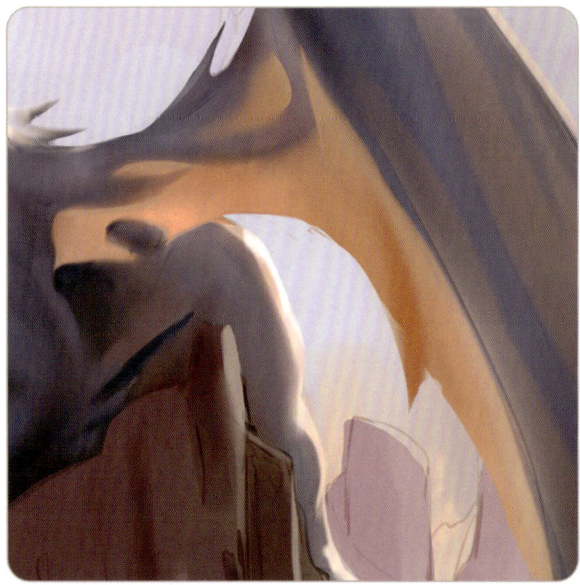

∧ Define the forms of the dragon's body and added areas of warmth

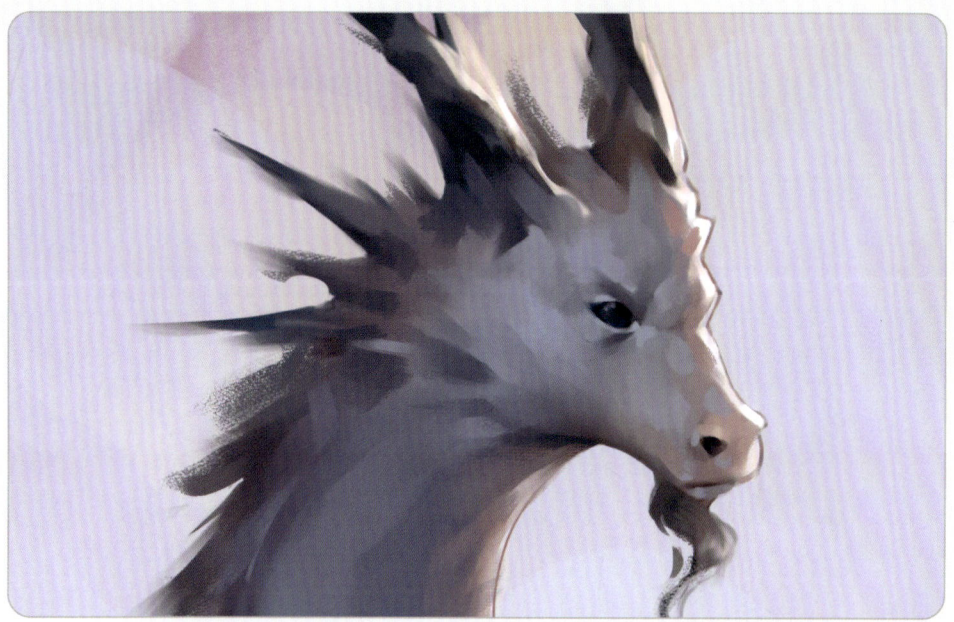

17 Now is the time to look at the medium-sized details. Since the face is the main focal point of the illustration, it's a good idea to start there. Feel free to zoom in on the face if this helps. Begin to define the more intricate details, including the form of the face and the horns. Always keep lighting in mind, taking care to paint in the strong, warm light from the side and a cooler blue reflected light from the sky. Take as much time as needed for this step and those that follow. Work at your own pace and don't rush. Rendering takes time!

< **Begin to render and add details to the dragon's face**

18 Turn your attention to the dragon's body and repeat the same process. Focus on creating the different textures and how to capture the look of fur. Some areas, such as the neck, can be kept less textured so that more textured areas, like the chest and arm, appear more pronounced. To create the fur texture, use the Gouache Brush for textured brushstrokes, and the Painterly Blender Smudge Brush for creating a blended painterly effect. The Gouache brush is located among the Thick Paint sub tools, while the Painterly Blender is found with the Blend tools or by pressing J.

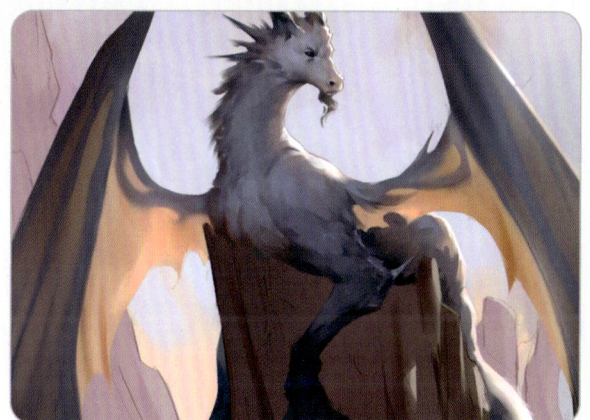
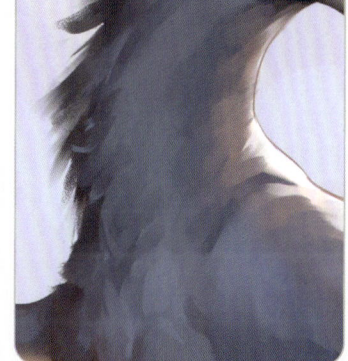

∧ **Render the body by introducing a painterly fur-like texture**

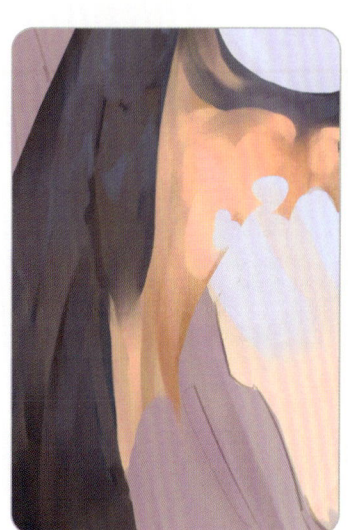
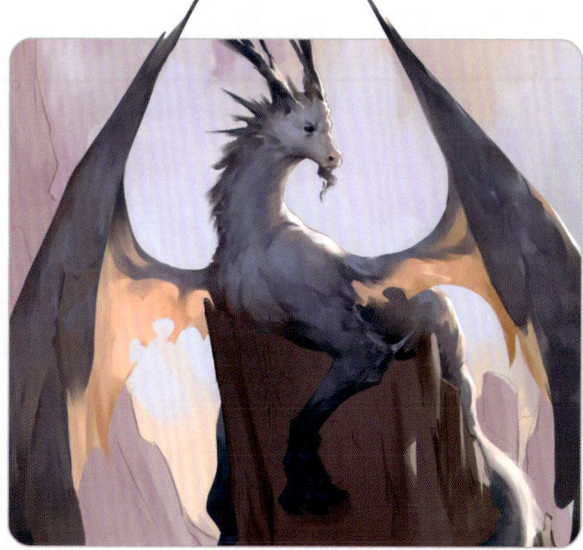

19 The wings are the only element left untouched so far. The method for rendering them is the same as previous steps: add a little texture while keeping the lighting in mind. Make sure the subsurface scattering remains clear and visible. As with the previous step, use the Gouache Brush for broad, textured brushstrokes followed by the Painterly Blender Brush to create that softness and a painterly brushstroke look.

< **Start to render the wings, adding texture while preserving the subsurface scattering**

20 Now the dragon is slightly more defined, you can move on to the other elements of the illustration. Start with the rock, since it's the closest element to the dragon. Applying the same rules and ideas, follow the lighting to establish lighter and darker areas. Rocks are hard forms, so make sure to paint in harder edges and sharp, angular shapes.

> **Add definition to the rock the dragon is sitting on**

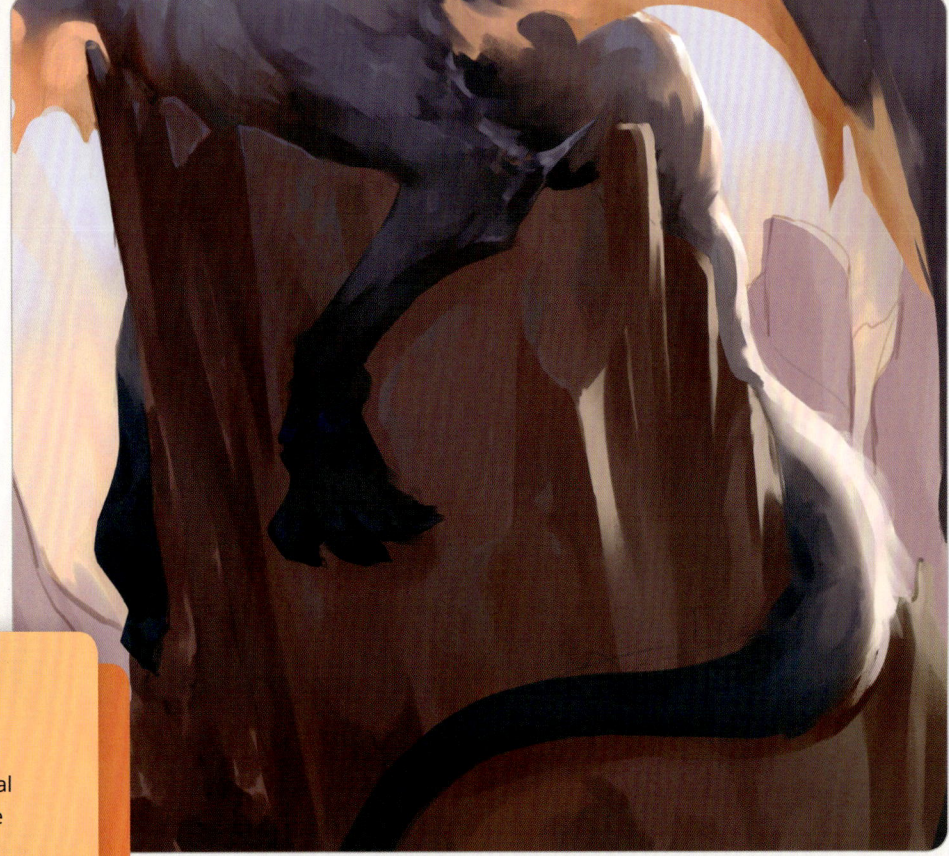

ARTIST'S TIP

Rendering is the longest stage of any digital painting. Don't get discouraged if the piece doesn't look correct right away. Be patient and work on one thing at the time. Keep revisiting and reworking until you're happy with the results. The great advantage of digital painting is that it's very forgiving. No matter what you've put down on the canvas, you can always go back and improve it later!

21 Move on to the mountains in the background and the sky. Apply the same rendering techniques covered in the previous steps, but remember that they should have significantly lower contrast than the dragon and rock. The further away the objects are, the less detail will be visible. Take care to only imply texture, rather than fully painting it in, and avoid darker colours.

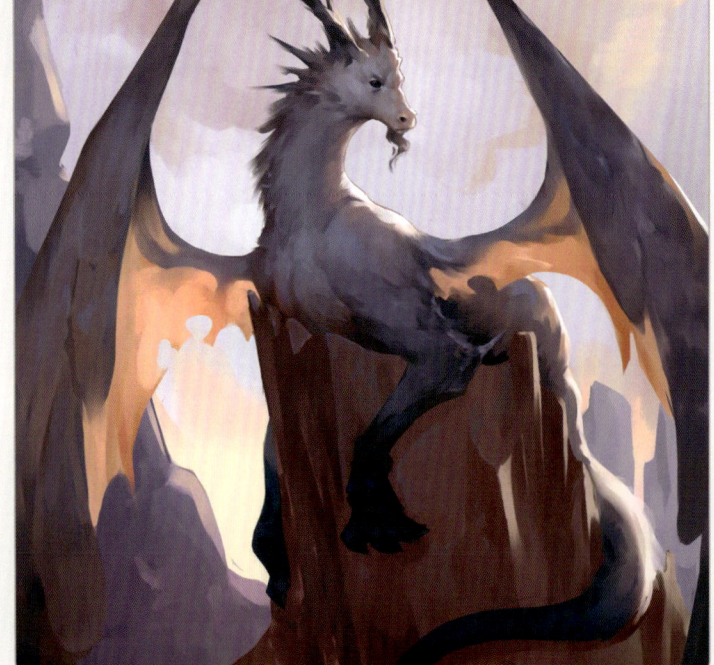

> **Introduce more definition to the sky and mountains in the background**

22 Now that the entirety of the image is rendered to a medium-level degree, you can work on all areas of the painting, improving anything that seems off. Make the darker areas darker, and lighter areas lighter. Imply stronger textures and fix shapes that look incorrect. Ensure the dragon's face remains the main focal point by giving it the strongest contrast, while lessening the detail in peripheral areas, such as the tail and feet.

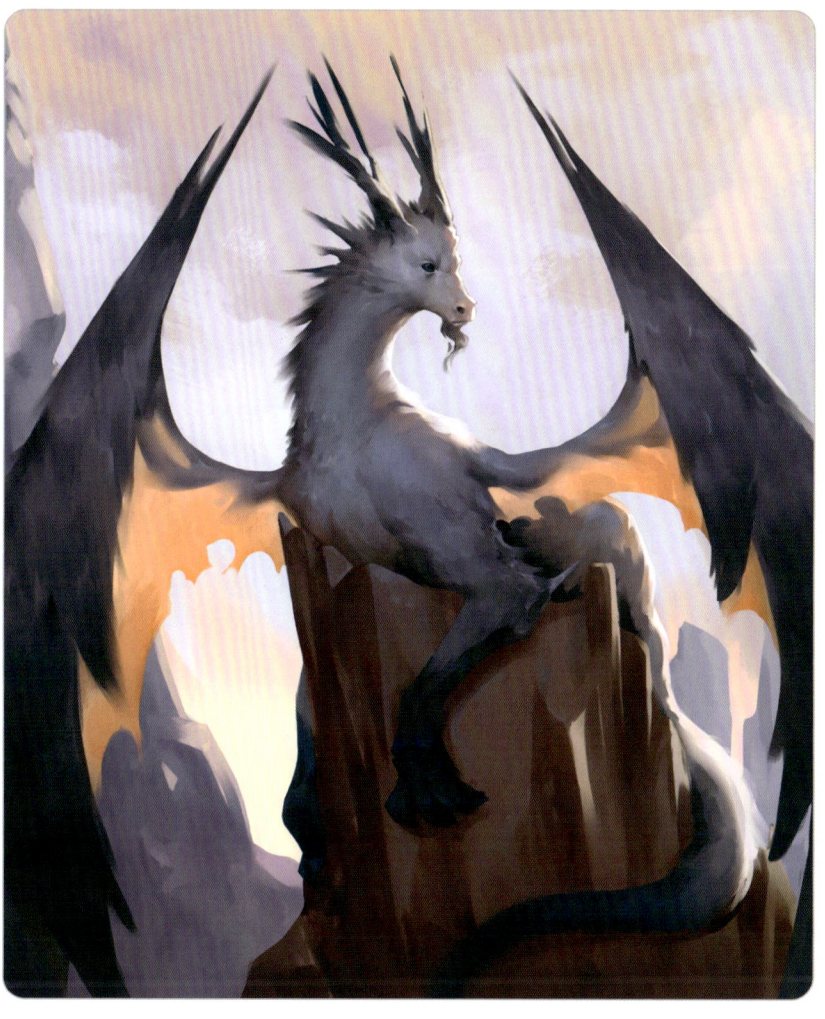

> Work on the smaller details, fixing any mistakes and painting in stronger textures

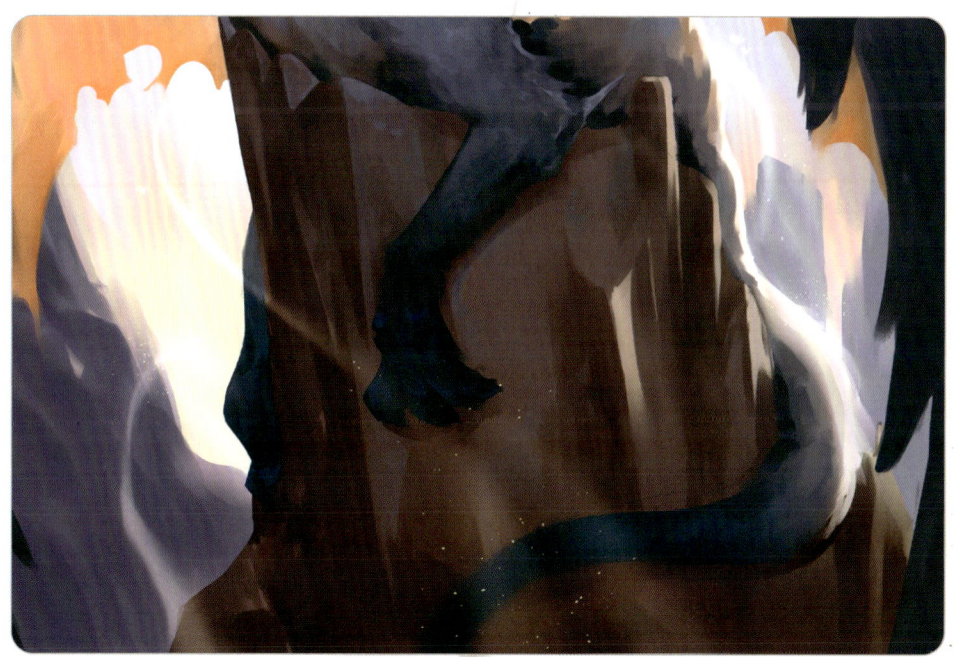

23 Before you finish rendering, you may wish to add some stronger visual effects to generate more of a fantasy feel to the scene. Create a new layer above the rest and set it to Add (Glow) mode. Painting on this layer will create a glowing effect, allowing you to add a magical radiance to certain parts of the image. Use this mode to further magnify the highlights, as well as add fantastical swirls of light circulating around the dragon and some sun beams streaming in from the side. Feel free to experiment with this layer to see what interesting effects you can create!

< Use the Add (Glow) mode to introduce magical glow effects

24 Now that everything is established, it's up to you how far you'd like to take the painting before you call it 'finished'. Focusing on the contrast, really darken those darks and light up those lights. Enhance the sun beams even further, add as much texture as needed, make sure the dragon clearly stands out from the background, and further define the dragon's form, as well as the rock, mountains, and background. Use everything you've learnt so far to add the final touches to the image, making sure to take your time and enjoy the process.

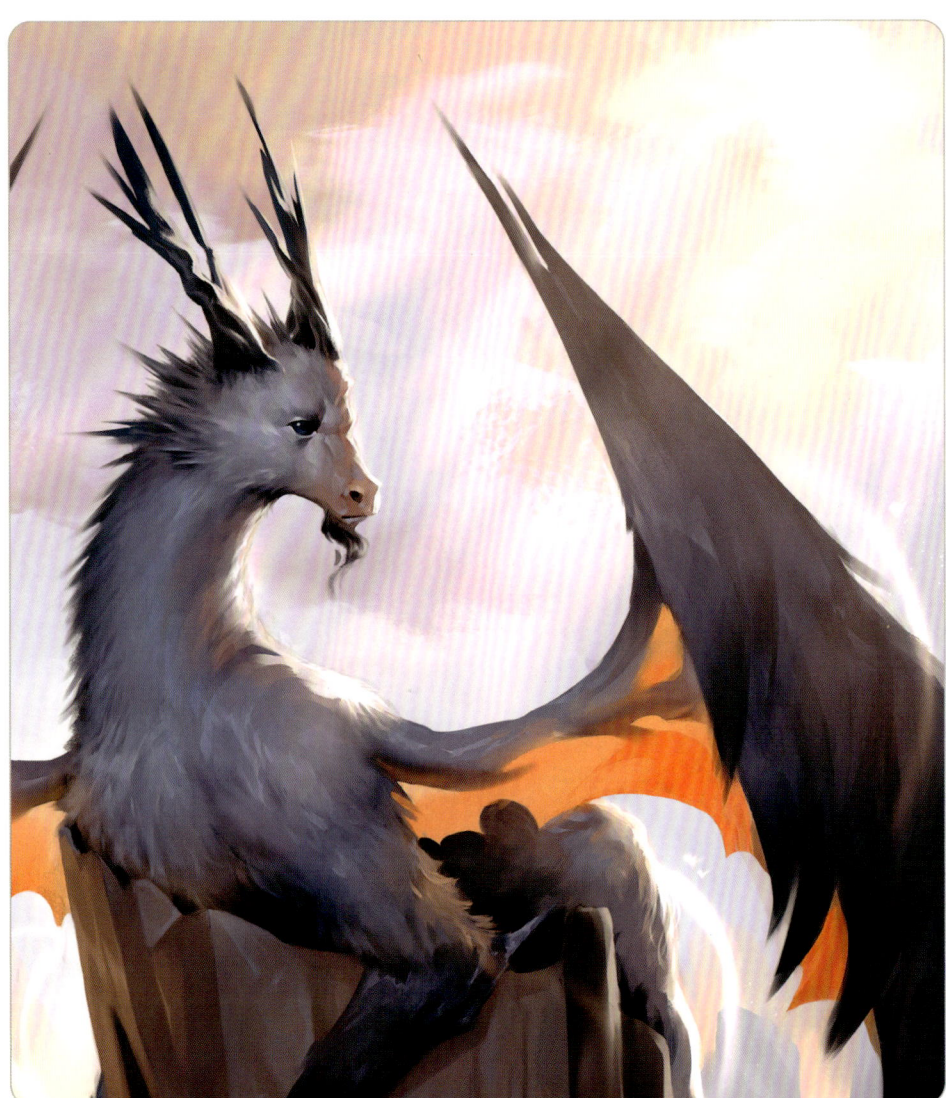

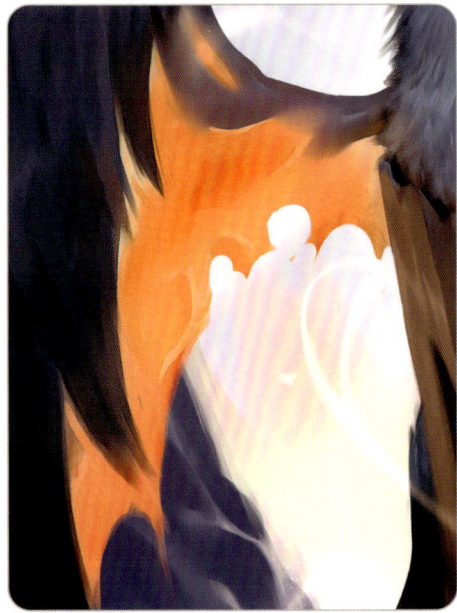

> **Add final touches to refine and enhance the overall design**

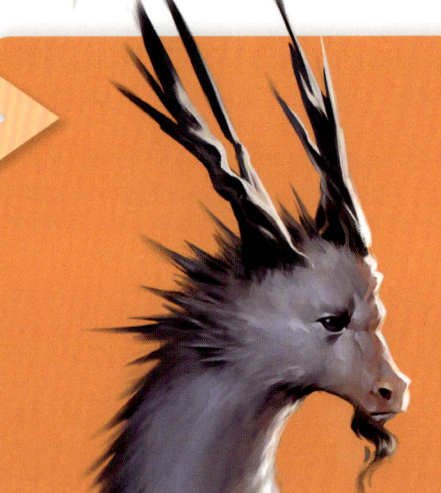

Conclusion

The final image depicts a majestic dragon in the wilderness. Strong lighting and high contrast evoke a sense of mystery around the creature, as well as the fantasy land it inhabits. Now the illustration is finished, take some time to imagine the story surrounding the dragon. If it has a family or clan it belongs to, what would they look like? How do they move and exist in their world? Now you've learnt how to build up a creature design using Clip Studio Paint's tools, you're free to bring these fantastical beings to life!

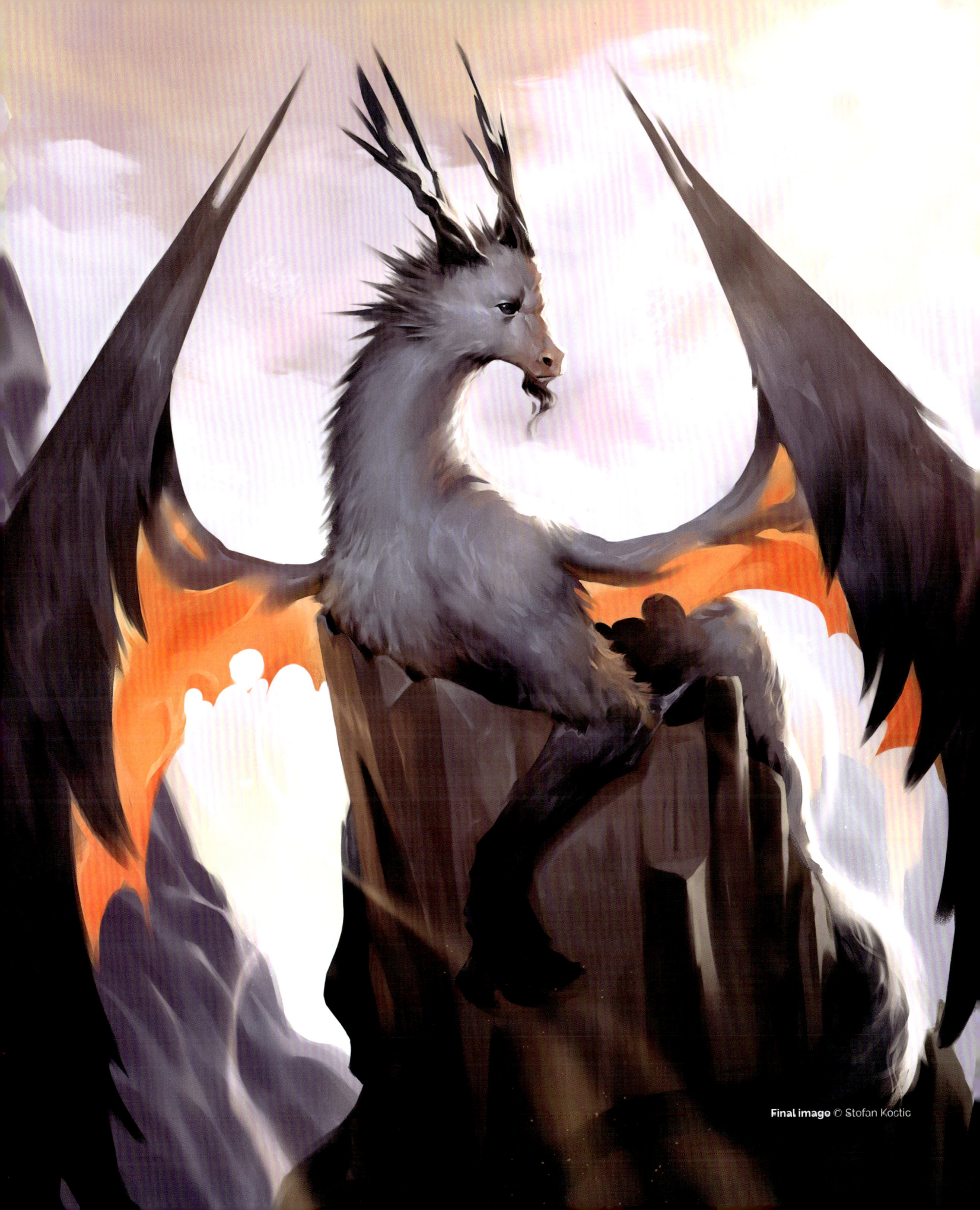

Gallery

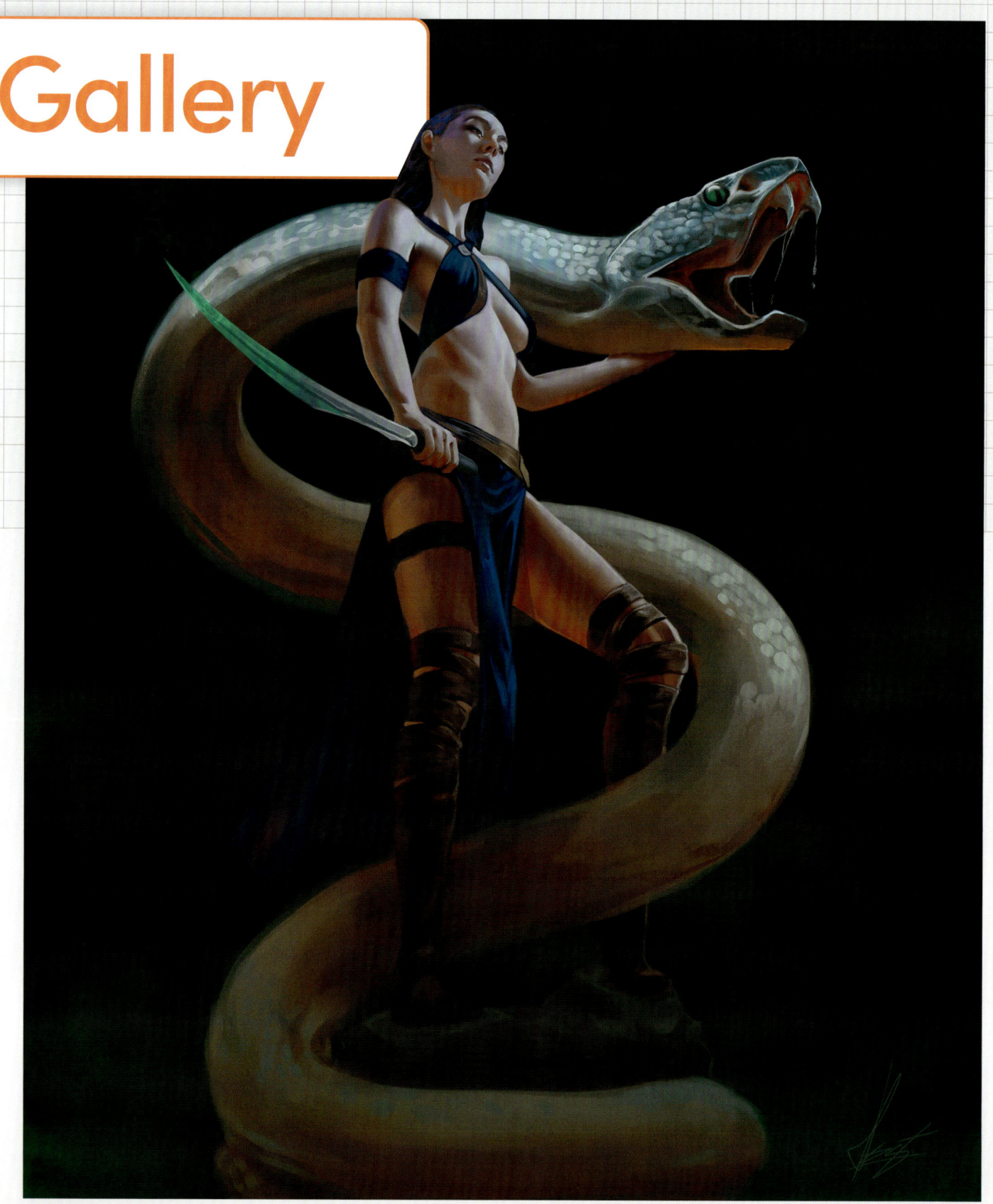

Snake Assassin © Stefan Kostic

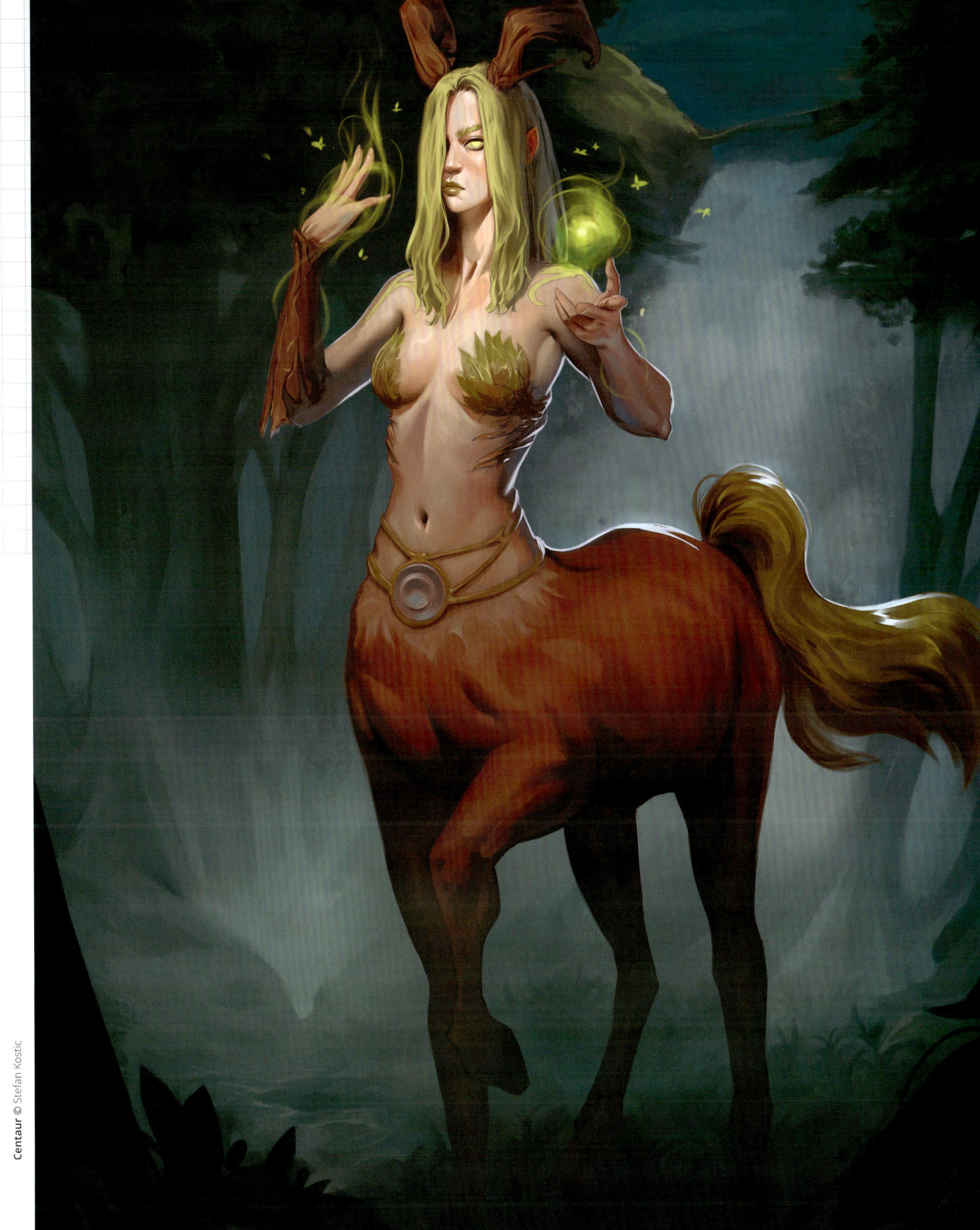

Character design

BY DADOTRONIC

In this tutorial I will share my illustration method and Clip Studio Paint tips, as well as the creative mindset that helps me to break free when I feel like I'm in a stagnant state with my art. The idea of being able to separate process from analysis is a very important step for me. While the challenges and requirements as an artist remain the same, a different mindset has enabled me to apply that common suggestion we so often hear from those more experienced than us: enjoy the journey!

I'm a self-taught artist and have been using Clip Studio Paint since the earlier versions of the software (known as IllustStudio). Perhaps that's why I know a lot of tricks and shortcuts for creating art in the program. The process I'm going to show you in this tutorial is as simple as possible, combining the best of the various methods and workflows I have tested over the years. This book contains many cool ways to make art and you should challenge yourself to try them all. Actual practice is what helps an artist evolve. You can learn a lot, even when the process seems very different from what you're used to.

Final image © DADOtronic

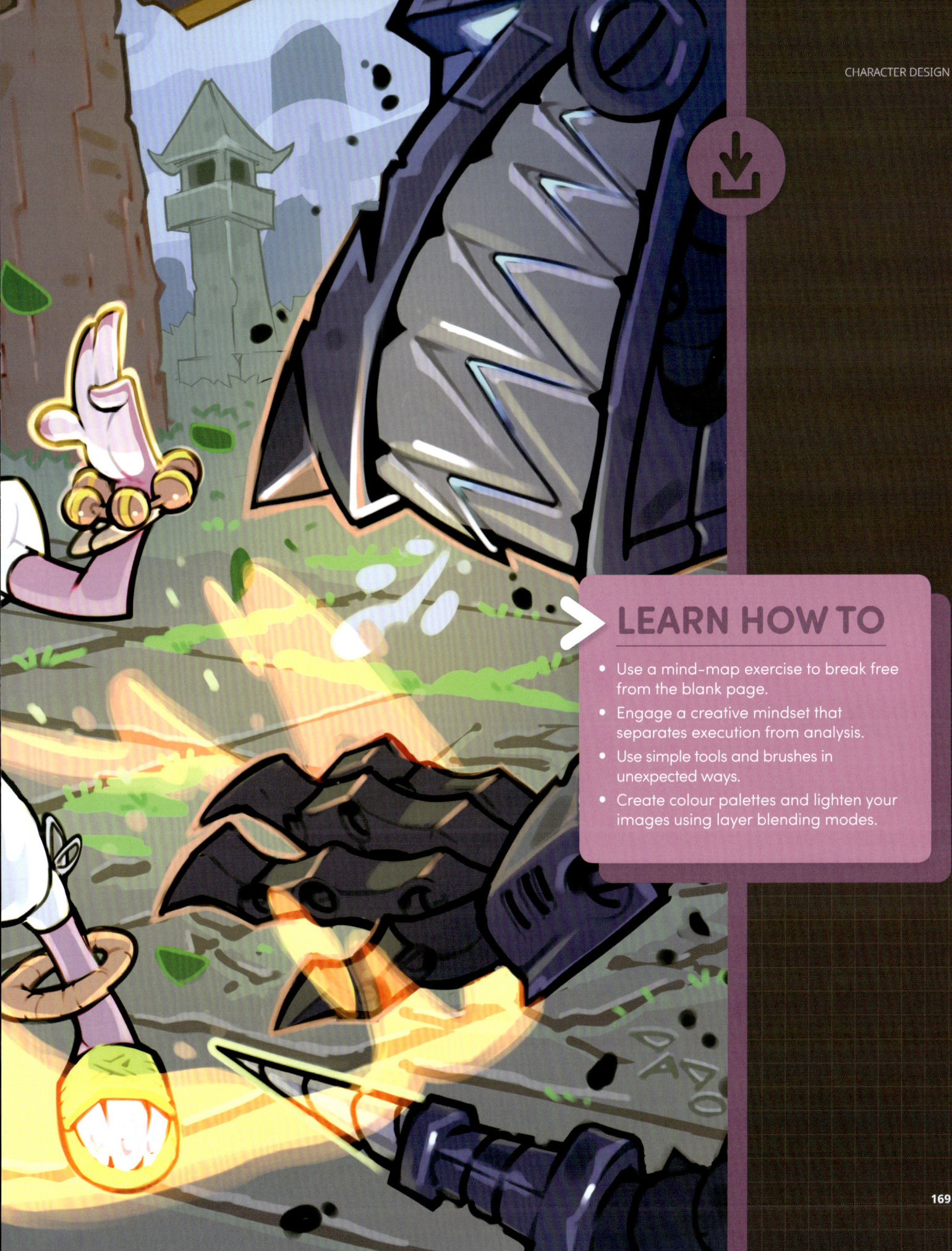

LEARN HOW TO

- Use a mind-map exercise to break free from the blank page.
- Engage a creative mindset that separates execution from analysis.
- Use simple tools and brushes in unexpected ways.
- Create colour palettes and lighten your images using layer blending modes.

01 How do you come up with a new and original illustration? If the blank page scares you, start by using a mind-mapping technique to generate ideas. Select File > New to create a new document, then enter a canvas resolution, approximately 1,920 × 1,080 pixels or higher, so you can make the canvas full screen on your computer monitor.

Typing out the mind map will allow you to edit the text and alter the order and placement of the nodes (text balloons). Create the central node using the Balloon > Rounded Balloon tool, then use the Text tool to write the main idea, brief, or purpose of the illustration you wish to create inside of it. Next, create new nodes around the centre one and write in ideas you want to explore. This example lists: animal, fantasy, class/job, and value/message.

Next, create new nodes from the existing ideas by duplicating the layer with the balloon and text. The aim for this tutorial is to create an animal character design. The first idea is to create a hare, but the scene could include other animals too, such as a turtle or fox. To rearrange the nodes on the page, use the Operation > Object tool. To create a line between two nodes, use the same Object tool to pull any of the vector points around the balloon shape. For this to work, make sure Mode is set to Move Control Points.

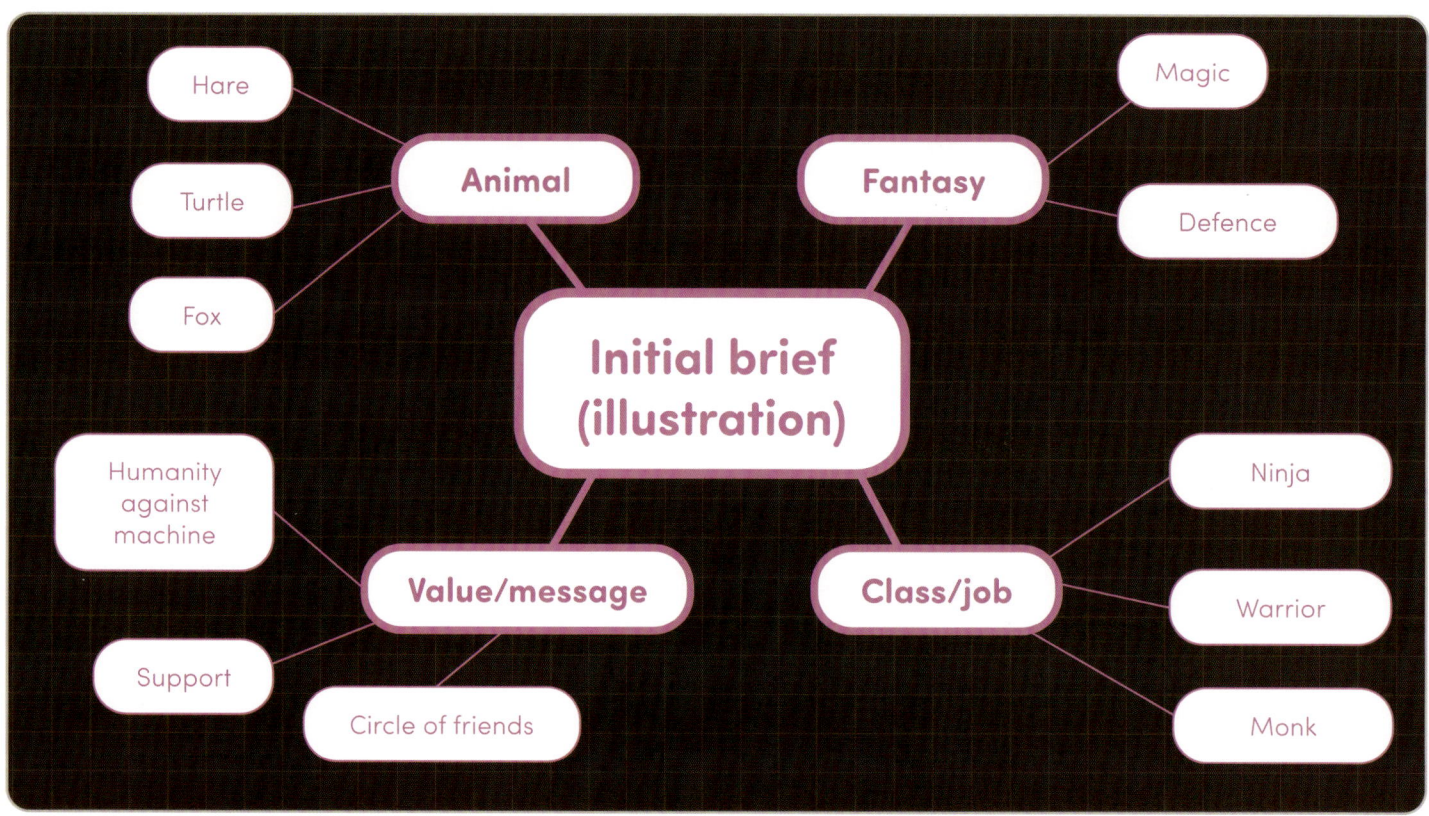

∧ You don't have to share this mind-map step with anyone – simply jot down all of your rough ideas and thoughts

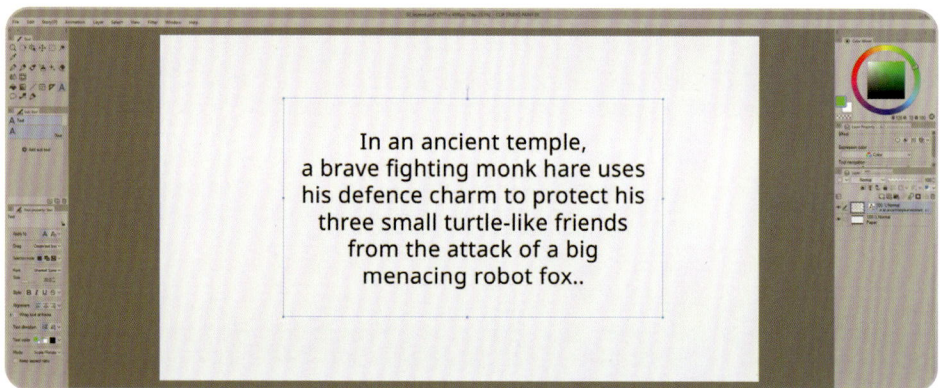

∧ Starting from words, instead of visual reference, should be a key part of your creative toolkit

02 Using the mind map you created in step 01, write out a short text description of what you plan for the illustration. Describe the characters it will include, where they are located, and, if possible, when the action will happen. This simple statement will help you enormously in the next parts of the process. It's up to you how descriptive you want to be. If you already have a strong idea, you can write down more specific details, such as the colour of objects, types of poses, and any details in the scene.

03 Let's start illustrating! Draw some studies using just one colour and a tool that makes it easy to draw large shapes, such as **Pen > Calligraphy** or **Lasso Fill**, which is found in the **Figure > Direct Draw** sub tool palette. To frame the thumbnails, use the **Rectangle Frame** sub tool, which is found under the Frame Border category. Click and drag to draw the frames in proportion to your final image. The **Frame Border** sub tool automatically creates folders and layers with masks for you to draw inside.

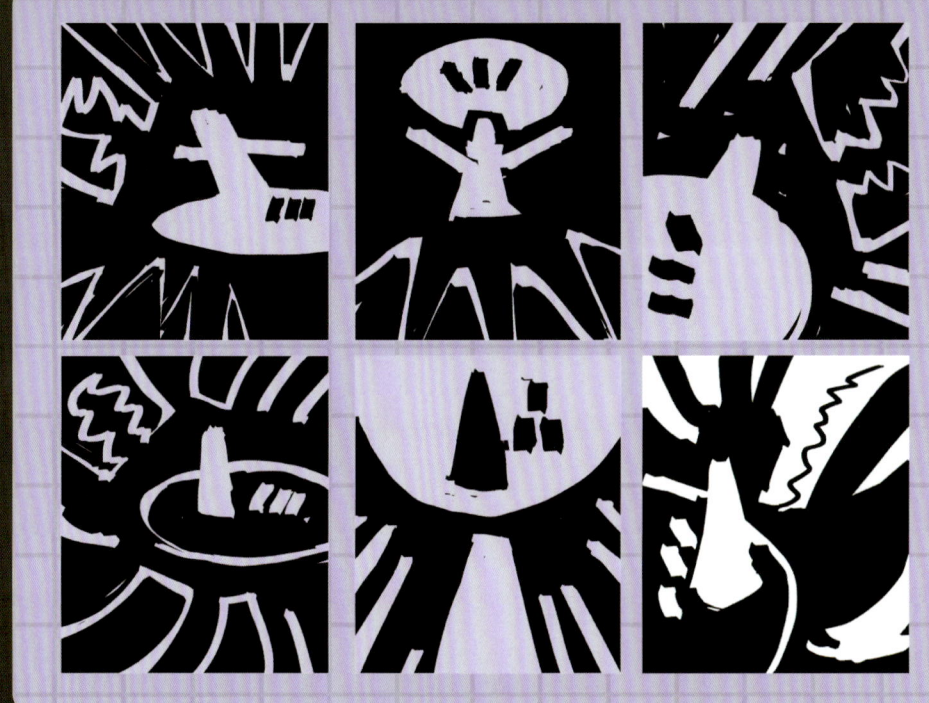

> Imagine that your task is to reduce the brief to the simplest possible image

04 You don't have to use lots of brushes to create a good illustration. As you learn to use Clip Studio Paint, all you really need is two types of brush: one with very soft edges, the other with slightly harder edges. Both should be without texture, dynamic colours, or blending effects. Here the image shows a simple method to build up your idea for a character. Let's try it now.

Create a new layer and name it Sketch, then use a soft-edge brush such as **Airbrush Soft** to sketch the overall shapes and features of the hare. Next, create a new layer called BreakDown on which to draw a clearer and more detailed version of each of the main shapes. Continue to use the soft-edged brush for this, but with a small size to produce darker lines. Finally, create an Ink layer on which to draw the final, cleaned-up line work. Use the harder-edged version of the brush for the line work and draw by 'sculpting' the lines.

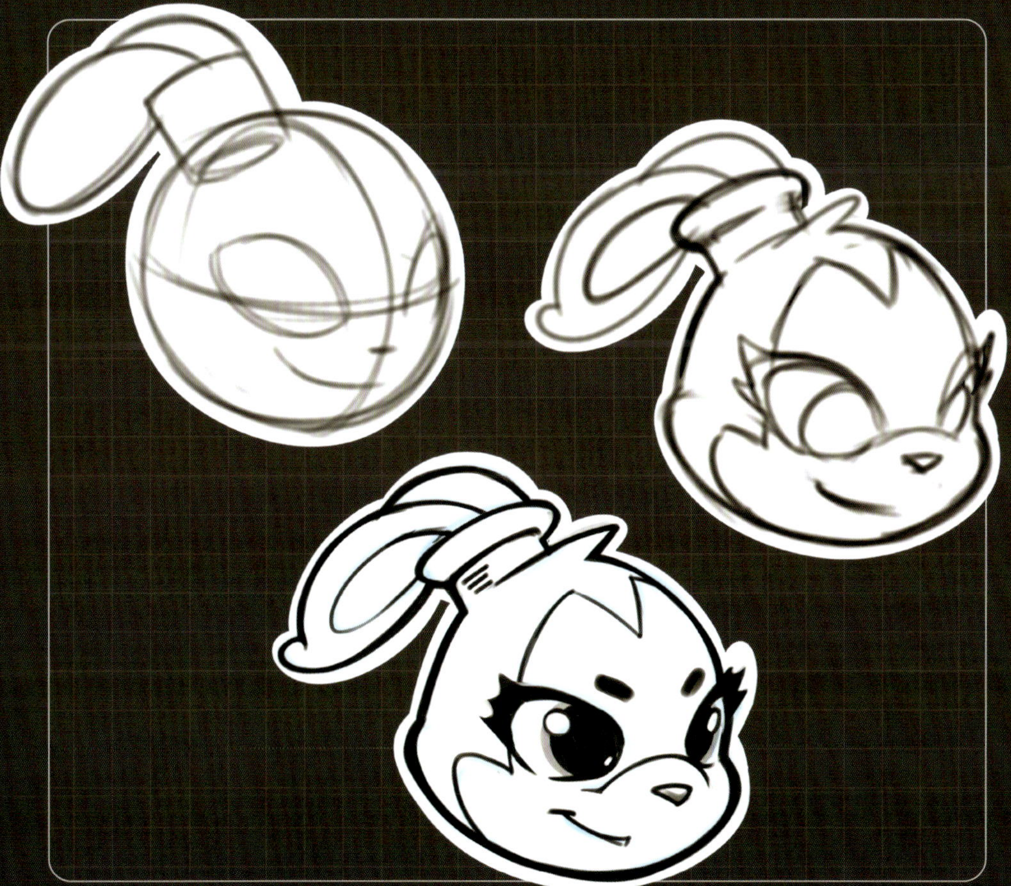

< Make the design more detailed on each pass

05 To create the brush for inking and painting, duplicate the **Soft Airbrush** and change the hardness, brush density, and pen pressure properties for a harder, opaque stroke with a fixed thickness. This will create darker, finer lines.

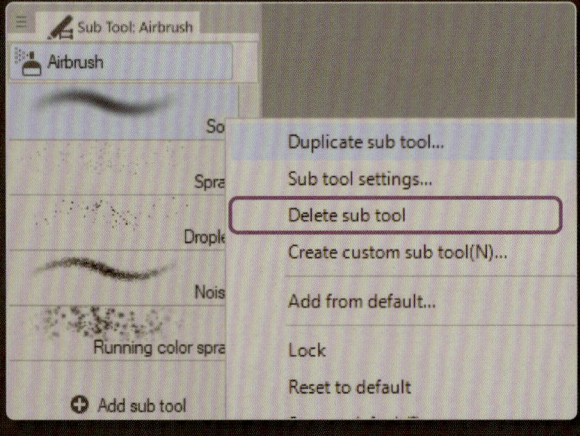

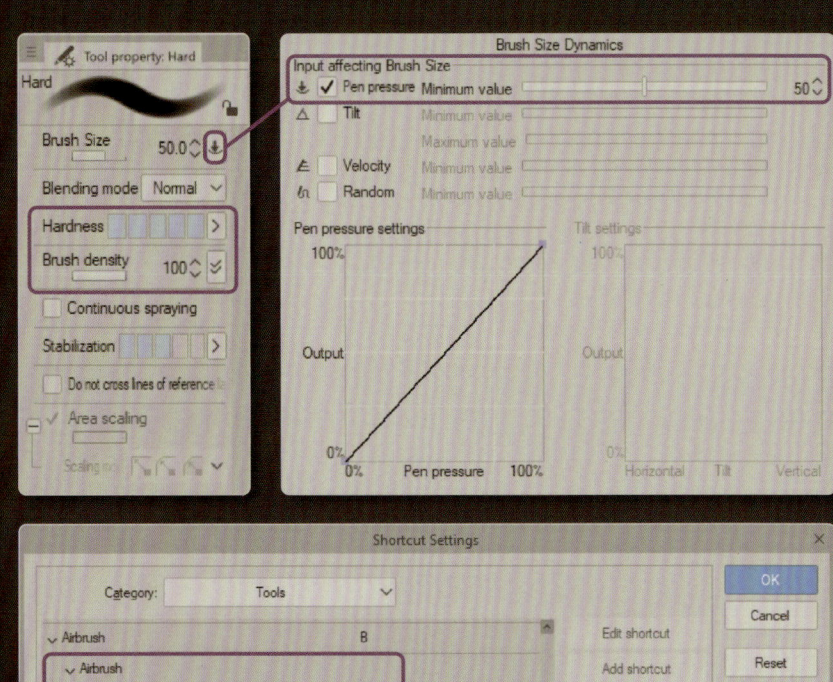

∧ **Set up the same shortcut for multiple tools and it will work like a toggle**

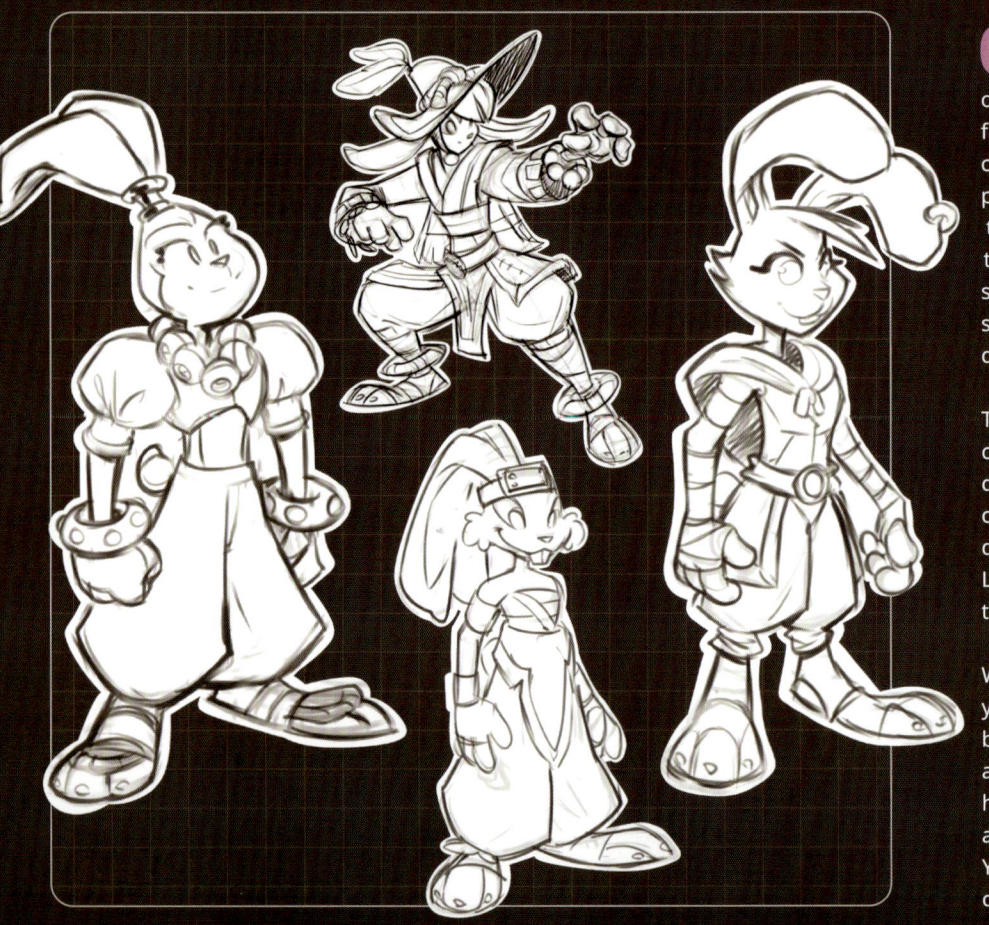

06 Sketch out a few different designs for the hare. While it can be tempting to seek out reference material straight away, try drawing from your imagination to begin with. Don't complicate the ideation process thinking about poses or details just yet. Here you can see how the three hare designs have the same upright three-quarter pose. Clothing and accessories are suggested, but not detailed. Try to think like a 3D sculptor, starting a model from the T-pose to lay down basic shapes and ideas.

To try a different method, draw the turtle creatures using the **Symmetry Ruler** tool. Simply drop the **Ruler > Symmetrical Ruler** onto your canvas before drawing, then delete or hide it after drawing by right-clicking on the ruler icon in the Layer panel. Next, draw the robot-fox in profile, as this is an easier angle for mechanical designs.

While drawing from the imagination can take you so far, references are invaluable for adding believable detail to your character designs. Create a mood board with various reference images of hares, turtles, and foxes, as well as monks, martial-arts clothing, accessories, and robotic machinery. You can then refer to this reference imagery when drawing the second line-up of hares.

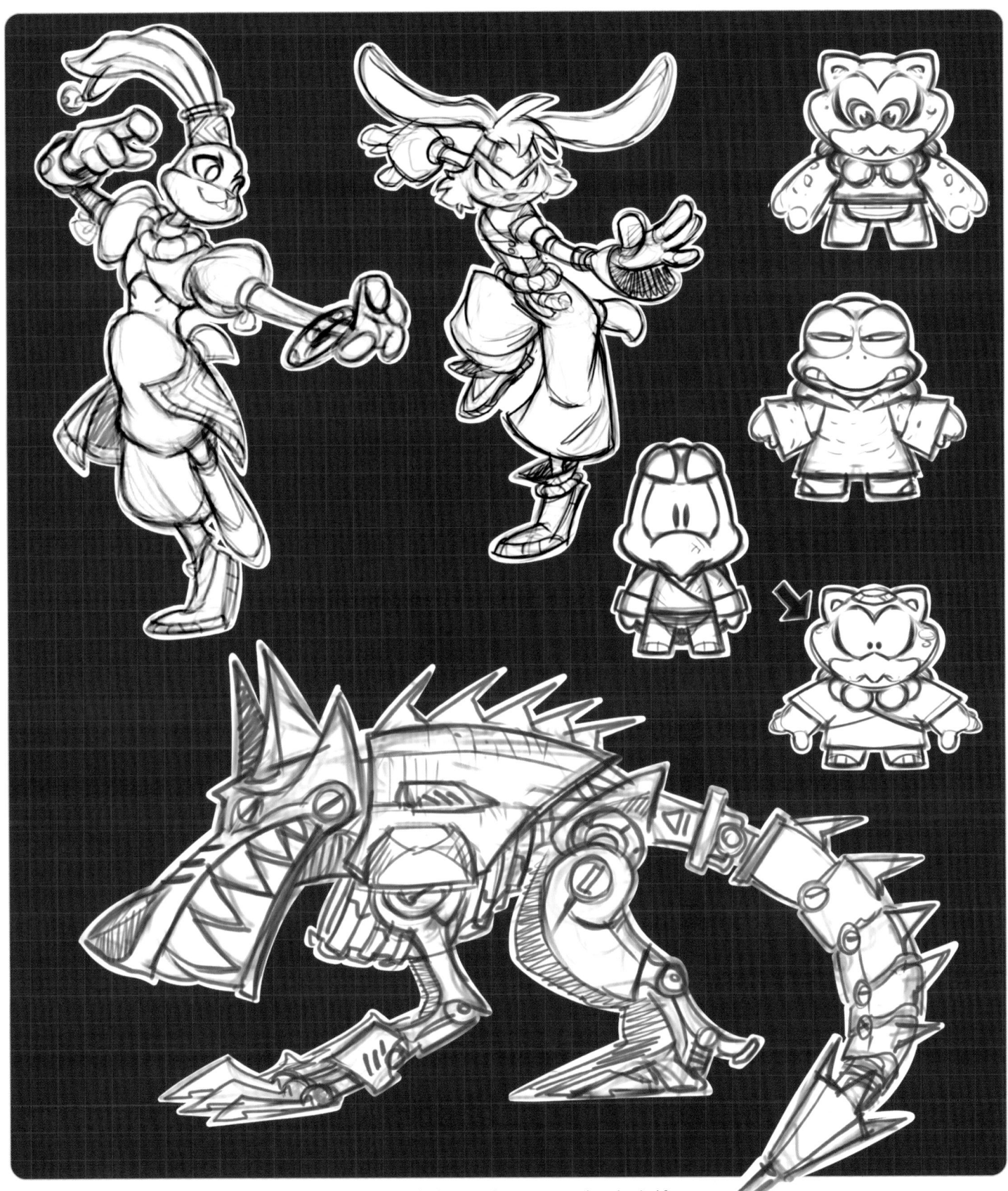

∧ Start by drawing from your imagination with different designs for the hare, turtle creatures, and mechanical fox

07 You can create your thumbnails, sketches, and concept art in lower (HD) resolution, as these will likely only be viewed on a computer screen or mobile phone. Using a lower resolution makes the software respond faster and generates smaller files. Create the final file (A4, 300 dpi, for example) when you start inking the final lines. You can set your monitor's dpi in File > Preferences > Canvas > Settings. Using a real ruler, adjust the slider until the canvas ruler and the real ruler are the same size. Now, when you click on View > Print Size, the page on the screen will be the same size as the real printed one.

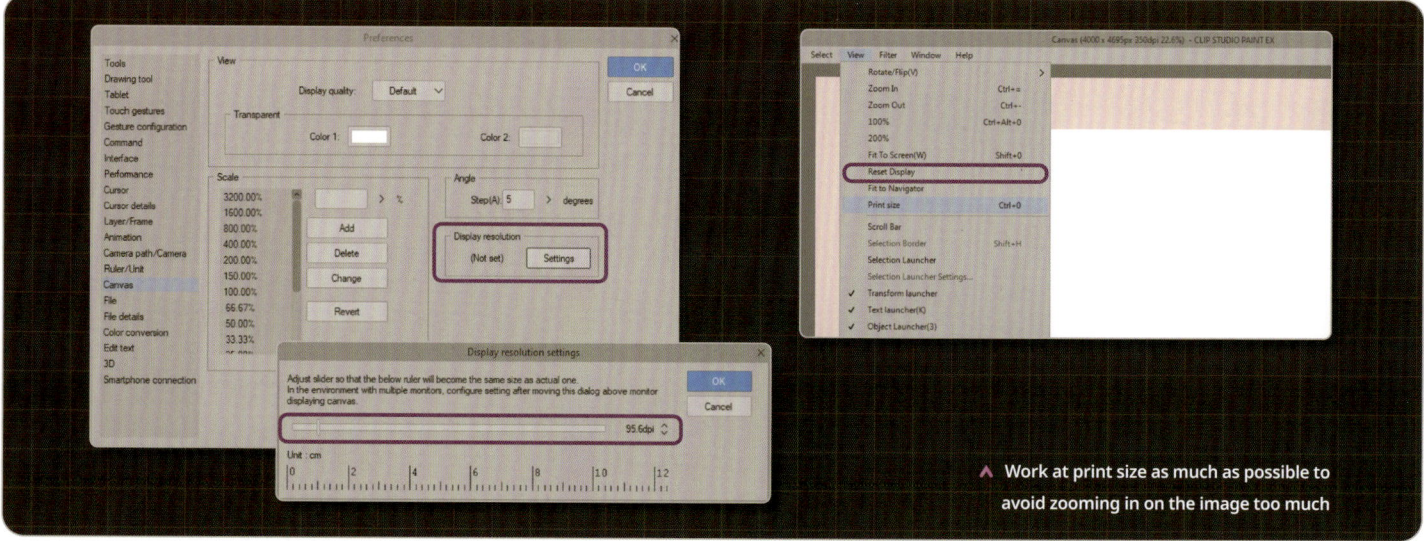

∧ Work at print size as much as possible to avoid zooming in on the image too much

08 The focus of this illustration is the characters, but adding an environment for them to exist in will make the artwork more interesting. Clip Studio Paint's materials library has a number of 3D primitives that you can drag onto the canvas to create a preview of the scene you want to illustrate. In game design, this technique is called 'grey boxing'. You can use this to find a camera angle that matches your vision, as well as experimenting with the size of the elements in relation to the characters. Make adjustments by changing the values on the General Primitive > Number of Divisions parameter of the selected primitive.

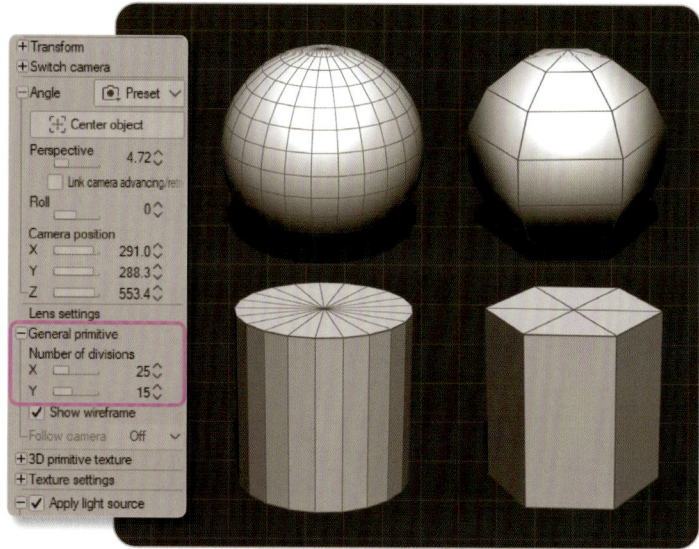

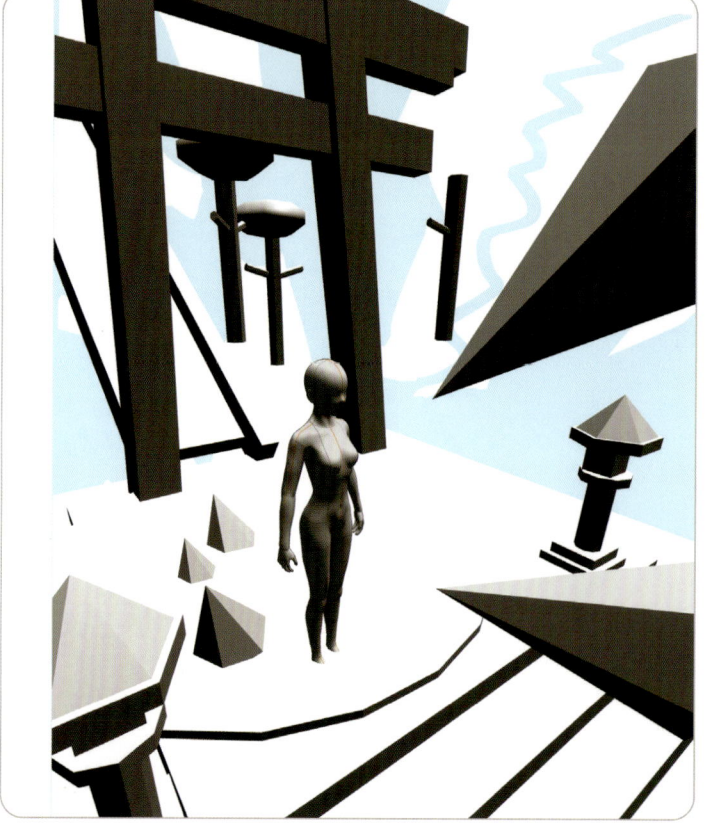

∧ Use 3D primitives from the material collection to roughly map out the scene

09

To reduce the contrast of the 3D layer and facilitate drawing over it, use the Layer Colour effect found in the Layer Property palette. This option colourizes the layer using the two swatches (Layer Colour and Sub Colour). Create a new layer and start drawing, very loosely, over your 3D references. Remember to start with the Soft Airbrush tool so you don't commit to any line or shape too soon.

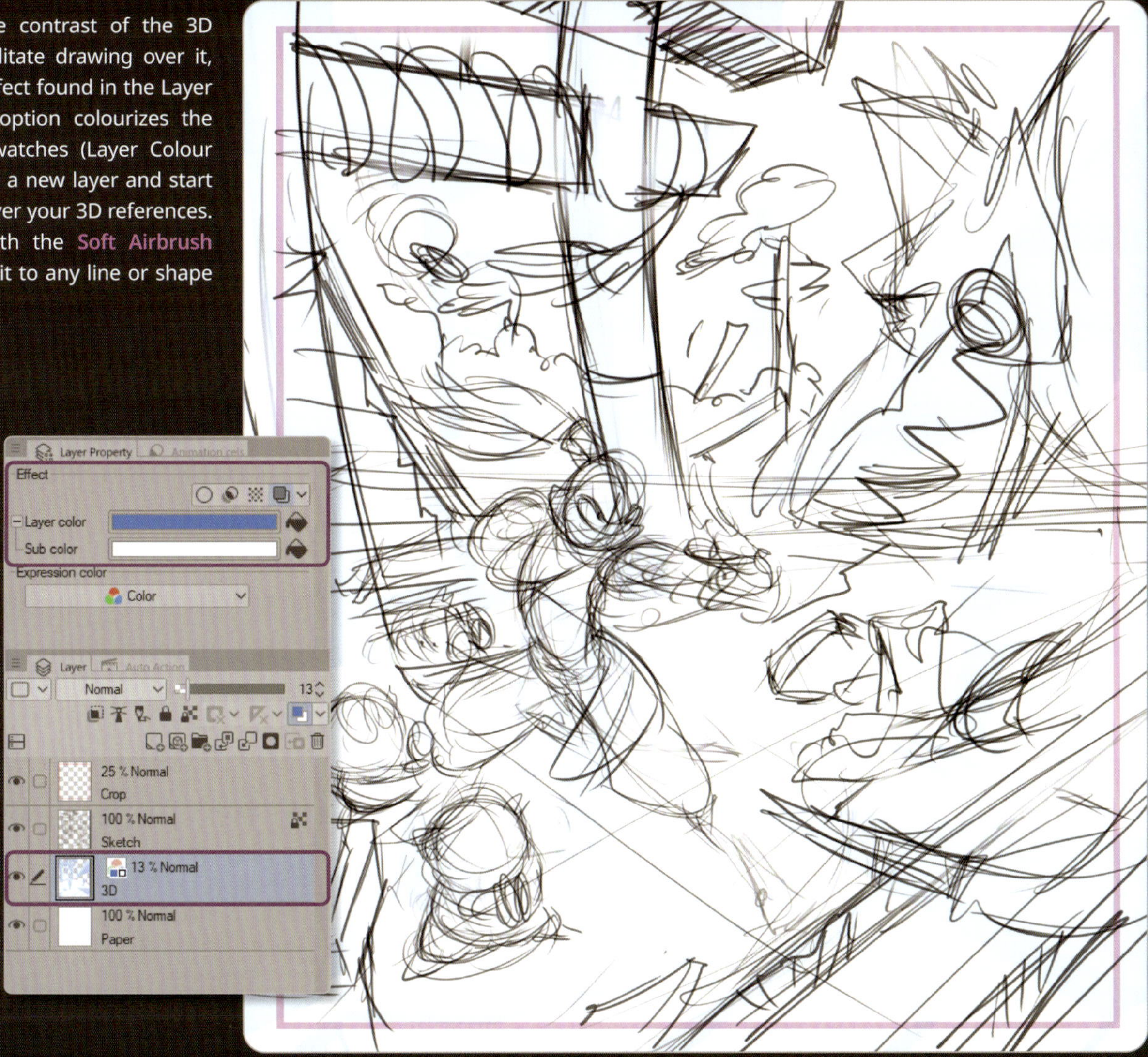

∧ Try to maintain the spiral composition of the thumbnail – gestural drawing is key to capturing the motion

ARTIST'S TIP

Split the process into two different personalities or approaches. At the beginning of the process, you are the artist: you can let loose and express yourself with all you already know about drawing. Somewhere down the line, however, you become the art director who analyses, provides structure, remembers the fundamentals, and adjusts the artwork to ensure it fits the original vision and story.

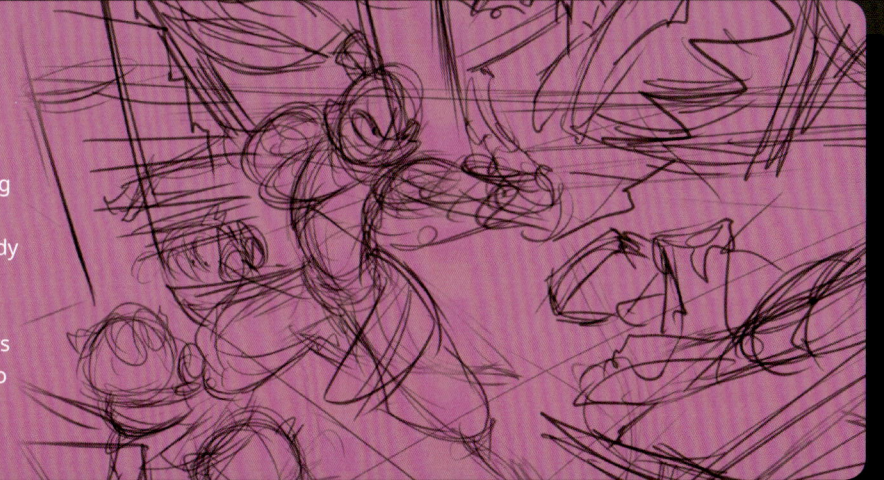

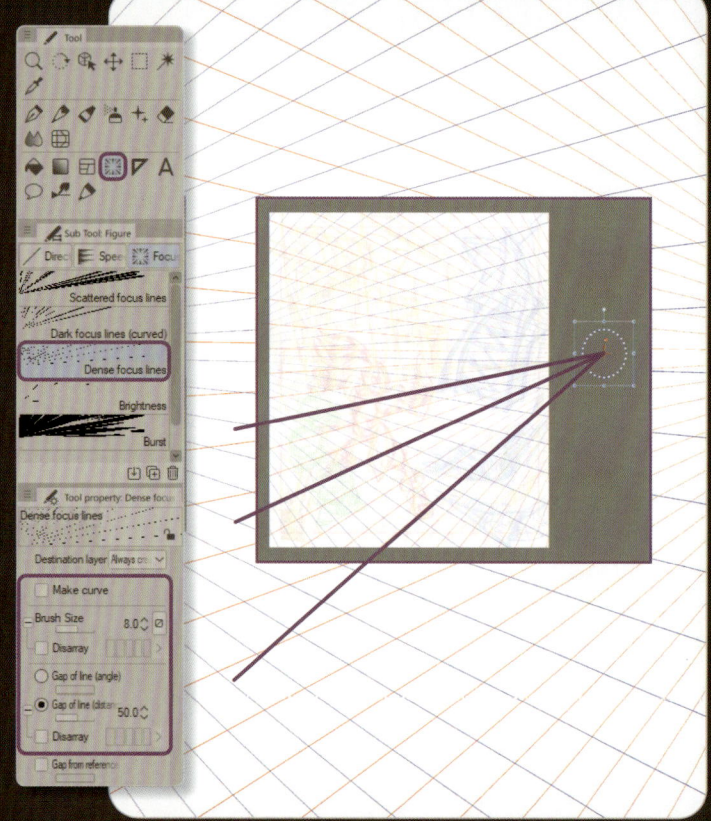

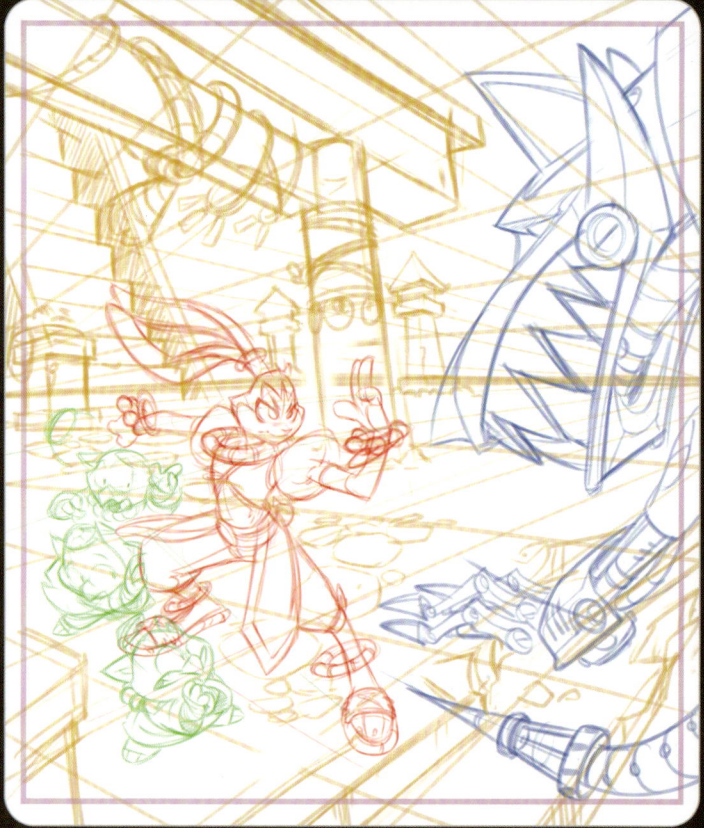

∧ Keep the Perspective Grids visible while you work

10 This stage – let's call it the 'breakdown' – is where you can start to put a bit more structure into your drawing. Build the characters using simple shapes and/or geometric forms. Use guidelines to remember the horizon and vanishing points in the scene. An easy way to set up perspective lines is to use the **Figure > Focus Lines > Dense Focus Line** sub tool, as this will enable you to radiate lines from a centre point. Turn off **Disarray** options and adjust the **Gap of Line** to create more or fewer guidelines. The trick is to position the Focus Line layer outside of the drawing area. Here you can see how the vanishing point is placed to match the perspective of the sketch. Duplicate this Focus Line and adjust it on the screen to create other vanishing points.

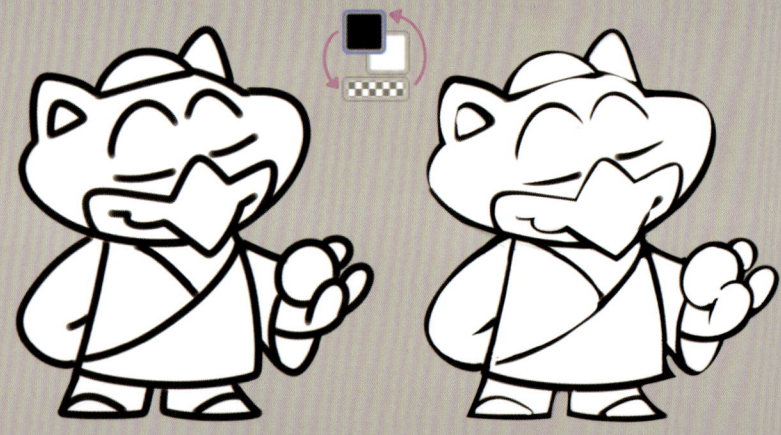

∧ Use the eraser or transparent colour to shape the line work

11 Sketch the final lines of the drawing using the customized **Airbrush** with harder edges. This brush is a good choice for converting a cel-shaded style into a more fully rendered look, with lines and shading made with the same style of brushstroke. This brush is also useful when you can't rely on pen pressure to vary the thickness of the lines. If you use the **C** key to quickly switch between between colour and transparency when drawing, you can sculpt your line weights to simulate pen pressure.

12 Start inking by making the contours. Create a hierarchy of thickness: draw objects closer to the camera using thicker lines and those further away using thinner lines. This may seem like a different way of creating the final artwork, but you will understand why in the next few steps. Make sure to close up all of the silhouettes to ensure you can use the Fill tool later to fill each shape with colour.

> Use thicker lines for objects closer to the camera, and thinner lines for objects in the distance

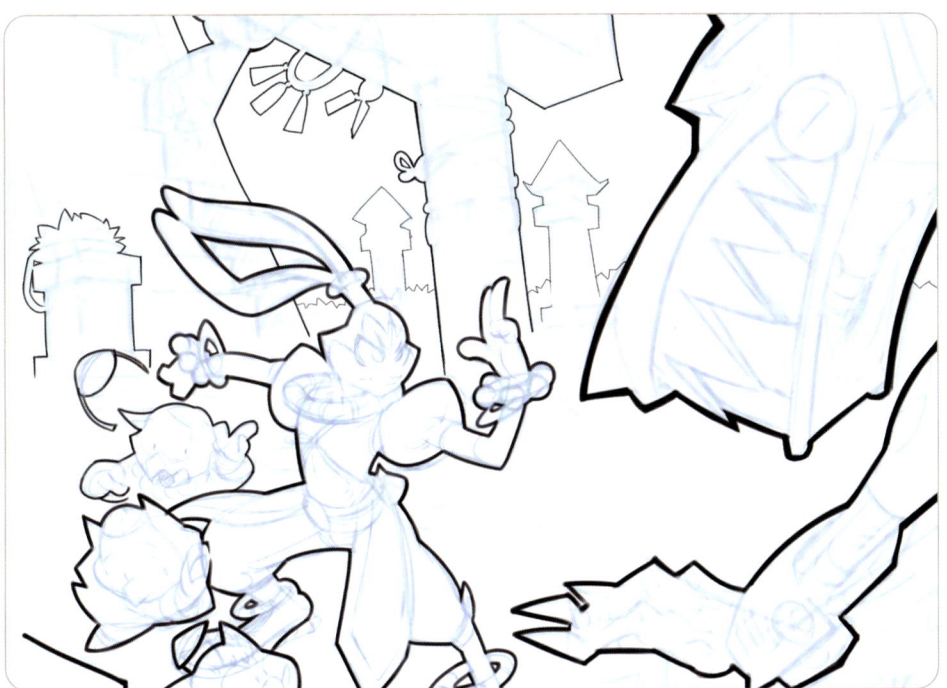

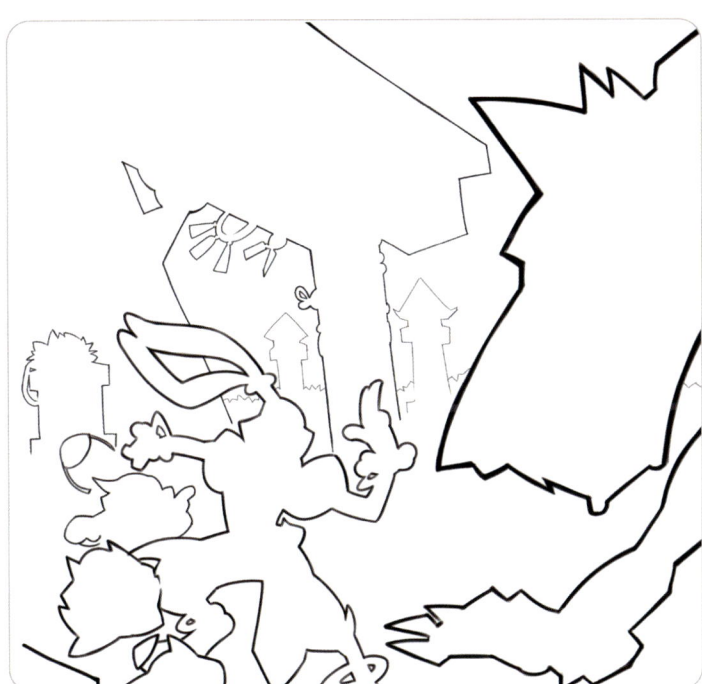

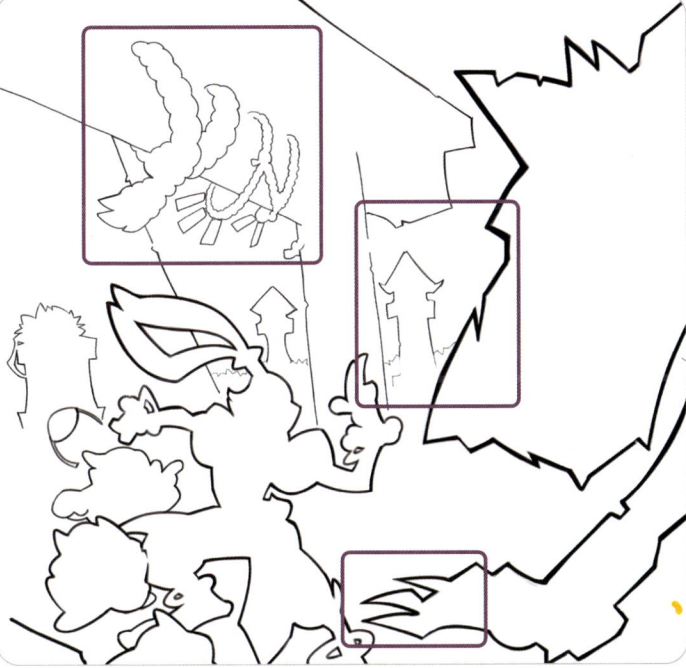

13 Beginning by inking the contours enables you to focus on the outlines alone. It can also make you quicker and more objective in your working, as you only need to outline the shapes. Another reason for this method is that, with the contours completed first, you can more easily examine the shape language of the silhouettes. This is particularly important for cel-shaded artwork, where the characters float above the background. Use this step to make the silhouettes more interesting with better-drawn curves or more distinct edges. Continue to use the transparency trick to sculpt the lines.

∧ Review the shape language of the silhouettes to see if they can be made more interesting

177

14 Once the outlines are complete, the next step is to draw the internal lines using a smaller brush size. Start by tapering the lines, usually at the T-ends of the drawing. Next, gradually draw the details inside each shape. The main advantage of this method is that it allows you to separate the process into more logical and practical steps that are less dependent on motor skills, fine brush control, and levels of pressure available on your hardware. This method of inking will work on any hardware or software.

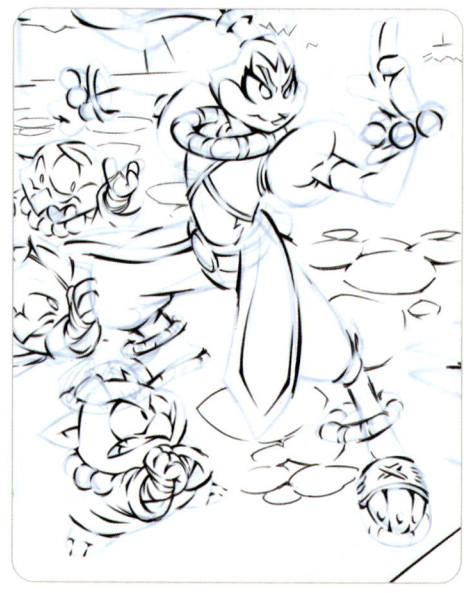

∧ Use a smaller brush size to draw the details inside of each shape

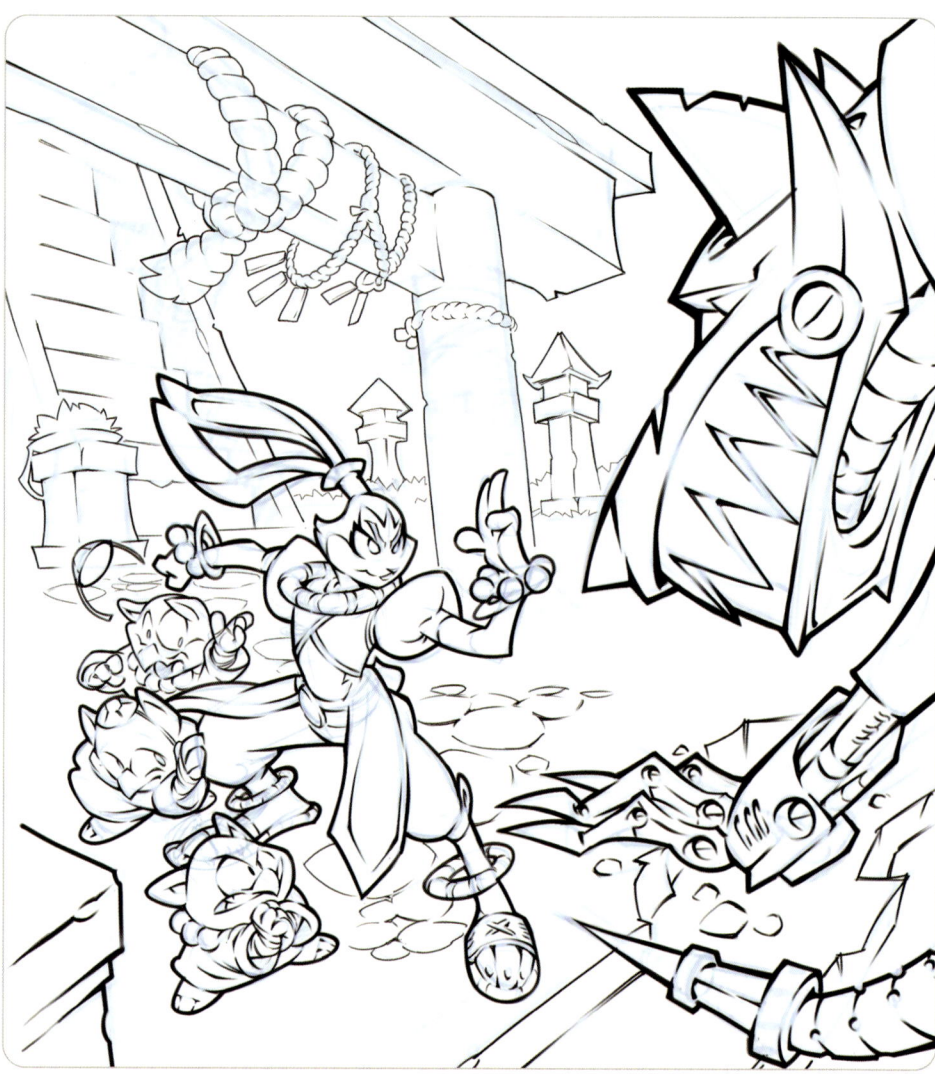

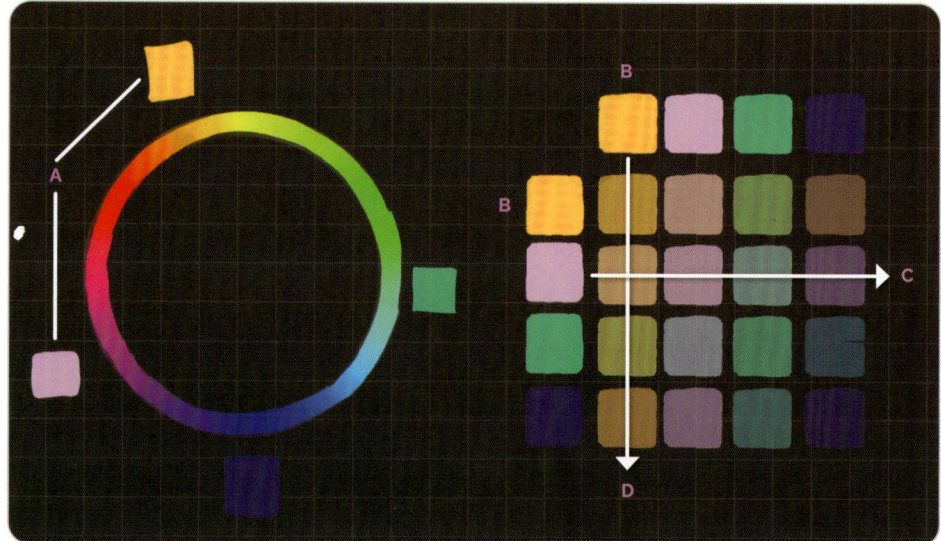

15 If you're struggling to decide on a colour palette, here is an extremely easy and effective method. Start by choosing a few random colours, or colours you know you want to include in the illustration (A). Here, pink is chosen for the hare and orange for the magic effect. Put these colours in a row and then repeat them in a column (B). Now, with the brush at 70% opacity, paint over the colours in the rows (C). Next, set the brush to 30% opacity and paint over the colours in the columns (D).

< Mixing the colours, with different amounts of transparency, results in a relatively harmonious palette that you can use as a starting point

16 Create a new folder and place all of your colour fills (each object separated by a colour) beneath the line work. Use the Fill tool (paint bucket) to fill areas, though this may need some configuration to work correctly. For example, you may need to set the line-work layer as reference. A simpler method is to fill the areas with colour by hand. Start by drawing the outline of the shape using a hard-edged tool, such as the default G-Pen. With the shape fully outlined, switch to the Fill tool and click inside it. You can increase the value of Area Scaling so the colour spills into the outline slightly. For a more advanced version of this technique, set a shortcut for the G-Pen sub tool by selecting File > Modifier Key Settings. You can then also set Alt+Shift to switch to the Fill tool, for example.

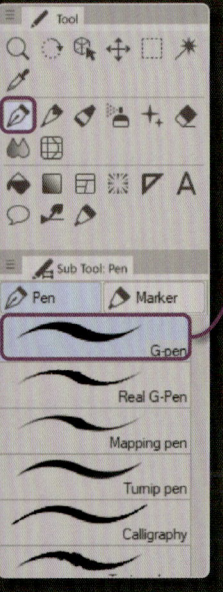

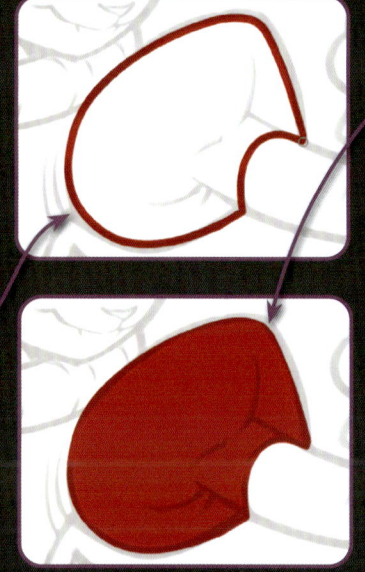

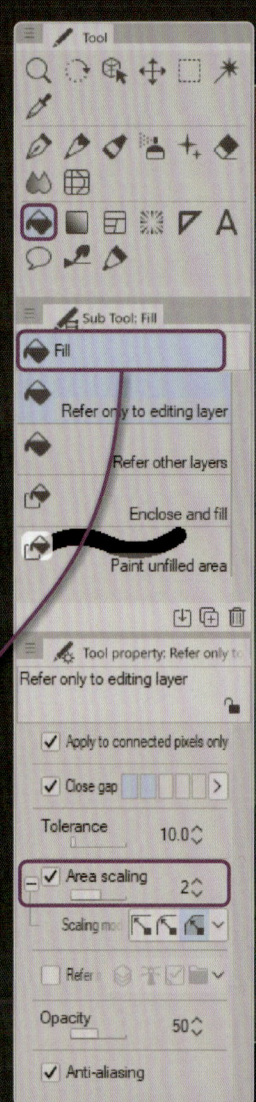

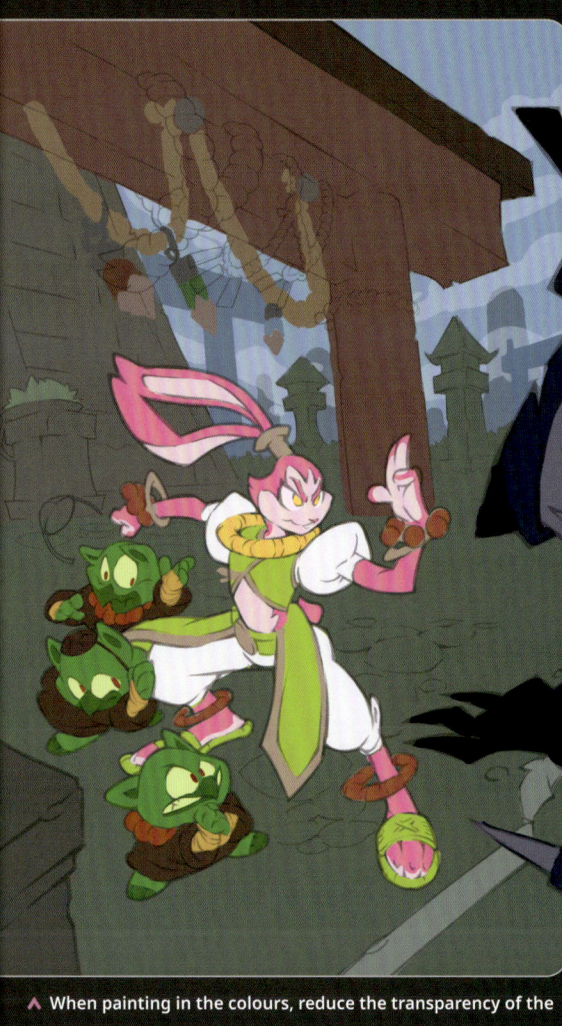

∧ When painting in the colours, reduce the transparency of the line-work layer to enable you to fill behind it

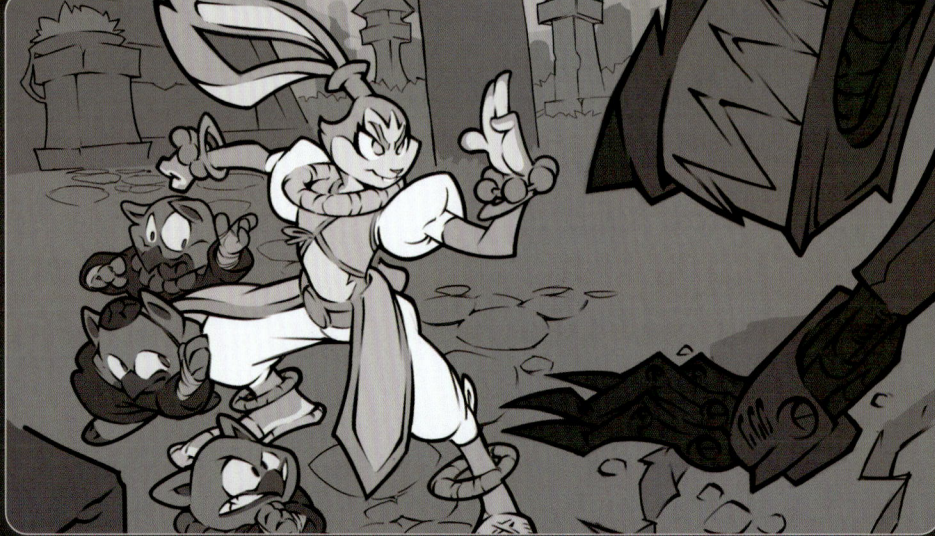

17 With all of the colours set, complete the following step to make sure they add to your composition. Create a new layer on top of everything, fill it with black, and set the blend mode to Colour. This will let you preview the brightness of each colour. Next, while the image is still in greyscale, select each colour (which is why it's important to paint them on individual layers!) and use Edit > Tonal Correction > Hue/Saturation/ Luminosity (or Ctrl+U) to make them brighter or darker. Adjust them to alter the contrast of the image, favouring the focal points.

< As you adjust each colour, remember to only introduce areas of contrast on the focal points

18 A final step before flattening the image is to colour the line work, especially if you want to give the illustration a more painterly look. Create a new layer on top of the line-work layer, right-click on its thumbnail in the Layer panel, and select Clip to Layer Below. Reduce the opacity of this layer and paint over the lines using nearby colours. By adjusting the opacity of this layer, you can make the lines more or less visible.

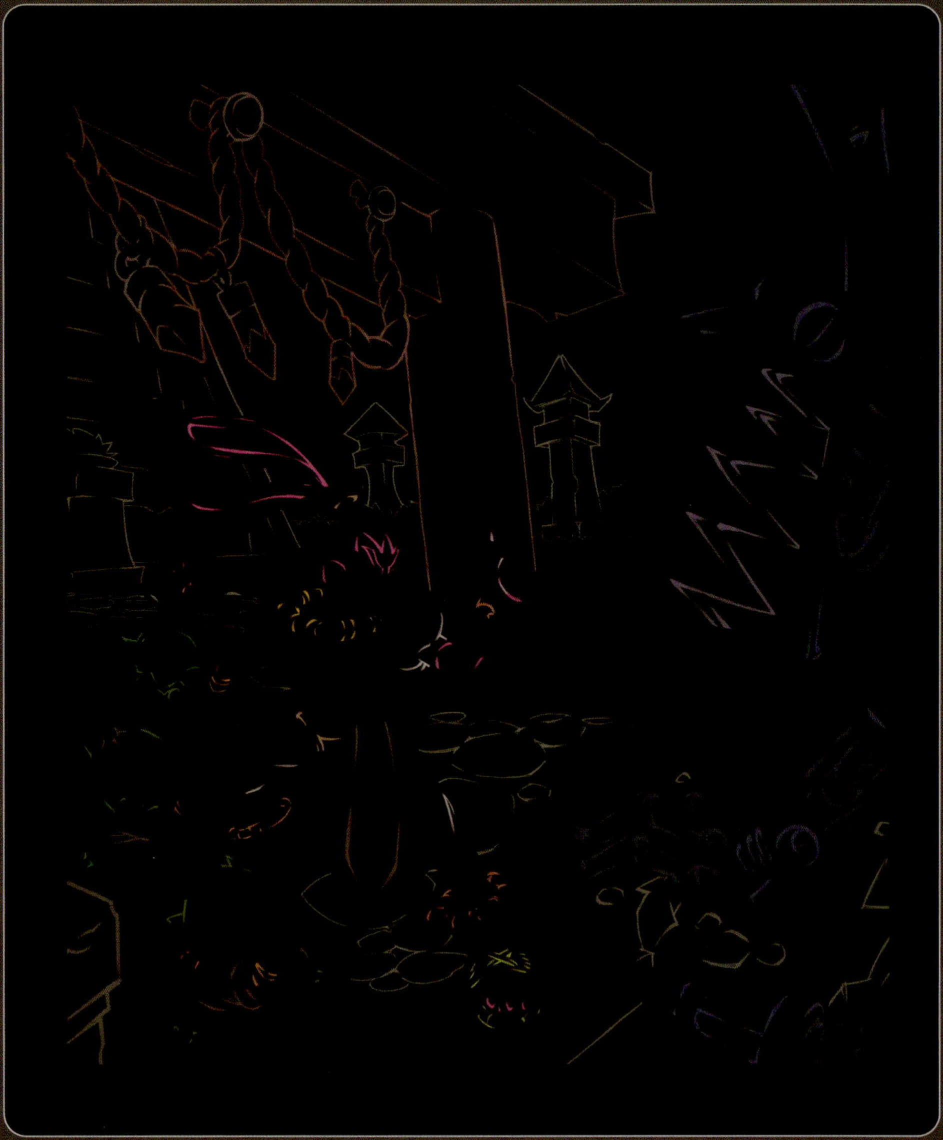

> Colour the line work to give the
painting a more refined look

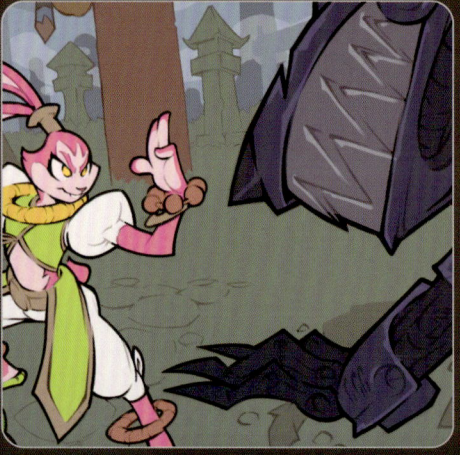

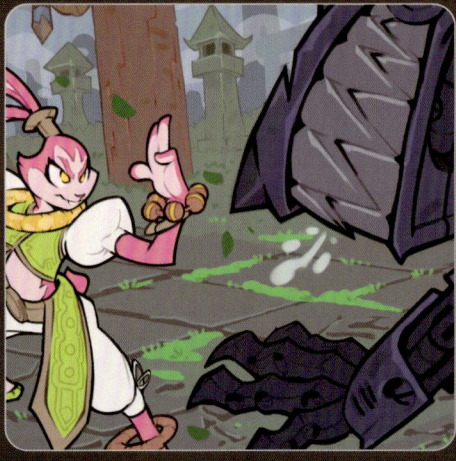

19 Remember that idea of splitting the tasks of illustration and art direction (see Artist's Tip page 175)? Now it's time to put on your director's hat to analyse the image as a whole. Right-click on the layer in the Layer panel, select Merge Visible to New Layer, and start painting from that point. Start by fixing the fox-robot's foot and changing the staircase idea to a tiled stone floor. Introduce other details to add to the spiral rhythm envisioned in the thumbnail stage.

< Add in details that help to tell
the story of the illustration

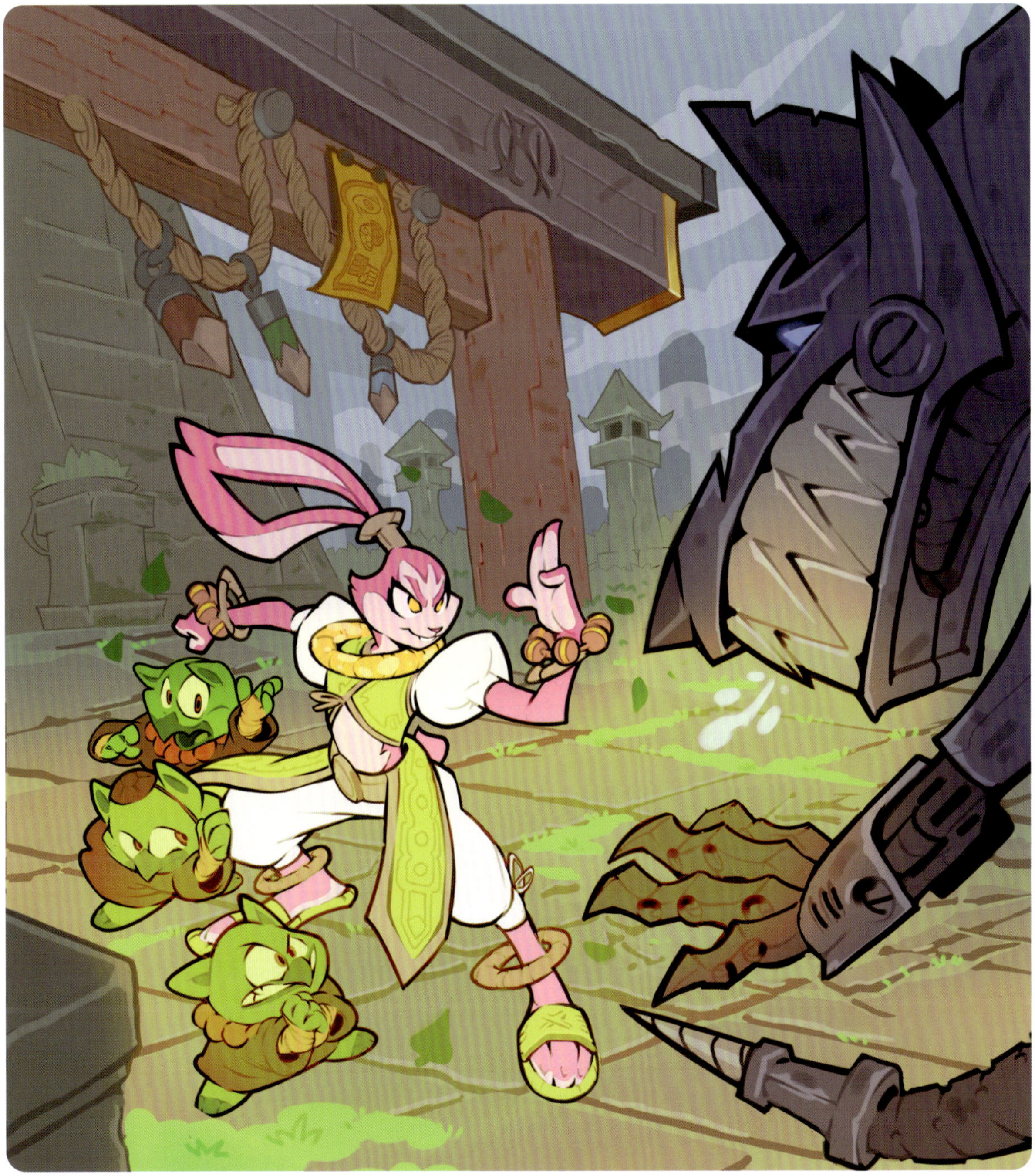

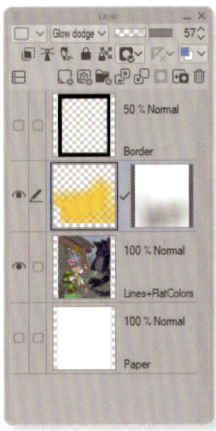

< Introduce a glowing light effect to emphasize the magic spell being cast

20 The intention is to keep a flat, cel-shaded look, as it favours the lines of the drawing and the local colours. However, the character is casting a defence spell, which will need to be emphasized with a cool light effect. Start by thinking about the main light in the scene. Where does it come from? What colour is it? Use the Soft Airbrush or the Gradient tool to tint the whole image with that colour. Experiment with a lightening blend mode – such as Glow Dodge, Overlay, or Soft Light – and different opacity levels.

Here, the main and strongest light of the scene is coming from the magic spell, which will be a glowing flame-like ring beneath and around the characters. The orange-yellow gradient is strong here, and then fades away towards the top of the image. While this may not be the way light works in real life, who says that every picture must have realistic lighting?

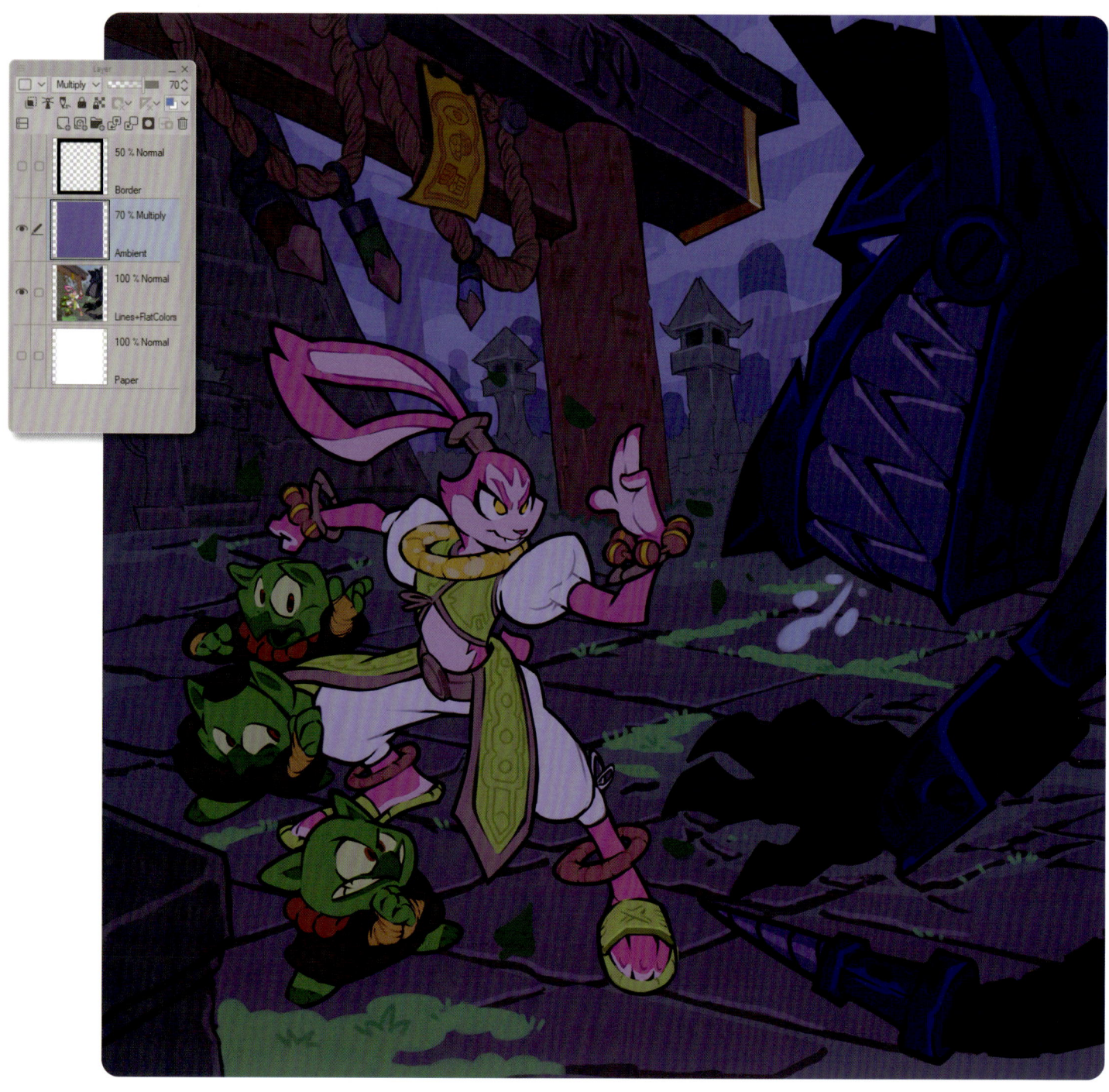

21 Create a new layer and fill it with a general colour for the shadows. When deciding which colour to use, think about the environment surrounding the characters. An outdoor scene with a clear, cloudless sky will cast a lot of blue on the shadow colours, whereas a night scene with a bonfire could have red and orange dominating the parts

not lit by the fire. Sunlight, daylight, overcast, fantasy, magic... Think about which lighting quality will affect your environment most and the story it will tell.

Here, a violet-blue, with the layer in Multiply mode, is chosen, as it creates a cold vs. warm contrast with the orange colour of the light.

▲ **Choose a cool violet-blue colour for the shadows to contrast against the warm orange of the magic glow**

22 On that same shadow layer, create a mask and use an eraser or transparent colour to paint the areas the light hits. Removing the shadow layer will reveal the colours underneath. As these will already be tinted with the colour of the light, this creates a combination of light and coloured shadows. It's almost like using transparent coloured gel papers over your artwork to give the impression of colour shifts.

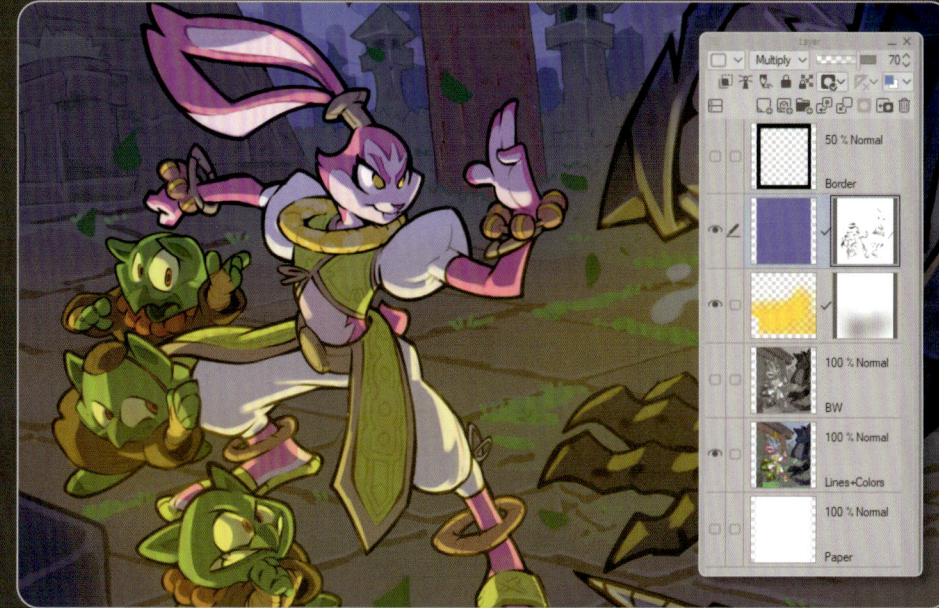

▶ Use a mask to remove the shadow layer on the highlight areas

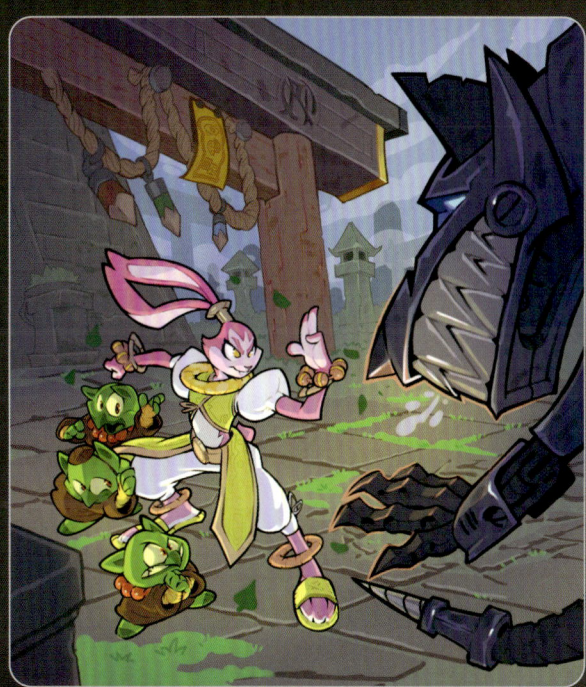

23 Merge the entire illustration into one layer by right-clicking the Layer panel and selecting Merge Visible to New Layer. You can now review the image with a more analytical eye, adding overpainted elements, fixing edges, adding details, and improving the flow of the composition. Use this step to paint in the magic effect – the glowing flame-like ring that protects the characters from the mechanical fox. This has been saved until last as it's a little more complicated than some of the other details and it can help to focus on the easier tasks first.

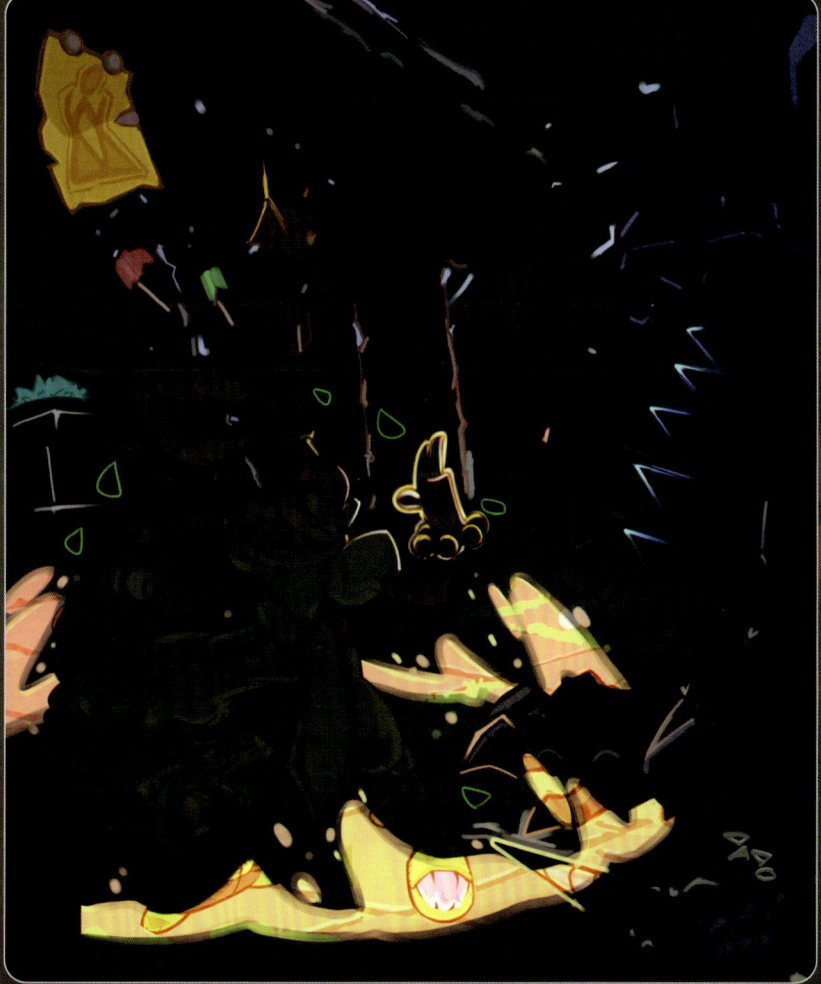

⋀ Pause between each step to take a step back and analyse the image, then fix any errors

24 Finish the illustration by making brightness, contrast, and colour corrections. If you do this using **Correction Layers**, none of the corrections will be permanent, making them easier to edit if you change your mind. Digital art is made of light. When printed, everything will look a little darker and desaturated, so you must learn how to compensate for this. You may wish to check the colours on another monitor or tablet. If so, create a new canvas window by selecting **Window > Canvas > New Window**, and then place the canvas on another screen. Once you're happy with how it looks, sign the image with your signature, even if it's hidden in a corner. To save the image, export it as a JPEG for the web or a PNG for printing.

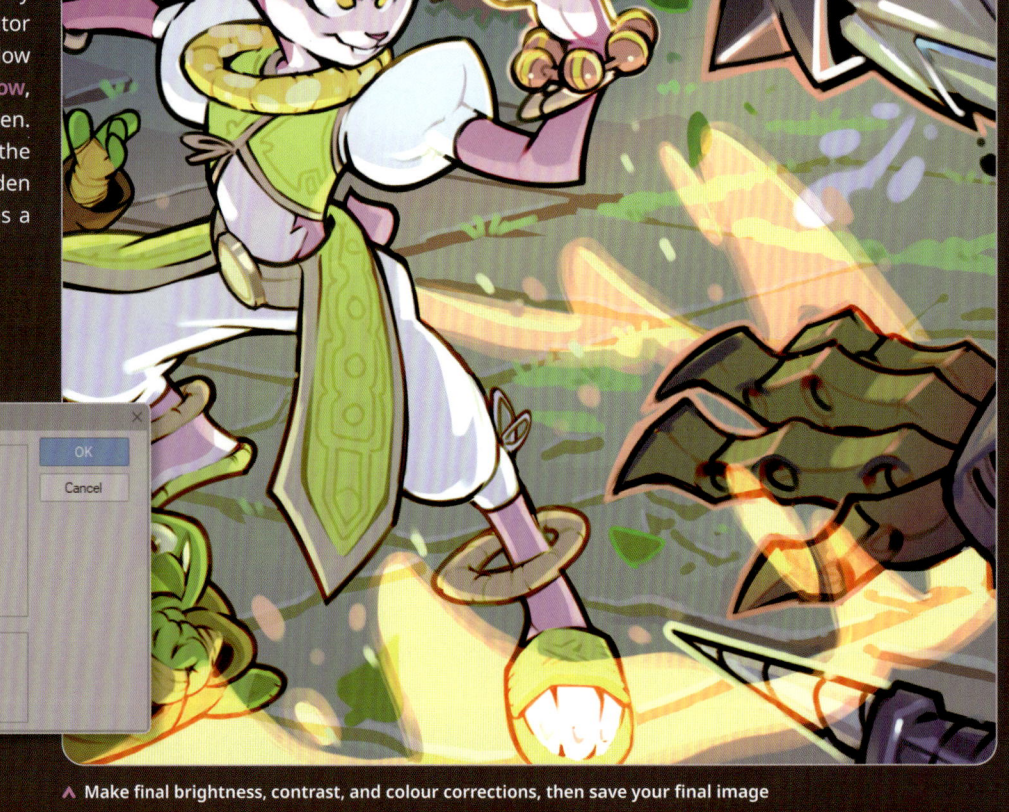

∧ Make final brightness, contrast, and colour corrections, then save your final image

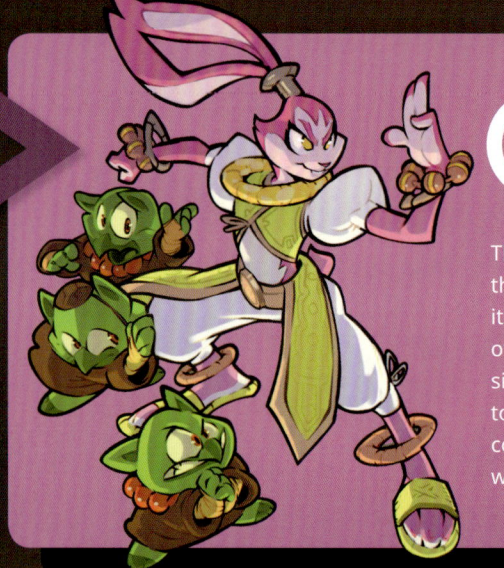

Conclusion

The illustration tells the story of a brave monk hare protecting his friends against the attack of a menacing robot-fox. It also acts as a metaphor for a larger message: it's a reminder to artists to keep defending their art and creativity against the threat of AI. Having a loyal circle of friends and the force of the art community on your side will help you to stand firm and remain focused on making the artwork you love to create. The final painting represents resistance, endurance, and hope. As you continue your artistic journey with Clip Studio Paint, think about what stories you want to tell. How can you use your artwork to communicate a message or narrative?

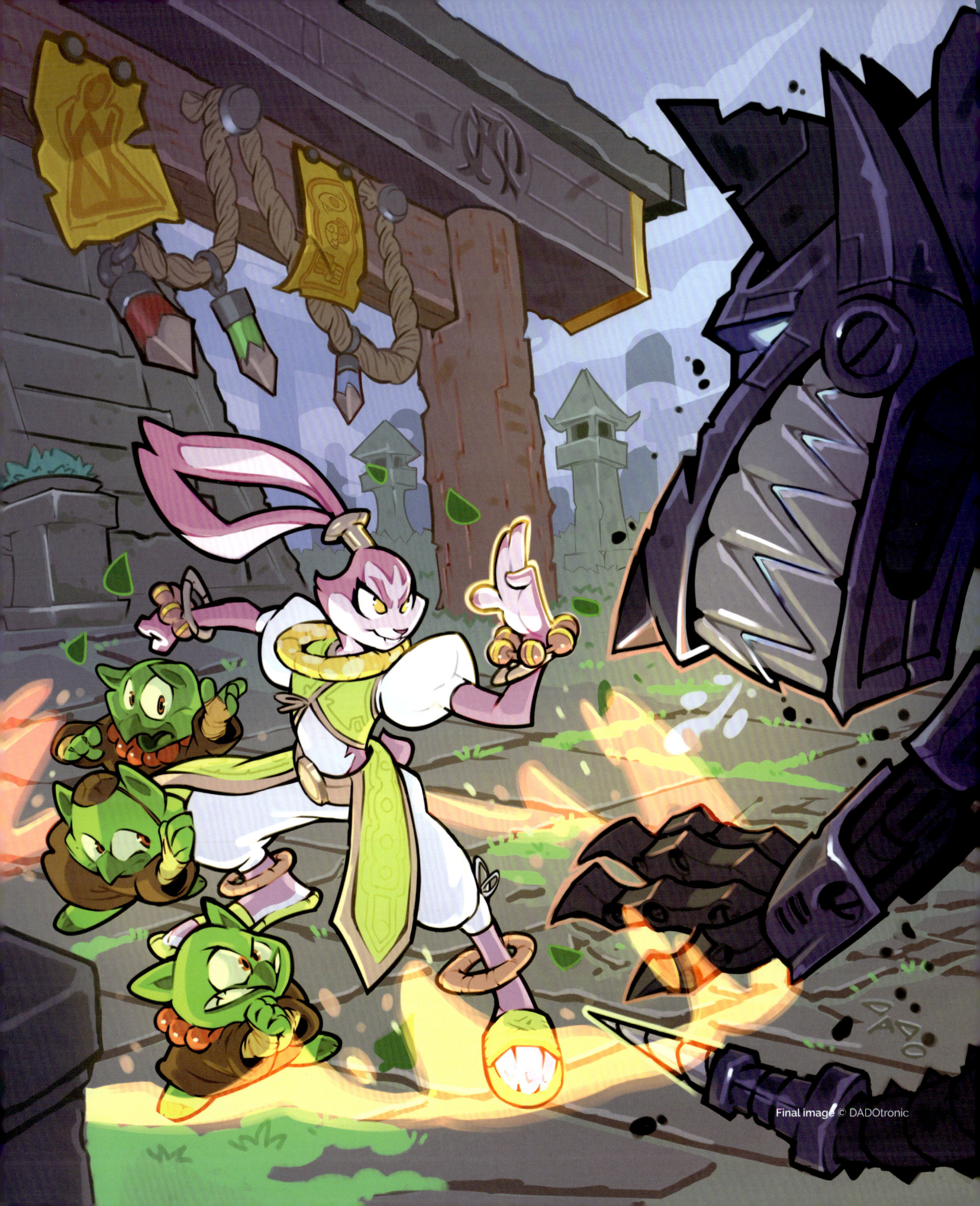

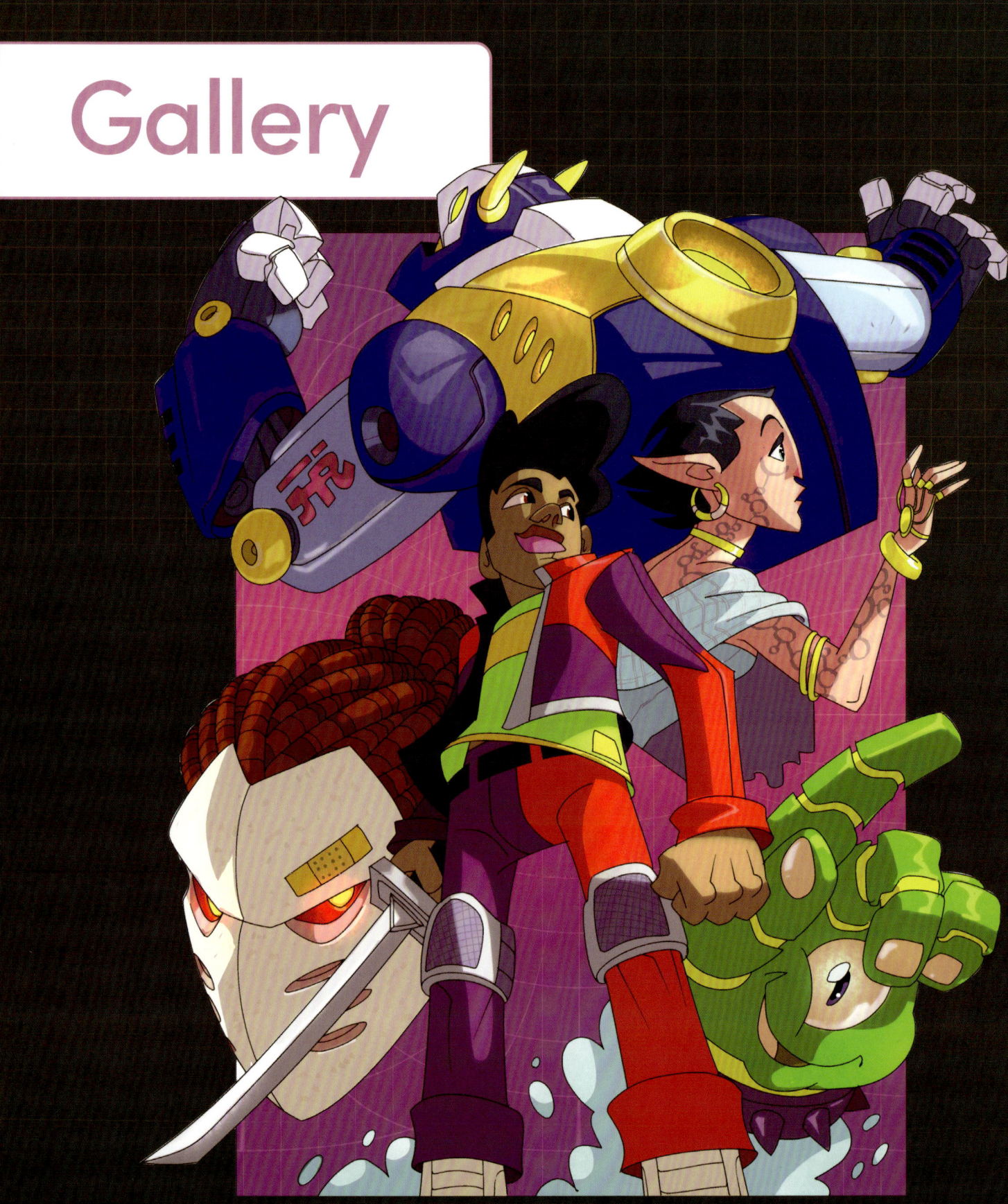

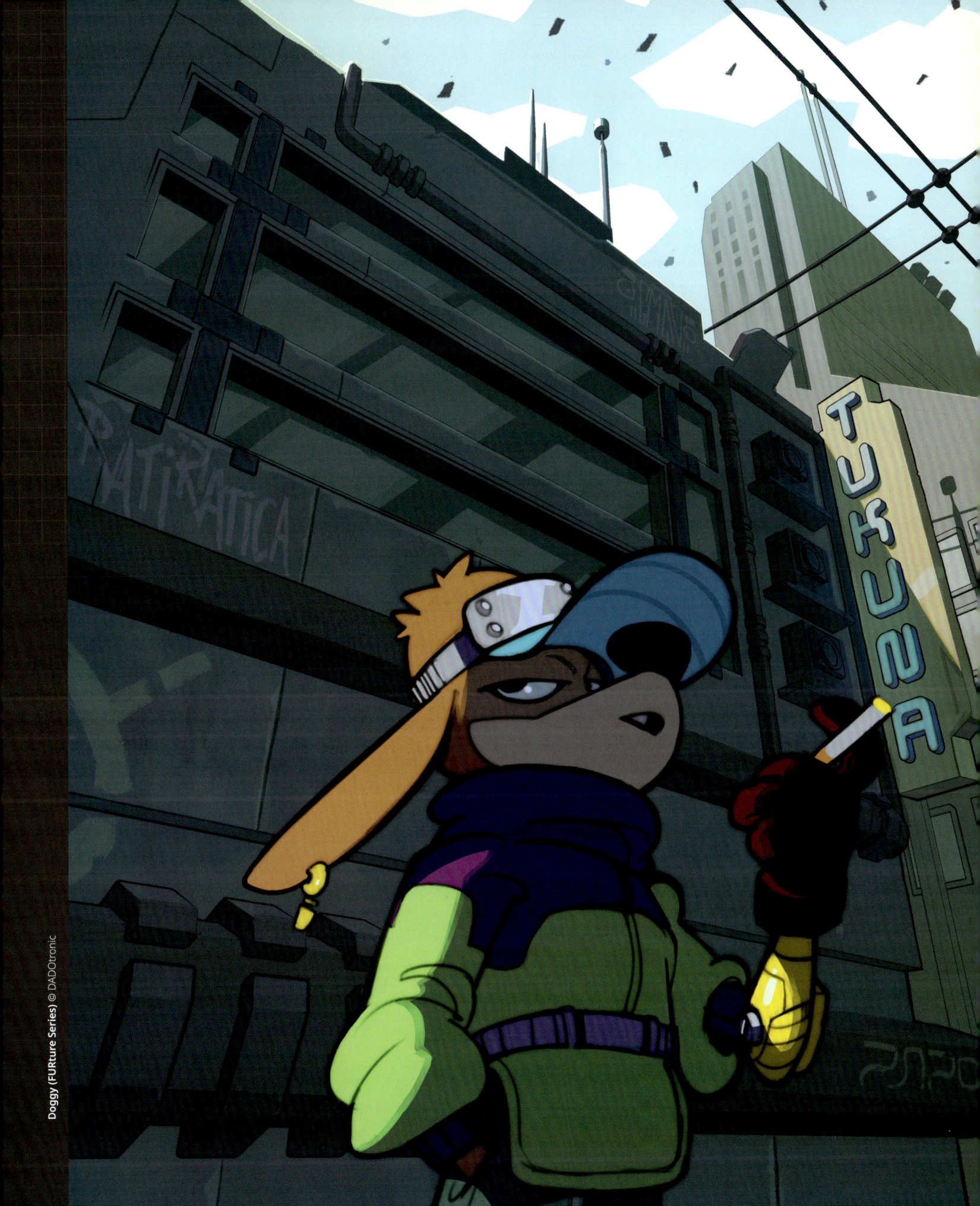

Real-world scene

BY ORENJIKUN

I will start my tutorial by teaching you how to begin a digital painting in Clip Studio Paint, from finding inspiration through to thumbnailing. I will then share a series of tools, techniques, and best practices to take an illustration from start to finish, resulting in a fully rendered, vibrant landscape scene.

The following pages will cover Clip Studio Paint's features for drawing and painting, diving into the intricacies of composition, colour, lighting, set dressing, perspective, and more! I will explore how and when to use layers, as well as brushes and perspective grids. You will also learn how to use digital tools such as blending modes, blur filters, and the Lasso Fill sub tool. By the end of the twenty-two steps, you will have peeled back the curtain to discover the thought processes behind my design decisions, all of which will help to turn your ideas into an engaging illustration full of depth and wonder.

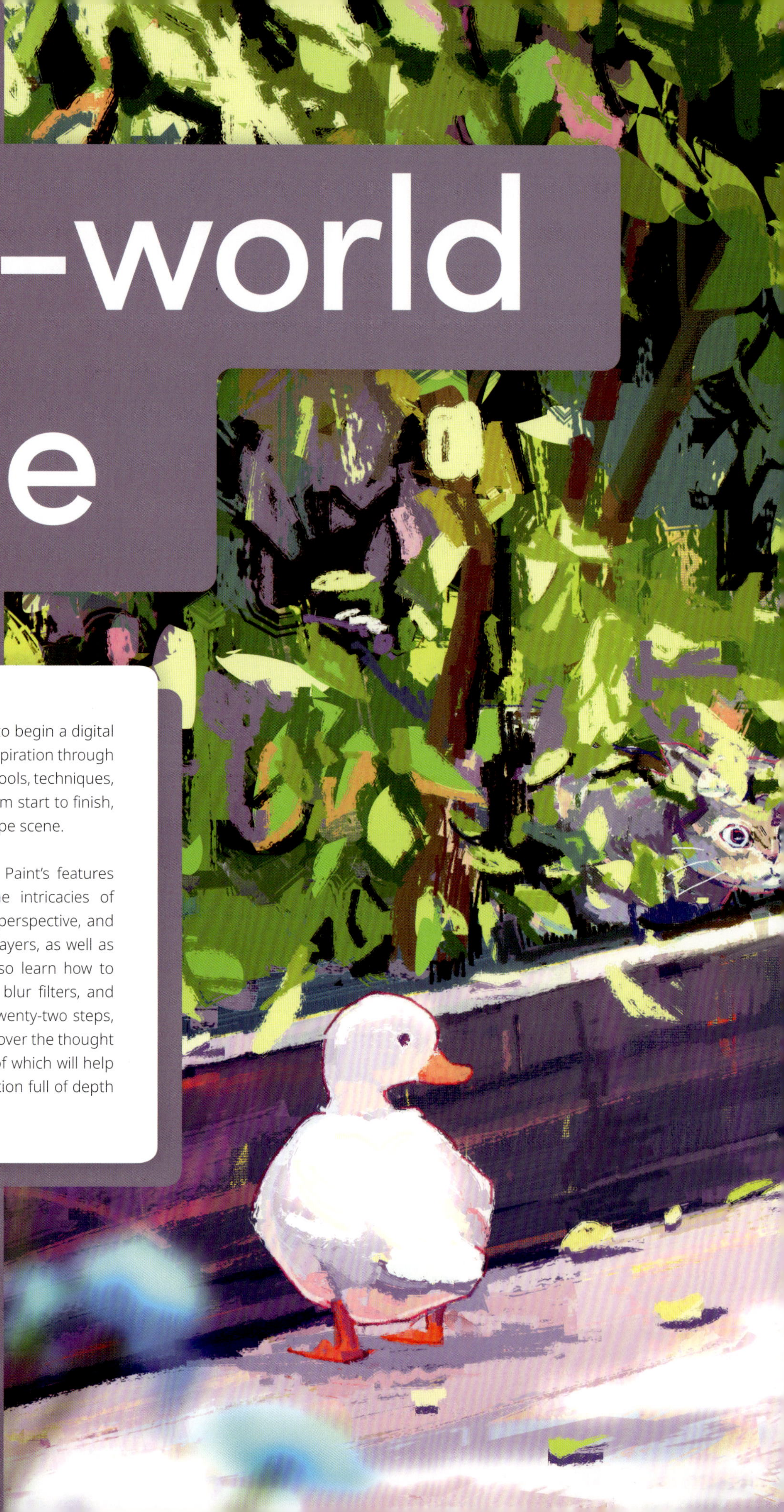

Final image © Orenjikun

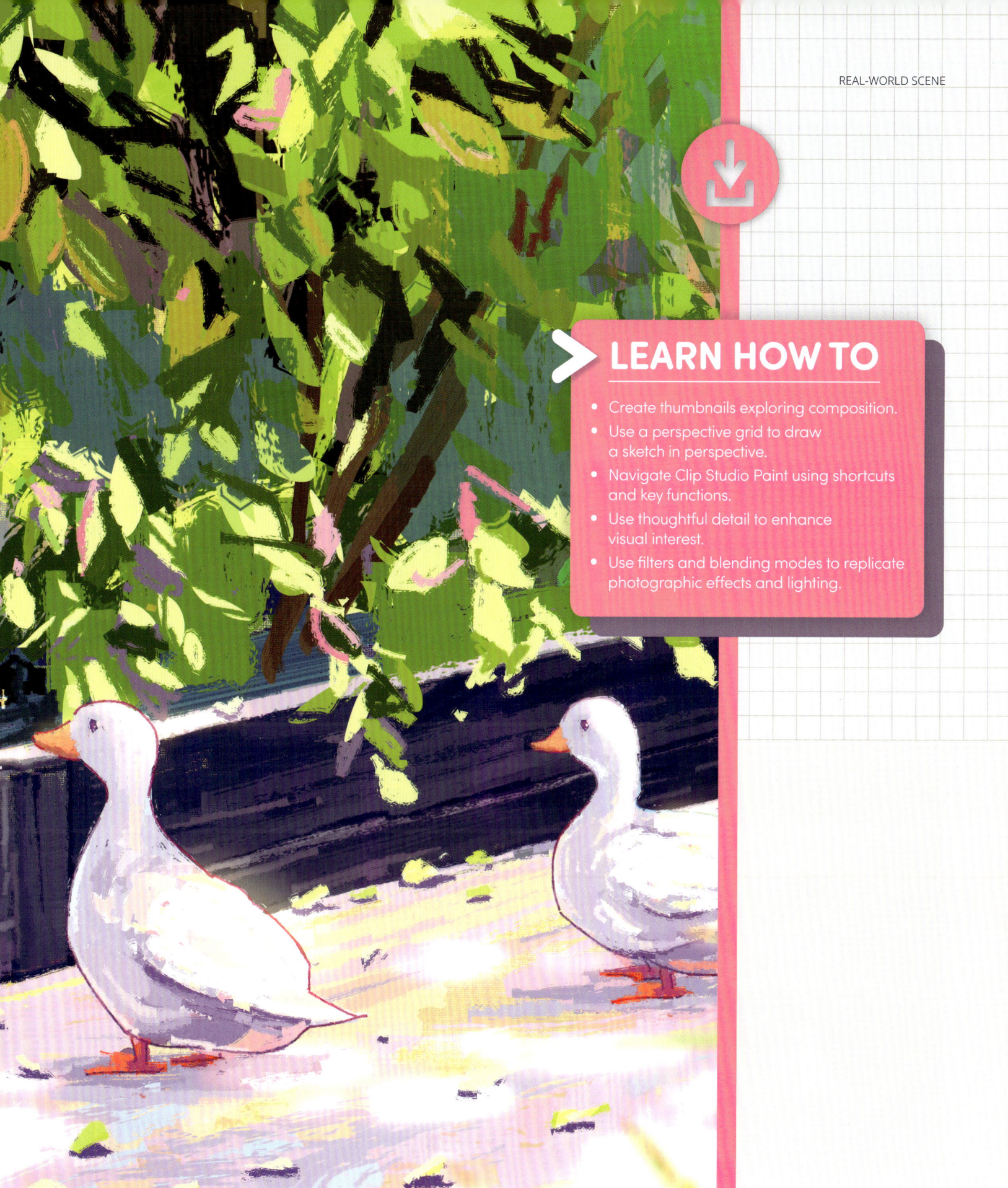

> LEARN HOW TO

- Create thumbnails exploring composition.
- Use a perspective grid to draw a sketch in perspective.
- Navigate Clip Studio Paint using shortcuts and key functions.
- Use thoughtful detail to enhance visual interest.
- Use filters and blending modes to replicate photographic effects and lighting.

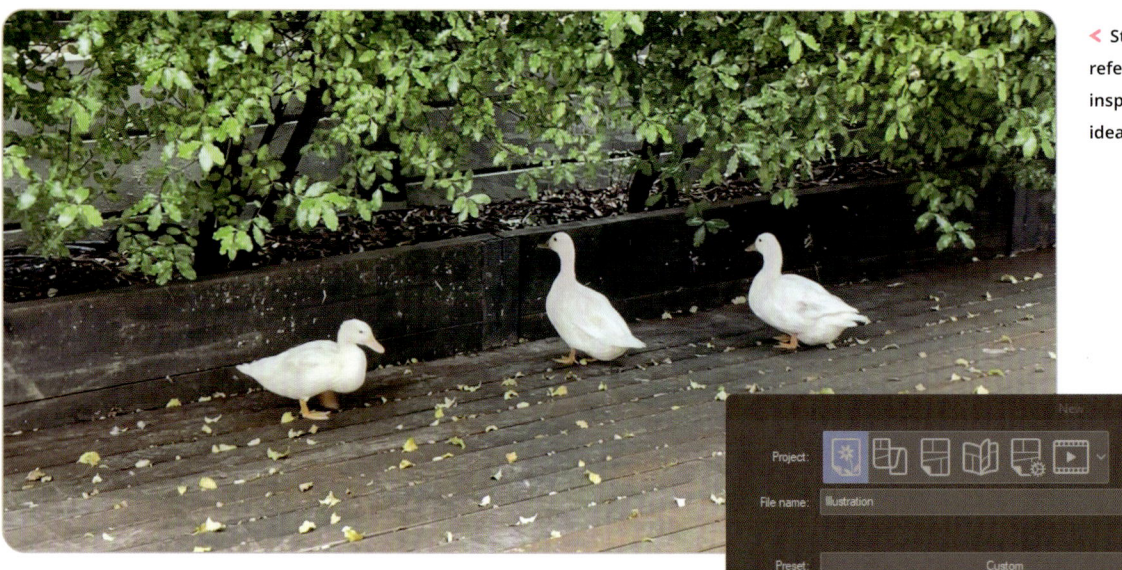

‹ Start by finding reference material to inspire and inform your idea for the painting

01 The creative process starts with research to find reference material. Using references will help to inform various elements of your artwork, such as subject matter, lighting, composition, perspective, colour, or anatomy. There's no limit to the amount of reference material you can draw from to inspire your design. This real-world landscape scene will be inspired by this photograph of ducks, while also aiming to include lots of foliage in the background. Once you've chosen your reference material, begin by creating a new document by selecting **File > New** or pressing **Ctrl+N**. Choose a canvas size from the drop-down menu or set a custom size, set your resolution to 300 dpi, choose your orientation with the arrows, and then select OK. The canvas size of this image will be 270 × 460 mm.

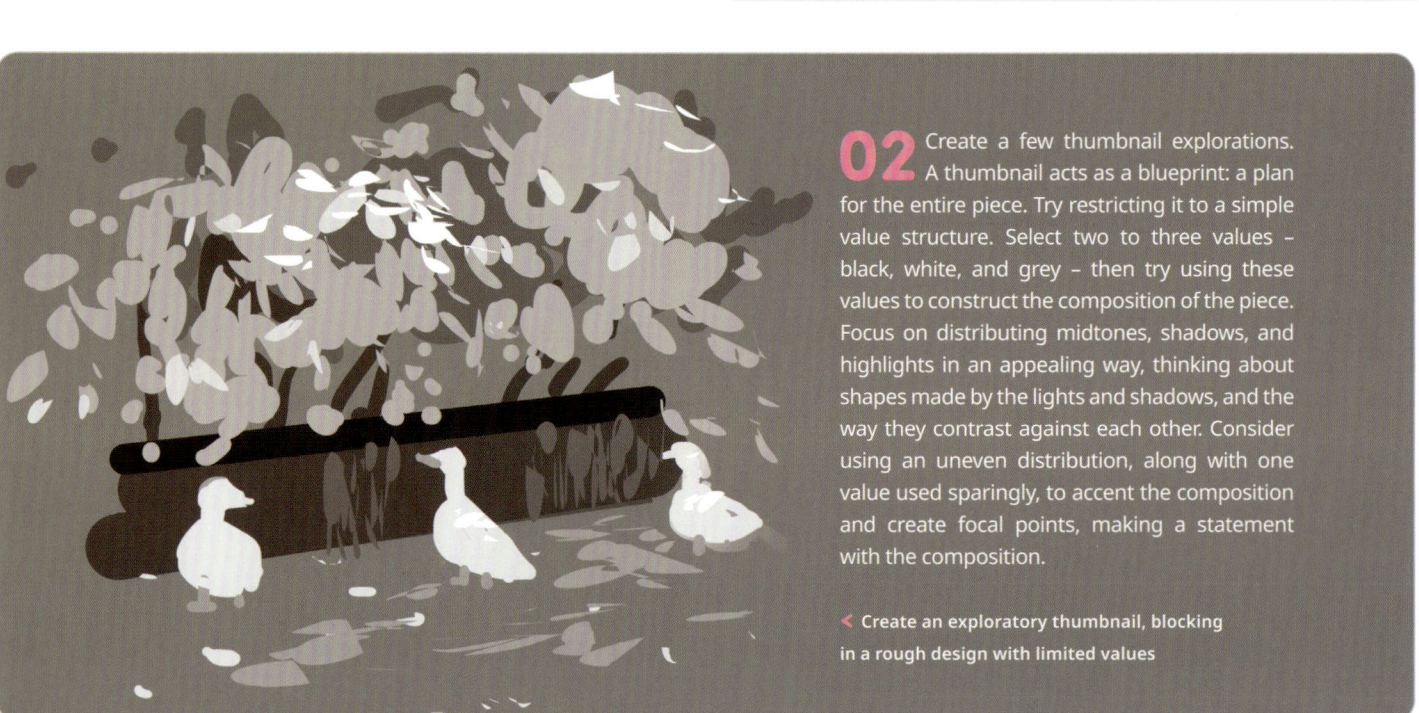

02 Create a few thumbnail explorations. A thumbnail acts as a blueprint: a plan for the entire piece. Try restricting it to a simple value structure. Select two to three values – black, white, and grey – then try using these values to construct the composition of the piece. Focus on distributing midtones, shadows, and highlights in an appealing way, thinking about shapes made by the lights and shadows, and the way they contrast against each other. Consider using an uneven distribution, along with one value used sparingly, to accent the composition and create focal points, making a statement with the composition.

‹ Create an exploratory thumbnail, blocking in a rough design with limited values

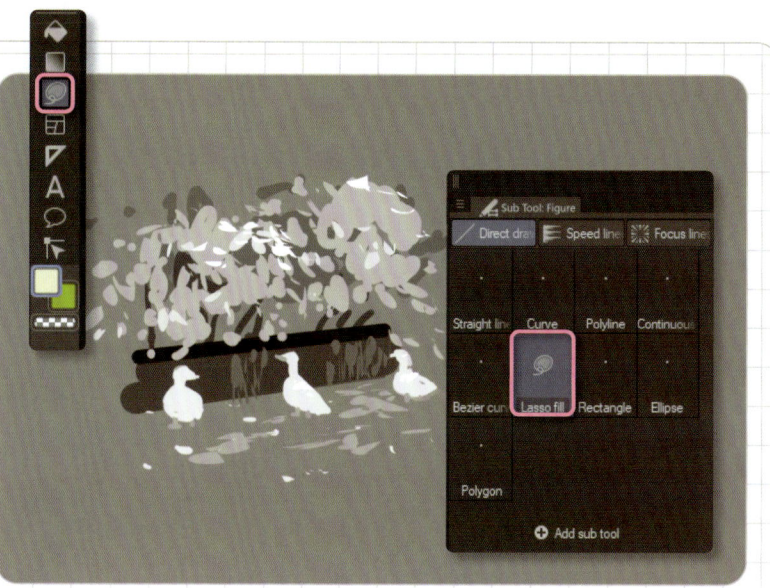

03 The Lasso Fill sub tool, found under Figure > Direct Draw, utilizes the lasso function to make selections, then immediately fills these selections. The tool forces an emphasis on shapes, tones, and colours, without getting distracted by details or trying to create perfect edges and textures. It's an ideal tool for the thumbnail stage, where the goal is to achieve broad statements with speed and efficiency. As thumbnailing is only focused on capturing an impression, rather than worrying about the final look, you can also use a hard round inking brush with solid colour, such as Pen > G-pen, to block in a rough composition.

> Use the Lasso fill tool and a hard round brush to explore a basic composition

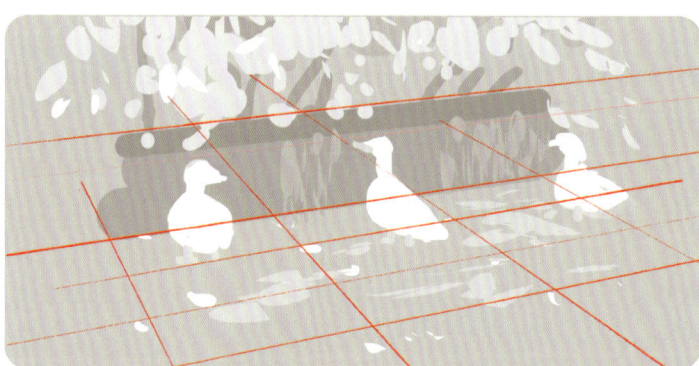

04 Perspective is an important aspect of drawing an environment or background scene. Much of experience in real life can be approximated by the rules of linear perspective in art. Depicting backgrounds in perspective helps to ground the piece in reality, even if the setting is fantastical, adding to a feeling of immersion and producing a richer experience for the viewer. Sketch in an indication of three-point perspective, or if you don't feel confident enough to attempt this freehand, create a new layer on which to set up a Three-Point Perspective Grid. (Perspective Grids will be covered in Step 19 on page 198.)

< Add a Three-Point Perspective Grid to help rough-in the environment

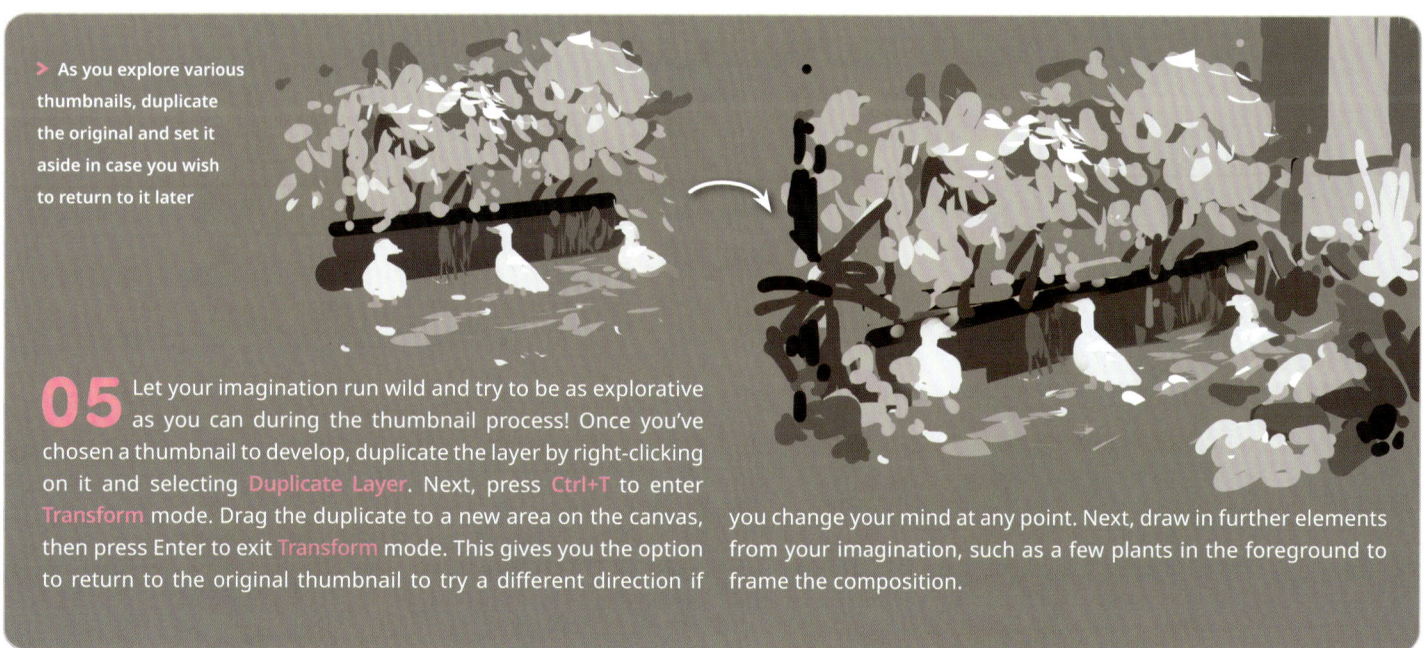

> As you explore various thumbnails, duplicate the original and set it aside in case you wish to return to it later

05 Let your imagination run wild and try to be as explorative as you can during the thumbnail process! Once you've chosen a thumbnail to develop, duplicate the layer by right-clicking on it and selecting Duplicate Layer. Next, press Ctrl+T to enter Transform mode. Drag the duplicate to a new area on the canvas, then press Enter to exit Transform mode. This gives you the option to return to the original thumbnail to try a different direction if you change your mind at any point. Next, draw in further elements from your imagination, such as a few plants in the foreground to frame the composition.

06 Once you're happy with the thumbnail, duplicate it onto a new layer and press **Ctrl+T** to enter **Transform** mode. Resize the thumbnail to fill the canvas by dragging the arrows in the corners. This thumbnail will act as the blueprint for the artwork.

The next step is to prepare the canvas with an underpainting. This is a technique used in traditional painting to prepare a canvas. It can be used to set the tone for a piece and can help unify your colour choices throughout the painting. Try using more saturated tones and more heavily textured strokes to lay down the underpainting. The saturation will give the painting a vibrant quality, while the texture will provide more visual interest and variation. As you paint over the top of the underpainting in the following steps, leave small gaps to let the underpainting peek through. Create a new layer by selecting the **New Raster Layer** icon in the Layers panel, or using the shortcut **Ctrl+Shift+N**. Next, select **Brush > Thick Paint > Oil Paint**, as its texture imitates the quality of a traditional fabric canvas.

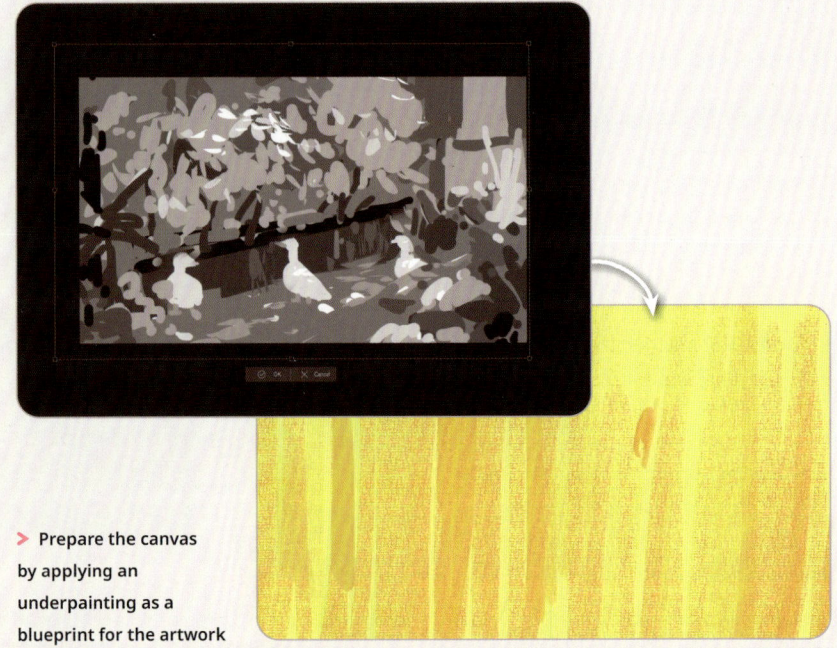

> **Prepare the canvas by applying an underpainting as a blueprint for the artwork**

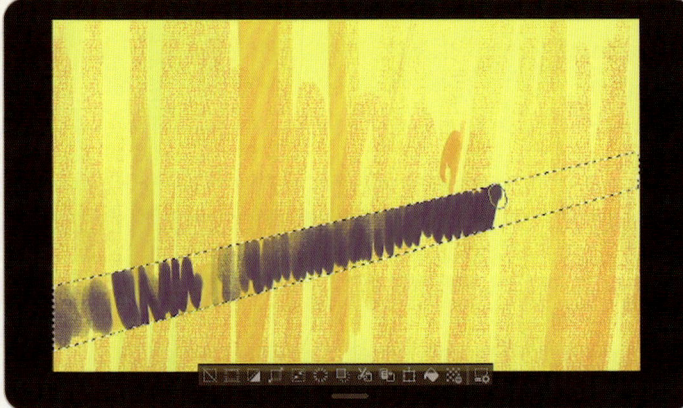

07 Making selections then painting within them is a great way to apply textured brushstrokes while keeping the edges clean. For example, this technique will allow you to apply texture for the wooden garden bed. This effect could also be achieved on a separate layer by using a layer mask, or with the messy edges erased away afterwards, but the selection technique will allow you to paint the whole painting on one layer to utilize the underpainting and colour pick from it. To make more flexible selections, hold **Shift** to add to the selection, and **Alt** to subtract from the selection. If you find the 'marching ants' of the selection too distracting, hide them by clicking the button on the Command bar, or selecting **View > Selection border**. You can also hold **Shift** when using the **Brush** tool to make a straight line.

< **Familiarize yourself with various tricks and shortcuts to aid the painting process**

> ARTIST'S TIP

Setting up shortcuts is a fundamental aspect of digital painting that will save you time in the long run. Customize the shortcuts for your most-used tools by selecting **File > Shortcut Settings** or **Ctrl+Alt+Shift+K**. For example, if you frequently use curves to finesse values, why not add **Tone Curve**, found under **Edit > Tonal Correction > Tone Curve**, to a shortcut. Similarly, you could make a shortcut for hiding the marching ants of a selection border by clicking **View > Selection border**, as seen in Step 07.

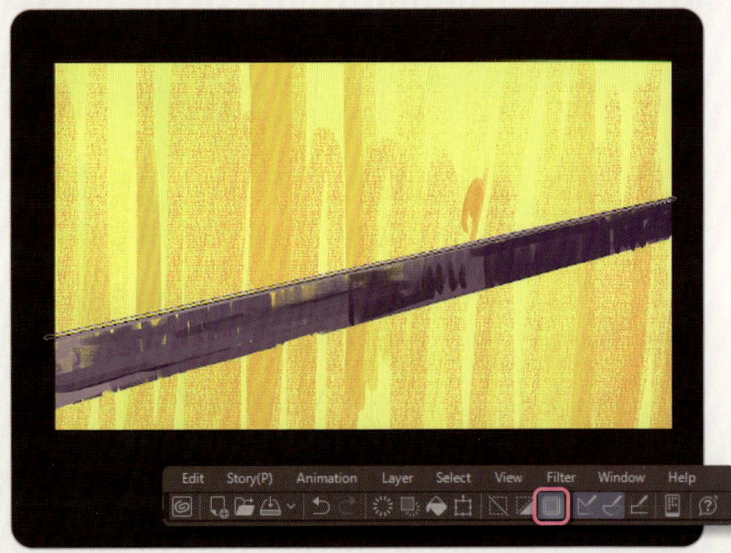

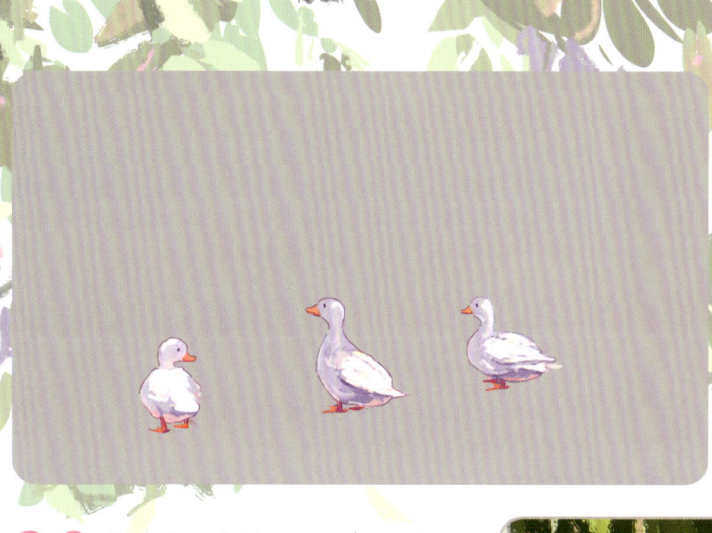

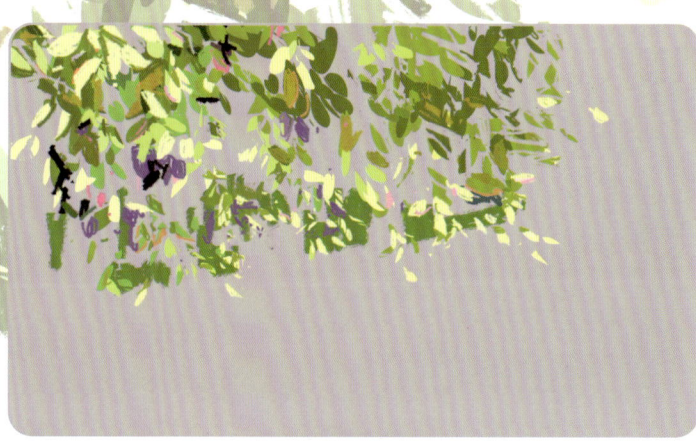

08 Begin to paint in more elements on separate layers. Layers are a useful tool and can be used effectively for separating various components of the painting, but not everything needs to be on a new layer. Paint the leaves on a new layer, as this will allow you to give them sharp, clean edges, while leaving rough, impressionistic brushstrokes behind them. Organize the layers as the elements would appear in real life, with the background elements below and the foreground elements at the top. For example, paint the bush and wooden plant beds on the back layer, and the foliage and ducks on separate layers towards the front.

> Paint the various elements on separate layers, working from back to front

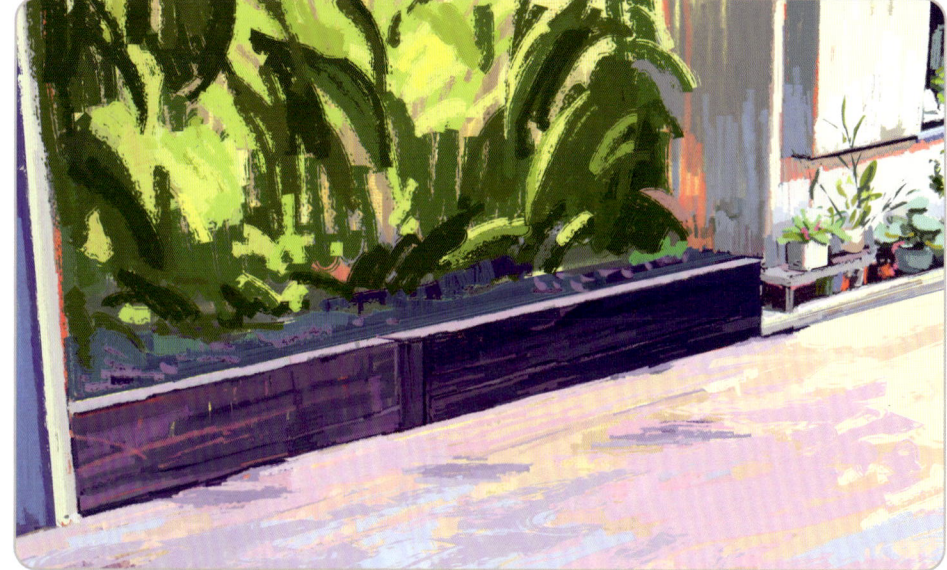

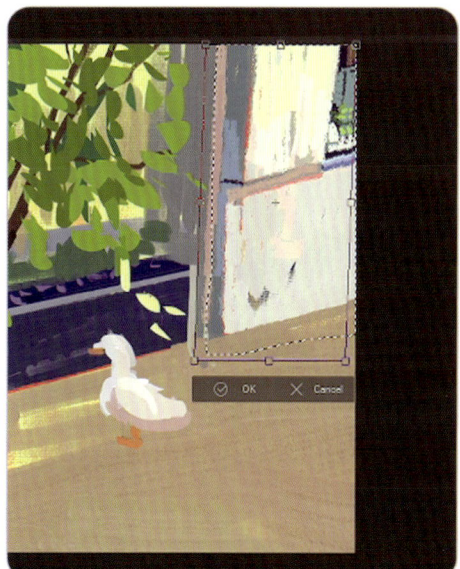

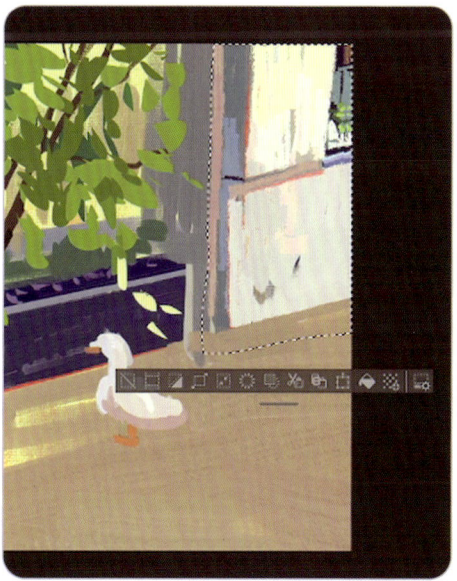

09 While in the early stages of a painting – before rendering, post-processing, and decorative enhancements – feel free to alter the illustration as you see fit. Don't be afraid to make bold choices to change the composition. Combining the Selection and Transform tools can be really useful in such situations. Try creating selections using the Selection Area tool in Rectangle, Polyline or Lasso mode, then transform them with Ctrl+T. Distort or resize a selection by holding Ctrl and dragging the corners. Here, the perspective of the building on the right side looks off. Using Select and Transform is a quick way to solve the problem.

< Make good use of the Transform tool and the flexibility a digital workflow allows to tweak the composition as required

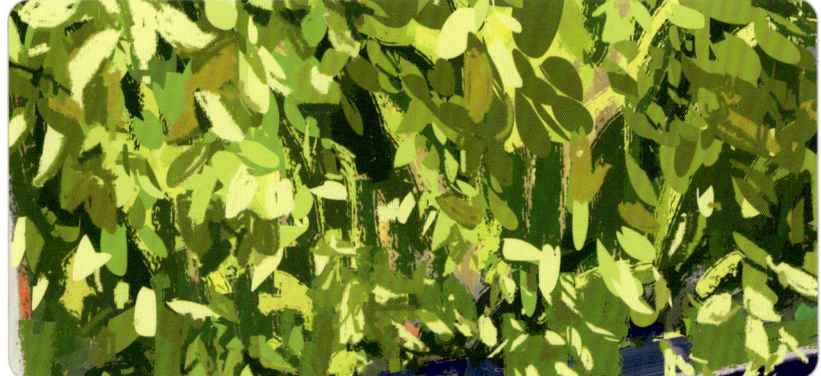

10 Once you've blocked in the shapes, lighting, and colours, the foundation of the painting is set. Move on to rendering, adding details as well as finessing edges and transitions. Look for opportunities to create interesting relationships between the colours, as well as ways to add colour variation. Use the hue cube and colour wheel in the colour wheel window to choose colours as you paint. Alt+click while using the Brush tool to sample any colour you are hovering over.

∨ Move onto the rendering stage, exploring ways to introduce colour variation

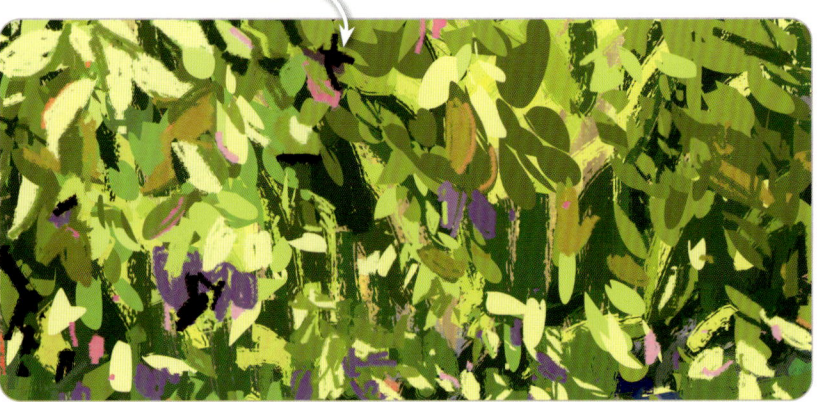

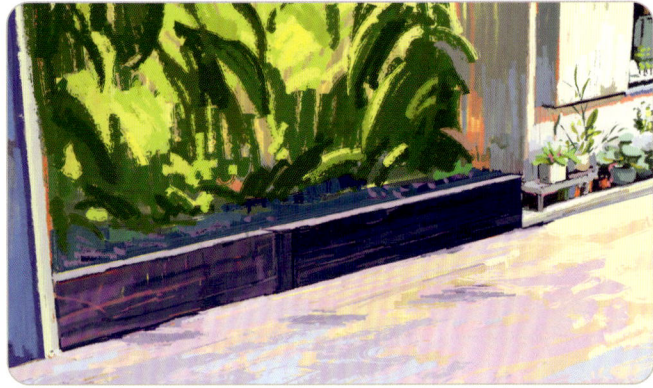

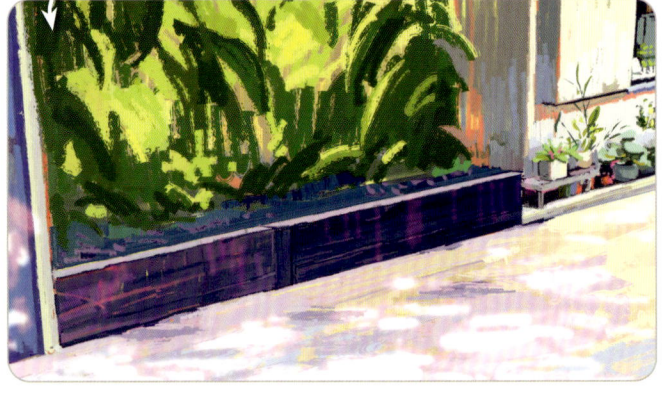

11 To create a dappled light effect on the ground, paint the dappled light on a separate layer set to Colour Dodge. Lighting is a good opportunity to work on a new layer, as you can alter the image in a non-destructive way. The image below the new layer won't be affected and the structure you've built will remain. To change a layer's blending mode, select the layer so it's highlighted, then change the blending mode in the Layer window. Colour Dodge mode is particularly well suited to depict strong direct light. When adding lighting, use a separate layer to allow for flexibility when adjusting the colour using the Hue/Saturation/Luminosity adjustment (Ctrl+U). You can tweak the hue, saturation, and value in real time to achieve the desired effect.

∧ Using separate layers and blending modes to depict lighting

Normal
Darken
Multiply
Color burn
Linear burn
Subtract
Lighten
Screen
Color dodge
Glow dodge
Add
Add (Glow)
Overlay
Soft light
Hard light
Difference
Vivid light
Linear light
Pin light
Hard mix
Exclusion
Darker color
Lighter color
Divide
Hue
Saturation
Color
Brightness
Normal

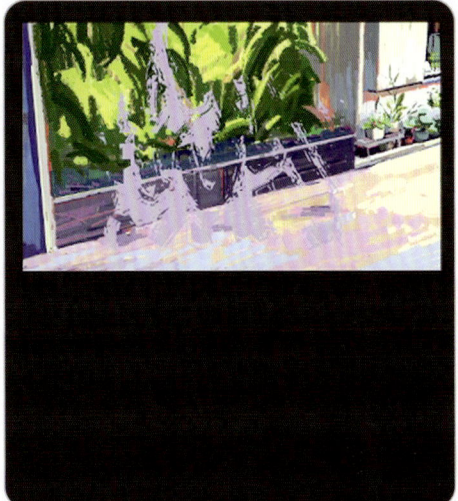
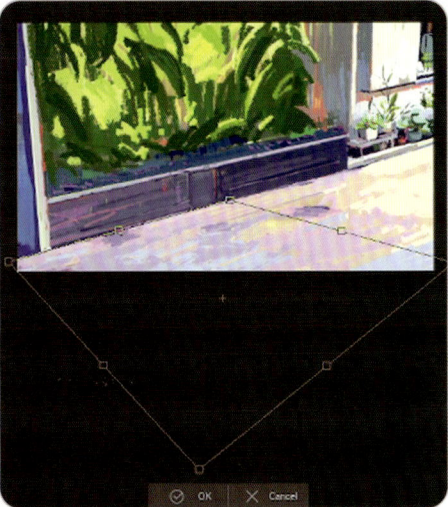

12 The next step is to paint in the texture of the ground, though depicting the subtle graininess of asphalt in perspective can prove tricky. While it can be achieved with precision and care, the process is time-consuming. There is, however, a time-saving method for painting texture in perspective. On a separate layer, create a flat texture using varied brushstrokes. Next, transform this layer using **Ctrl+T** and align it with the perspective of the object – in this case, the ground plane – and erase any unwanted parts.

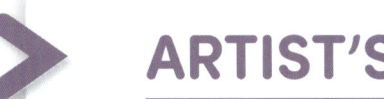

> Texturing and transforming flat assets in perspective

ARTIST'S TIP

When painting in colour, it's crucial to check values along the way. Value is an important aspect of colour and any muddiness in value can adversely affect the perception of your composition. To check your values, use a greyscale layer set to **Colour** to turn the composition black and white.

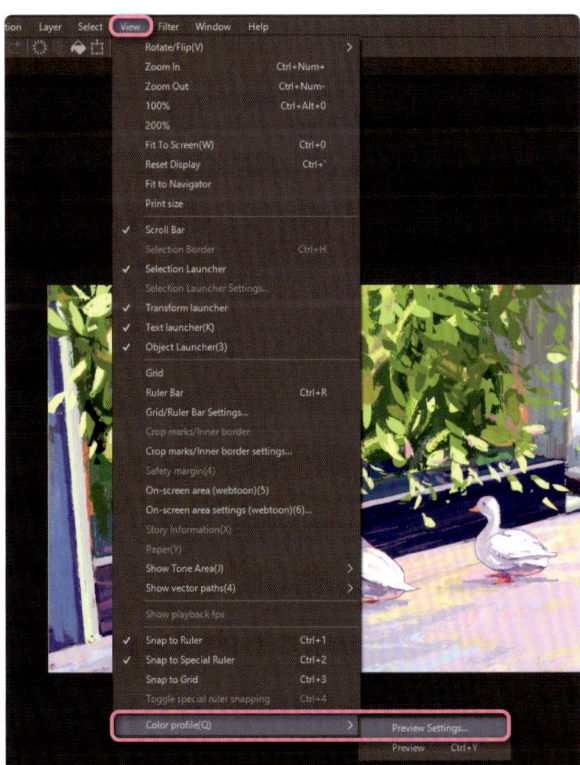

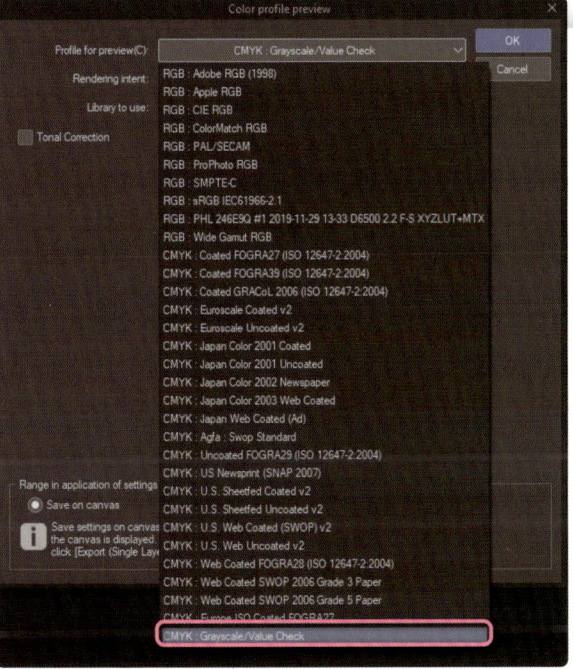

Alternatively, you can use a custom proof set-up. Set this up by copying the Greyscale&ValueCheck.icc file included in the downloadable resources (see page 257) to C:\Program Files\CELSYS\CLIP STUDIO 1.5\CLIP STUDIO PAINT\Settings\PAINT\ICCProfiles. (An ICC profile is a colour profile that converts the colours of an image into a particular colour space. This ICC profile accurately converts the image into greyscale, so you can check your values as you work.) Next, go to **View > Colour Profile > Preview Settings** and set the preview to **CMYK Greyscale/Value Check** in the drop-down menu. This will enable you to toggle between greyscale and colour previews, whenever you press **Ctrl+Y**.

13 The indication of bevels on turning forms can be a subtle way to enhance the read of volume and lighting. For example, where there is a larger crack or crevice on the wood grain, a brighter sliver of highlight should accompany it to indicate the edge of the wood reflecting light towards the viewer. This provides more emphasis on describing the dimensionality of the wood, adding texture and a sense of volume to help make your landscapes more lively and believable.

∨ Indicate lighting and form through the use of bevels

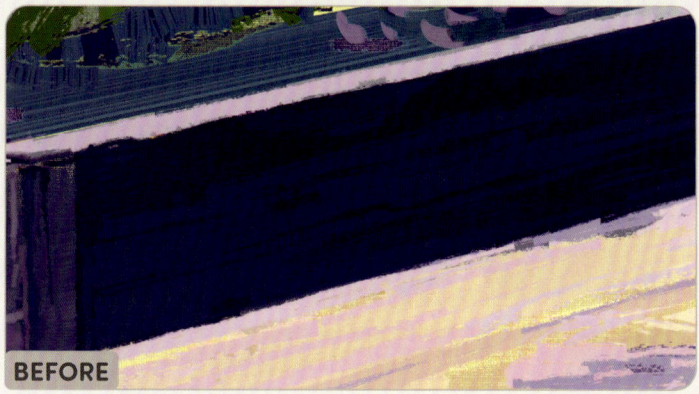

BEFORE

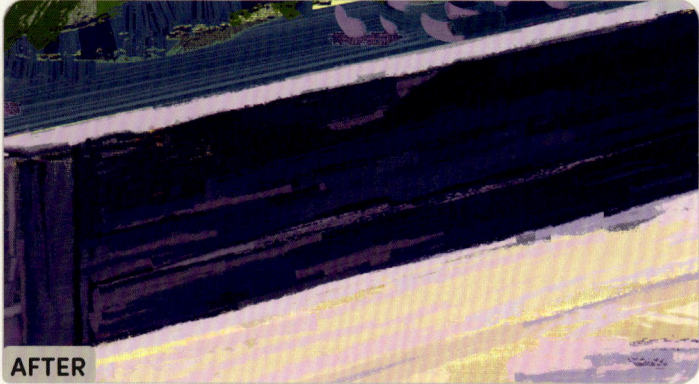

AFTER

14 Once you have the foundations set, the addition of global lighting can add a spark to the image and take it to the next level. Enhance the lighting by airbrushing a soft yellow gradient on a separate layer in **Add (Glow)** mode to introduce more brightness and yellow saturation. The **Airbrush**, as opposed to the **Brush** tool, has a scattering effect. You can select it from the toolbar directly under the **Brush** tool. Its **Soft** sub tool is perfect for creating subtle, soft transitions. As the lighting and shadows of the piece indicate the light source is coming from the right side of the canvas, that is also where the soft lighting should radiate from. This will replicate an effect similar to a sun flare.

> Airbrush subtle transitions and global lighting layers

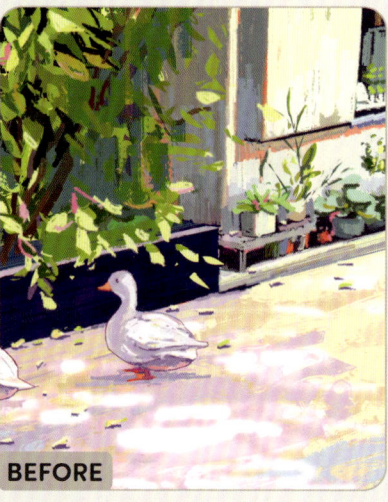

BEFORE

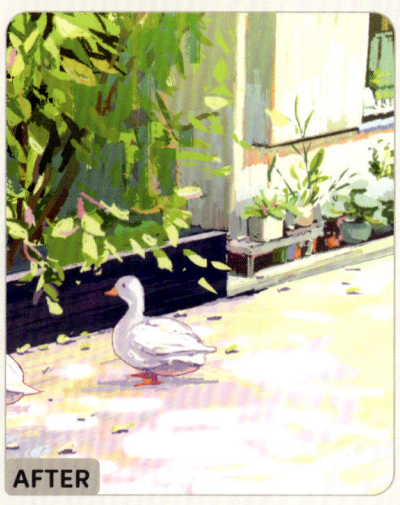

AFTER

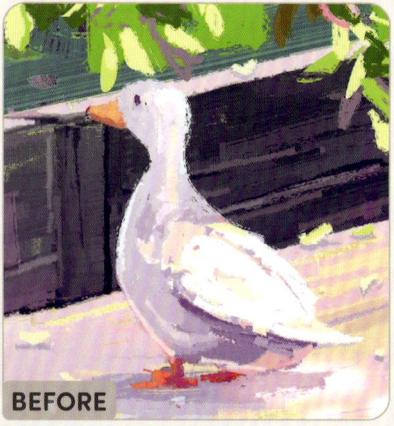

BEFORE

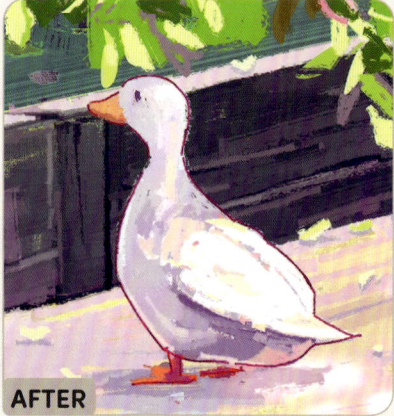

AFTER

∧ Graphic approaches to contrast adjustment and adjusting composition

15 It's easy to feel boxed into realistic rendering when tasked with depicting lighting realistically, but don't be afraid to explore more graphic approaches. You're creating art after all! Although the ducks have been rendered in a way that successfully indicates the environmental lighting, they still don't stand out enough as the subject of the image. An easy fix is to outline the ducks using a pink colour. This approach suits the vibrant and graphic nature of the existing background well. Draw the lines on a separate layer to preserve the underlying layer for the ducks, then once you're happy with the lines, merge the ducks layer and lines layer together using **Ctrl+E**. For convenience and to reduce clutter, consolidate layers in this way as you go.

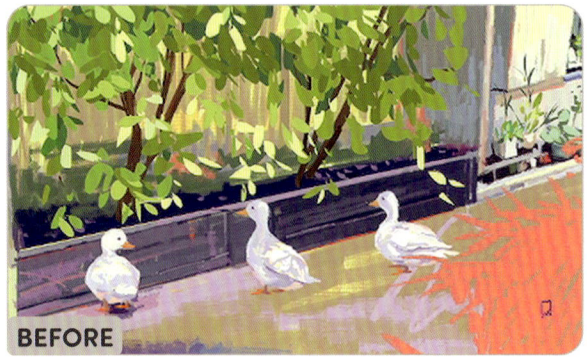

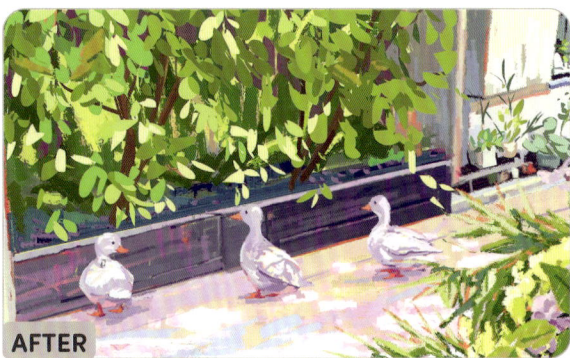

< **Introduce foreground
elements to create
depth and focus**

16 Foreground elements are a great way to add depth to a painting, as well as to aid the flow of the composition. This piece features a larger area of rest towards the bottom of the canvas. To lead the viewer's eye to the centre, fill in the empty space and frame the composition by painting some plants and flowers previously indicated in the thumbnail sketch. On a new layer, start by underpainting a rough shape for what will be the silhouette of this new foreground element. Next, on a new layer on top of that, paint the final look of the foliage over the top. This will preserve the underpainting and any information on the layers below.

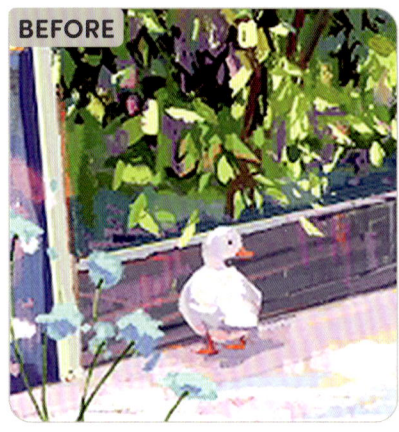

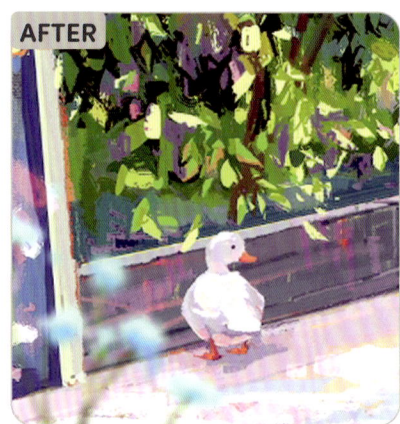

17 Depth of field is an element of photography and film created by adjusting the aperture of the lens. A wider aperture reduces the depth of the area in focus, blurring the subjects nearer than or further beyond it. This photographic effect can be replicated in art for cinematic appeal. First, determine the area of focus. The foreground plants are extremely close to camera and will therefore be very out of focus. It is the ducks in the middle of the image that you want to capture the viewer's attention. Simulate the depth-of-field effect using Filter > Blur > Gaussian Blur. Use the slider in the dialogue box to adjust the blur to your liking.

< **Apply the Gaussian Blur filter to alter the depth of field**

18 When painting digitally, it can be useful and highly efficient to duplicate and reuse an existing asset in another part of your painting. For example, to save time and preserve consistency, you can duplicate the blurred flowers at the front to help fill out more of the foreground and to frame the composition a little more. If looking for more opportunities to reuse existing assets, you could duplicate the ducks and then paint over them to create individual poses. You could also duplicate parts of the foliage. Don't be afraid to make use of the capabilities of digital art to aid consistency and save time.

> **Duplicate existing assets to save time**

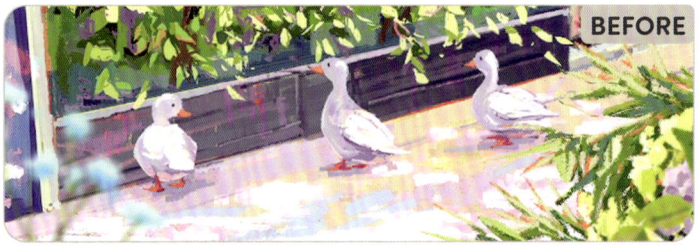

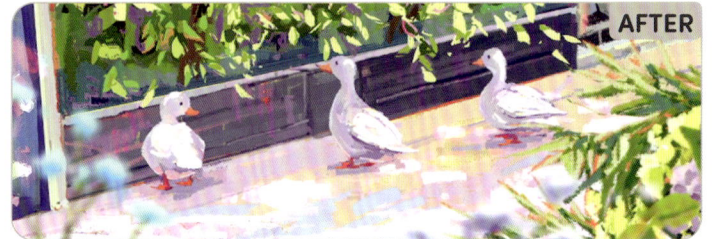

19 Clip Studio Paint features a brilliant in-built perspective grid function, found under Ruler > Perspective Ruler. Line up two perspective lines to two parallel objects in the illustration to find the first vanishing point. Do the same for another set of lines to find the second vanishing point on the horizon line. Then line up verticals to find the third vanishing point. Next, select the Operation tool, and in Object Select mode, select the triangle icon to select the perspective grid. In the Tool Property palette, select the planes you want to construct the grid on and adjust the grid size to your liking. Setting up perspective guides will allow you to draw more accurately, creating a more believable environment.

> **Create a perspective guide**

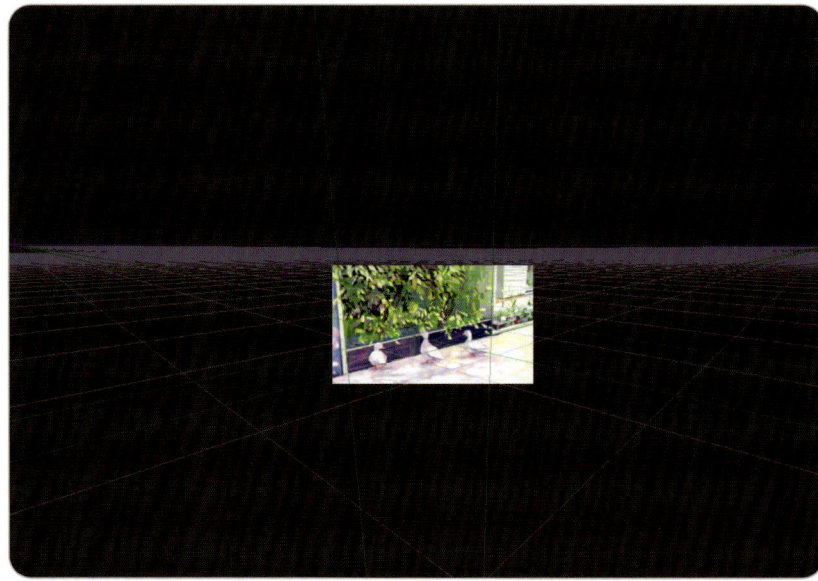

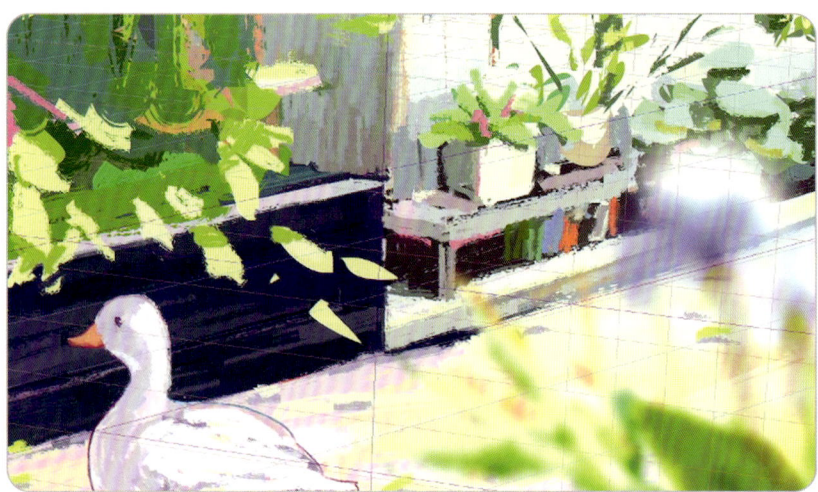

20 Perspective is an important fundamental to get right. As it's something you see every day, you can easily spot when something looks wrong, whether consciously or subconsciously. Introducing perspective guides too early in the process, however, can restrict the exploration of the composition. As the foliage-heavy background of this painting isn't so perspective-intensive, most of it could be painted freehand. But by adding in a perspective guide at this late stage, you can see that part of the building towards the right of the canvas looks incorrect. Use the grid as a guide to correct any elements that aren't quite right.

< **Use the Perspective Grid to correct any errors in the painting**

21 Upon the introduction of a perspective guide, take the opportunity to add another set-dressing element to give the scene more personality. Paint in a small seat on the right-hand side of the composition. This will help to frame the composition and focus the viewer's eye on the centre of the image. Using the perspective guide to help guide the construction, but not adhering strictly to it, you can retain a sense of realism by placing the chair slightly askew to the orthogonal arrangement of the building and plant bed.

> **Flesh out the scene with set dressing for added character and framing**

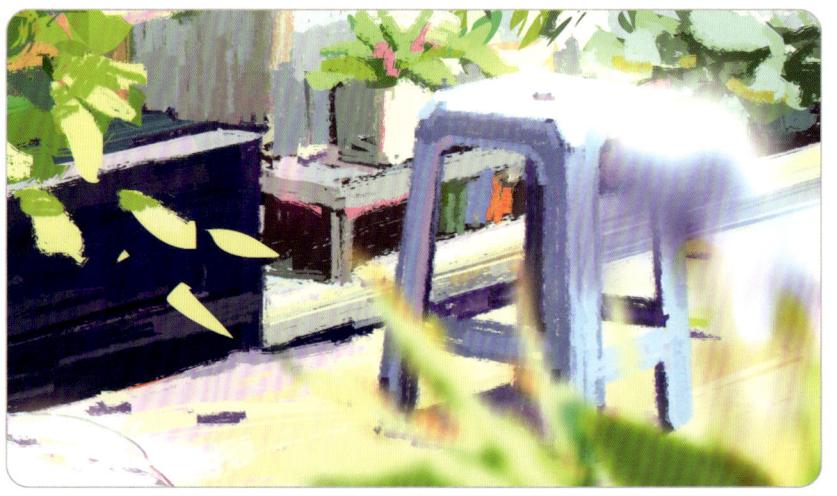

22 Creating a digital painting is not always a straightforward process. After applying finishing touches to the piece, take a step back and review the painting as a whole to see if you feel there's anything missing. The centre of the image, with the ducks as the focal point, could use something extra. Return to your reference photos to provide inspiration. Something as simple as painting a cat hiding in the bushes will add some interaction with the ducks, as well as a further story element. The cat also provides a link between the ducks and the background.

> **Review the image, make final changes, and add any additional storytelling elements**

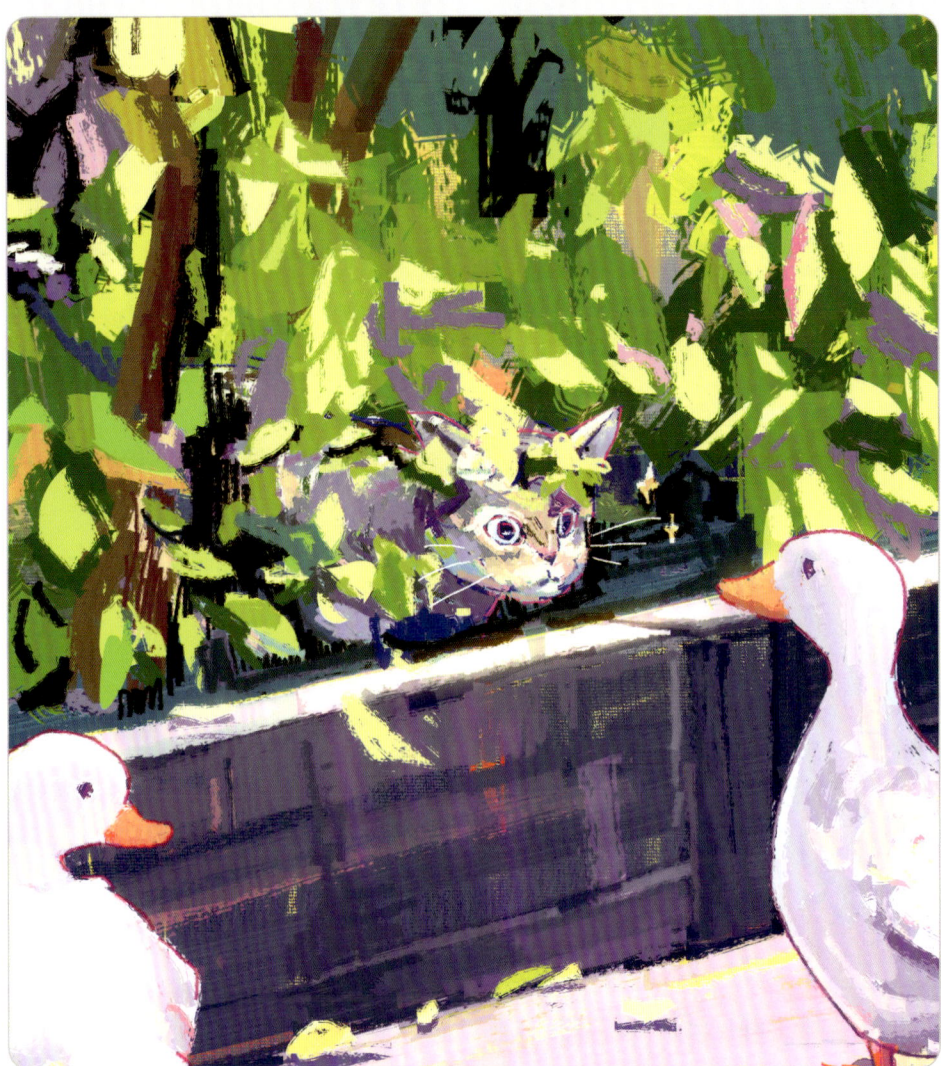

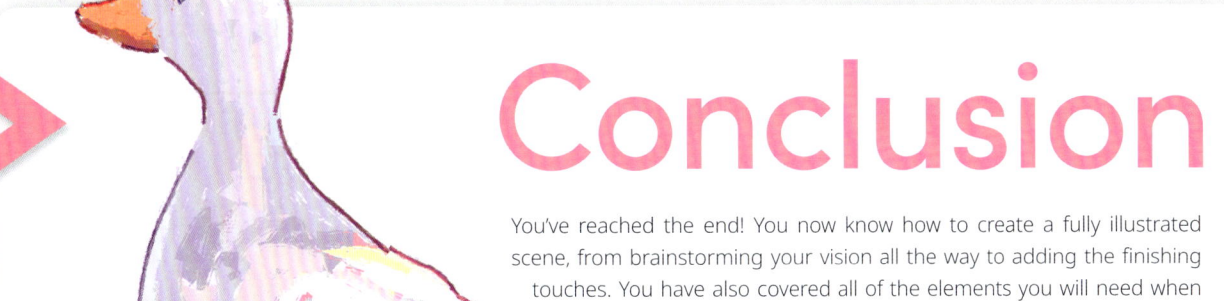

Conclusion

You've reached the end! You now know how to create a fully illustrated scene, from brainstorming your vision all the way to adding the finishing touches. You have also covered all of the elements you will need when creating your own piece, including composition, lighting, set dressing, storytelling, perspective, and more. Coupled with the technical knowledge of the various tools Clip Studio Paint has to offer, use these insights to kickstart your own journey into digital art and help turn your ideas into reality!

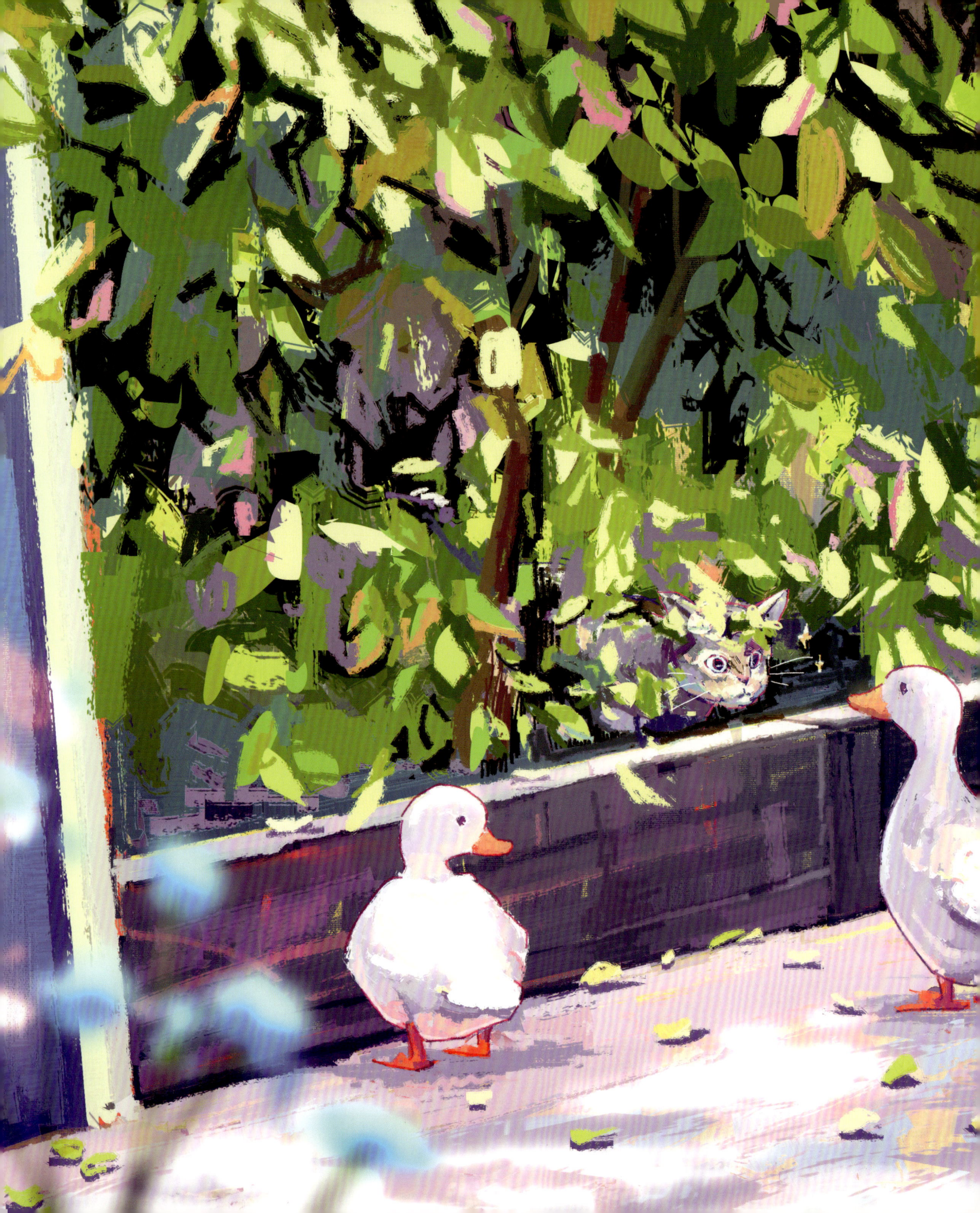

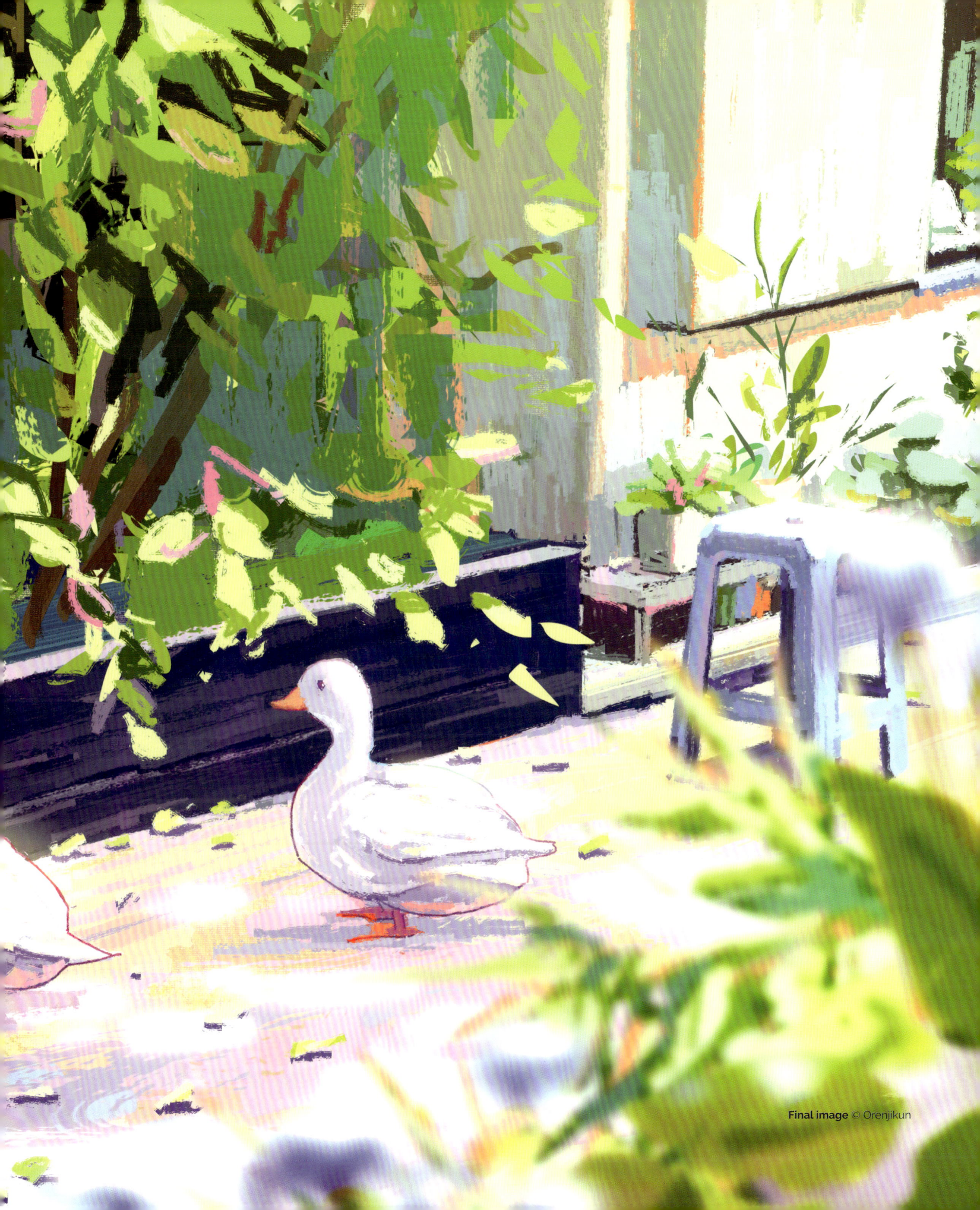

Picture-book illustration

BY SIMONE FERRIERO AKA SIMZ

> In this tutorial I'll show you how to create a digital illustration featuring multiple characters in an environment with an element of storytelling. The steps will cover a range of topics, including composition, colour theory, and perspective. Following along, you will learn how to create a composition that's both engaging and readable, plus the importance of perspective in composing a solid background. I'll also demonstrate how character design, body language, and rhythm can inject an illustration with a spark of life.

With the aim to create a picture-book illustration, the style will be a combination of painterly digital watercolour and outlined sections. I'll show you how to use a pencil-like brush for the lines, and share how the use of paper texture and a round brush with opacity control can be used to create a watercolour effect. Overall, the approach will be reminiscent of a painting created using traditional media. The colouring stage will draw on colour theory to put together a limited palette and muted tones, followed by how to create masks and clipping masks for easy, efficient painting. As you grow in confidence, I'll demonstrate how to use blending tools to create a visually pleasing transition in each brushstroke.

Final image © Simone Ferriero AKA Simz

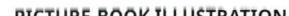

LEARN HOW TO

- Compose a scene that invites the viewer into the story.
- Use colour theory to your advantage.
- Paint with digital watercolours to create the look of traditional watercolour.
- Use Clip Studio Paint's various blending modes to enhance the illustration.

203

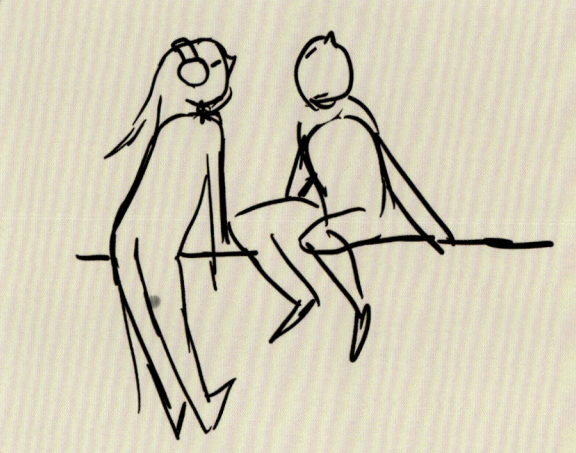

01 Everything starts with an idea. Before you begin drawing, spend some time thinking about what concept, story, mood, or emotion you want to illustrate. Explore your idea and how you can best depict it on the canvas. For example, a stationary character staring straight out at the viewer isn't very interesting or dynamic. Consider how your characters can interact with the environment around them and how you can tell the story of the scene. When you have a few ideas, create a new canvas – this can be any size – and sketch out some rough ideas and compositions using Clip Studio Paint's default **Round** brush.

‹ Sketch out a few simple lines to describe your idea for the illustration

02 Once you've chosen an idea, create a new canvas that is 4,000 × 5,000 pixels, and 300 dpi on which to draw your initial sketch. Set the background to a cream, almost yellowish, paper colour. Create a new layer, select a round ink brush without any opacity control, and start doodling. Use this step to experiment with shapes and compositions, making as many mistakes as is necessary in order to find a composition you like. Once the sketch is finished, click on the **Layer Colour** button in the Layer Property palette and turn the layer blue. Making the sketch a different colour will help when drawing over it in the next step, allowing you to see what is the initial sketch and what is the new line work.

› Open the Layer Property window to alter the colour of the layer

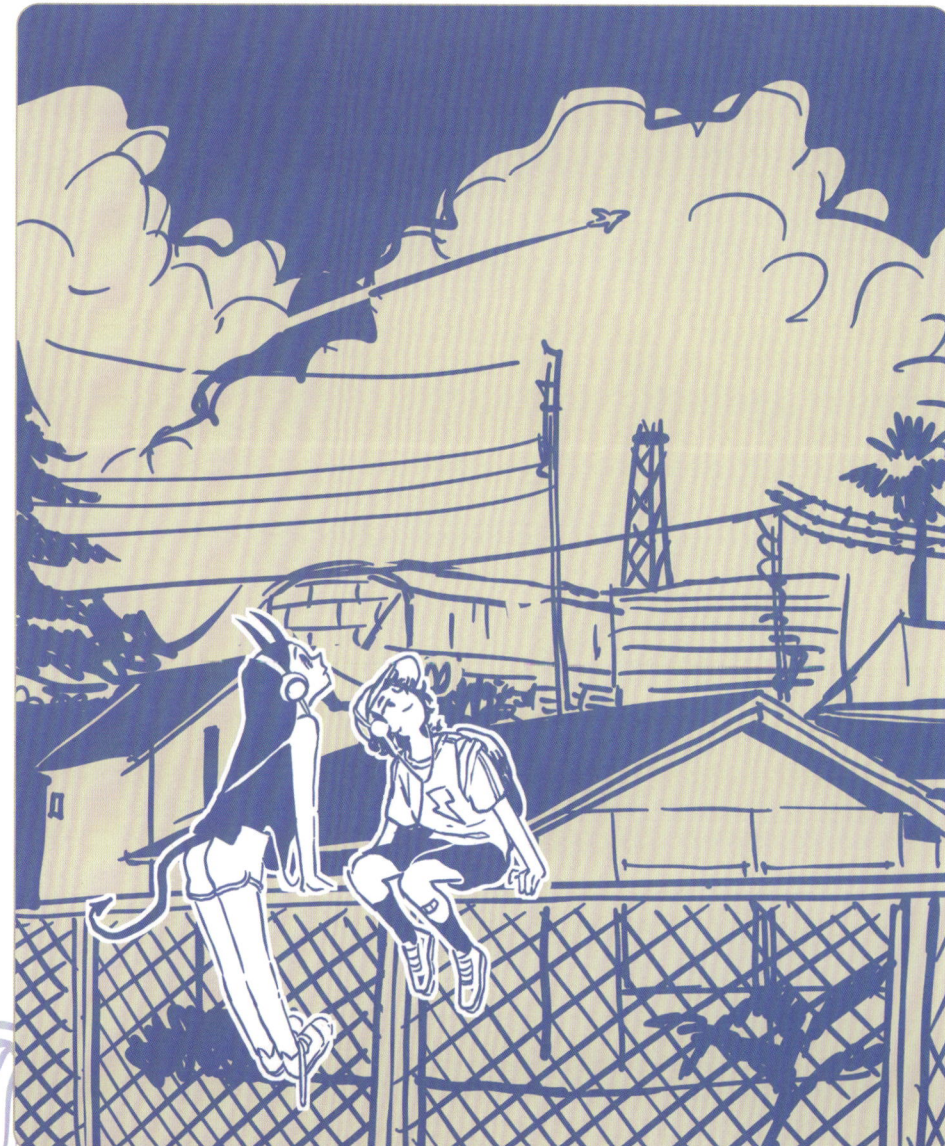

03 Composition is one of the most important aspects of an illustration. Look closely at the placement of the main elements in your sketch and see if there are any distracting shapes that will confuse the viewers' gaze. Try to ensure that the houses, power lines, roofs, and all other elements work together to frame the characters to avoid them blending into the background. Remove any distractions to make sure the characters are the focal point and the first element viewers look at.

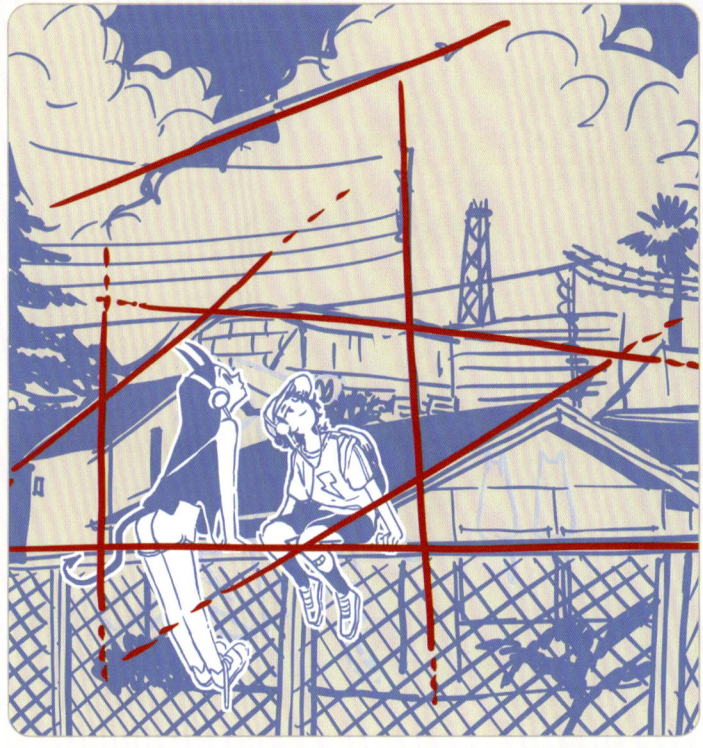

> Tweak your composition to make sure the lines of the background visually frame the main subjects

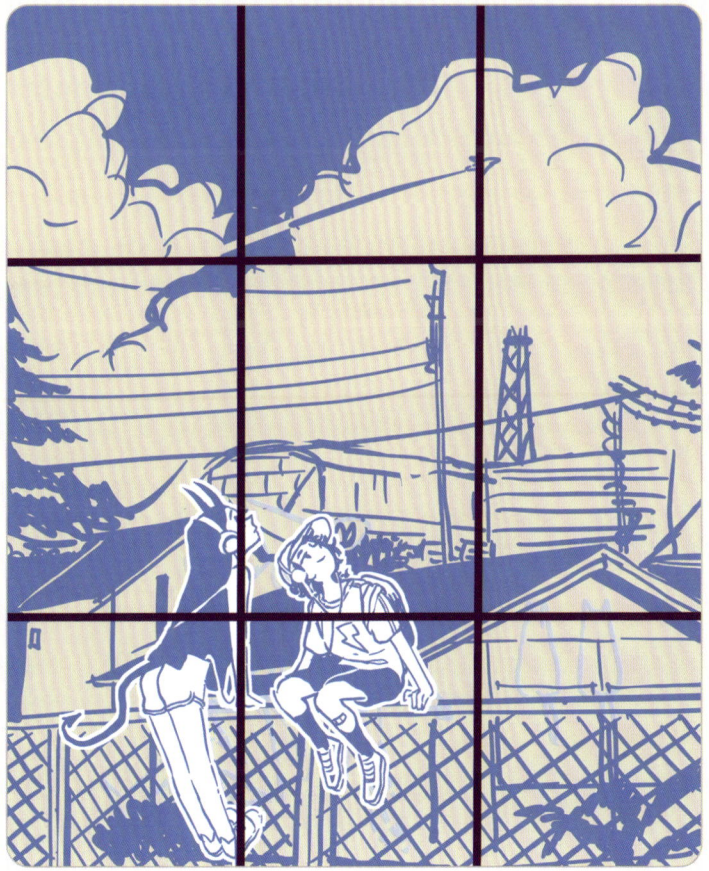

04 There are many different techniques for composition, but the rule of thirds is one of the most popular. Using a rule-of-thirds grid can help you to place the main elements of the illustration in visually pleasing positions. You can either create a rule-of-thirds grid on a new layer by adding two horizontal lines plus two vertical lines, or alternatively, you can find a rule-of-thirds grid on the internet and paste it onto your canvas, again on a separate layer. This will allow you to adjust your illustration as needed to ensure the main elements sit at the intersection of lines, directly on the lines, or just to the side of the lines. A third option is to imagine where the rule-of-thirds grid would be and to place elements by eye. The rule of thirds is just a guideline; not everything has to sit exactly on the lines or intersections.

< Use the rule of thirds to inform the composition of the scene

> **Design the characters separately from the background – this will allow you to focus on them more easily**

05 Once the draft is complete and everything is in the right place, start to think about designing each element in detail. Hide the initial sketch and create a new layer on top, then begin to define the characteristics and body language of the two characters, separate from the rest of the image. One will be a human boy, while the other will be a little demon with red skin, horns, and a tail. Both will be curious and adventurous, their attention caught by something in the sky above them. Unhide and hide the background at regular intervals to make sure the design of the characters fits with their environment. It's important to define each element clearly in this early stage, before moving on to a cleaner version of the image.

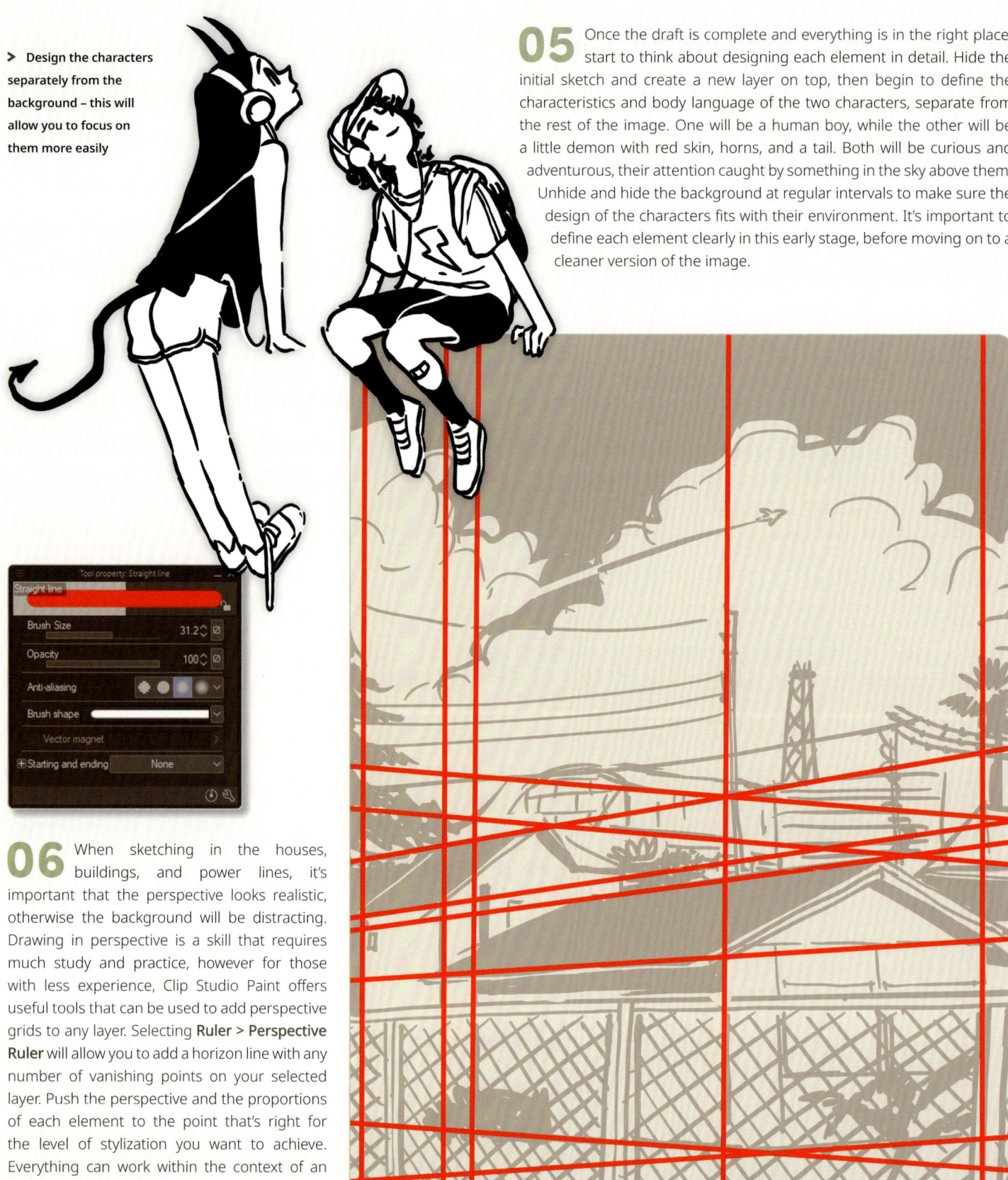

06 When sketching in the houses, buildings, and power lines, it's important that the perspective looks realistic, otherwise the background will be distracting. Drawing in perspective is a skill that requires much study and practice, however for those with less experience, Clip Studio Paint offers useful tools that can be used to add perspective grids to any layer. Selecting **Ruler > Perspective Ruler** will allow you to add a horizon line with any number of vanishing points on your selected layer. Push the perspective and the proportions of each element to the point that's right for the level of stylization you want to achieve. Everything can work within the context of an illustration as long as it relates to the rest of the elements in a seamless way.

∧ The red lines show the perspective grid used to align all of the architectural elements

07 Once the draft is complete, it's time to clean up the design. Create a new layer, then with the blue draft layer set to low opacity – around 20% should be about right – draw the line art using a pencil brush. The **Mechanical Pencil** brush will be perfect for this task. The goal is to create clean, readable line art that doesn't appear obviously digital, but rather has more of a traditional-media look. This will help to give the piece a more organic feel.

> Draw clean line art on a new layer
set on top of the draft layer

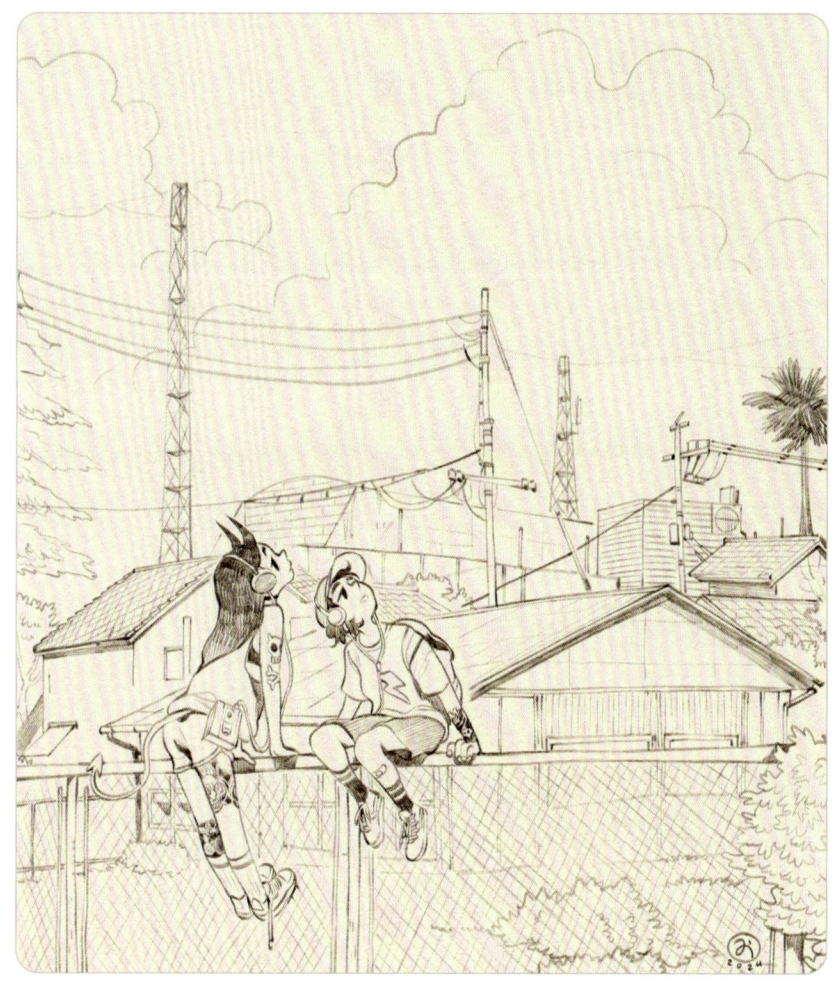

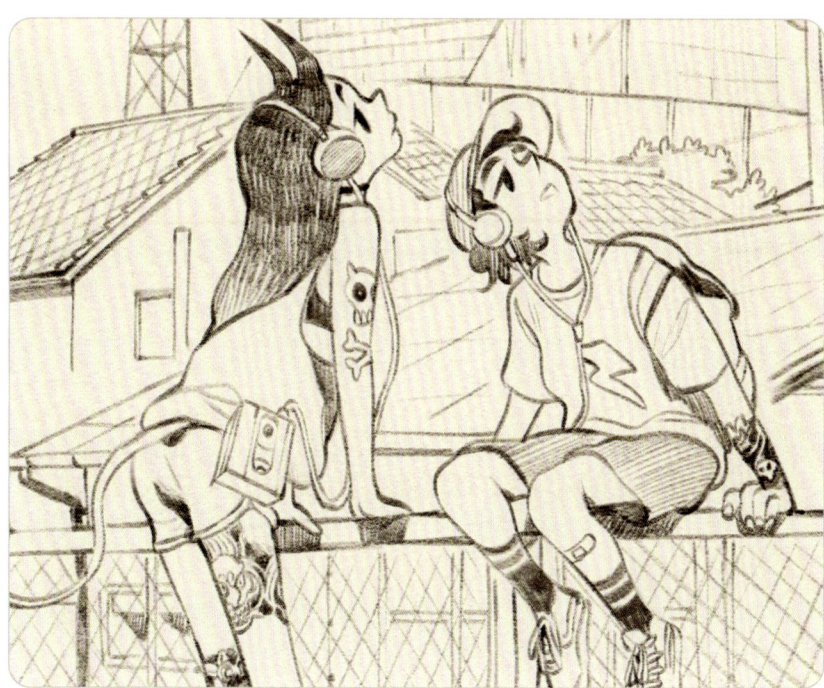

08 The rhythm of each line can dramatically change the perception of a drawing. When drawing the line art, vary the pressure and size of the brushstroke to inject the design with a sense of weight and dynamism. Use thick lines when drawing areas that are darker or in shadow, and thinner, lighter lines for areas that are close to the light source. Cross-hatching is also a useful technique that can convey darker areas and enhance the quality of a drawing.

‹ Vary the line weight and make use of cross-hatching
when drawing areas in shadow

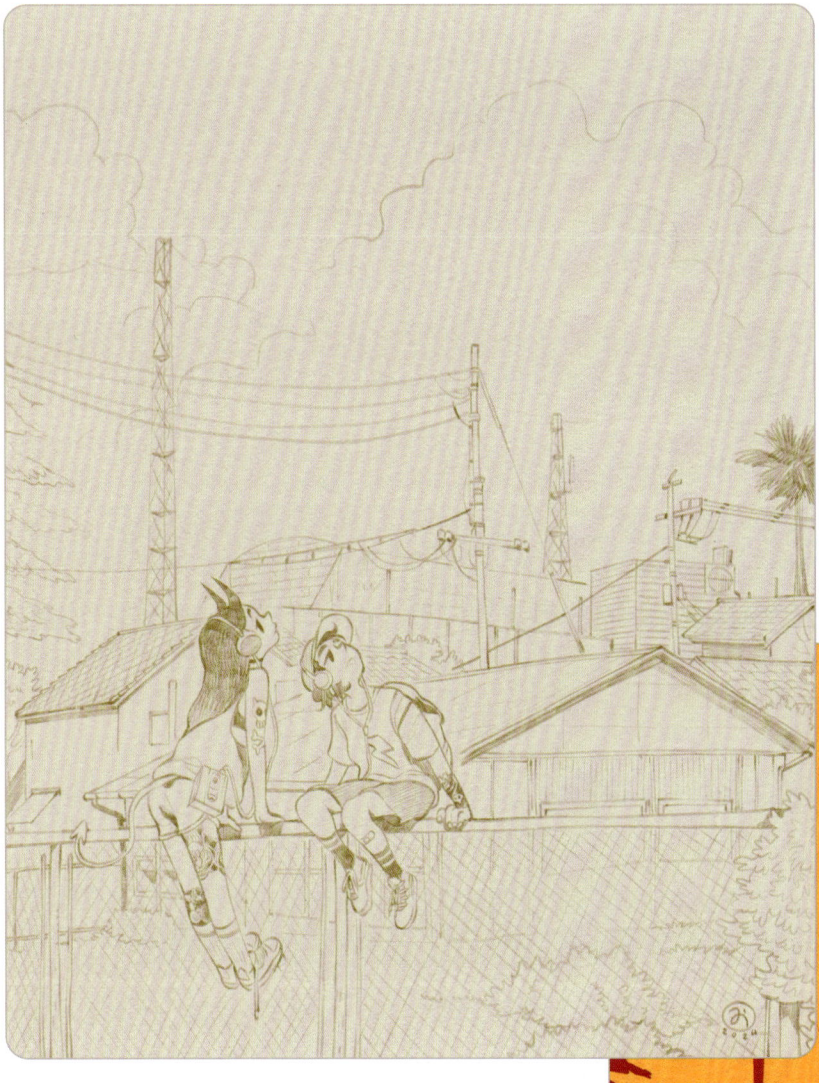

09 Clip Studio Paint has many colour blending modes. Take some time to experiment with them to learn how each one works and when you could use them to help create the look you're after. Change the line colour to dark grey, avoiding full black or white. Next, set the blending mode of the line-art layer to **Colour Burn**. This will make sure any colour added later will change the appearance of the line art accordingly. Similarly, take care to ensure that the canvas is bright, but never full white or black.

< **With the Colour Burn blending mode, the grey line art will appear coffee tinted when sat on the cream-coloured background**

10 Before the colouring stage, select the main areas or important elements of the illustration to separate on different layers. Do this by choosing any solid brush, then on a new layer located under the line art, fill the whole area with the same colour. The colour chosen for each selection isn't important as long as they aren't similar – what matters is that you're able to make a reasonable distinction between foreground, middle ground, and background. Do this for the characters, fence, buildings, and sky.

> **Fill each key element with a different colour**

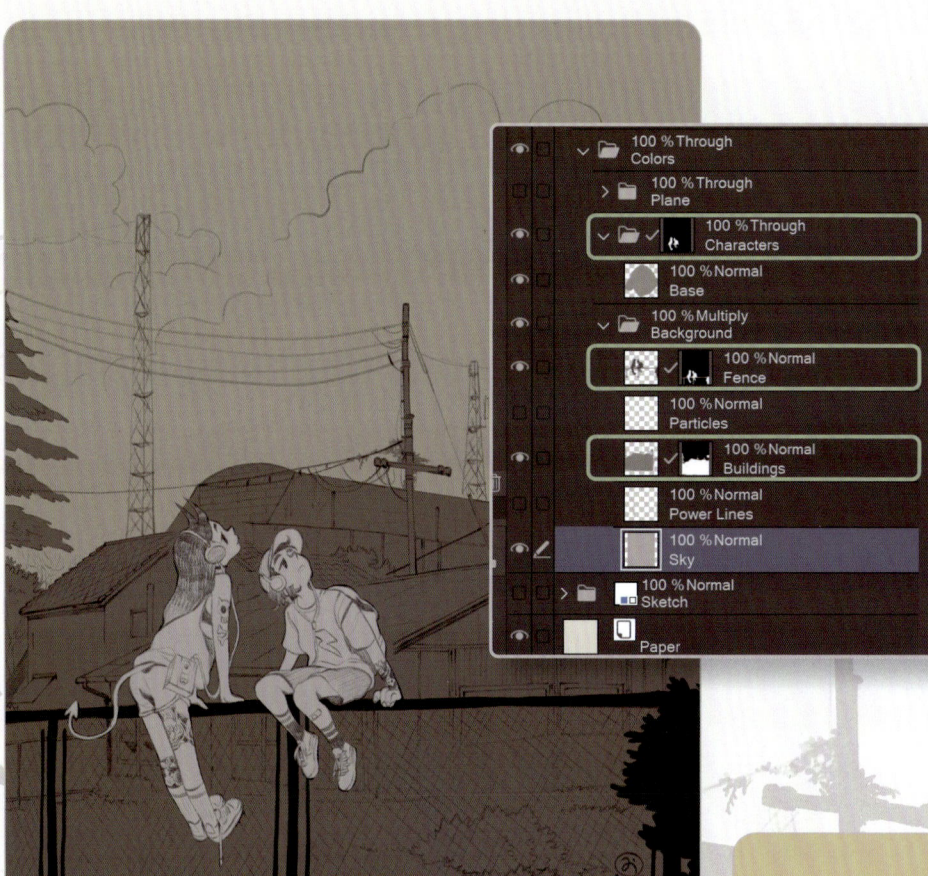

11 Some elements, such as the two characters, will need colouring over several different layers. This is where layer masks can be extremely helpful. With the **Selection Area** tool active, **Ctrl+click** on a layer's thumbnail to create a selection of the entire layer's contents. Then select any layer, or folder, from the layer window and select the **Mask** icon. This will attach a layer mask in your chosen place, in the shape of the selected area. Now anything inside the layer, or any layer inside a masked folder, will be limited within the area defined by the mask. This trick can make the colouring process much quicker and can be used for every section, if needed.

< When colouring the illustration, make use of layer masks to speed up the painting process

12 The colours chosen for an illustration can define its mood and set the atmosphere of the whole piece. Before you start to paint, set each of the main-area colours to various shades of a harmonious palette; for example, ranging from light to dark green. These undertones will influence any colour you paint on top. A basic understanding of colour theory will help to inform your colour choices, yet the mood that you want to convey and your personal taste should still play an important part in the selection process. Forcing principles or pre-made colour palettes into an illustration can make it feel soulless.

> Fill each of the main areas with a different shade of the same colour

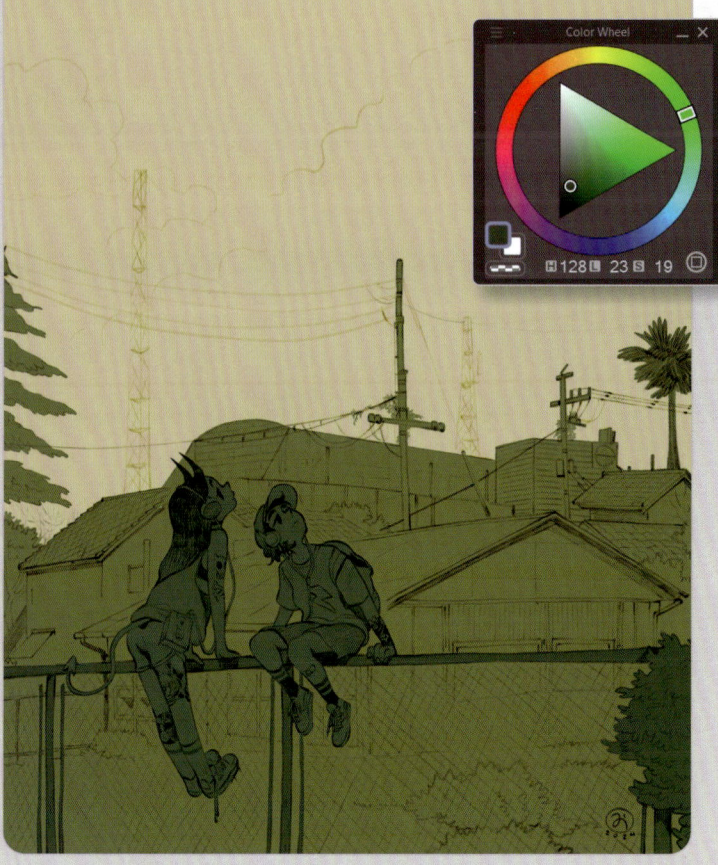

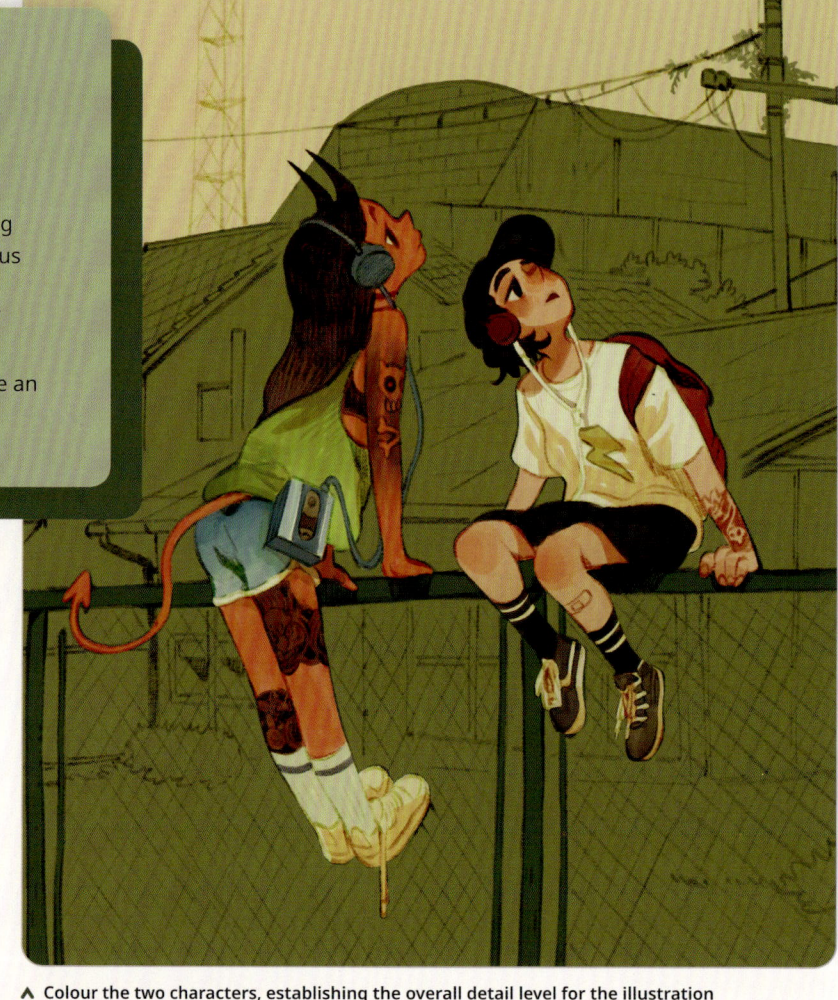

ARTIST'S TIP

Using too many layers or colour blending modes can risk the painting becoming overwhelming and result in poor beginner-looking designs. Conversely, working mainly with basic tools and a few important layers, plus a thoughtful use of filters or special effects, is what will create a solid design. Some of the core aspects of an illustration are shape language, composition, and colour choices, but overuse of special effects can have an unpredictable effect on them. Keep it simple.

13 Start to colour the characters. Do this by applying a **Layer Mask** to the character folder, making sure the Layer Mask is a selection of the two characters. Add a new colour layer inside the folder and begin painting using an opaque, hard-edged round brush. You'll see that the colours stay within the boundaries of the layer mask, so you can easily paint within the characters' shapes. Select a few simple colours to define each one and avoid over-detailing unless necessary. Use a round brush with opacity enabled to create an even but slightly textured effect on each surface and material. Small imperfections will add character and personality to the piece, so make sure to keep any interesting brushstrokes. Blending all the colours together can be useful at times, but over-blending risks removing unique aspects from the painted areas.

⌃ Colour the two characters, establishing the overall detail level for the illustration

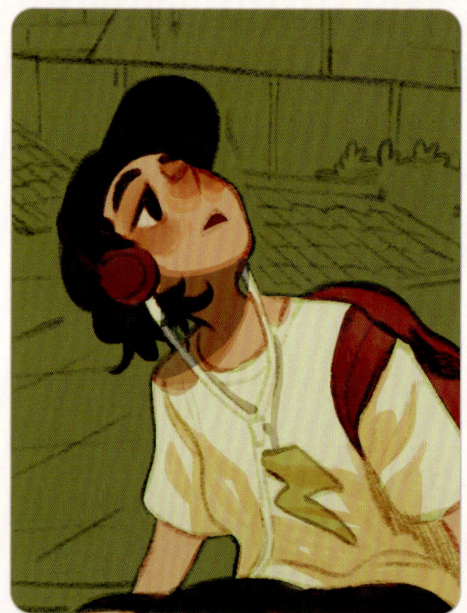

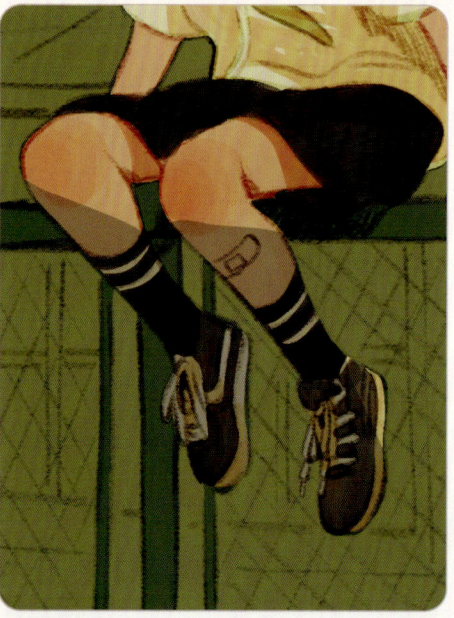

14 When colouring, leave the base layer set to **Normal** mode in order to overlap with the layers located below without blending. However, on top of the **Normal** layer, add new layers with different blending modes to shade the colours. Use a new layer set to **Multiply** to create shadows, or **Overlay** or **Screen** to add a brighter colour. Add **Normal** layers to paint over with a new opaque colour, or use the opacity slider to blend new layers, making them slightly transparent. Finally, when you're happy with how the colouring on the characters looks, merge all the colour layers of the section together.

< Once the characters' colours are complete, merge the layers together – just remember, once this is done, you won't be able to make changes later

15 For the middle-ground section of the illustration, use the **Selection** tool and lock the layer transparency. This will allow you to paint freely without risking painting outside of the selected area. Use a painterly approach that gives the shapes a more organic look. Keep the colours limited to the palette you selected initially. If needed, create new layers and set them as different blending modes (such as **Multiply** or **Overlay**) to create shadow areas or highlights. Once finished, merge the entire middle-ground section down to a single layer.

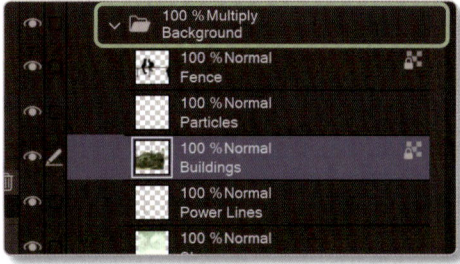

> Paint the middle-ground section using a painterly technique

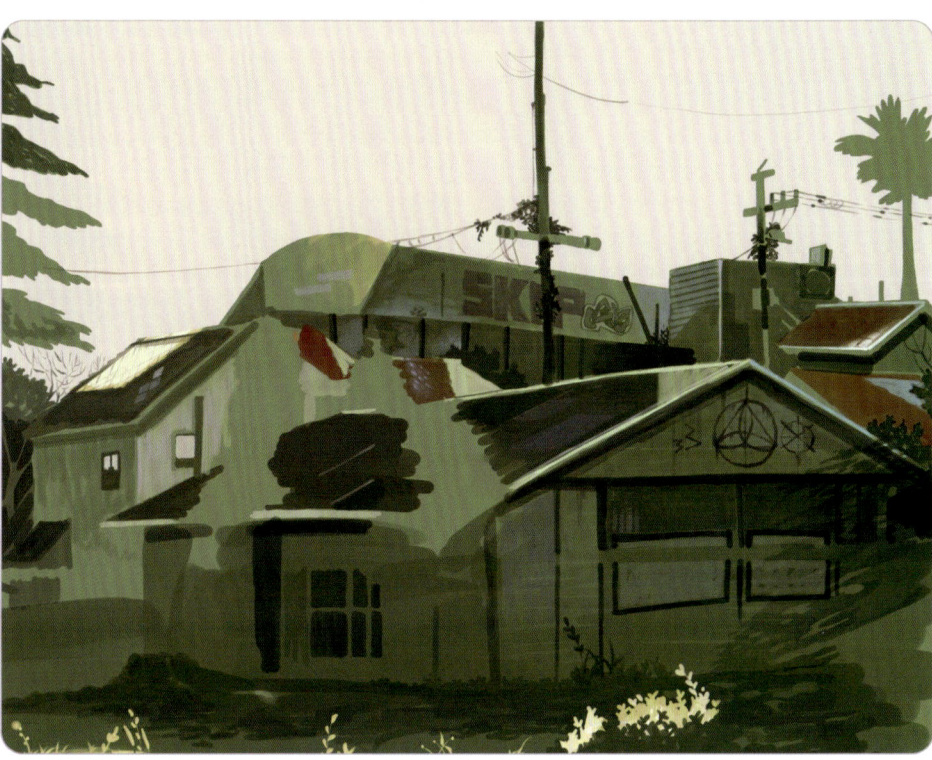

16 Paint the background with colours that respect the vision of the original colour palette, while making sure to retain contrast. This illustration will use greens, from dark to bright, mixed with a pinch of red as an accent colour. From time to time, pause to check the values of your illustration to avoid a painting that's too flat (the values blend into each other) or too contrasting (the values change from very bright to very dark), damaging the readability. There's no formula for an optimal contrast besides studying other artists, but with experience and time, contrast will become just like any other tool. To quickly check the values, simply create a new layer on top of your layers, set it to **Colour** blending mode, then fill the layer with black.

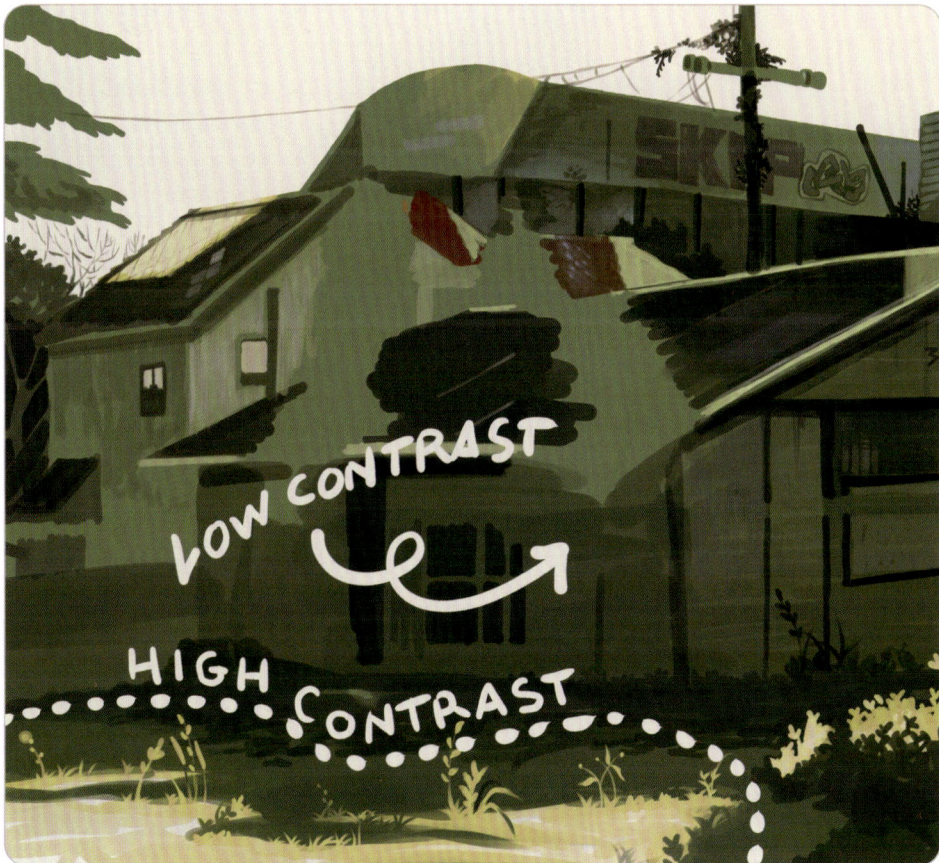

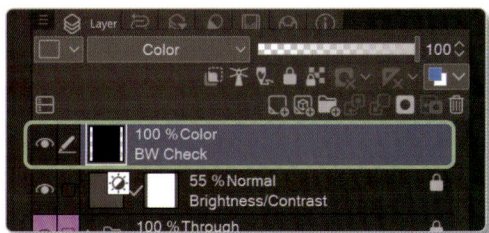

< Make the grass a high-contrast area, then use low-contrast values to shade the house

17 Details should always have a purpose. Adding lots of details for the sake of it will lead to an uninteresting final image. On the contrary, adding a few interesting details can generate a sense of depth and story. Though you should avoid changing brush size too often, in order to maintain a sense of consistency through the rendered areas, try adding small details here and there to create a greater sense of immersion and provide the eye with a few points of interest to explore. Scatter the details across the whole image, without drowning every corner. Paint details like leaves, roof tiles, graffiti, and power lines – anything that adds information, without which the illustration wouldn't communicate the same story. For example, the plants growing on the power line indicate that no one is taking care of them. The graffiti suggests the area is frequented by graffiti artists and there's no interest in repainting.

> The foliage-covered power line and the graffiti on the building help to make the environment appear 'lived in'

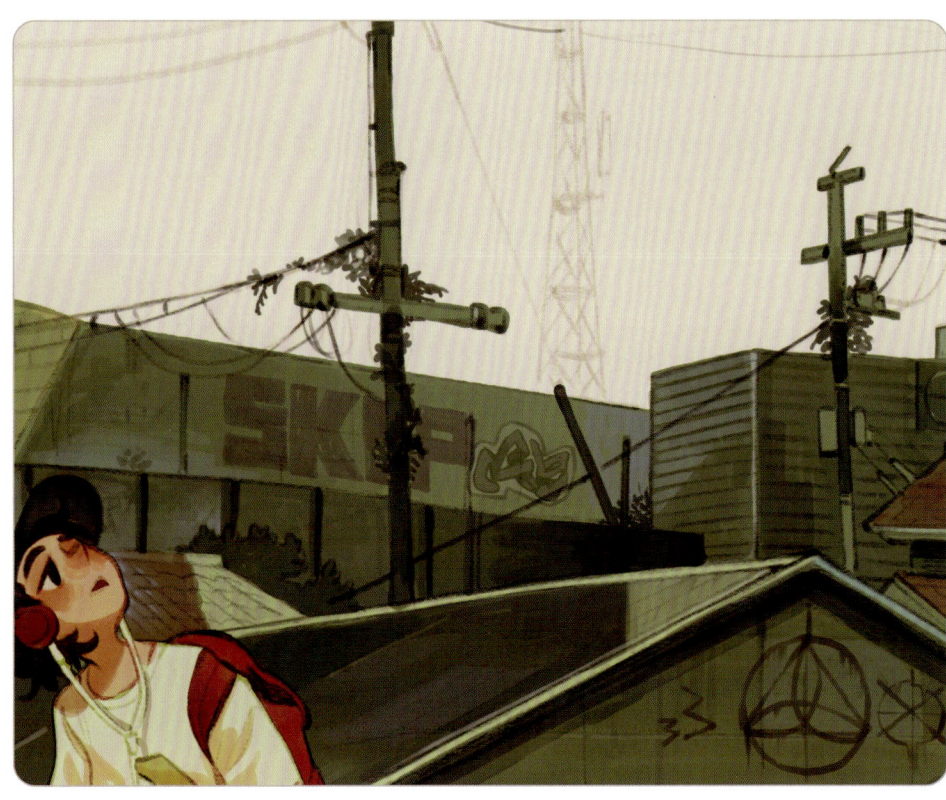

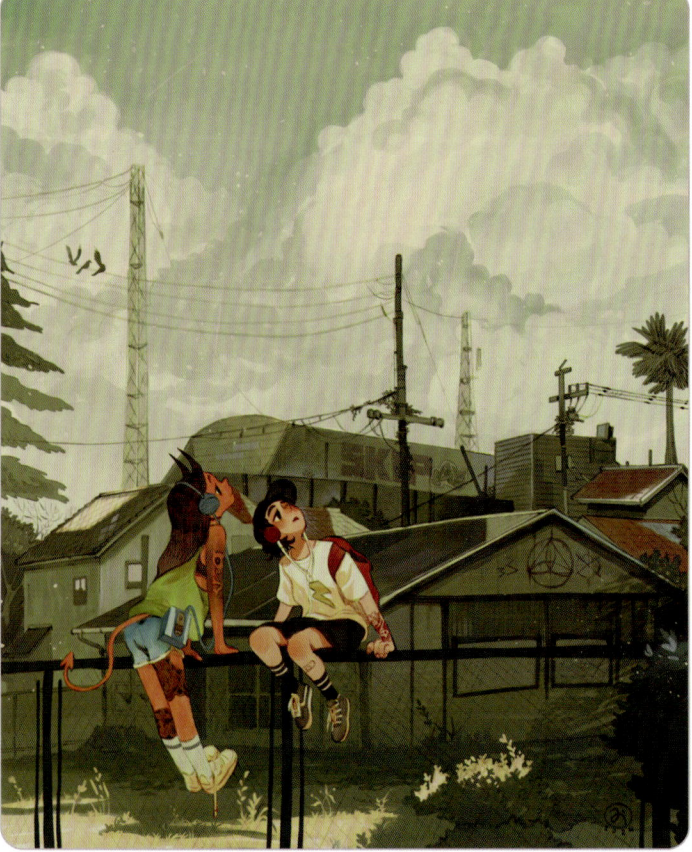

18 Working on separate areas of the illustration, masking the parts you don't want to colour yet, is a painting technique that can be applied both digitally and traditionally. Always overlap all of the separate layers and pause at regular intervals to review the image in its entirety. This will allow you to tweak the colours to ensure the illustration is readable as a whole. Intensely focusing on single elements can easily give you the illusion that you're doing a good job, but when compared to the rest of the image, the reality could be the opposite. Render the illustration as one whole image that is more than the sum of its many parts.

< Turn the layer visibility on and off to check how the piece looks as a whole

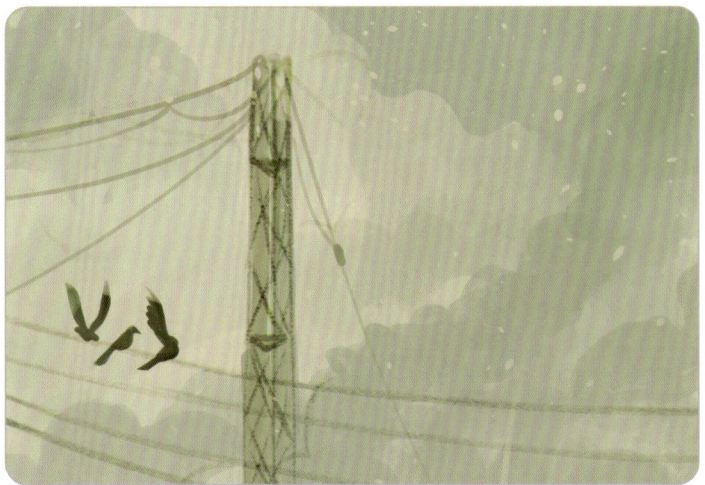

19 Paint the clouds in the background, remembering that there's still a focal point to be added in the sky. Keep the contrast low to create a sense of lightness, concentrating the weight of the image at the bottom, where the ground is. Use references to help you paint the sky and clouds with a more appealing and realistic look. Keep the values in check between the sky and ground to maximize the contrast of the illustration.

> The clouds are an important aspect of the illustration, creating a sense of space and air

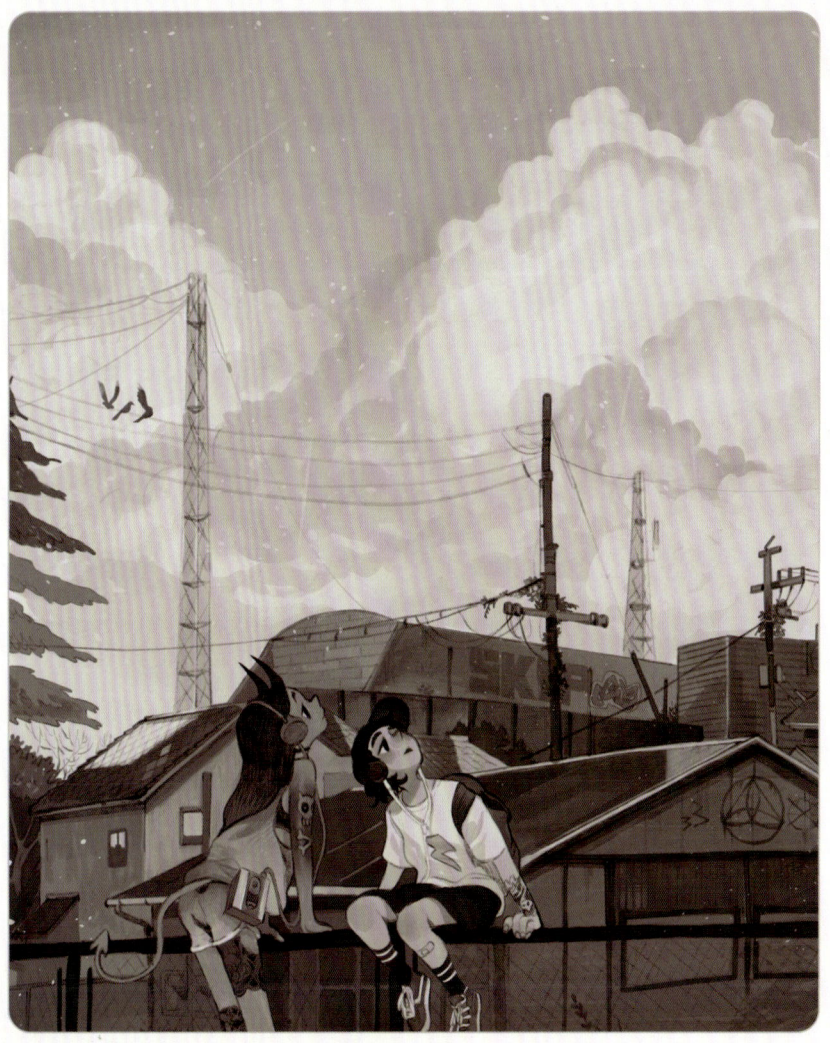

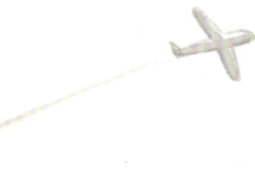

20 Now the illustration has a colour palette, background environment, characters, and details, it could easily be considered finished. However, every image needs a focal point within a focal point – something that will make it truly engaging and unique. Two characters hanging out on a fence is an interesting enough event, but with a plane flying over their heads, the illustration tells more of a story. Paint the plane on a new layer, using the **Move** tool to adjust its position until it looks right. This is the final element that will either make or break the image.

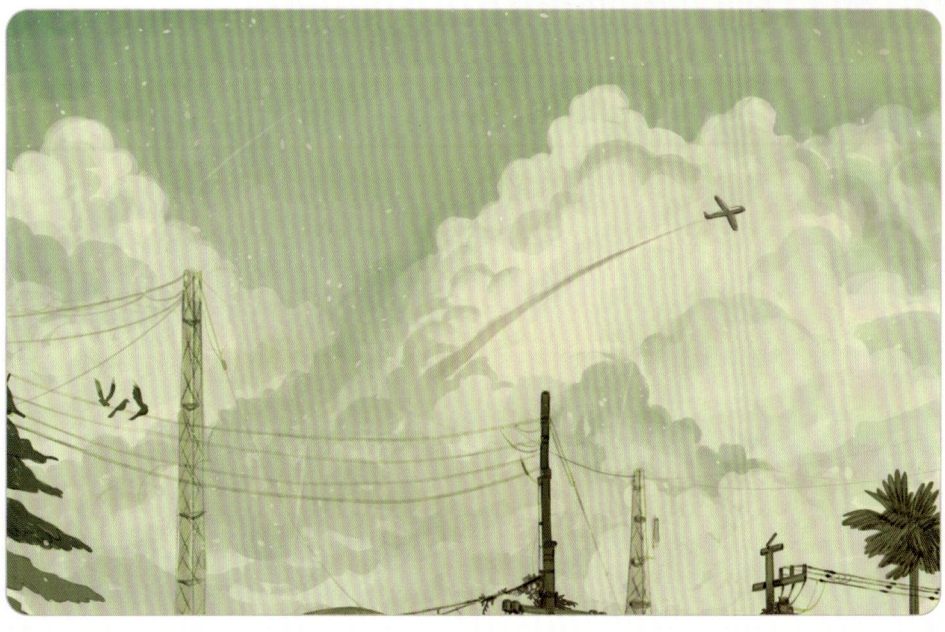

< Paint in a plane flying overhead – this is what the characters are avidly watching

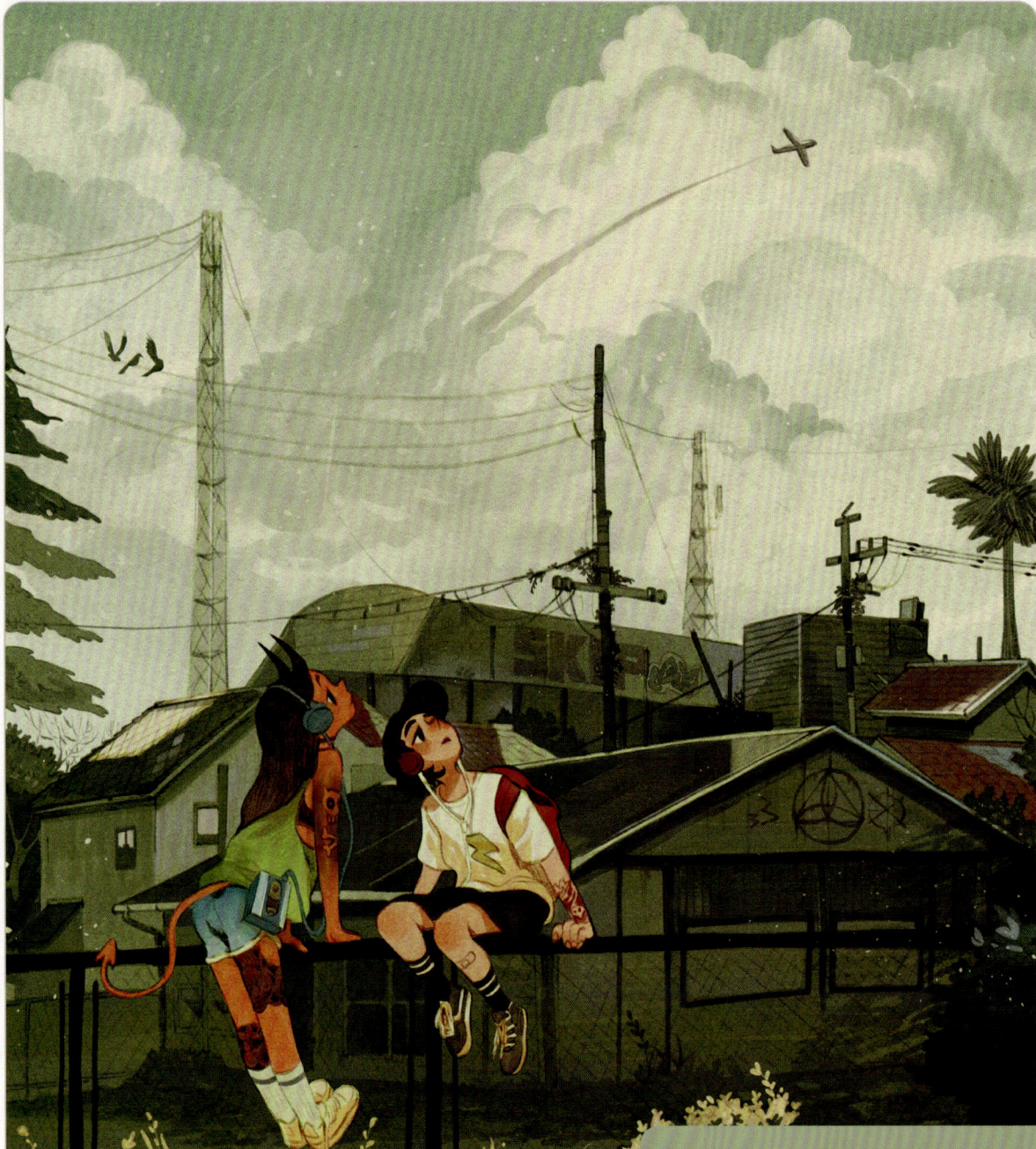

< Add a textured grain to give the illustration a slightly dated look

21 Adding grain or textures on top of your illustration can make it look more organic and like it was created using traditional media. It's possible to create your own grain texture by creating a new layer, filling it with white, and then selecting **Filter > Render > Perlin Noise**. Make sure that the scale is set to a ratio small enough to become grain-like. Once the noise has been generated, simply set the layer blending mode to **Overlay**. You can try this with paper textures, or any kind of texture that fits the style, in blending modes such as **Overlay** or **Multiply**.

ARTIST'S TIP

Don't get lost in Clip Studio Paint's many features and possibilities. The power and flexibility of a tool doesn't mean that every aspect of it has to be used at all times. It's possible to create great illustrations by sticking to a few select tools and techniques. Only use new tools when the task calls for them.

22 The last few steps involve looking at the illustration as a whole and working on all of the different sections at the same time. Create new layers on top of the others on which to add details and unify the various elements into one image. There's no right or wrong way to do this, but try to keep an eye on the characters when adding details in. Paint in particles, textures, and glow effects where it feels right, but don't get too carried away. Floating dust particles and sparkles can be painted one at a time using any brush and a delicate hand. The **Overlay**, **Glow Dodge**, or **Colour Dodge** blending modes are perfect for quickly creating a glowing effect. Alternatively, you may prefer to select a luminescent colour to paint in the glow by hand, as has been done here. Add the bright dust particles and glow effects over the entire image, but keep in mind that they will be more impactful when painted over dark areas where they will create a striking contrast. If something doesn't look right, don't try to cover it with details or special effects. Backtrack a few steps and fix that element of the painting.

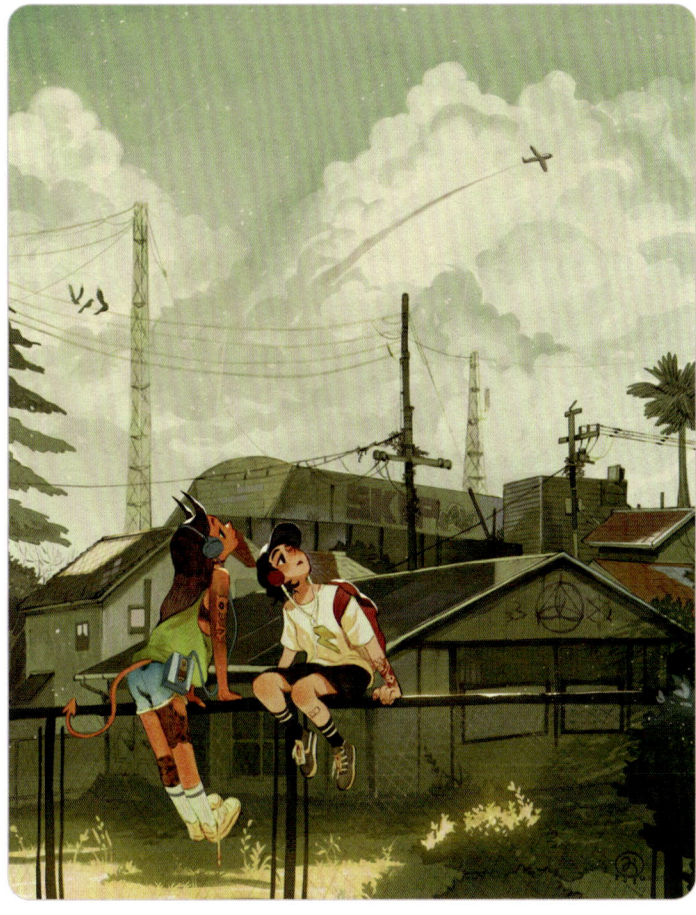

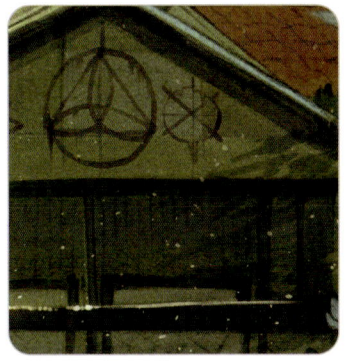
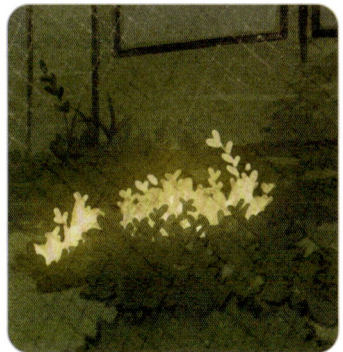

∧ Finalize the artwork by adding special effects, viewing the image as a whole

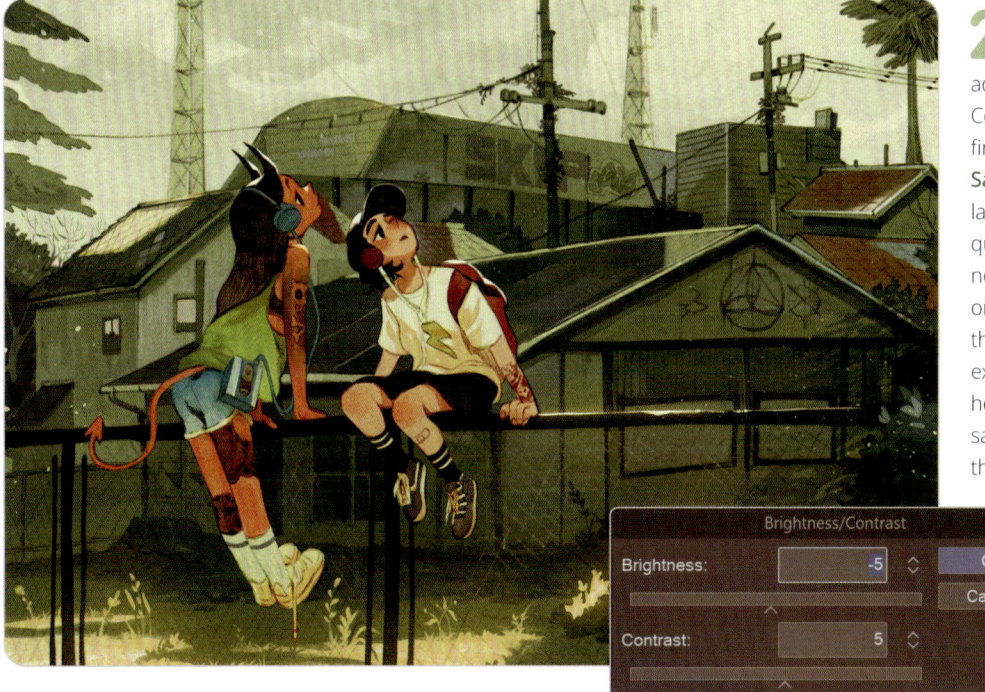

23 Right-click on the **Layer** panel and go to **New Correction Layer** to view the adjustment options. Use a **Brightness/Contrast** Correction layer to adjust the illustration before finalizing it. You can also consider using the **Hue/Saturation/Luminosity** or **Level Correction** layers to tweak the image. Correction layers quickly inject a little more punch, and are usually non-destructive, meaning they can still be edited or deleted if you change your mind. Don't overdo the corrections – this is an obvious sign of a poorly experienced artist. Use a light hand to slightly heighten the contrast and adjust the colour and saturation. Next, use a **Tone Curve** layer to adjust the contrast in a more granular way.

< Brightness/Contrast is an essential adjustment layer – tweaking these things can make an illustration 'pop'

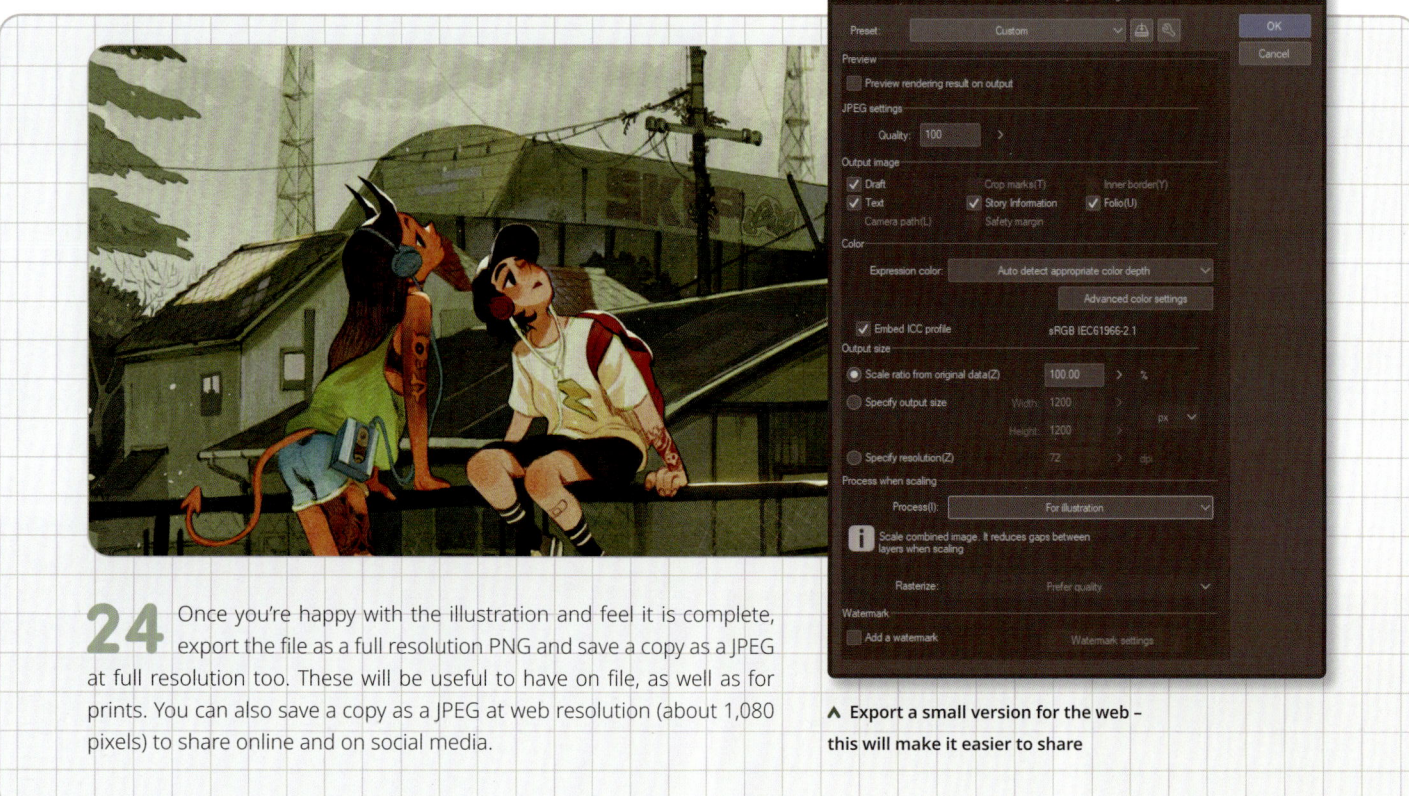

24 Once you're happy with the illustration and feel it is complete, export the file as a full resolution PNG and save a copy as a JPEG at full resolution too. These will be useful to have on file, as well as for prints. You can also save a copy as a JPEG at web resolution (about 1,080 pixels) to share online and on social media.

∧ **Export a small version for the web – this will make it easier to share**

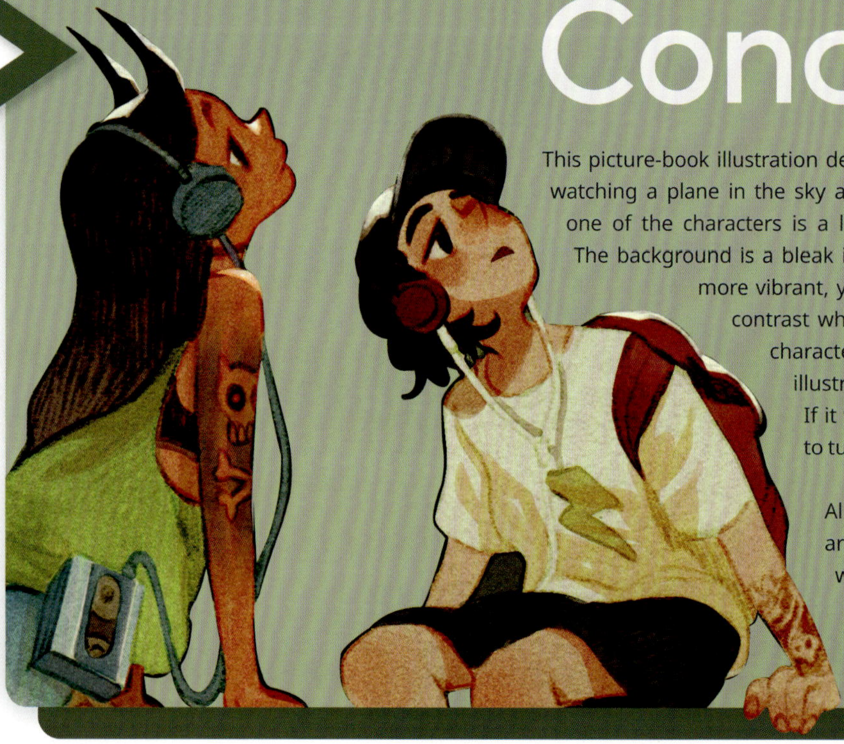

Conclusion

This picture-book illustration depicts two young characters sitting on a fence, watching a plane in the sky above. This would be a normal scene... except one of the characters is a little demon, with red skin, horns, and a tail! The background is a bleak industrial environment that contrasts with the more vibrant, youthful characters. This creates an interesting contrast which invites the viewer to wonder who the two characters are and what brought them to that spot. The illustration invokes a sense of nostalgia and mystery. If it were in a book, it would encourage the reader to turn the page to see what's going to happen next!

All of the techniques demonstrated in this tutorial are just the tip of the iceberg. Time and practice will allow you to master them, giving you more control over your craft and allowing you to tell the stories you want to tell.

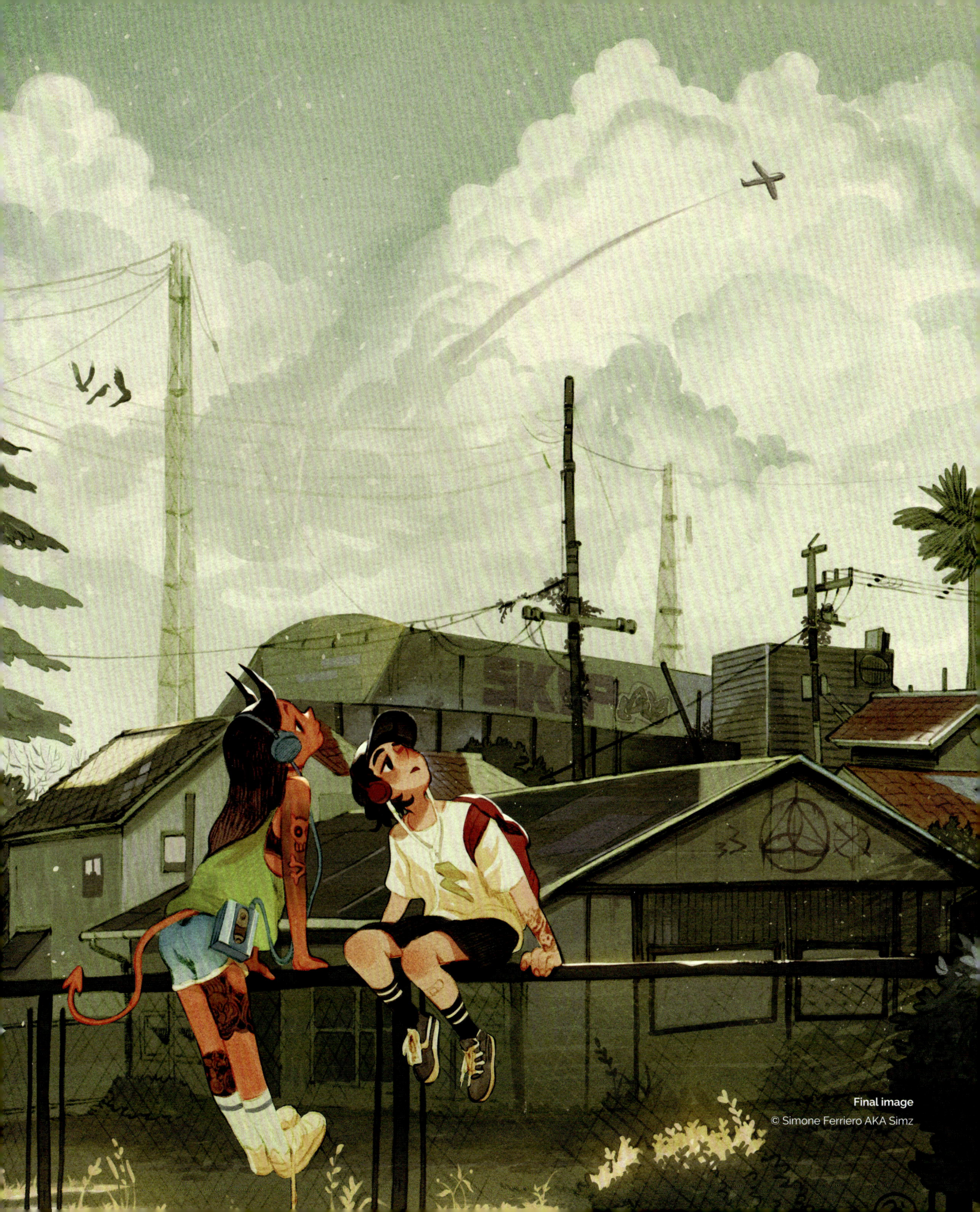

Gallery

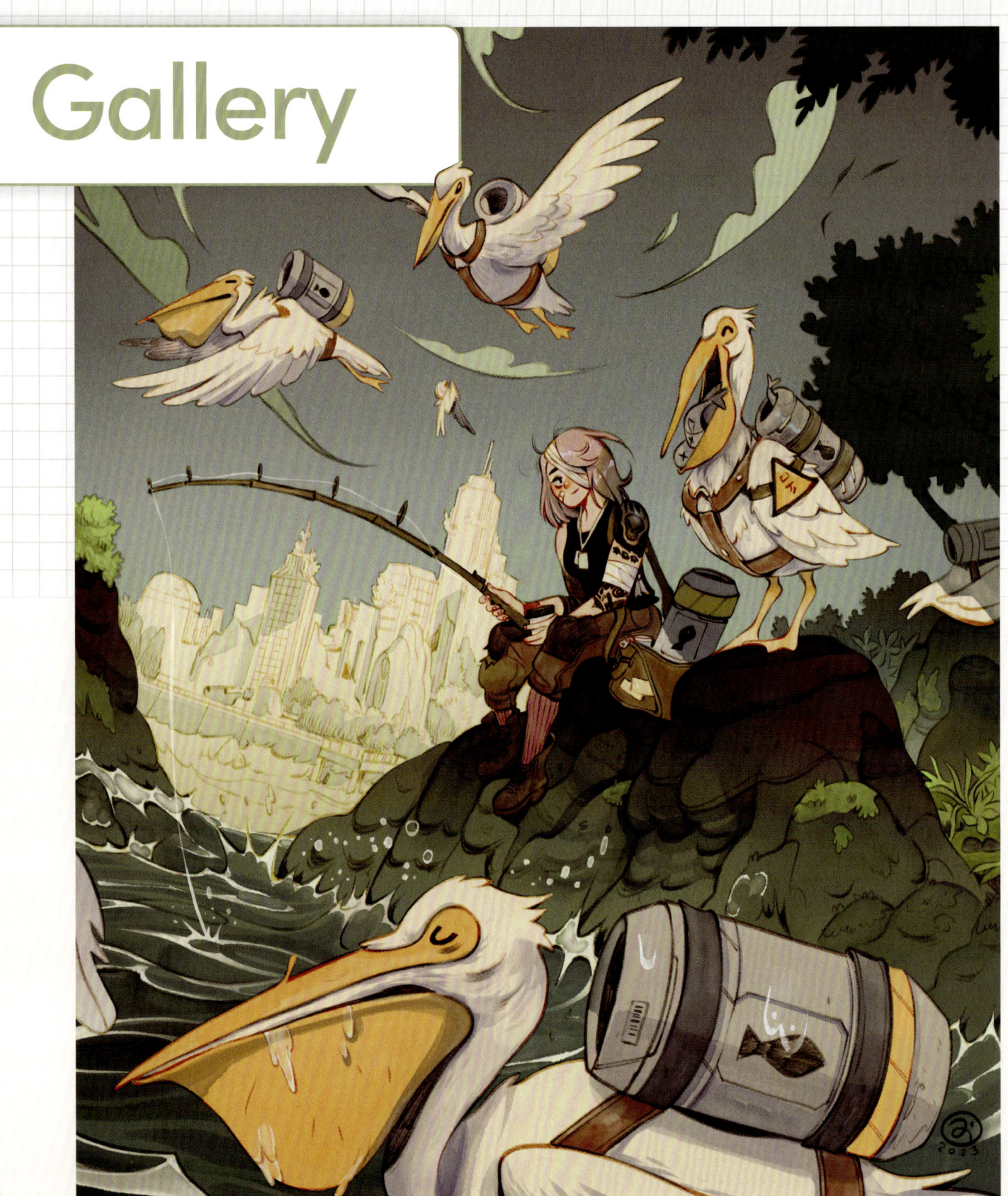

Pelicans © Simone Ferriero AKA Simz

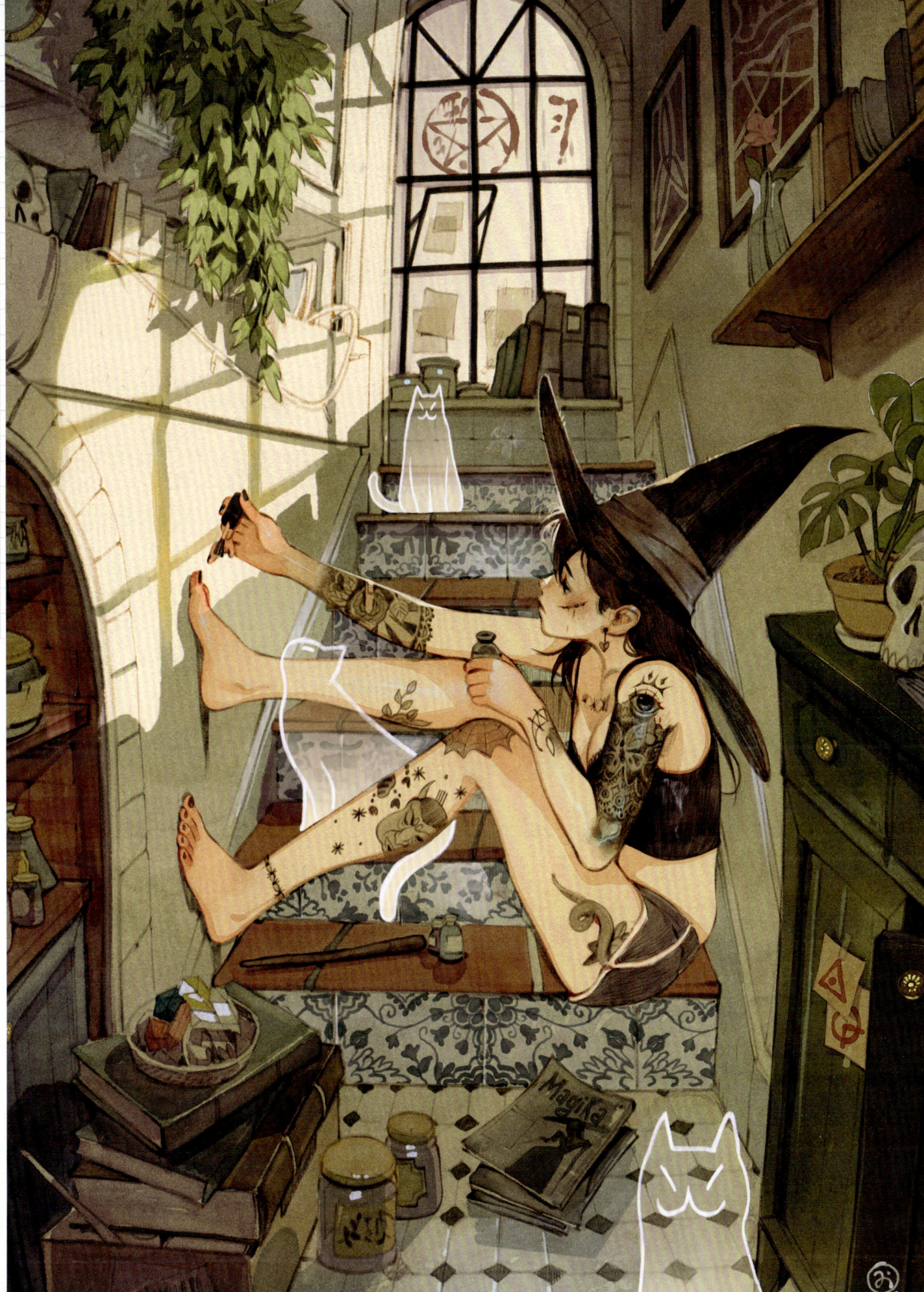

Comic

BY DYLAN TEAGUE

Ever wished you could use your artwork to tell a story? Interested in building a narrative into the imaginary worlds you dream up? Well, strap in, because this tutorial will take you through the creation of a comic page in Clip Studio Paint. I'll guide you through each step of my process, starting with creating the initial rough, setting up panel borders, and using perspective rulers. Following this, I'll cover pencilling, refining the pencils, and finally colouring.

I will create a page from a sci-fi comic strip I draw for myself in between paying jobs. It combines all of the things I enjoy drawing, with a storyline to tie it all together. The page will show the arrival of a large enemy spaceship on a rundown, war-torn planet. Our protagonist will spot the ship arriving and quickly make her escape.

I work with a format similar to old newspaper strips. It's a fun shape to work in, as it's not quite as intimidating as a full comic page, but still provides the opportunity for wide establishing shots. I've always been a big fan of sci-fi comic strips from the 1980s by artists such as Enki Bilal, Mœbius, and Juan Giménez. Even though I'm working digitally, I enjoy incorporating influences from these artists. I try to replicate the watercolour-type feel to the colouring, avoiding making the artwork appear overly digital. Working with Clip Studio Paint allows me to jump from frame to frame, and back again, rather than one frame at a time. For this tutorial, however, I will break down each panel individually.

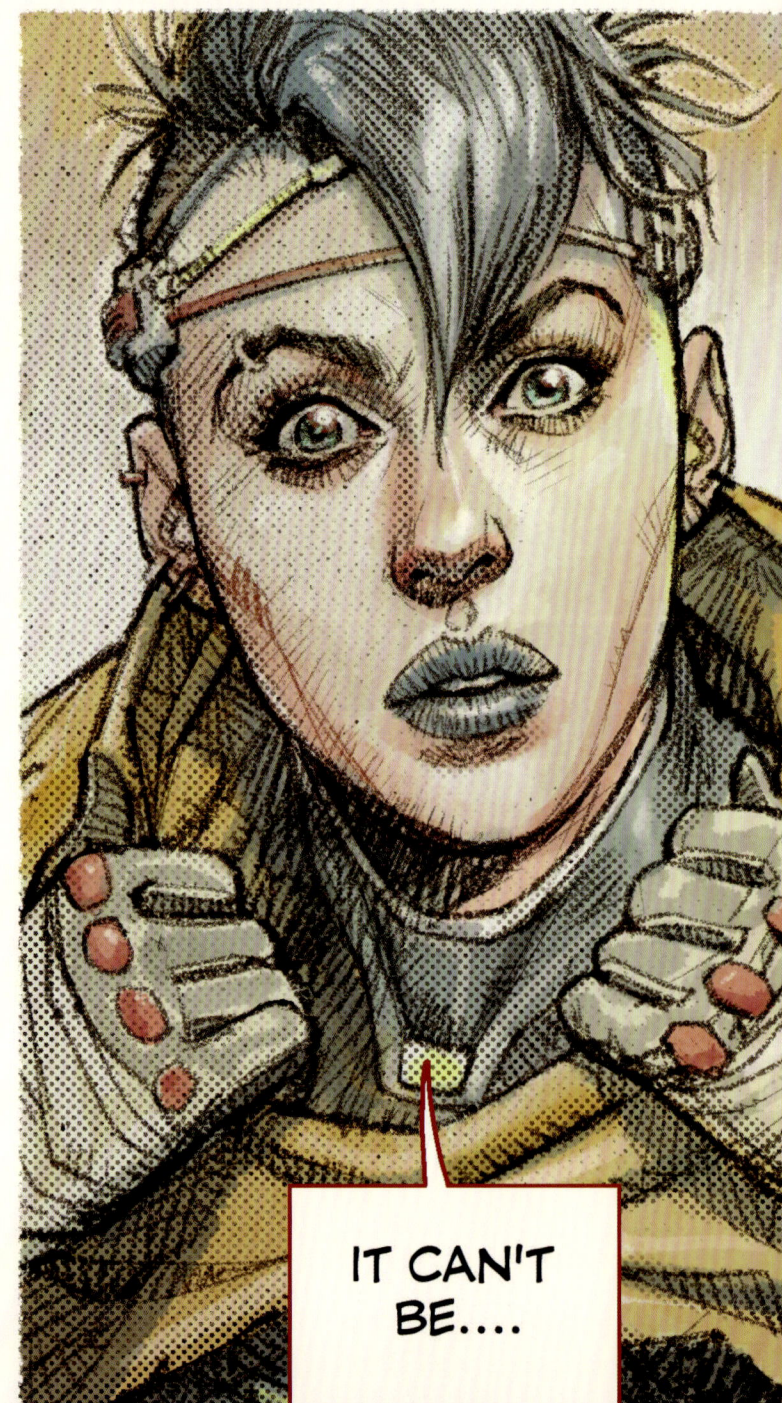

Comic © Dylan Teague

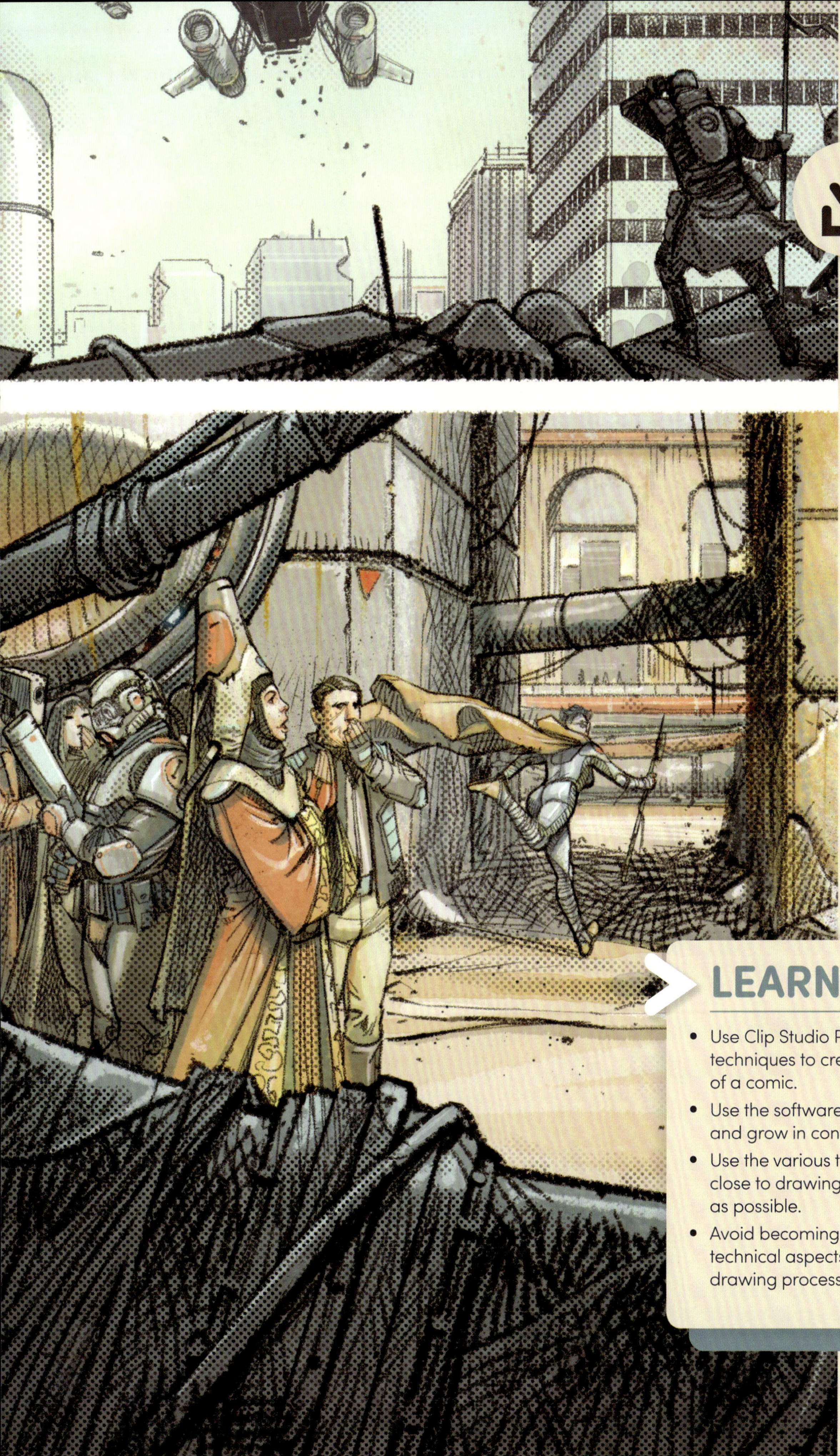

LEARN HOW TO

- Use Clip Studio Paint's tools and techniques to create a dynamic page of a comic.
- Use the software in a straightforward way and grow in confidence using it.
- Use the various tools in a way that's as close to drawing with traditional tools as possible.
- Avoid becoming bogged down with the technical aspects so you can enjoy the drawing process.

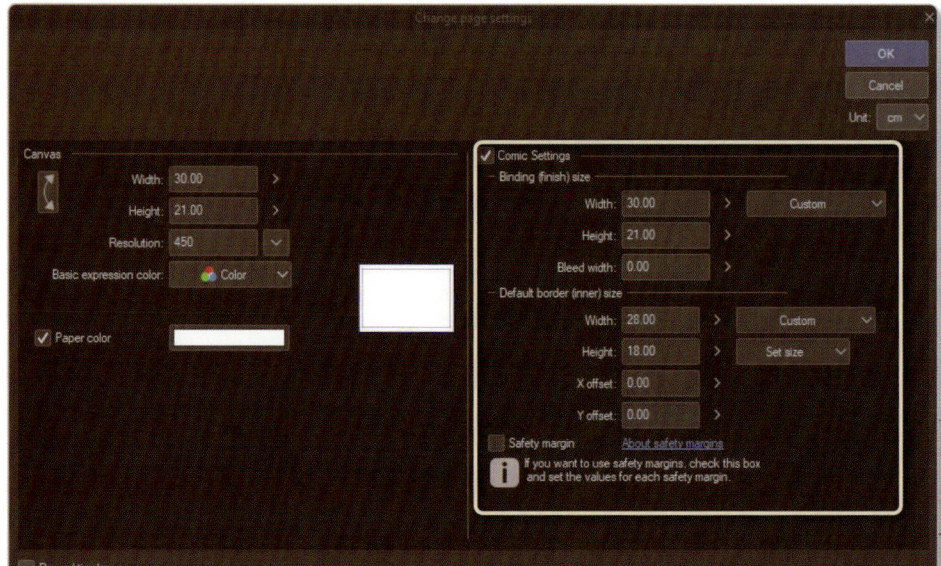

< Set the page dimensions to 300 mm wide × 210 mm high, with margins 280 wide × 180 mm high

∨ The initial rough – keep this stage loose, merely blocking out shapes for the composition

01 The first step in creating a comic is the rough. You can draw this in a sketchbook, or straight into a new file in Clip Studio Paint. As the purpose of this stage is simply to set up the composition of each panel, you can keep it very loose. If you plan to make this page part of a longer project, set up a Clip Studio Paint story file (see step 02) that contains all of the pages in one file. (Multi-page settings are only available in Clip Studio Paint EX.) Start by sketching a big establishing shot of the enemy battleship and fighters arriving above the city. Next, cut to a closer overhead shot of the main character looking up at the looming ship. The final frame will then zoom out to show the character making her escape.

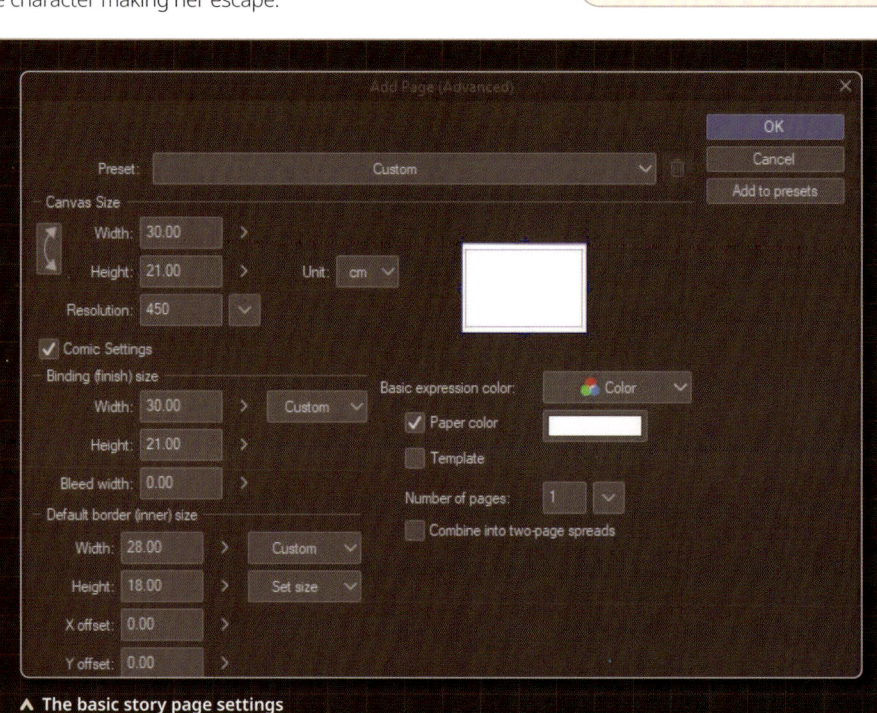

∧ The basic story page settings

02 Clip Studio Paint's story file allows you to organize all of your pages into one file, as well as setting out the content you wish to be the same across all of the pages for consistency. This includes settings such as the resolution of the images, page dimensions, and margin sizes (as detailed in the screengrab). You can even go into binding and other print set-ups, but for this tutorial it can simply be used as a way to organize the pages. If you wish to create a hand-drawn look, you don't need to use the Panel Border tool. This would make the border snap to the margin you have set up, whereas for this tutorial, you can just use it as a rough guide. A useful feature for users of Clip Studio Paint EX is the ability to add a page at any time by selecting Story > Add Page. Depending on the page you're currently on, a new page will be created next to it with all of the same settings.

03 Reduce the opacity of the initial rough in the Layer palette and create a new layer to work on. Begin to refine the initial rough on this new layer and start thinking about Perspective Rulers, though these will be introduced later. Add in another panel with a close-up of the character – this will help to introduce the character to the reader. Next, start to refine panel three and the new panel four.

> On a new layer above the initial
rough, begin to refine each frame

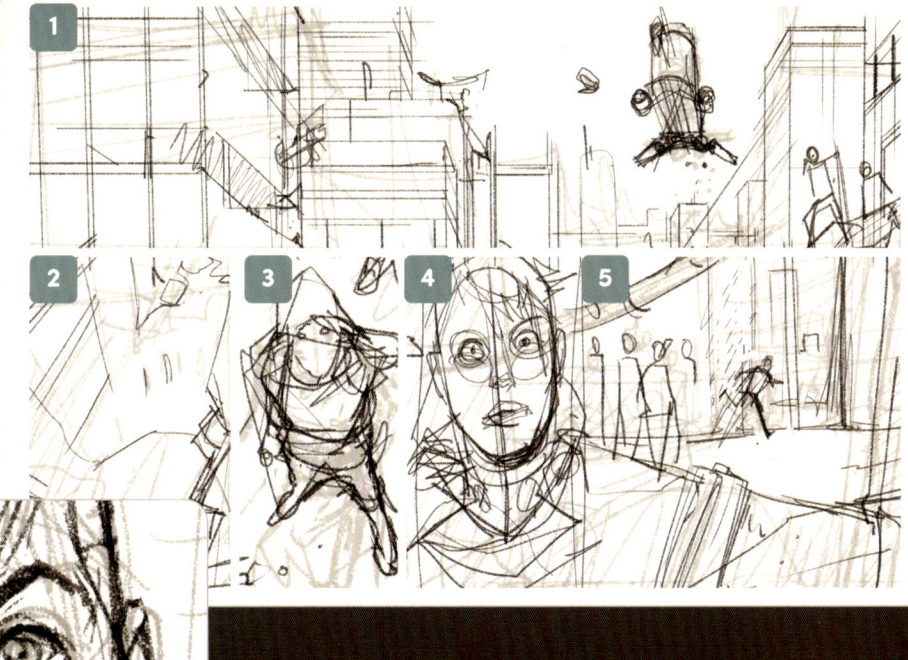

04 Zoom in on the close-up of the character's face to refine it further, though still keeping it pretty loose. This is the trick to making a digital artwork look as analogue as possible. Lower the opacity of the previous layer and create a new layer above it to draw this more refined sketch on. Rough-in her facial features and start to think about her costume and hairstyle.

< Refine the close-up sketch of the character – this will act
as a foundation on which to draw the neater line work

< Set a shortcut
to make the ink
switch between
drawing colour
and transparent

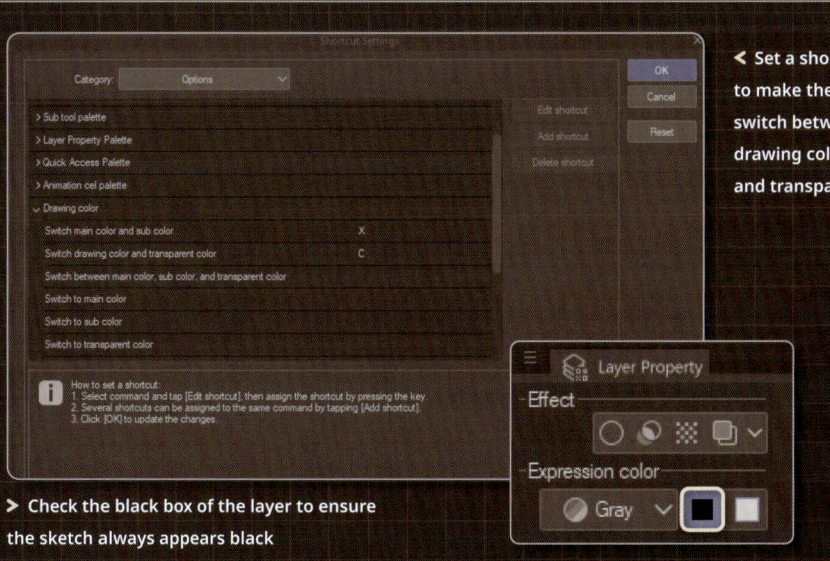

> Check the black box of the layer to ensure
the sketch always appears black

05 Take a moment to look at the set-up of your layers. Set the layer's Colour Expression to Grey, with only the black box checked. This means it doesn't matter which colour you draw with – it will always come out black. To set up a shortcut to make the 'ink' cycle between black and transparent, select File > Shortcut Settings. This will open a dialog window where you can assign keyboard shortcuts for various different actions. Select Options > Drawing Colour to assign a shortcut to switch between drawing colour and transparent. This is useful if you find you need a quick method to erase what you've done and it will also make it easier to colourize your line work at a later stage.

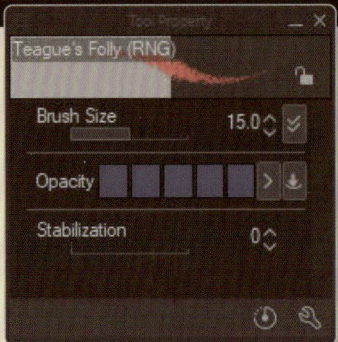

ARTIST'S TIP

You don't need to use an abundance of brushes to create a quality image. In the downloadable resources you will find a brush called Teague's Folly RNG (see page 257) created by artist Ray Frenden. You can use this brush almost exclusively until you reach the colouring stage. It's also possible to modify the Line and Ellipse sub tools to have the same tip as the pencil, which you can use for more technical aspects of the image. Simply open the Tool Property menus of the Line and Ellipse sub tools and, under Brush Tip, set the Material to match the tip of Teague's Folly. You won't need to use them much if Perspective Ruler snapping is turned on, however, as the pen is great for creating straight lines.

∧ The settings of Teague's Folly RNG brush

> Draw a near-final sketch of the close-up character on top of the previous sketch

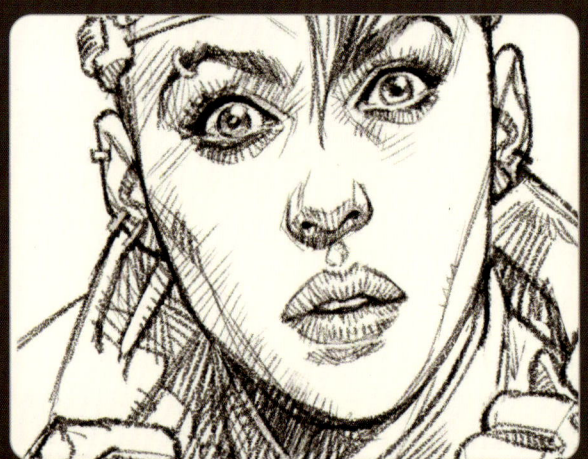

06 Use this step to refine the close-up sketch of the character and work towards the final line work. There will be a lot of back and forth at this stage as it's where the bulk of the work is done. As with the earlier steps, lower the opacity of the previous sketch layer, then create a new layer above it on which to draw a refined sketch. Make regular use of the Navigator panel's Flip Horizontal button at this stage – this will provide you with a fresh look at the drawing and help you to spot mistakes. In addition, it can be useful to turn off the rough layer from time to time to check the progress of the new sketch.

07 Changing the layer colour is one of the most useful features of Clip Studio Paint. You may find it helpful to change the colour of the previous layer to blue so it doesn't interfere with the lines on the new layer. Sometimes lowering the opacity of the rough layer alone isn't enough and it can become a bit distracting when drawing the neater final lines. There's a tab in the Layer Property palette that will turn the layer blue by default. This replicates the use of 'non-repro blue' typically used in comics created by hand, as the blue colour would not show up in print. You can choose whatever colour you like though!

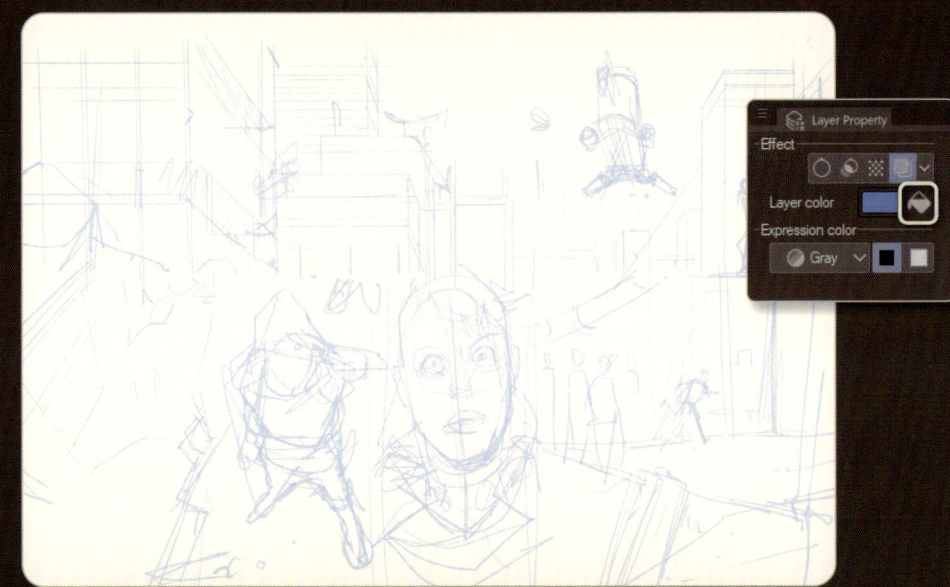

∧ Click this icon in the Layer Property palette as well as the blue rectangle next to the bucket – this will let you choose any colour you like

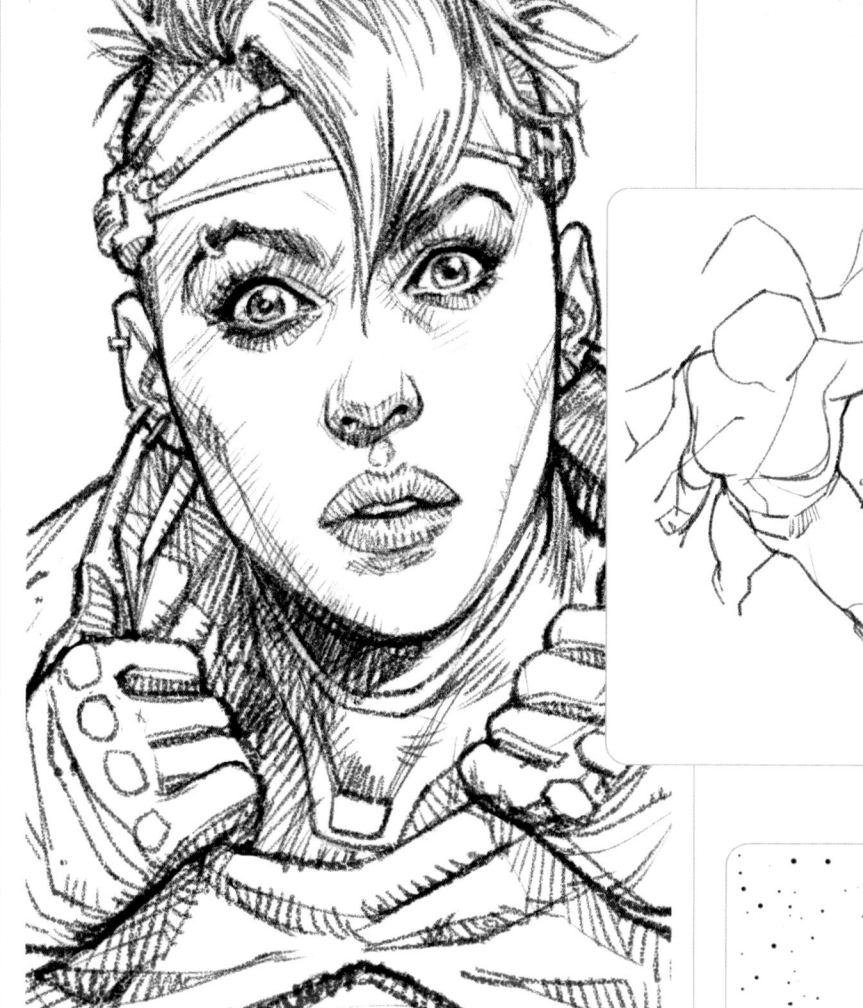

08 Panel four is now complete, though it's inevitable that you'll return to it for minor tweaks. This is one of the perks and perils of working digitally! Next, move on to panel three. This is a full-figure shot of the main character, so you can take this opportunity to design her costume. Before this, however, you need to rough-in the figure. The panels are gradually zooming in from the initial wide establishing shot of panel one, but it's still a high overhead angle. Though this can sometimes be a little tricky, it can look very effective when done well, so keep the perspective of the shot in mind when sketching the figure.

< **Start to refine the figure and work out the costume**

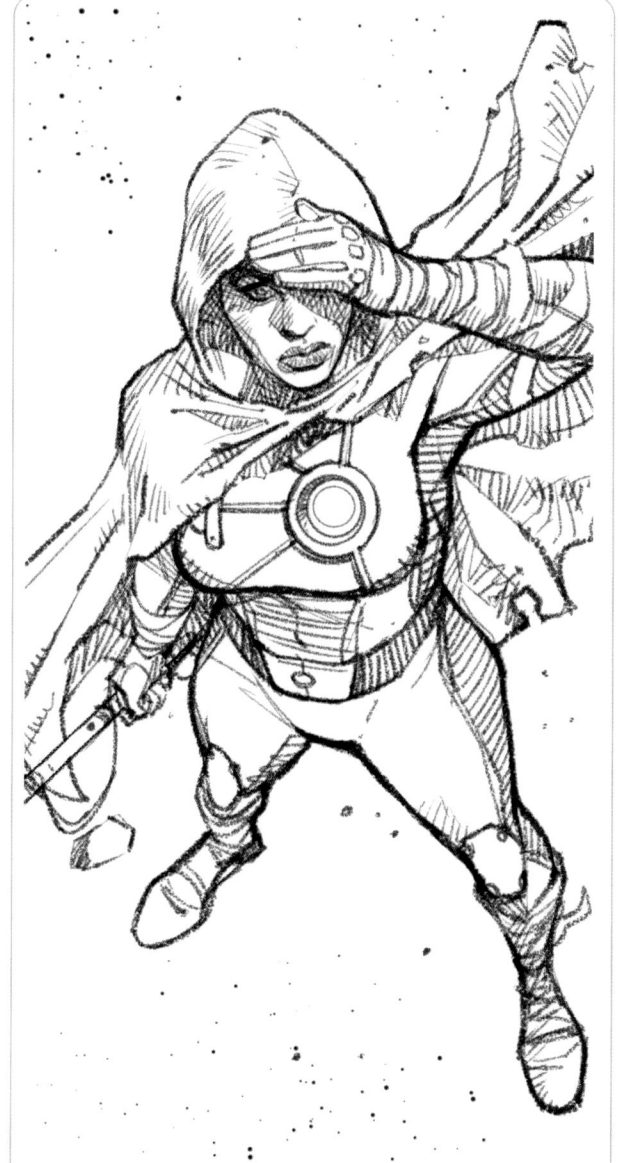

09 As with panel four, lower the opacity of the rough layer and draw a refined sketch on a new layer on top of it. As this is a sci-fi comic, you can use your imagination when designing the character's costume. Here she wears a circular control panel harness top with some piping detail. The addition of a hooded cloak will make the reveal in the next panel more interesting, as well as implying movement between the panels. It's a good idea to show your characters performing some kind of action to keep the narrative moving and ensure the reader is engaged. You can also add some indication of the ground, plus a bit of debris.

> **Draw the final lines over the rough sketch**

10 While you have already set the story page settings (dimensions and margins) in step 01, at this stage you can now consider adding the panel borders, as it's unlikely you will be making many changes to the layout from here on in. If you want to create a hand-drawn look, you don't need to keep strictly to these – go with what you feel suits the page. As the 'paper' is set to a slightly off-white colour, you can use this for the borders too. Select the Marquee tool near the margin box, then invert this selection and fill it with the off-white colour. After turning off the selection, use the Figure > Line sub tool, to create the panel borders. Hold down Shift when drawing to create exact horizontal and vertical lines. As mentioned on page 224, you can even apply texture to the Line tool by going to Tool Property > Sub Tool Detail palette > Brush Tip > Material to assign a rougher texture.

> Here the background is set to red to allow you to see the borders more clearly, but for the final page it will be off-white

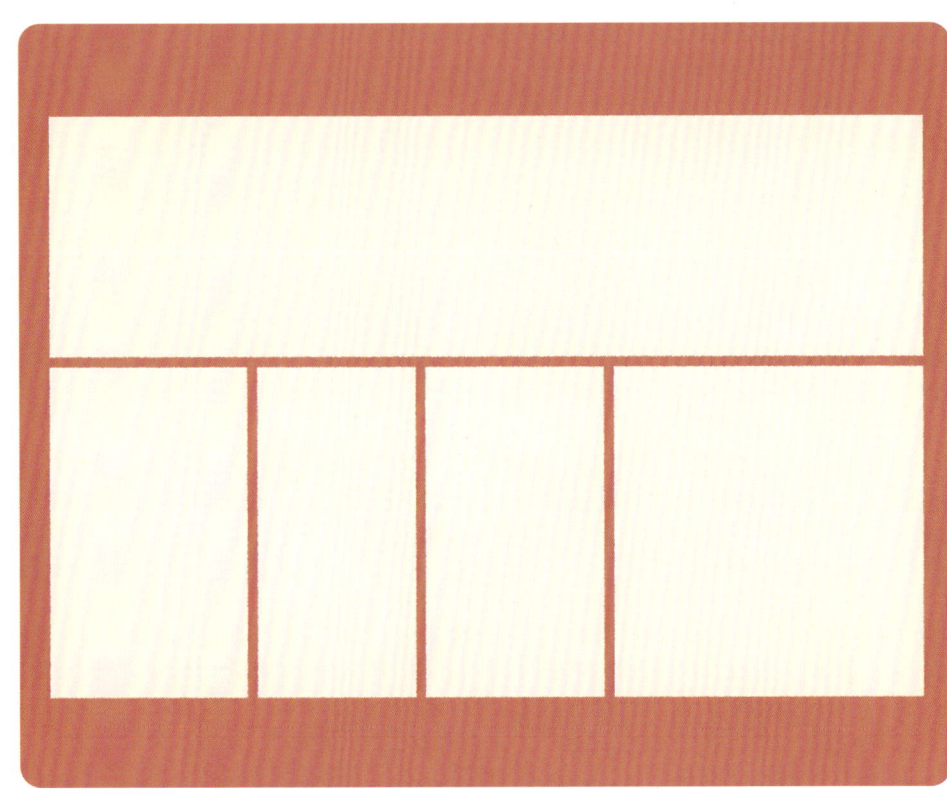

11 Tone layers can add visually pleasing texture to your work. To create a tone layer, select Layer > New Layer > Tone. A window will appear where you can enter the settings shown in the screengrabs. The Tone layer functions just like a mask: remove areas of tone with an eraser or by painting with transparency, and add tone back in using a drawing tool with any colour.

∨ Use these settings for your Tone layer, then apply and remove tone to achieve interesting textures

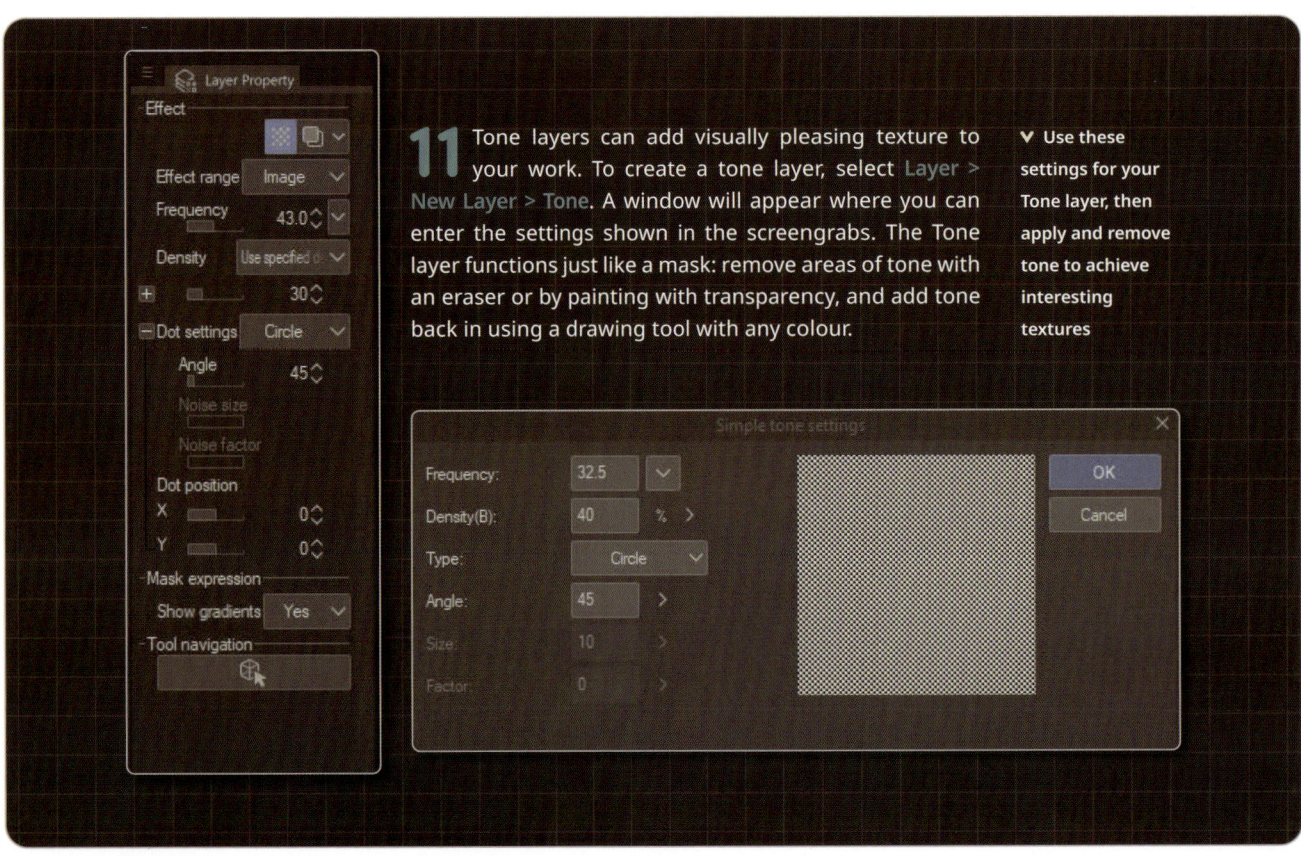

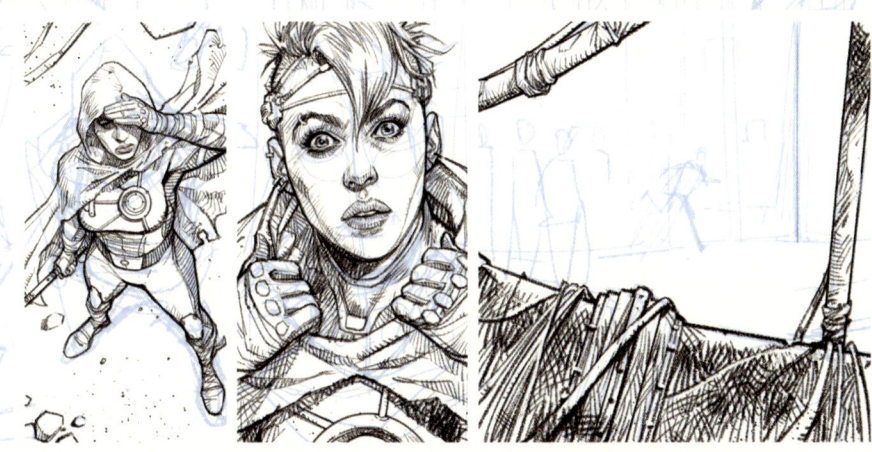

12 Take a look at the final panel. Try using a foreground element of simple pipework to frame the action. You don't need to set up Perspective Rulers yet, but keep where the perspective will be roughly in mind. Shade the pipe shapes quite heavily so they are almost like a silhouette that will outline the more detailed action taking place in the rest of the panel.

< Draw the line work and tone for the foreground elements of the final panel

13 Set up a Perspective Ruler for the final panel by selecting Layer > Ruler/Frame > Create Perspective Ruler > 2-Point Perspective. Adjust the settings to ensure the ruler will be visible on the layer you're working on. Right-click on the ruler to make it show up on all layers. Using the anchor points, drag the horizon line to the desired height (around neck level) and drag the right vanishing point to the desired location. To get the vanishing point to the correct position, use the anchor points to move it until it aligns with your sketch. By holding Shift and dragging an anchor point on the left vanishing point, you can make them become horizontal, so all those lines will become parallel. Clicking the semi-circle icon on the top toolbar will lock all of your tools to the Perspective Rulers.

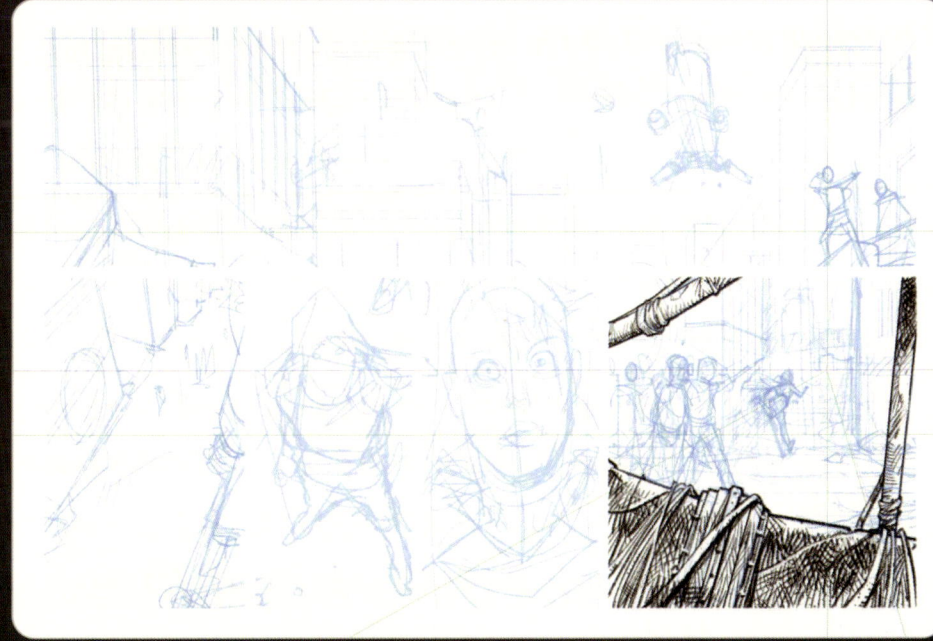

∧ Set up a two-point perspective ruler that is visible on all layers

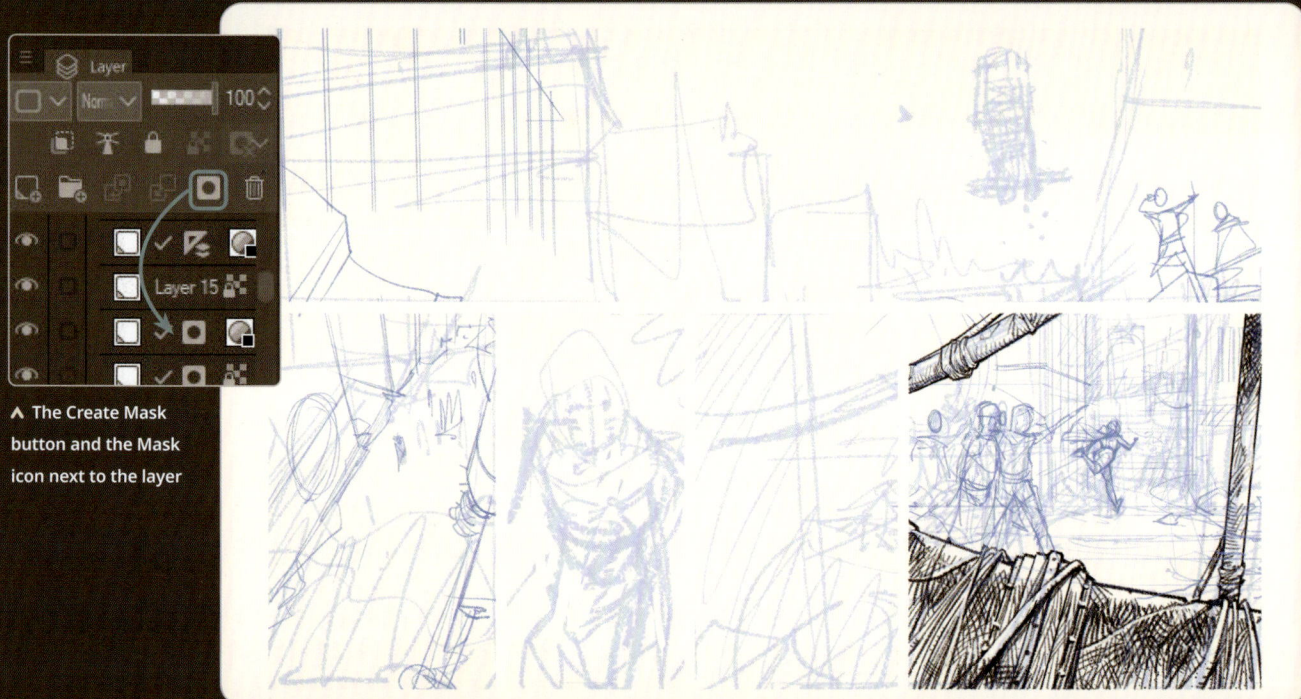

▲ The Create Mask
button and the Mask
icon next to the layer

14 Once you're happy with the background, it's time to move on to the figures. Working digitally allows you to draw the whole of the background and then make a new layer for the figures. While this may take a little longer, it means you can change your mind when drawing the figures without damaging the background. Once you're happy with the figures, plus any other foreground elements, you can return to the background and mask off what you don't need. Simply select the background in the Layer panel, click the Create Layer Mask icon, then use an eraser on the mask to get rid of anything you don't need. This is a really useful function, as the erased content can be brought back at any time (if you have the mask thumbnail selected) by drawing with any colour. Using a solid tool like the G-Pen with full opacity will bring the masked areas back fully, though using different tools can create some interesting effects.

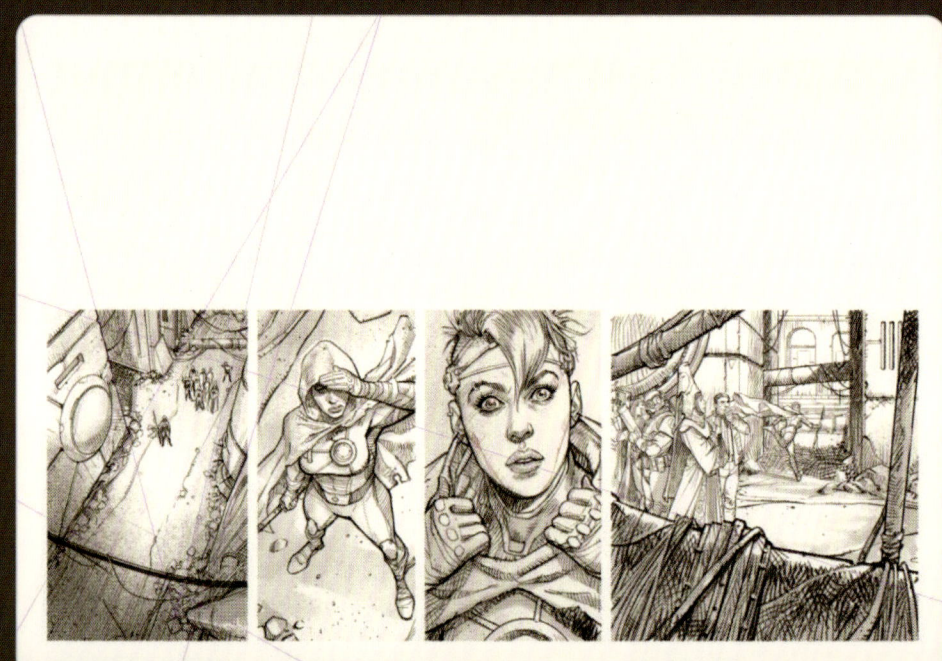

15 Now turn your attention to panel two. Rough-in the basics of the frame, then create a more refined sketch on a new layer above. As it has a high-angle view, this panel will need a 3-Point Perspective Ruler. Like before, select Layer > Ruler/Frame. It can be useful to develop the image until it's almost what you want before adding a Perspective Ruler, and then shift the rulers to make them fit with the current scene. The more you practise, the more confidence you will gain. The Perspective Rulers are a great help, but you need to build a good understanding of perspective to get the best out of them. Another great feature of Clip Studio Paint is the ability to snap shape tools to perspective. This is useful for creating the ellipse on the wall at the correct angle.

‹ Draw in the refined line work for panel
two with visible perspective rulers

16 You will likely have a lot of layers by this point. The Story settings have Colour Expression set to Colour by default, so when you select the New Layer icon in the Layers palette, a Colour layer will be created. Earlier in the tutorial you created lots of greyscale layers, with just the black box ticked. It's possible to create a time-saving Auto Action for this. Using the Auto Action panel (tabbed behind the Layer panel), add a new action, hit the record button, perform the steps you want to automate, then stop recording. Drag the action to the Command bar to make an icon that you can easily click. Custom actions that a comic artist might find useful to automate are a one-click copy-and-paste, displaying the outline of the current layer, and merging a layer with the one below. You can add default behaviours as shortcut icons, too. Just right-click on the Command bar, go to Command Bar Settings, and add your most-used tools to save time. Simply right-click on the Command bar and go to Command Bar Settings to add or remove buttons.

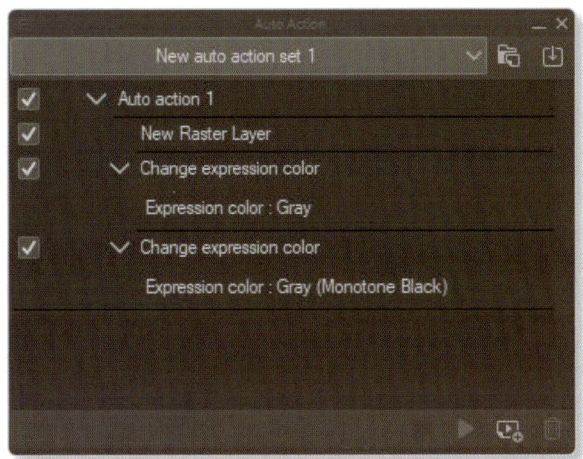

▲ The settings for Create Layer auto action

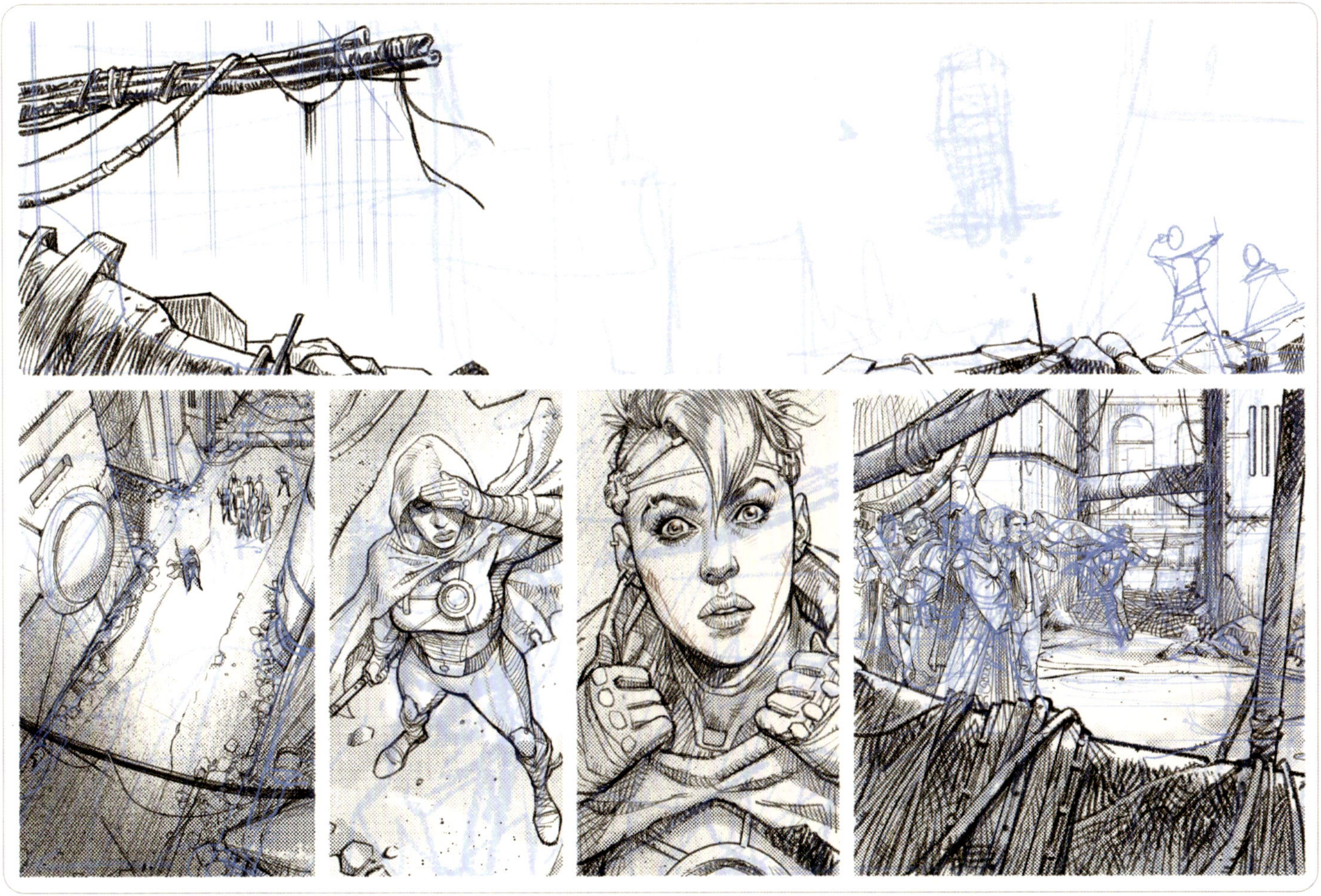

17 Finally, turn your attention to panel one. Create a new layer for this. Sometimes it's worth locking the layer, or layers, you've previously workedon, as it's all too easy to accidentally merge down your current layer and possibly erase what you've already created. Draw in the foreground elements first, as these won't need the Perspective Rulers, but still keep in mind where the horizon line will be. Once you're happy with the basic outline, sketch in quite heavy hatching to create an almost silhouetted feeling to the foreground. You can add the two small figures later.

▲ Sketch in the foreground elements, using hatching to add shading

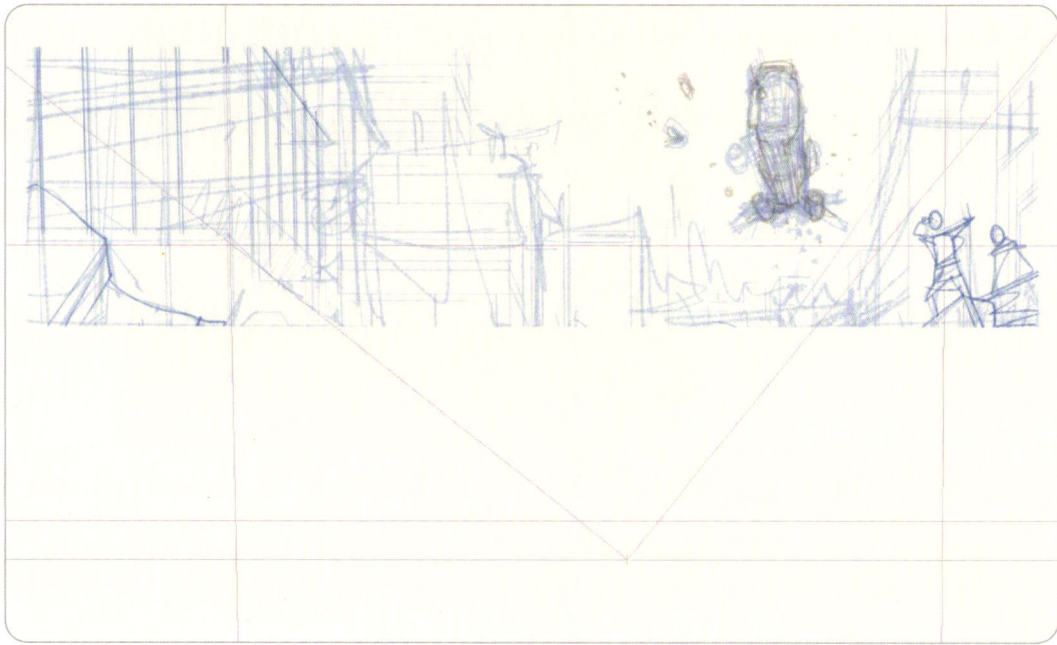

18 Using the same method as before, set up **3-Point Perspective Rulers**. Make one set horizontal and position another around the focal point of the panel. Set the third point low below the horizon. Though this is very subtle, it will give a slight lean to the buildings. Next, draw simple rectangle shapes and work into them, using hatching to shade the building on the left. Use the masking trick covered in step 14 to remove some of the hatching and create the look of broken glass.

◄ Set up 3-point perspective rulers, then sketch simple rectangle shapes to rough-in the buildings

19 Create a hatched look on the buildings on the left and right. You can use a brush from the **Decoration > Hatching** sub tools to save some time, or try creating a hatching brush of your own. To do this, create a greyscale layer and draw small parallel strokes with the pen. Next, use the **Selection Area** tool to make a selection around the strokes, then go to **Edit > Register Material > Image** to save your new brush tip file. You can then open the **Sub Tool Detail** palette for a pre-existing **Hatching** brush and change the tip for the new one you've created.

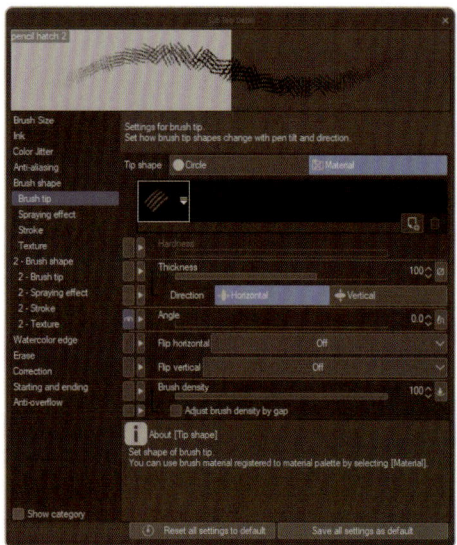

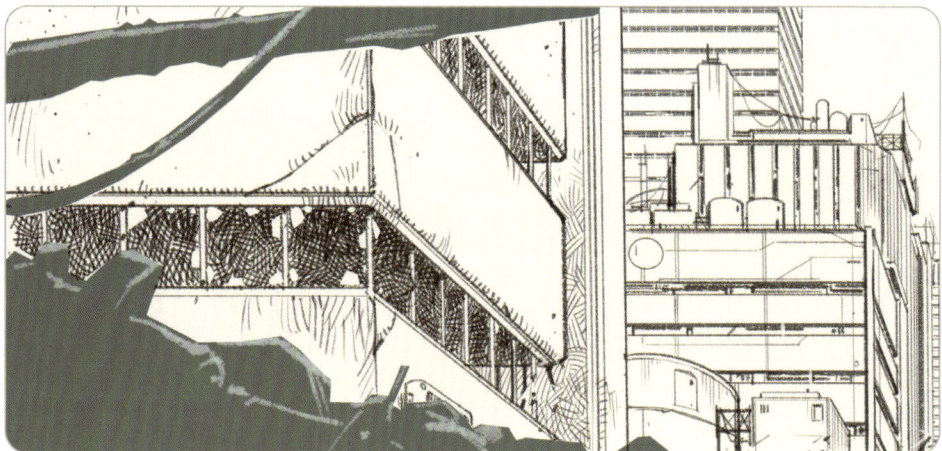

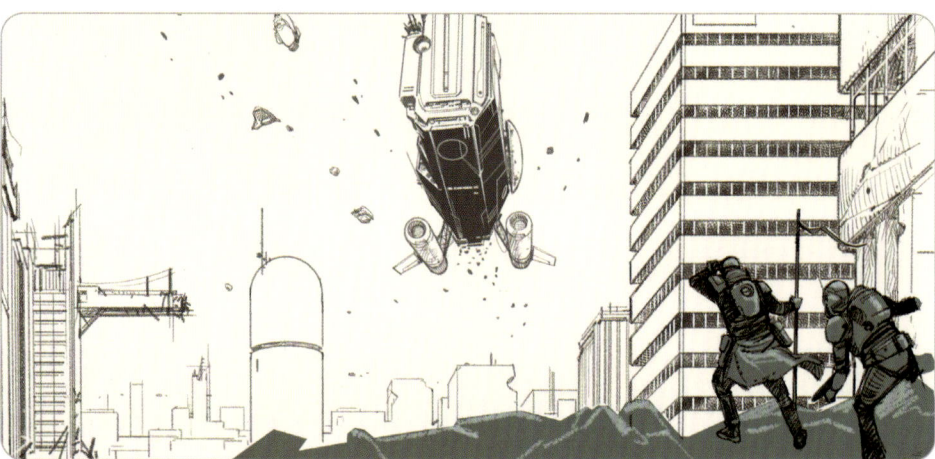

◄ Try creating your own hatching brush to speed up the shading process

> ARTIST'S TIP

At this point in the tutorial you have more or less finished the line work and are almost ready to start introducing colour. It's important to ensure the image can stand on its own at this stage. This page shows how important a good knowledge of perspective can be to creating believable settings for your comics. Learn how to draw freehand before bringing in Perspective Rulers – this will prevent the image from looking too stiff and clinical.

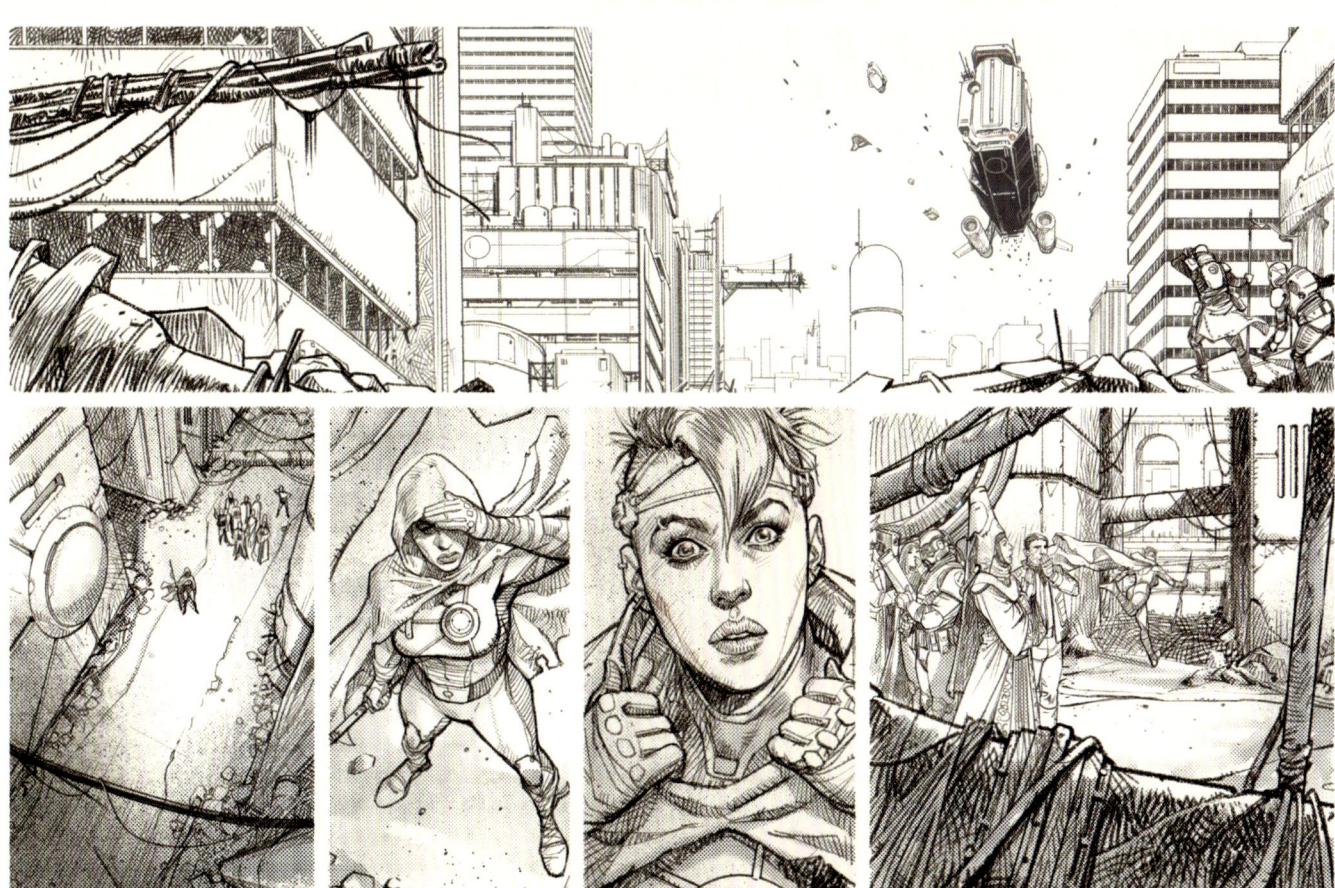

20 Now the line work is complete, it's time to prep for colour by applying some settings to the layers. A big benefit of working digitally is that you can keep different elements of an image separate. For example, positioning the foreground and background

the layers for colour, change their Colour Expression from Grey to Colour in the Layer palette, then lock the transparent pixels on each layer. This will allow you to change the colour of the lines to whatever you wish. Try using lighter blue colours as elements move

∧ **Prep the line work ready for colouring**

21 Make a new Overlay layer and insert a paper texture onto it. Any type of high-res paper-texture image can be used for this - it's helpful to build up a collection of your favourites.

Place this layer almost at the top of the layer stack, above the line work but below the panel borders, so it will affect all of the layers beneath it. It will be barely noticeable on the line-work layers, but will add a slight texture if you have a large area of solid colour. You can reduce the texture's opacity and adjust its colour with Edit > Tonal Correction > Hue/Saturation/Luminosity until you are happy with the effect it produces."

> Create a paper texture layer to overlay the rest of the comic, giving it a hand-drawn, paper look

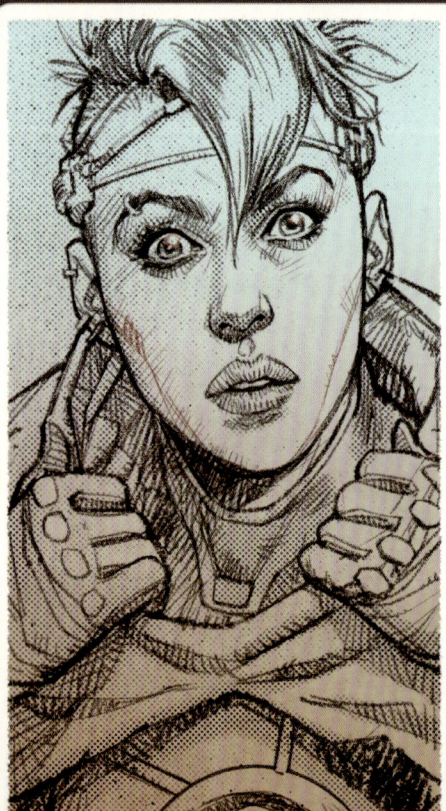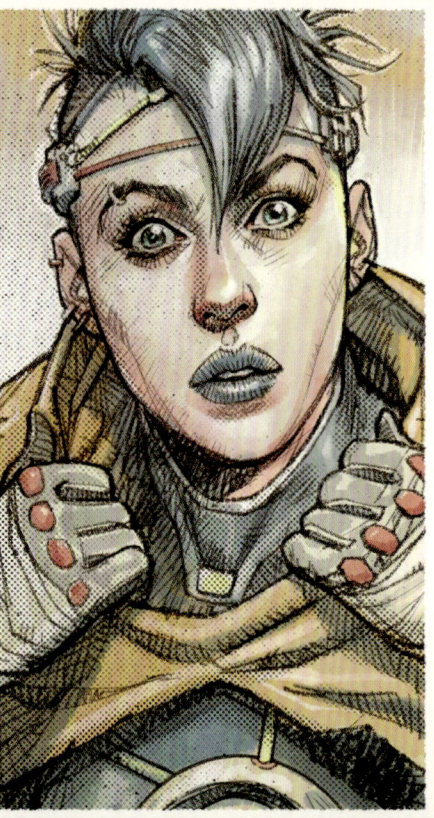

22 Now to colour each panel in turn. Select the panel you wish to colour first, create a rectangular selection of the whole panel area, and fill it with a base colour. Next, lock the layer's transparent pixels to make sure you're only colouring within that area. Using the Gradient tool, paint a slightly blueish hue at the top of the panel and a warmer brown/yellow coming up from the bottom. From this point onwards, work on each element of the panel in turn, picking colours to suit the scene and overall sci-fi genre. When applying the colour, try using texture brushes, such as an airbrush and a textured pen, as well as the Teague's Folly RNG brush mentioned on page 224. Once you're happy with the colouring of a panel, create a new layer and repeat the process on the next panel. If numerous layers slow your computer down, merge any layers you have finished painting. Working with as few layers as possible can help you to stay organized.

< Paint in colour, panel by panel

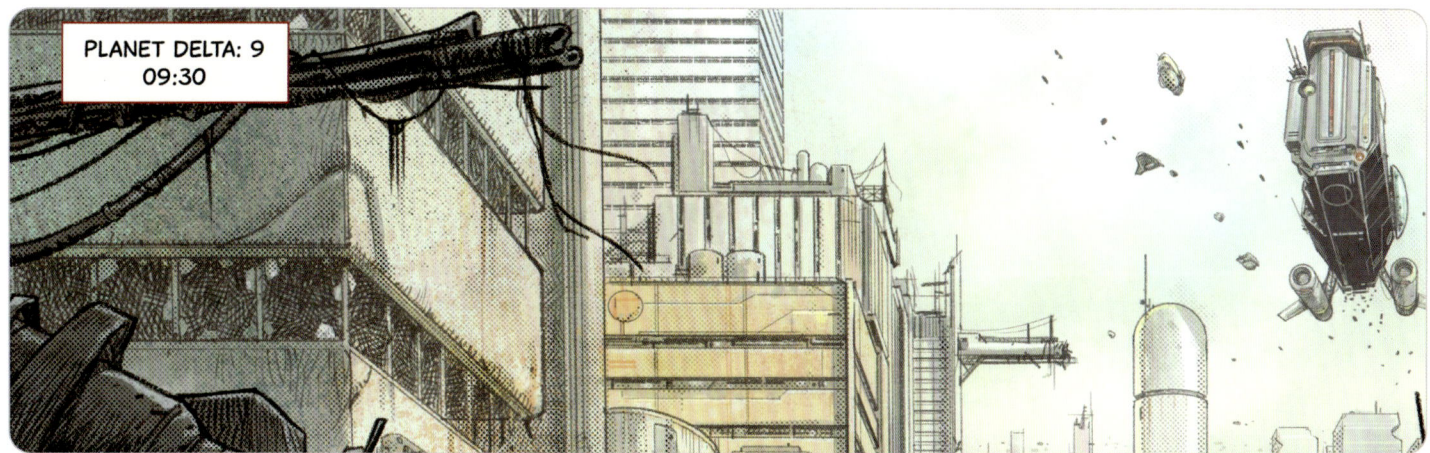

23 When creating a comic, you may want to add text, such as narration or speech bubbles. Once you've selected your font, select the Text tool and type in your text. This comic uses a font bought from Comicraft, but Clip Studio Paint has plenty of fonts to explore. Once you're happy with the text and the shape of it, you can add a box or balloon – either one of Clip Studio Paint's defaults or your own creation. To make your own, create a new layer below the text, create a selection in the desired box or balloon shape, then fill this selection with white. Select an outline colour, create a selection from the layer by Ctrl+ clicking on its thumbnail, then go to Edit > Outline Selection to set the placement and thickness of the border. To make the tail, create another layer and draw the tail shape with a pen. Erase a bit of the border on the previous layer, make a selection under the tail, and fill with white. Finally, merge the tail layer with its balloon layer.

> The final comic page with text

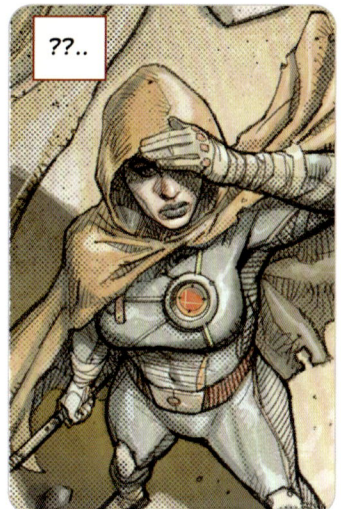

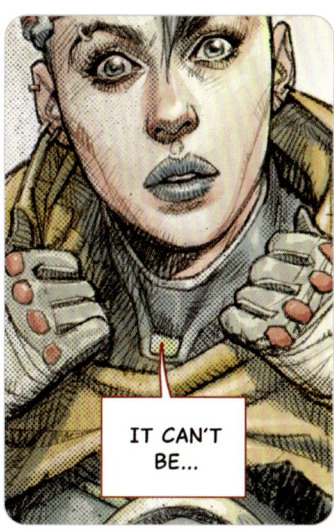

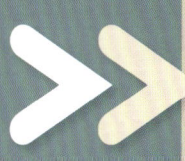

Conclusion

This page of comic is now finished! The panels successfully establish a new character and location. While it's fairly simple in terms of storytelling, the angle choices and flow from panel to panel add a dynamic quality and convey the information in a clear and engaging way. The tutorial has demonstrated how you can use Clip Studio Paint in a similar way to how you would approach traditional materials. Experiment with creating your own brushes, text boxes, and borders, especially if you want to recreate that hand-drawn look. There are many benefits to working digitally, including the way each panel can be drawn on a separate layer, or even multiple separate layers for optimum flexibility. Use the guidance in this tutorial to find the comic creation workflow that works for you!

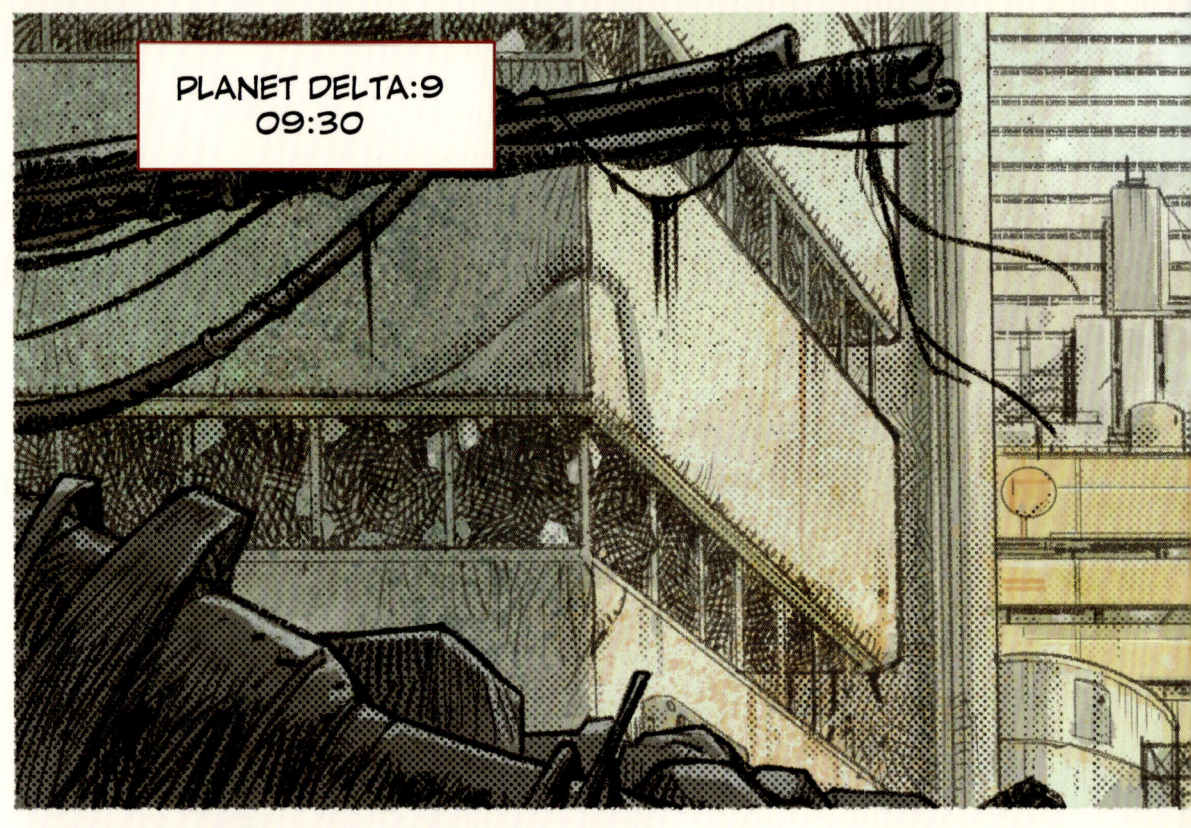

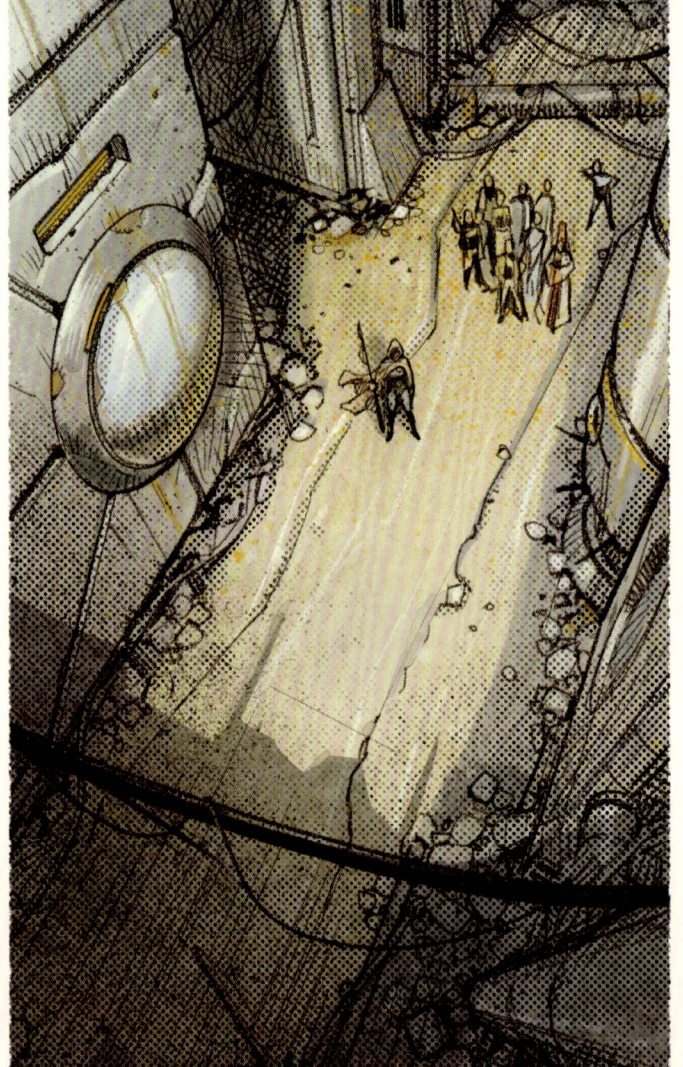

Final image © Dylan Teague

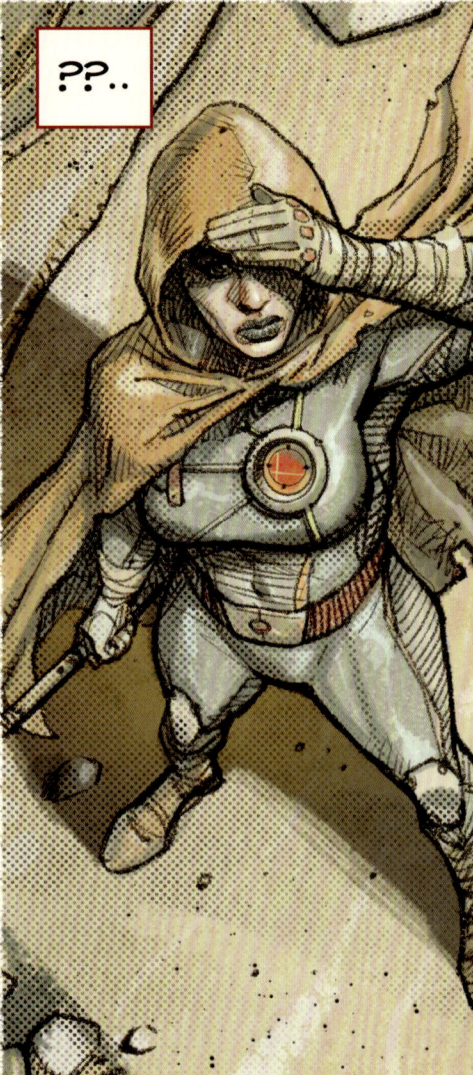

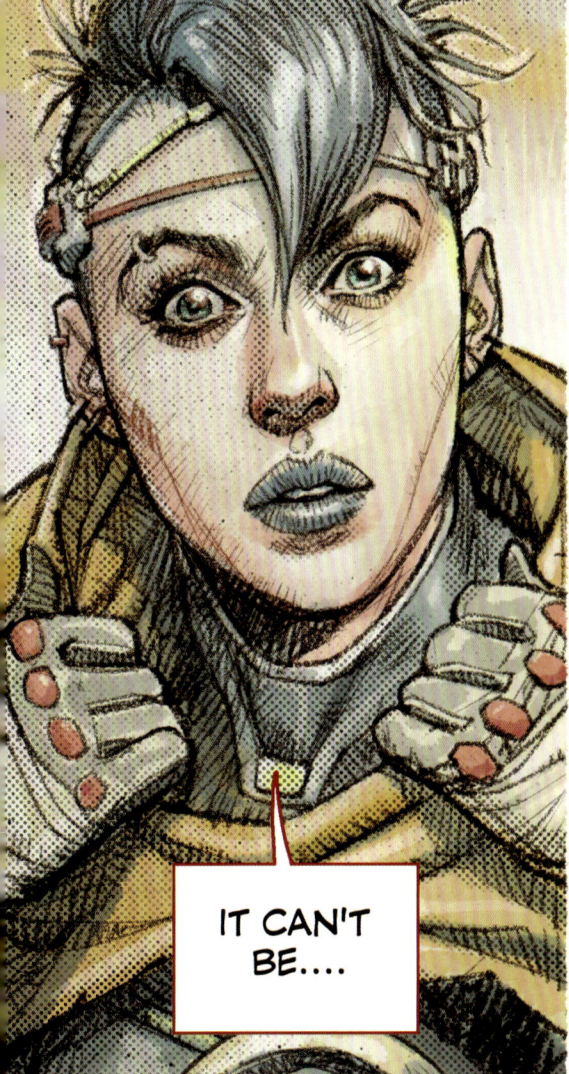

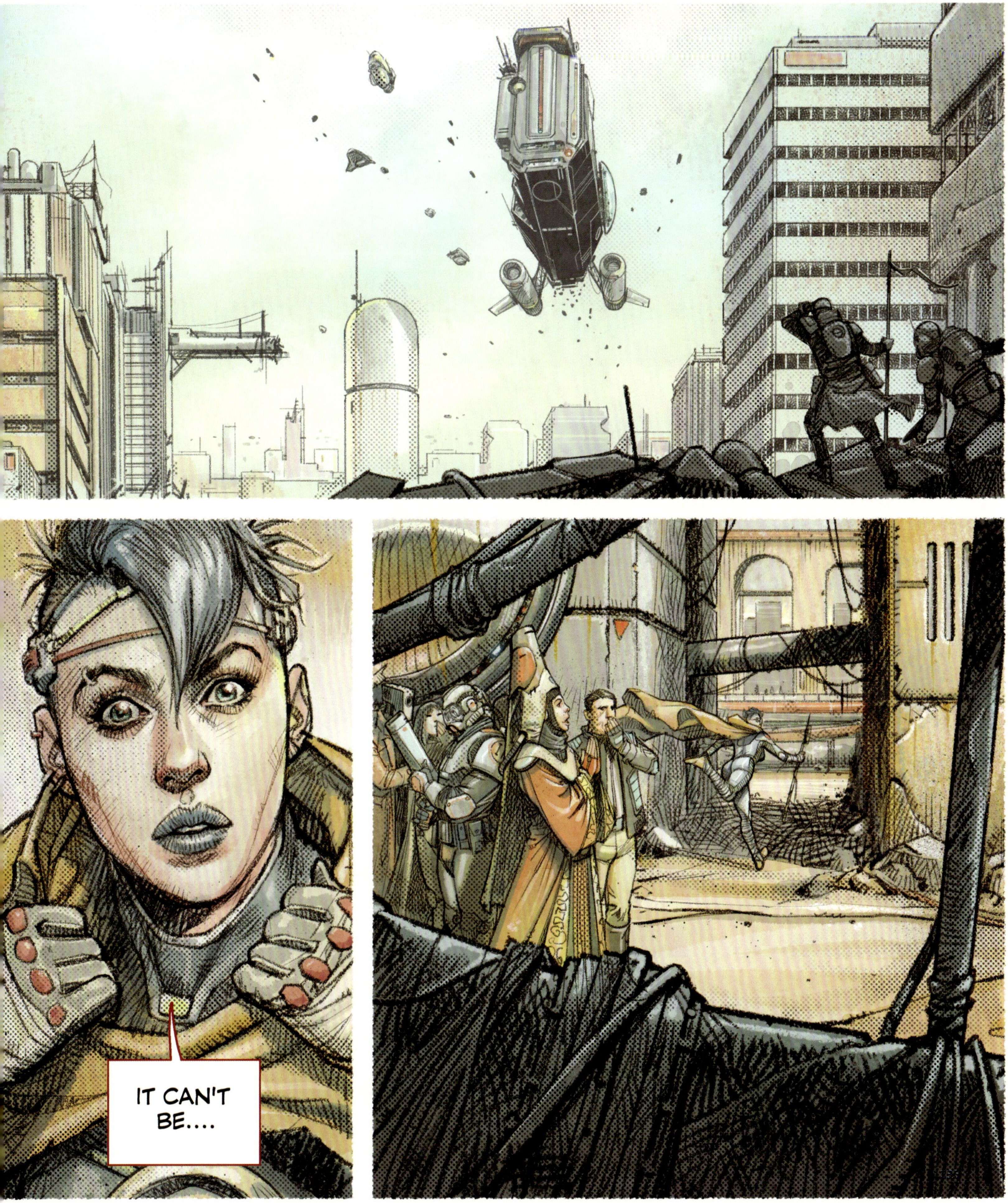

IT CAN'T BE....

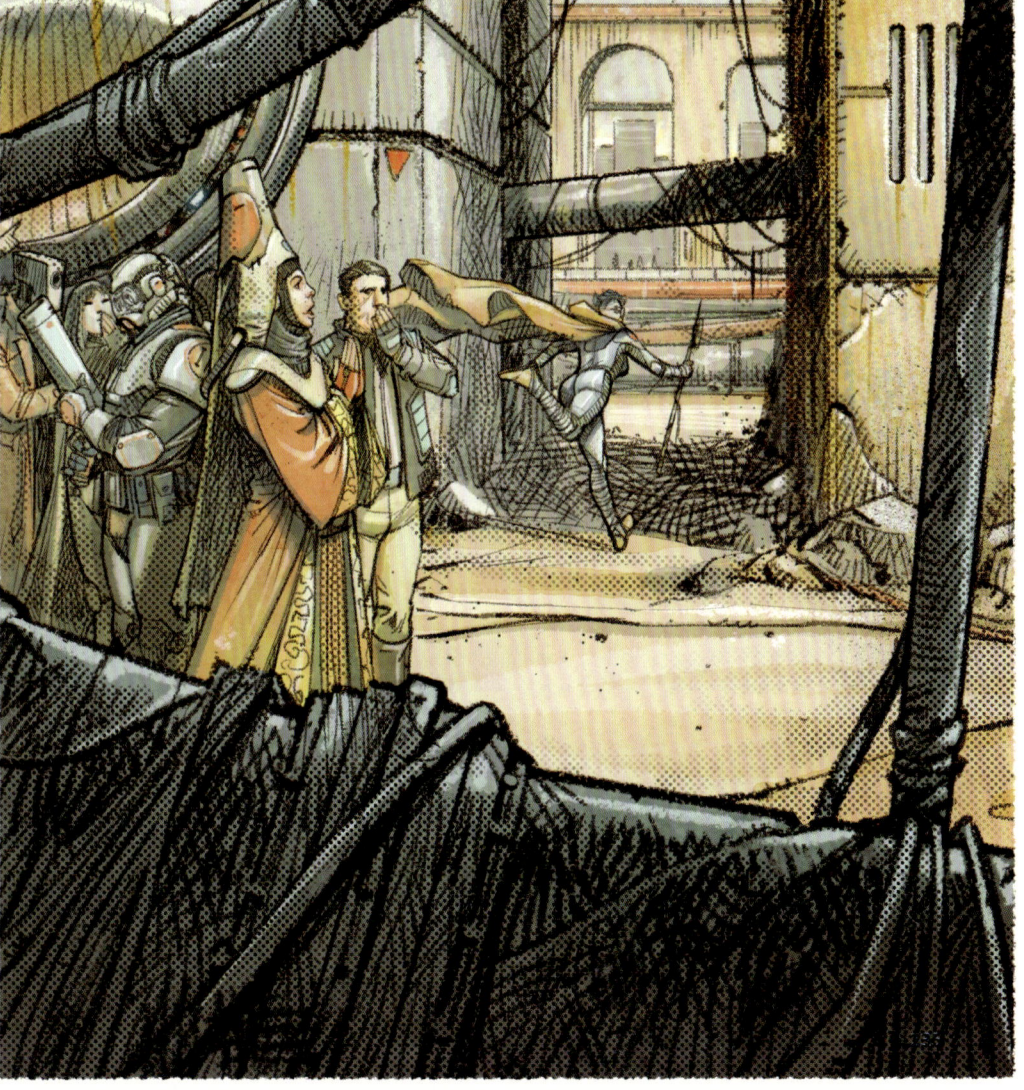

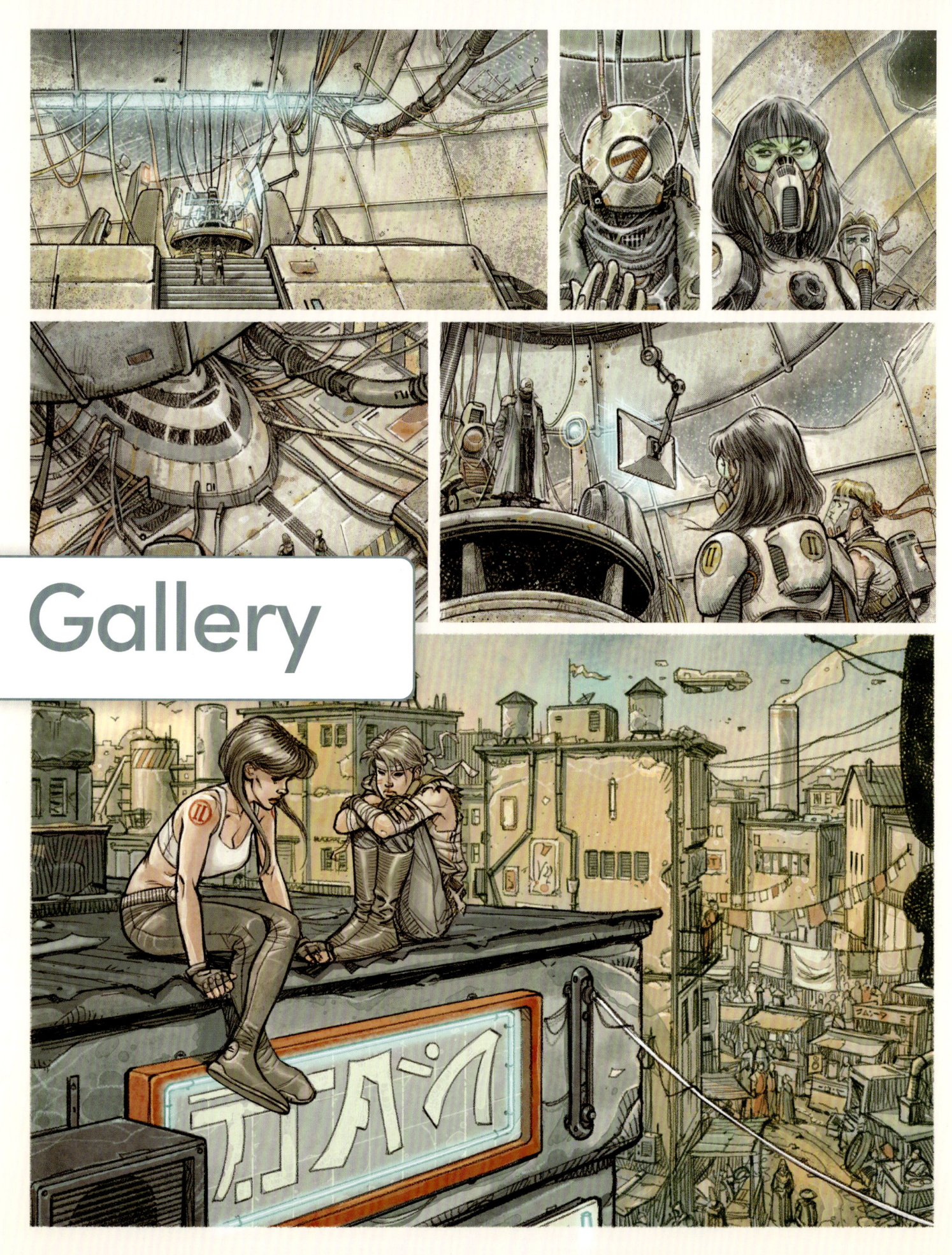

Gallery

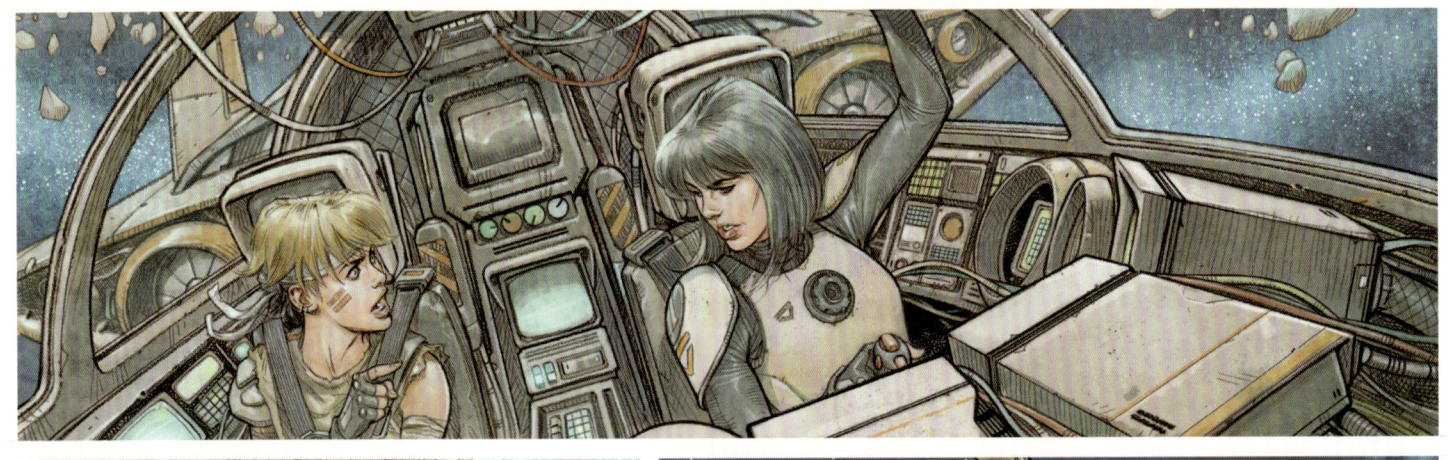
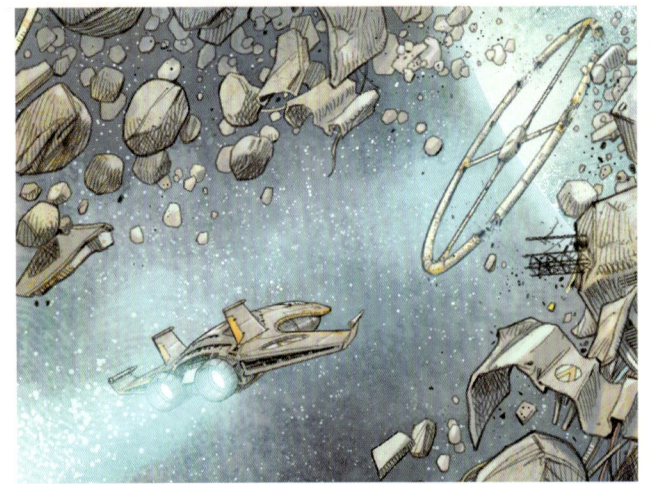
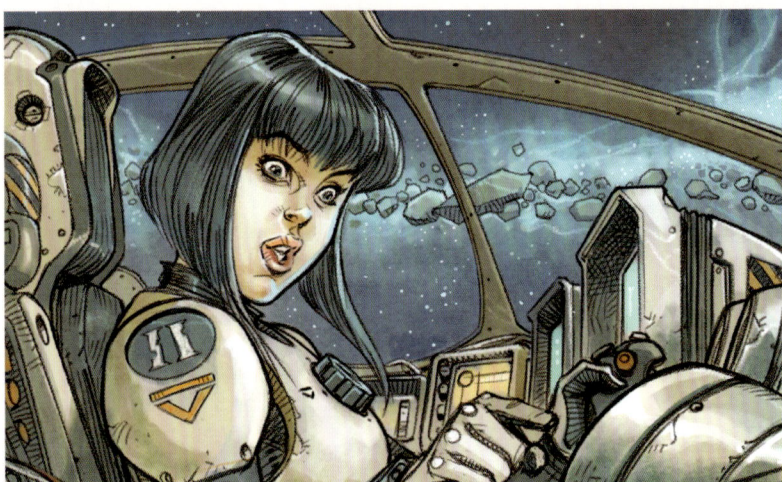
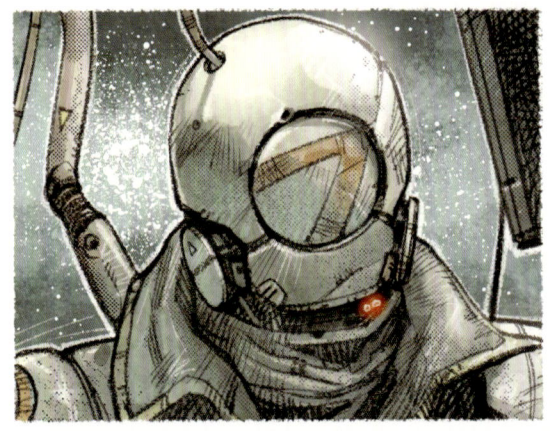
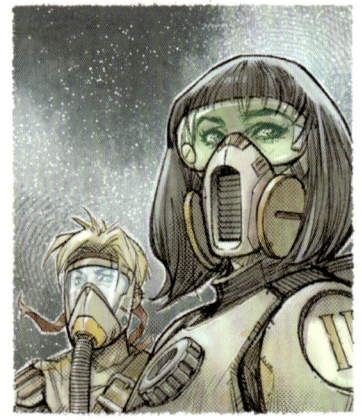
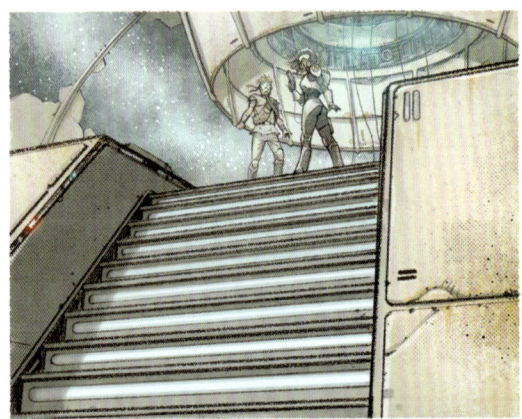
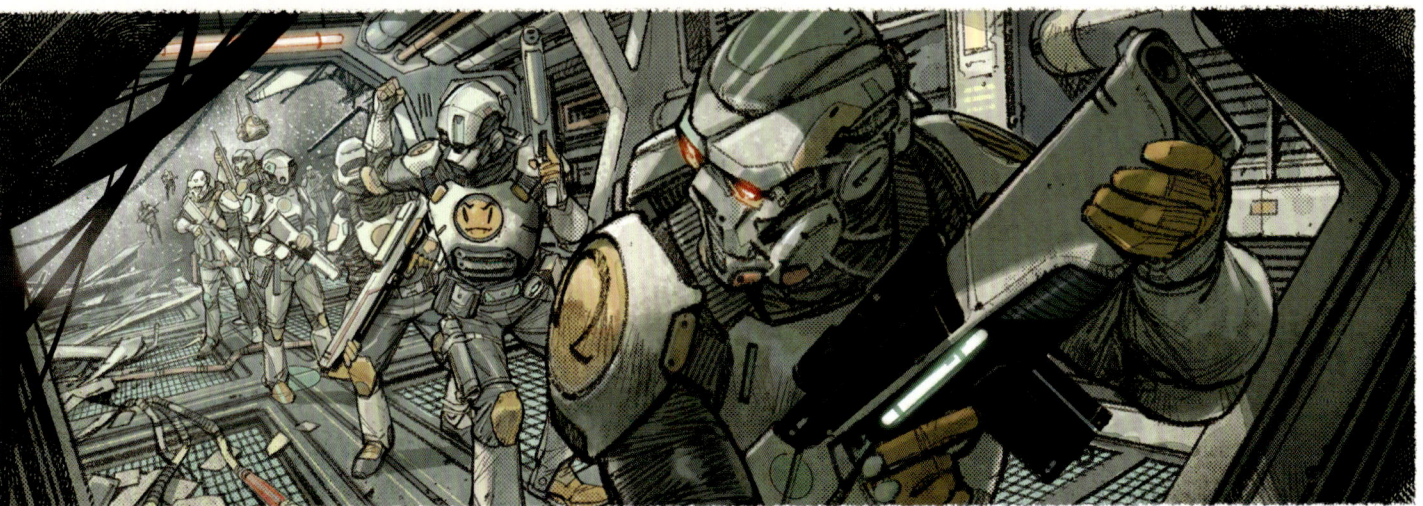

Manga webtoon

BY LOLITA ALDEA

Do you believe in magic? My tutorial will guide you through the steps that go into creating a manga webtoon in Clip Studio Paint. Set in a magical fantasy world, the webtoon will introduce the beginning of a dynamic action scene. You will learn how to introduce multiple characters and create an engaging environment; in this case, a fierce warrior under oath battling a powerful witch who wields arcane magic, all taking place amid eerie historic ruins!

Over the following pages I will walk you through the initial ideas and sketching process as you start to build the narrative. The steps will teach you how to guide the eyes of the reader, use basic tools to create panels, customize text bubbles, introduce onomatopoeia, and add complex kinetic lines. I will then share how to ink the line art, colour the panels, and add final touches. Additionally, I will explore a variety of techniques for adding vibrant lighting effects, as well as teaching you how to use blur and motion filters to enhance the impact of the action.

Keeping in mind the fast-paced nature of webtoons and the need for efficient chapter production, the aim is to create a series of engaging panels that grip readers and leave them eager to see more. You will learn simple, tried-and-tested methods, all while taking into account the specificities of working with the webtoon format, which is intended to be read vertically on mobile phones.

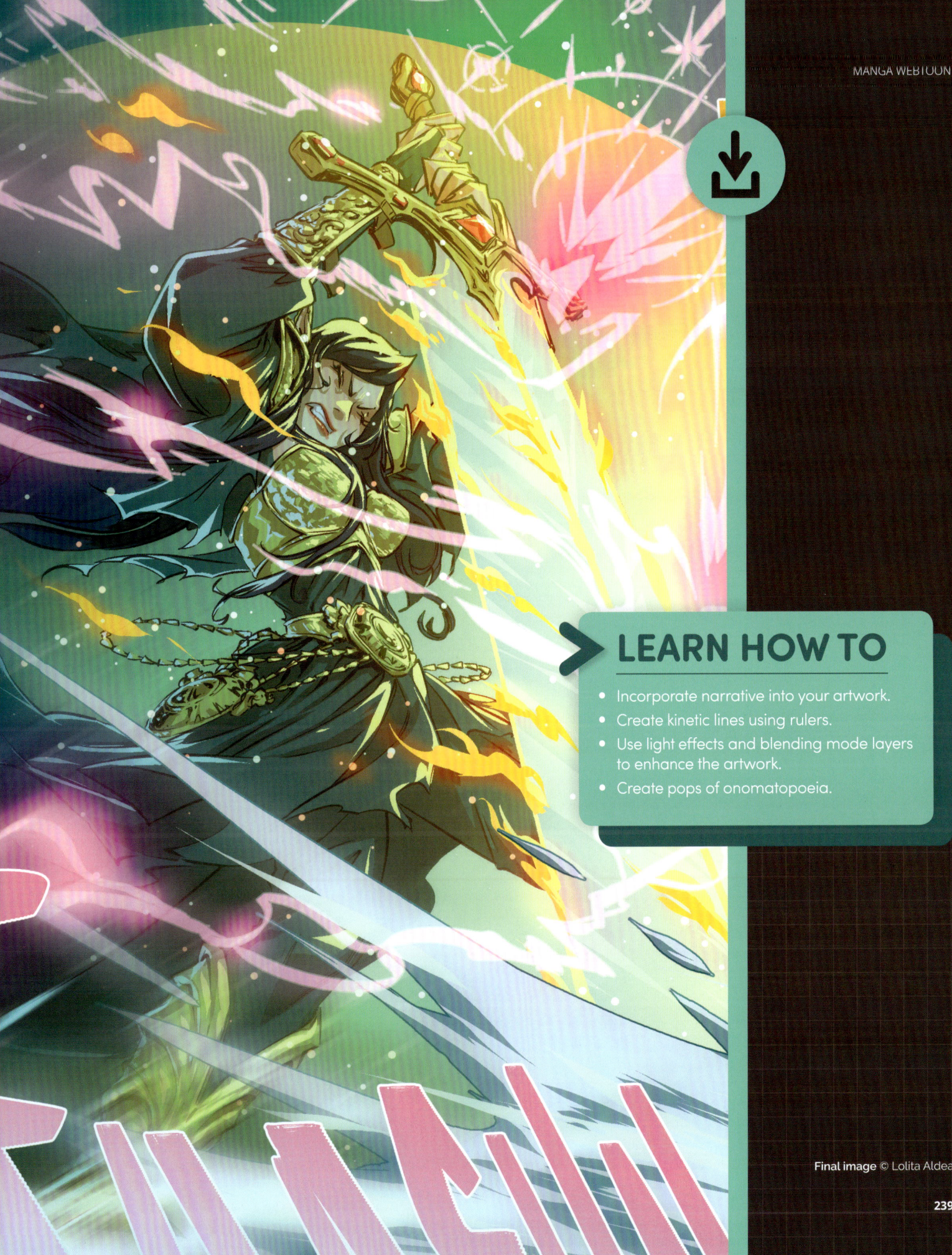

LEARN HOW TO

- Incorporate narrative into your artwork.
- Create kinetic lines using rulers.
- Use light effects and blending mode layers to enhance the artwork.
- Create pops of onomatopoeia.

Final image © Lolita Aldea

01 As webtoons are intended to be read on a phone with a vertical scroll, you need to design the canvas accordingly. Additionally, keep in mind that webtoons are typically uploaded to platforms such as Webtoon, so make sure to check the size requirements of your intended platform. To create a new canvas, select File > New and select the webtoon icon. Choose a file name and select a preset that best suits the length of the story you want to tell. (You can preview the canvas on the right side of the screen.) A resolution of around 350 pixels is suitable for both phone and desktop screens, but you can go up to 600 pixels if you want a crisper original file. If the canvas ends up being too long or too short, select Edit > Change Canvas Height to extend or crop it as needed.

< Open the Webtoon Canvas Settings window and Canvas Adjustment window

02 To understand how the webtoon appears on a phone screen, it's important to activate View > On-Screen Area. This setting allows you to clearly visualize the portion of the webtoon that will be visible on the screen at any given moment, greying out the other areas. It can also help when planning the narrative, panels, and dialogue. You want to avoid making a panel so long that the viewer needs to scroll to view it fully, or leaving so much space between panels that readers think something is missing.

> Activate the On-Screen Area setting to see how the webtoon will be viewed on a phone screen

03 Now everything is set up, you can begin sketching out each panel. This webtoon will depict the start of a duel between two powerful characters: a warrior and a witch. You need to introduce the characters, show the location of the clash, and illustrate the beginning of the action scene, incorporating some dialogue too. The initial sketches only require a pencil-type pen brush and a Normal layer. The intention is to create a simple skeleton sketch that captures the key elements the webtoon requires. This simplicity will allow you to fine-tune every detail as needed.

∧ Using a thick brush will prevent you from focusing on small details too soon, forcing you to capture key elements instead

04 Start sketching an environment that sets the scene, and the first close-up of the protagonist. On a layer below the sketch lines, you can roughly block in some lighting in greyscale tones, just to help visualize the mood. Start thinking about text placement, too, by adding a layer on top for a rough speech balloon.

The action should pick up speed as the viewer scrolls through the webtoon. You can somewhat control the tempo by adjusting the distance between panels. The closer the panels are together, the faster the viewer can read through them. Conversely, if the panels are further apart, it will take the viewer slightly longer to scroll through them. You can also perform two simultaneous actions by superimposing two panels, or placing them side by side at the same height.

For this battle scene, large panels will help to set the scene and introduce the characters. Smaller panels are better suited for showing details, while large spaces are best used to capture the action.

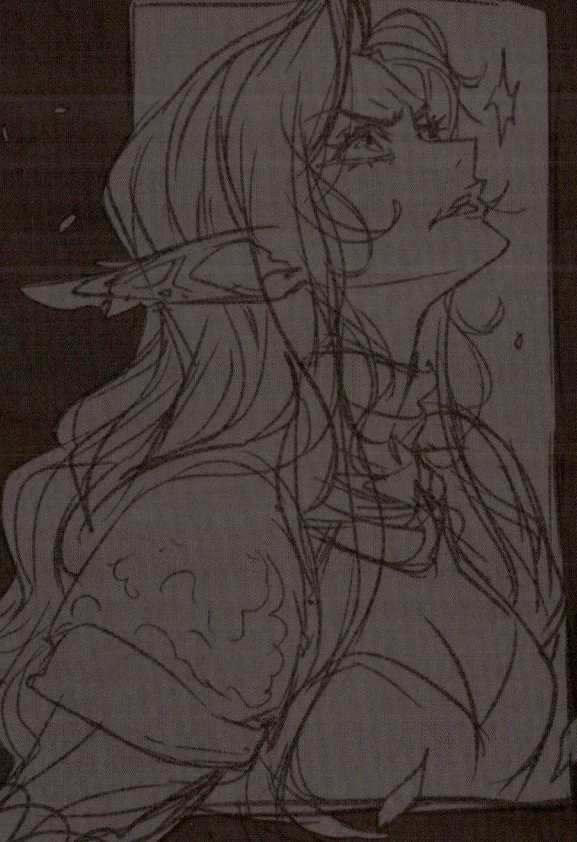

∧ When planning out your webtoon, make sure a panel isn't too large to view in its entirety on a phone screen

05
Continue sketching your way down the canvas, roughly mapping out the unfolding narrative in greyscale. Depending on how you wish the webtoon to be read, the narrative that weaves through the panels could be drawn as a zigzag line down the canvas. The first panel will be used to present the location, with the warrior character as the central focal point. Placing the first speech bubble just below the warrior leads the viewer to the second panel, which will show a close-up shot of the warrior looking off-camera. These two panels are slightly further apart to create a slower pace, which can build suspense.

The next panel is from the warrior's point-of-view. The viewer sees what the warrior is seeing: a magical attack coming directly at her. It then quickly shows the impact of these attacks, as well as the warrior trying to protect herself with a magic shield.

In the next panel, the witch character speaks off-camera and the warrior looks up, searching for the source of the voice. These two actions occur in the same panel, but by placing the dialogue bubble higher than the warrior's face, you can dictate the reading order.

The witch will be introduced in the next panel. The placement of the first dialogue bubble, the character's face, and the angle of her clothes can be used to guide the gaze towards the next dialogue bubble, which leads to a close-up of the witch's face.

The final panel shows the warrior charging at the witch. The battle begins!

Try to provide a clear guide for how you want the viewer's eyes to move through and rest on each panel and moment. The human eye will typically focus on eyes, faces, and text first. You can also use visual lines to lead the viewers' gaze to the next point of interest, such as clothing, hair, or arms.

> Consider how you can connect the starting point of the first panel to the desired destination in each panel that follows

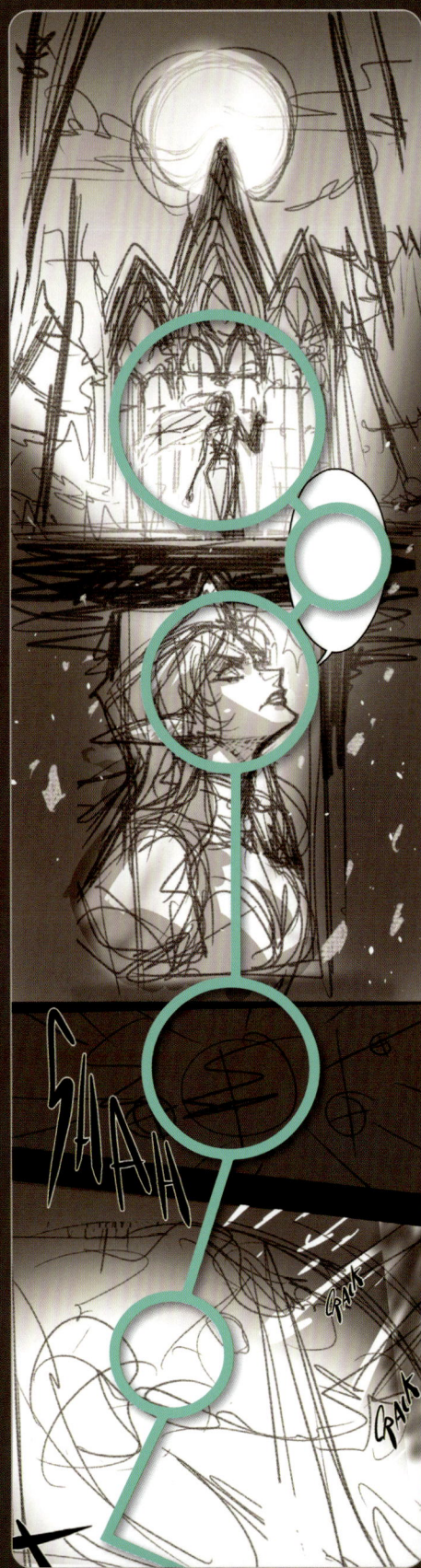
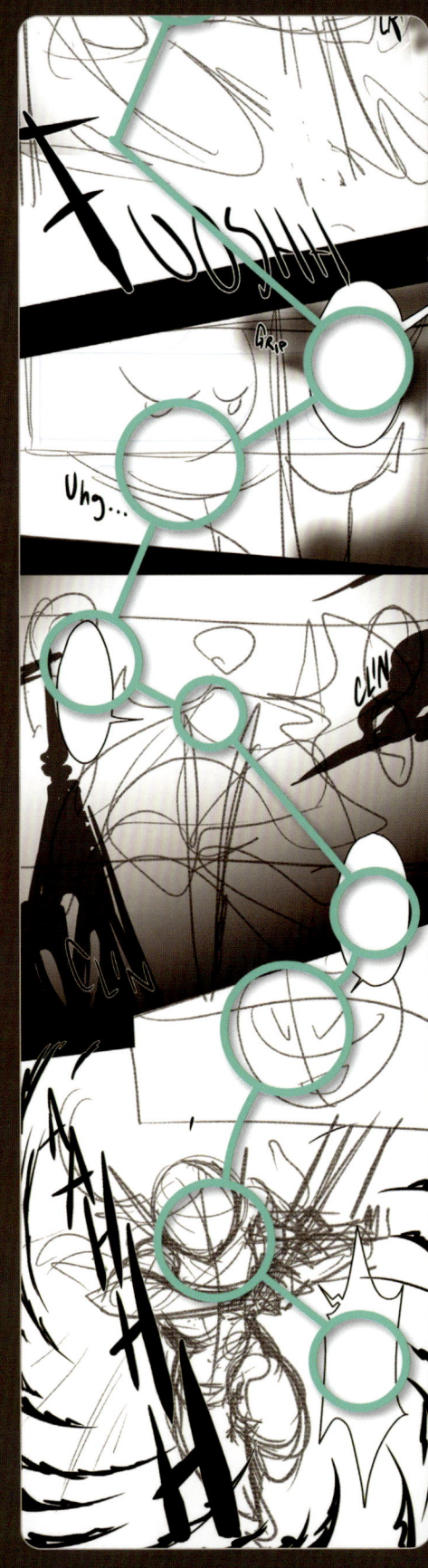

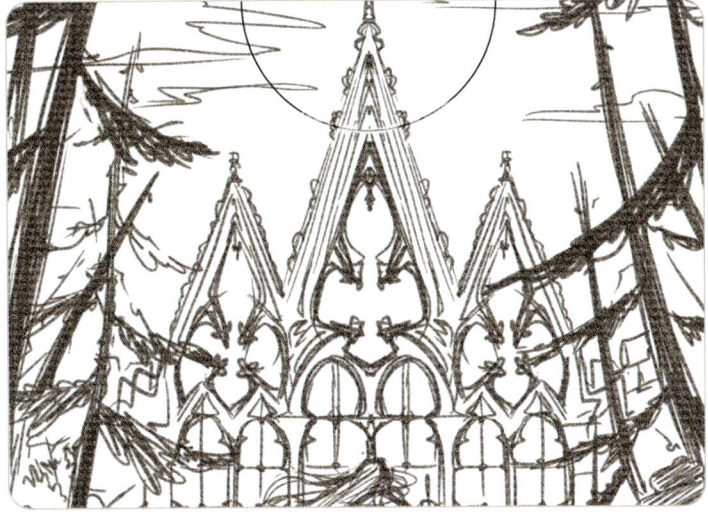

06 Backgrounds in webtoons are sparse but crucial, as they establish the setting for the action. There's no need to include excessive details in them, due to the small screen size and fast reading pace. Try to choose an environment that's easily readable, with a simple but strong silhouette. If you wish to emphasize a specific detail in the background, create a new panel specifically for that detail, such as a close-up shot. To create this background of a ruined abbey, use the Symmetry Rulers by selecting Ruler > Symmetrical Ruler and set the number of lines to two.

< Make use of Symmetry Rulers when drawing manmade structures in an environment, though it's generally recommended to avoid using them for natural elements, such as trees

07 Once you have a rough sketch of each panel, step back and assess whether you need to make any changes to improve the narrative. Type out the dialogue to ensure each speech bubble is the right size. Adjust the positions of your characters, panels, and speech bubbles as needed. Use the selection and transformation tools to adjust the size and position of each element. You can also use the Liquify tool to adjust the size of various elements, such as tweaking the shape of the characters' faces. Once satisfied with this rough sketch, begin to draw in all of the details that are missing. Take time to draw a more detailed sketch of both characters, their outfits, and facial expressions. Sketching in such details now will make the line-art process much easier.

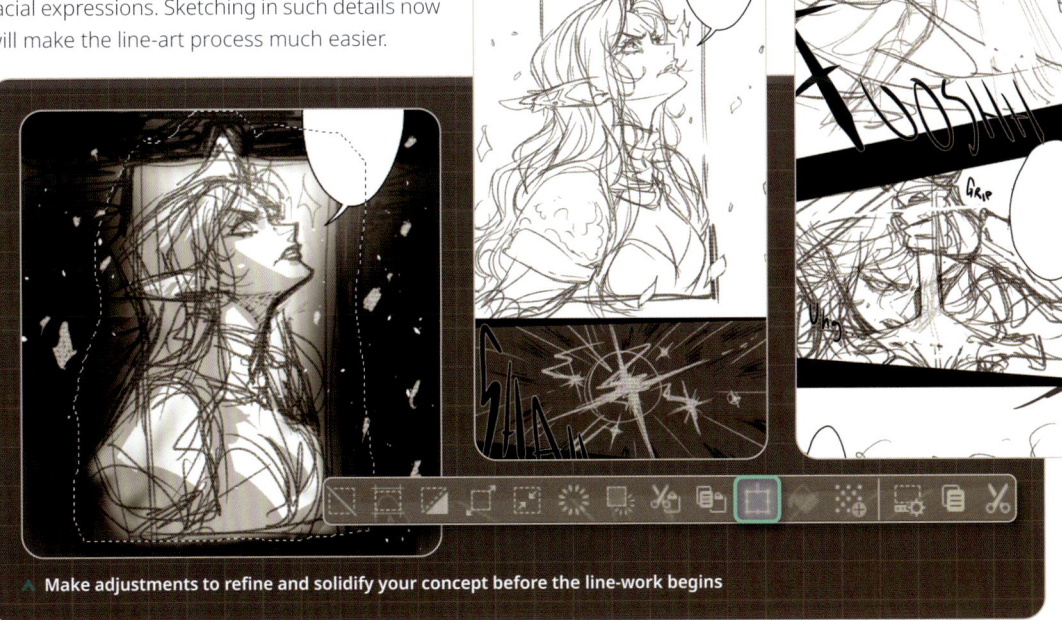

^ Make adjustments to refine and solidify your concept before the line-work begins

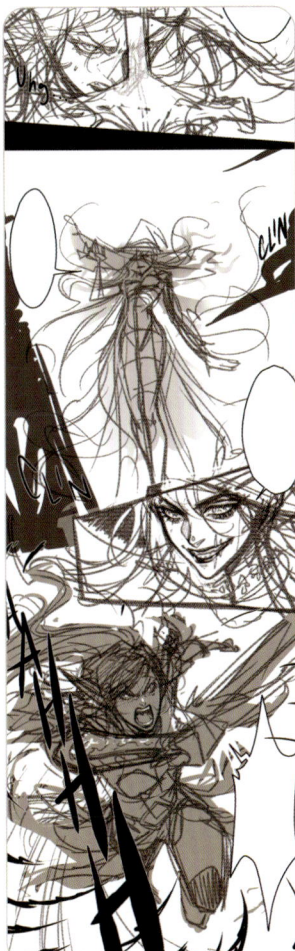

^ Once you're happy with the rough sketch of each panel, draw a more refined sketch and add missing details

08 When you've finished the final sketch and are happy with the placement and detail level of each element, it's time to ink the line art. But before that, you need to create panels using either the Rectangle Frame tool or Polyline Frame tool. These are sub tools of the Frame Border tool, found in the lower section of the Tool bar. This will automatically generate a new layer type that will contain all of your panels. If you need to make adjustments to specific panels, use the Operation > Object sub tool (near the top of the Tool bar) to interact with the lines in the panels.

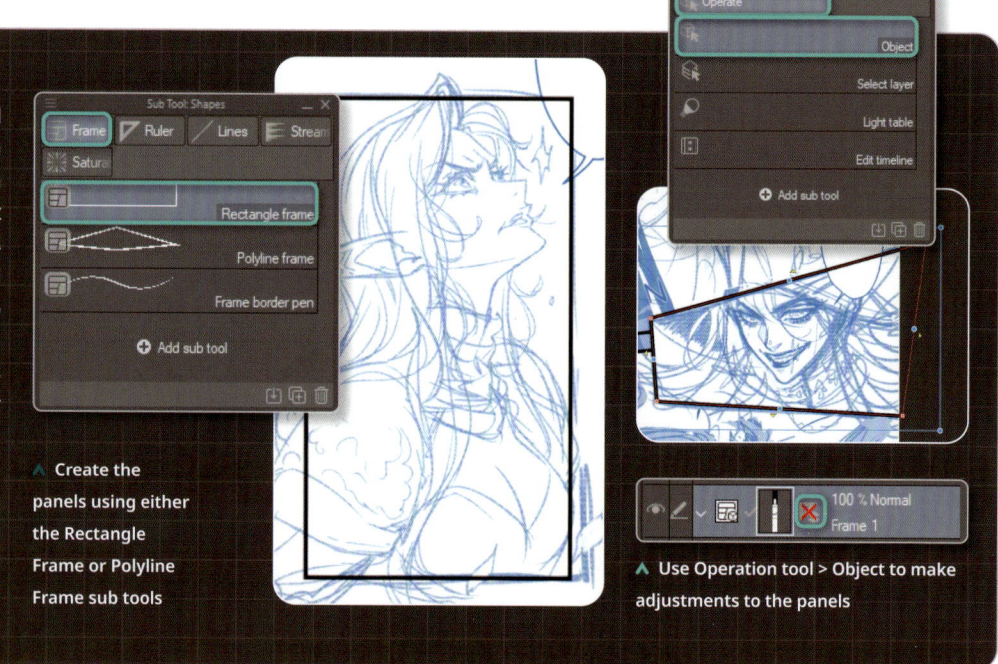

⌃ Create the panels using either the Rectangle Frame or Polyline Frame sub tools

⌃ Use Operation tool > Object to make adjustments to the panels

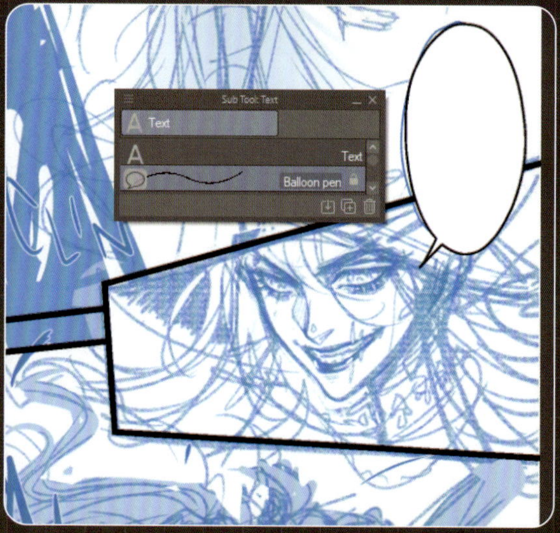

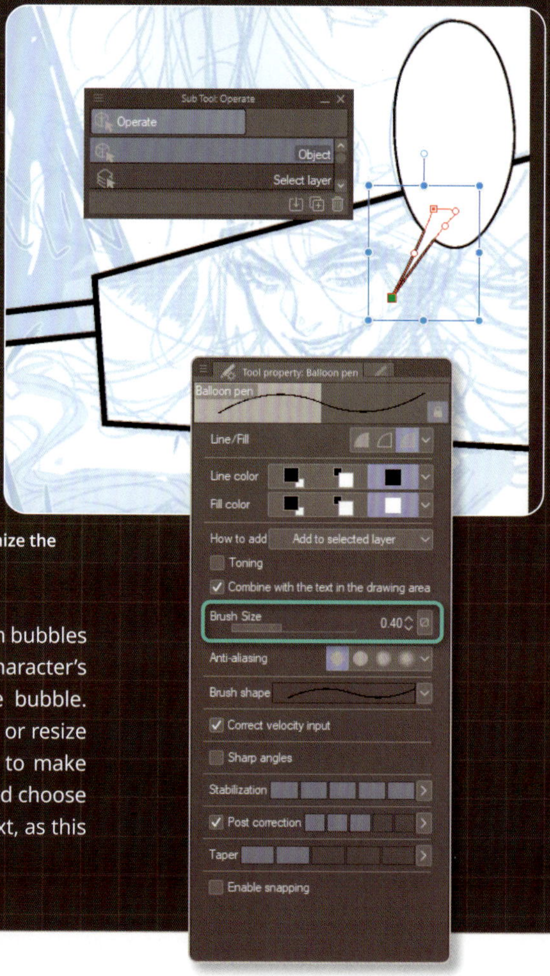

> In the properties menu of Balloon Pen, there is the option to customize the thickness of the line for the balloons, as well as many other settings

09 Use the Text > Balloon Pen sub tool to draw speech bubbles large enough to accommodate all of the character's dialogue. You can also use this to create the tail of the bubble. Select Operation tool > Object to correct any imperfections or resize the bubble, using the handles that appear on the screen to make adjustments. Next, use the Text tool to type the dialogue and choose a font from the menu. Always use capital letters for your text, as this will make it easier to read.

10 Intense, kinetic action lines are common in manga. You can draw them by hand, or using the Special Ruler tool. Select Special Ruler > Radial Curve. Under Tool Property, set the drop-down to Radial Curve and set the Curve icon to Quadratic Bezier to draw twister-like lines, creating a sense of fast forward movement. Use the focal point of the movement/panel as the origin point, placing the curve points in the direction the lines should follow. Double-click and the ruler will be set. You can then use the Operation tool to adjust it, if needed. It can be helpful to make the ruler much larger than you think you will need. This will save you from having to remake it if it's too small.

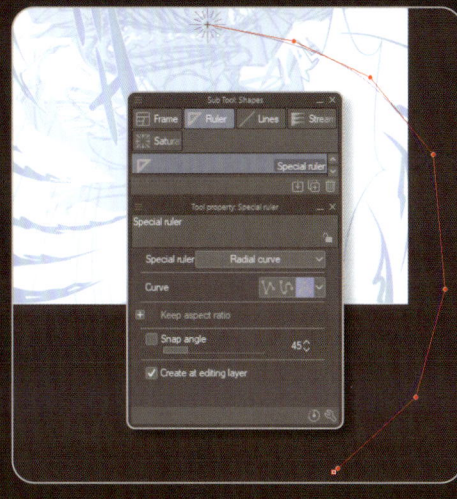

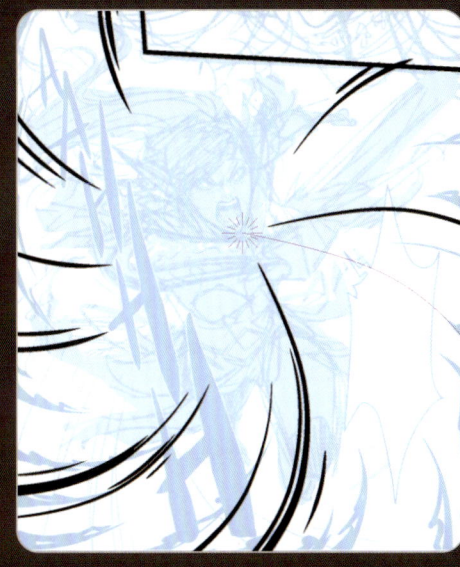

∧ Use Special Ruler > Radial Curve to draw curved lines that spiral out from the focal point

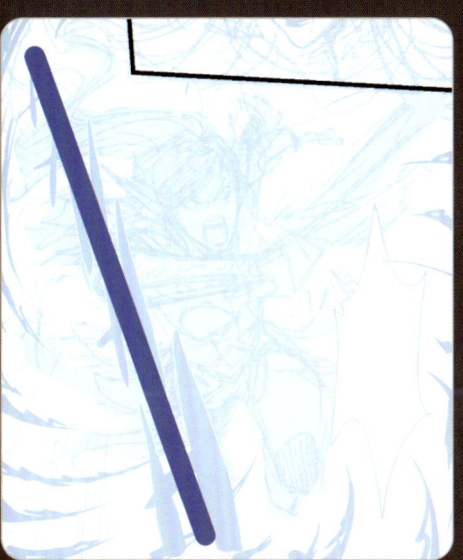

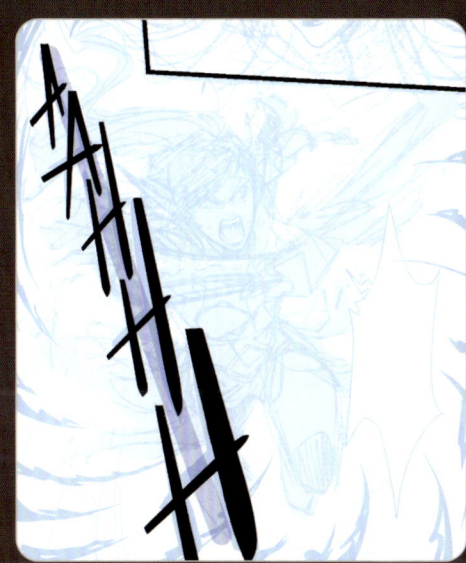

11 To create onomatopoeia, start by transcribing the sound into a word and deciding where it should go in the panel. Choose a location that is visually appealing and complements the action of the scene. Next, draw a line as a guide and write in the letters using a pen brush. Experiment with different stroke thicknesses and styles to achieve the desired visual effect. Select all the letters and use Ctrl+T to transform, resize, or rotate to adjust the position and shape.

< Add onomatopoeia to enhance the action and drama of a scene

ARTIST'S TIP

Sketching is the most important part of the process, as you're putting together all of the pieces that will make up the webtoon. Take your time during this stage; after you have completed an initial sketch, take a step back and let it breathe until the next day. Returning with fresh eyes, you can then make any adjustments you think it needs. Once you feel it's ready, you can start on the line art and colour!

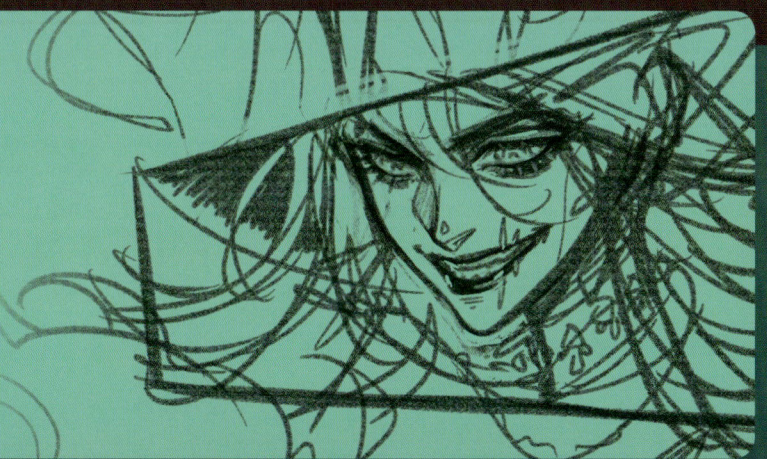

12 When it comes to creating line art, one of the key factors is the choice of pen. Consider what type of pen best suits your art style and the tone of your webtoon. In this case, choose the default Ink > G-Pen. This offers the perfect balance between thickness variation and subtlety. It allows you to create lines with personality, adding depth and character to your artwork without overpowering the colours that will be added later on. Inking these lines in a Vector layer can make it easier to erase excess lines using the Vector Eraser tool.

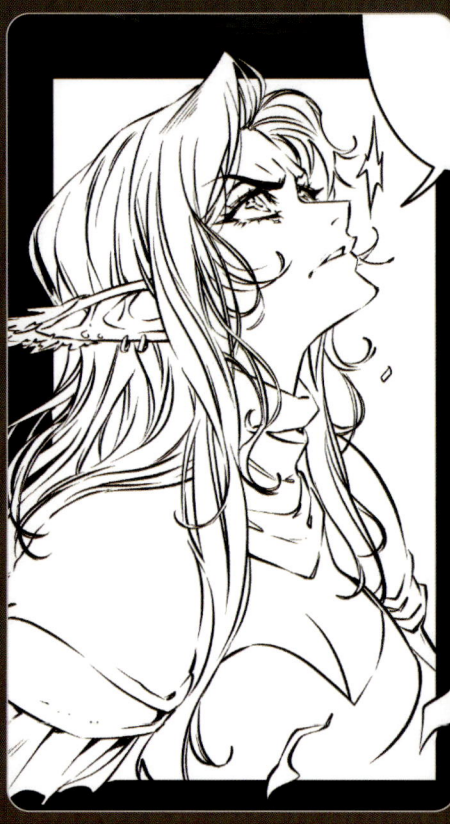

> Use the G-pen brush to draw neat line work on a new layer on top of the sketch

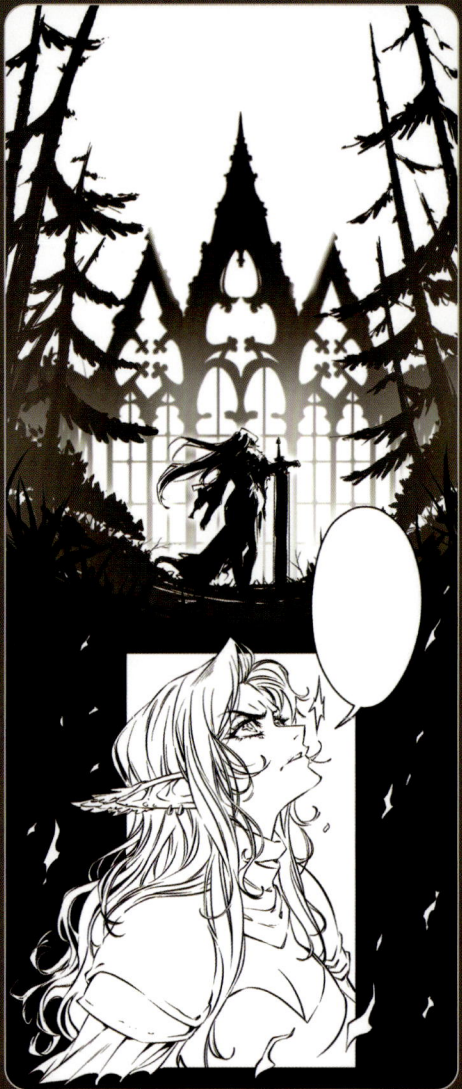

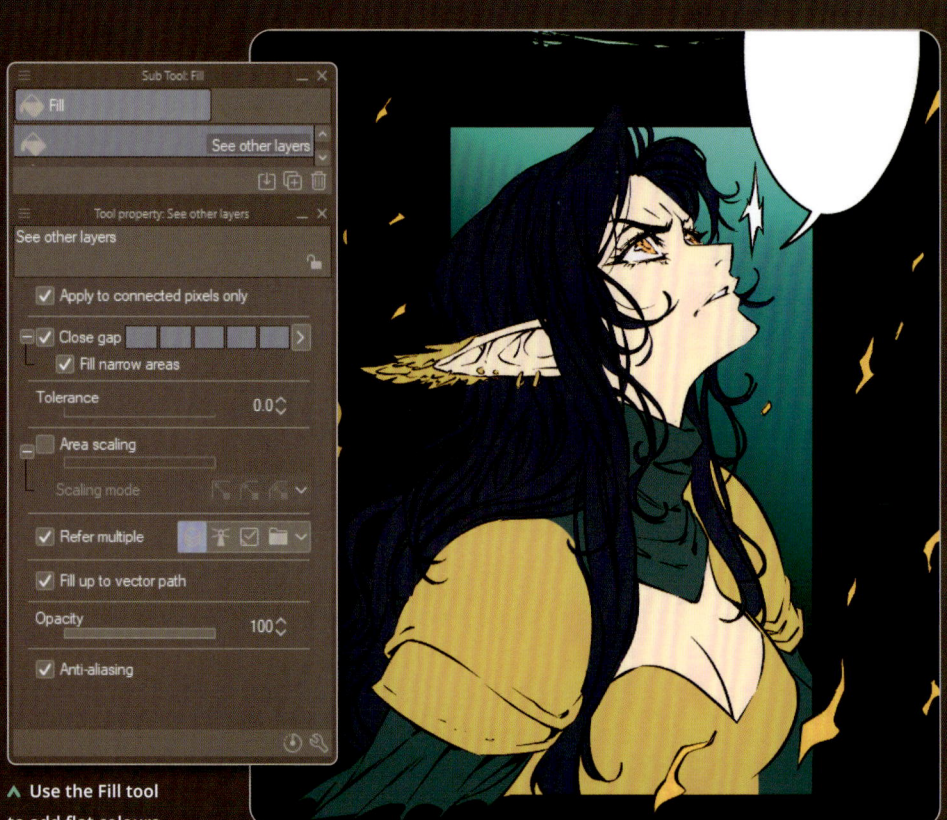

∧ Use the Fill tool to add flat colours

13 Now it's time to add colour. To improve navigation, create separate layer folders for the character flats (base colours) and backgrounds. To speed up the process, use the Fill tool for the flat colours , set to the Refer Other Layers sub tool, with the Close Gap and Refer Multiple options enabled. Don't worry about choosing colours tinted with ambient light during this phase; you will use other resources later to achieve that goal. Colour each character and background on a single layer and try to keep the number of layers to a minimum for now, as you will need to add more layers later on. Also, remember to name each layer.

14 To add shadows, create a new layer and set the blending mode to Multiply. Next, below the Blending Mode drop-down menu, select the option Clip to Layer Below. Place this shadow layer just above your flats layer in the layer list. To create the shadows, use a relatively dark and saturated colour that falls between red and purple. This will help to create contrast, highlight certain elements, and add depth and dimension to your illustration. Adjusting the layer opacity to a value between 50% and 60% will allow the shadows to blend with the colours underneath. For the best results, use a solid pen brush with the anti-aliasing set to none. You can also use this type of layer to add additional colours, such as the tone of the lips or the skin.

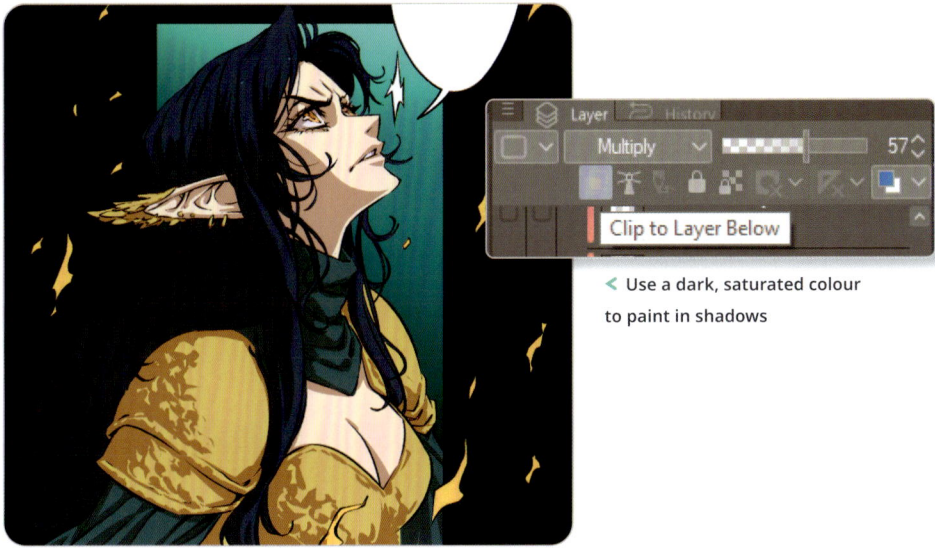

< Use a dark, saturated colour to paint in shadows

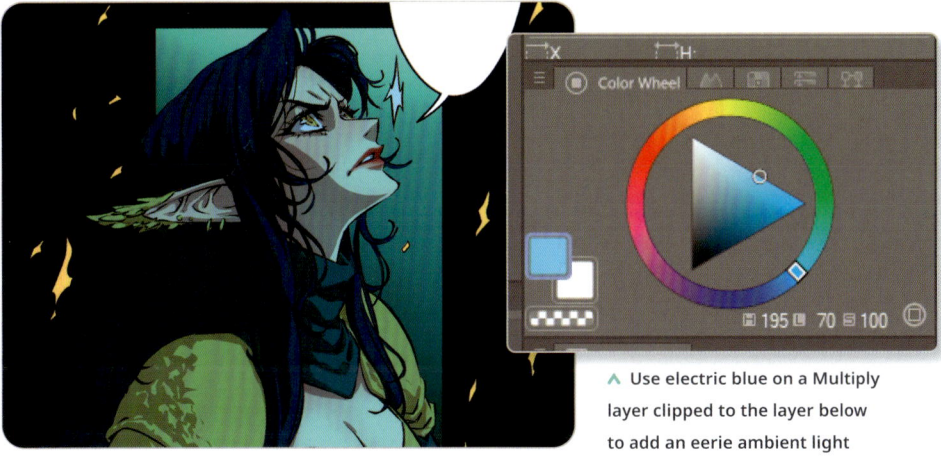

∧ Use electric blue on a Multiply layer clipped to the layer below to add an eerie ambient light

15 To add ambient light, create a new layer and set it to Multiply at around 80% opacity. Next, choose a super-saturated colour, in this case, an electric blue, and fill the entire canvas. Set the layer to Clip to Layer Below (it will clip to the flats layer). This will produce an eerie feeling to the colours, perfect for historic ruins illuminated by moonlight. This method will work with any other ambient light you wish to add. For example, you can use orange for a sunset, or purple for a dark night. Just remember to use saturated colours. The most saturated colours are always found near the right vertex of the triangle.

16 Use secondary lighting, with a different tone to ambient light, to give the colour more drama and impact. These colours emanate from the weapon, the magic the characters use, and the moon. Create a new layer and use the Clip to Layer Below option above the folder that contains the line art and the rest of the character's colours. This layer will overlap the line art and create the sensation of the light tinting the character. By applying a couple of intense, almost backlit light sources to the character, we can define some of the contours using coloured light instead of lines. You can see this occurring on her hair and the back edges of her armour.

> This lighting effect allows you to depict the details of the armour without the need to outline specific shapes

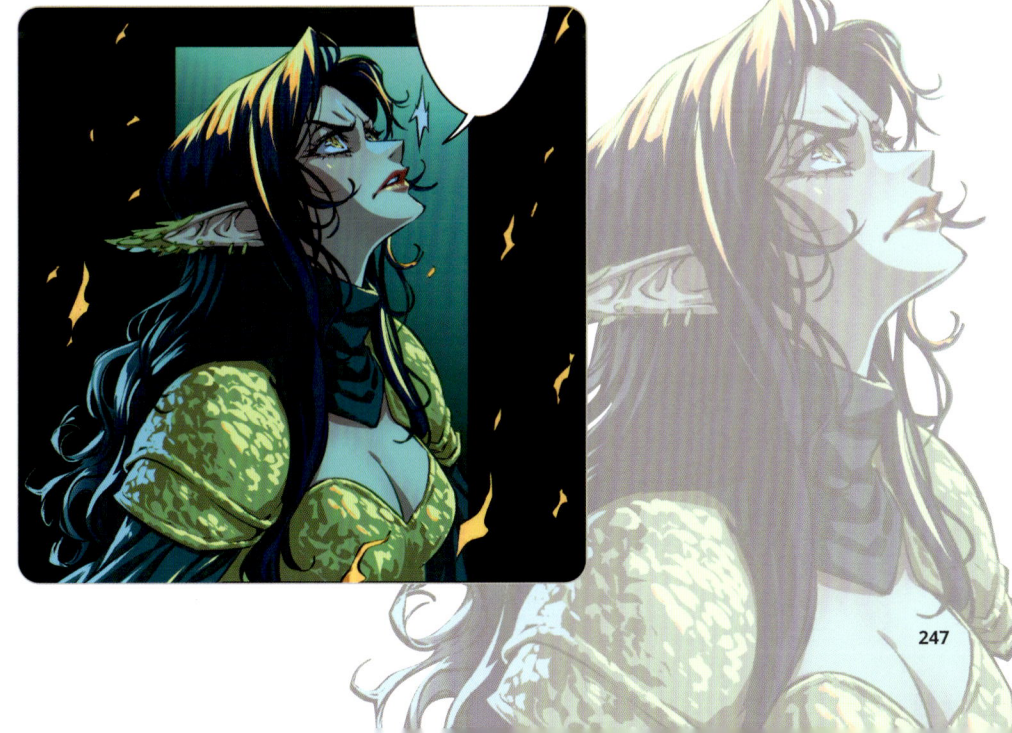

17 To create the background, paint simple outlines of the ruined abbey arches, as well as trees and plants in various sizes. You can also include plants in the foreground to help frame the panel. Place each element on separate layers to create different depths. For example, separating the foreground trees from the background ruins. For added depth, create additional layers in Clip to Layer Below mode and apply gradients to them. Next, use the Gradient tool and Airbrush to paint in colours that represent the influence of ambient light. The objective is to clearly differentiate each layer of depth.

> **Add elements of the environment in the background, middle ground, and foreground**

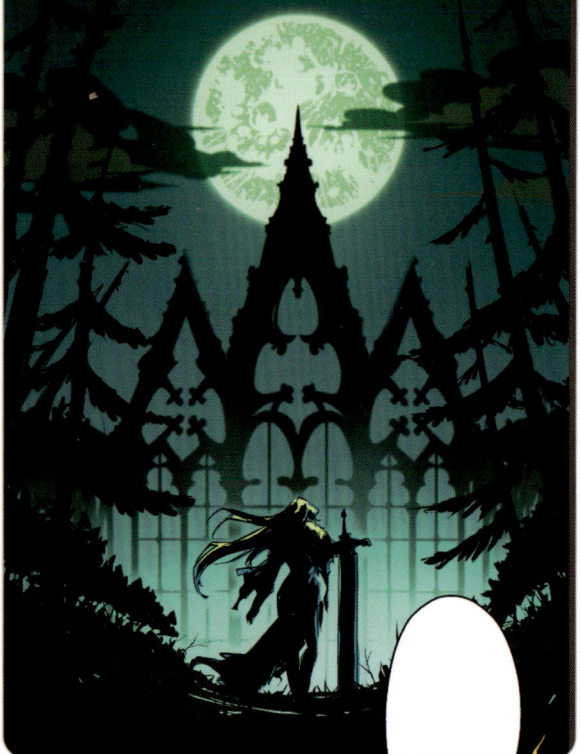

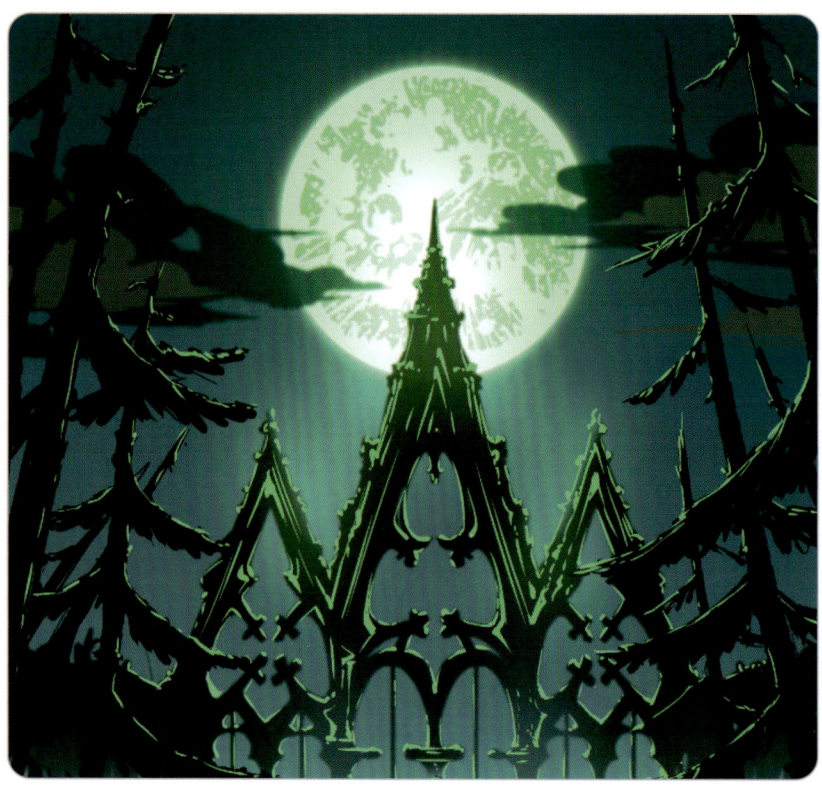

18 The entire environment is bathed in an eerie greenish moonlight. Add a new layer above all of the other background layers and then, with a Pen brush and a creepy pale-green colour, use small brushstrokes to define the edges of the background elements. The goal is to provide definition to the shapes without overpowering the characters, allowing the viewer's attention to focus on the warrior while still adding details to the surroundings.

< **Use a Pen brush and green colour to paint fine lines that mark the most intense light points and reflections**

19 Now it's time to create general lighting effects. These types of effects can inject illustrations with a spark of life, creating new depths of tone and providing the opportunity to make certain elements more eye-catching. You will need two types of blending mode layers, one in Overlay and another in Add (Glow). Overlay allows you to add colours that will enhance those already on the canvas. It lets you dye the scene with any colour to produce a chromatic richness. Add (Glow), as its name suggests, is used to give brightness to the colours already present on the canvas. Try this using a bright colour and you will see how everything seems to shine. You can use the Pen brush to apply highlights, as well as the Airbrush for more general lighting. Experiment with the tools to find out what you prefer working with.

> **Use the Overlay and Add (Glow) blending modes to create extra lighting effects to enhance the artwork**

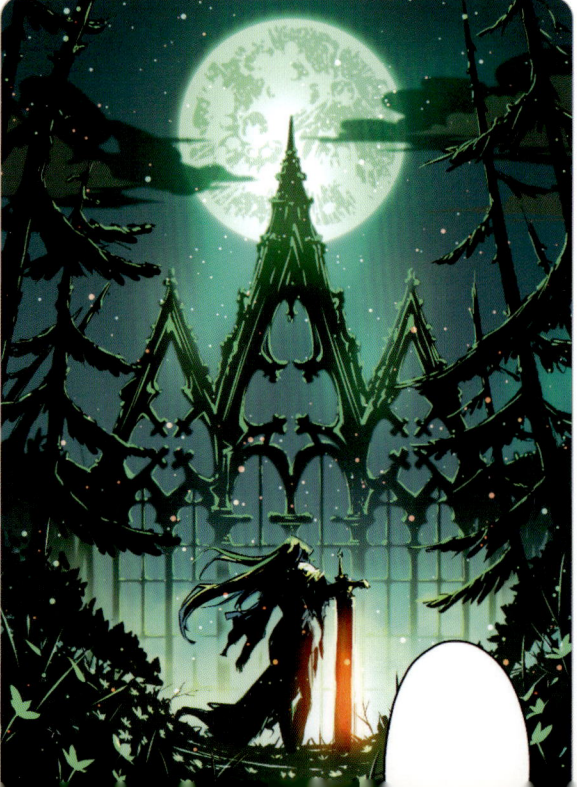

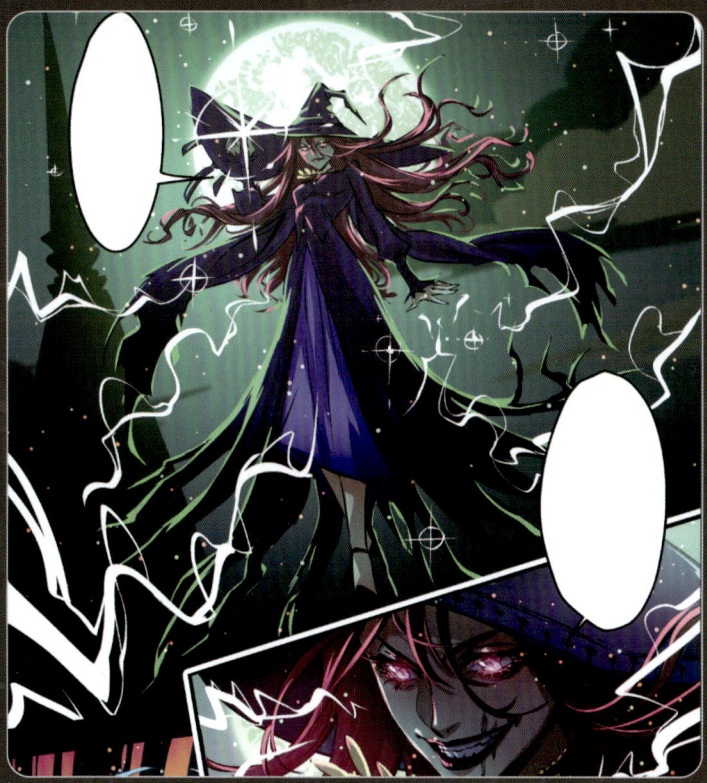

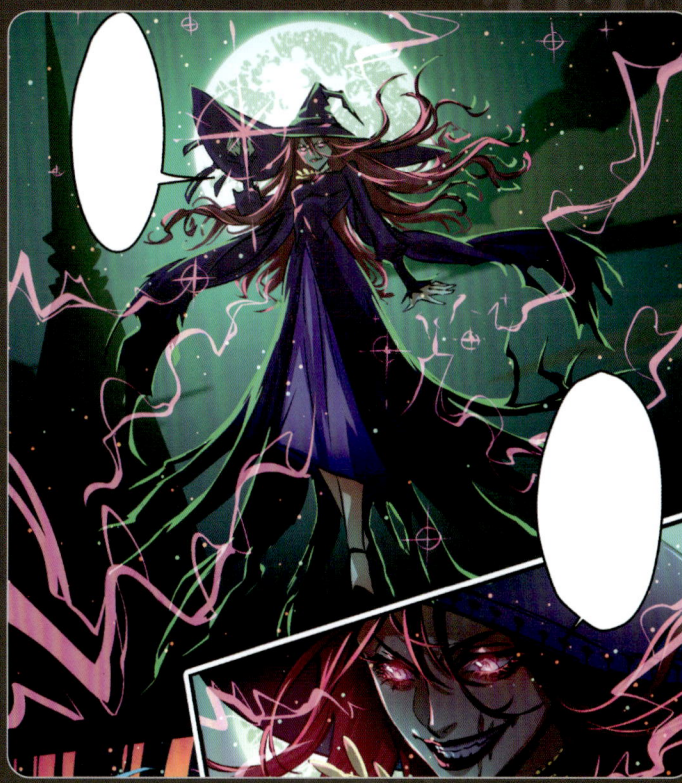

20 To paint in the magic, create a new layer and draw the silhouettes of the spells in white. Once they are well defined, choose a primary colour, such as electric pink, to stand out against the darker background colours. Duplicate the spell layer and fill the drawing with the pink colour. Set this new layer to Overlay and duplicate it. Next, lower the opacity of the original white layer to 40%. Doing this will allow the Overlay layers to retain the intensity of their colour without being 100% transparent. Apply Filter > Blur > Gaussian Blur to one of the Overlay layers to create a magical halo effect.

> ## ARTIST'S TIP
>
> With the use of lights and shadows, it's very easy to bring life to panels, simply by choosing four colours. One general colour for shadows and light, plus two effect colours, can make a seemingly simple image increase in complexity. These are just some of the possibilities that blending modes offer, so play around with them and different shades of colour. It's possible to completely change the atmosphere of a scene using only a few adjustments. Colours evoke sensations, meaning they can be used as part of the narrative to convey mood and emotion.

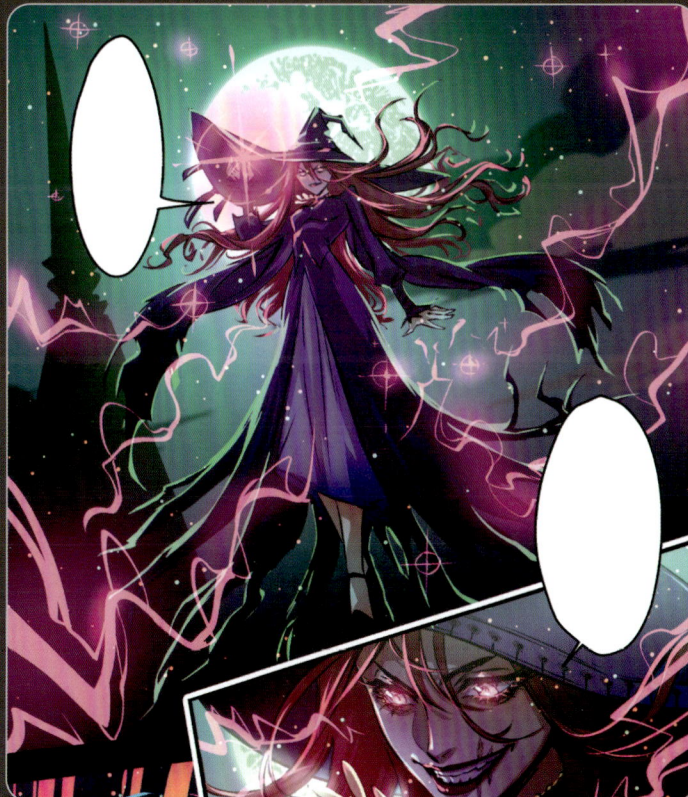

∧ Add Overlay layers on top of the other layers and use the Airbrush tool to paint touches of colour in different places to give the spell an even more spectacular look

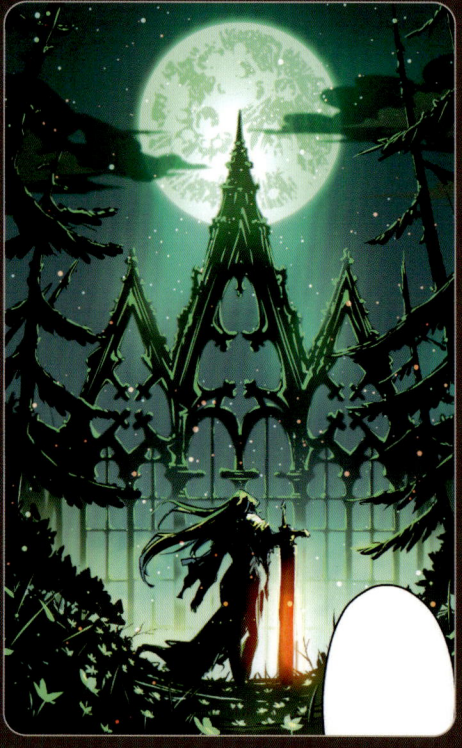

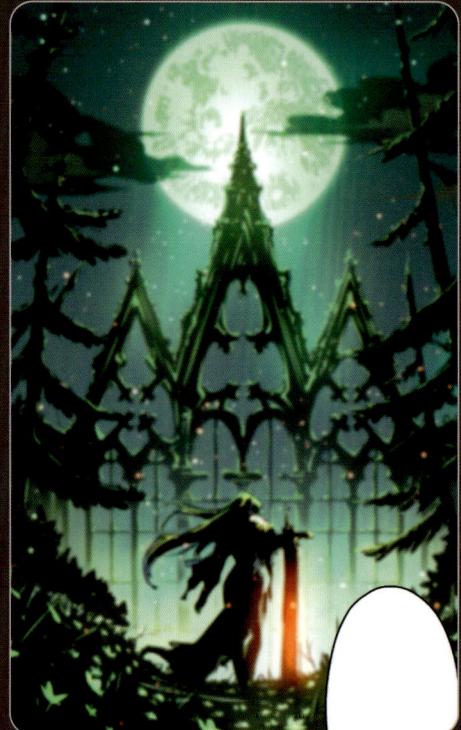

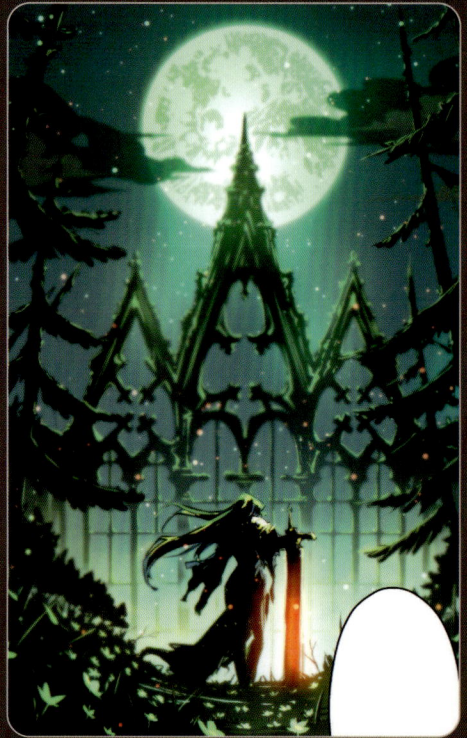

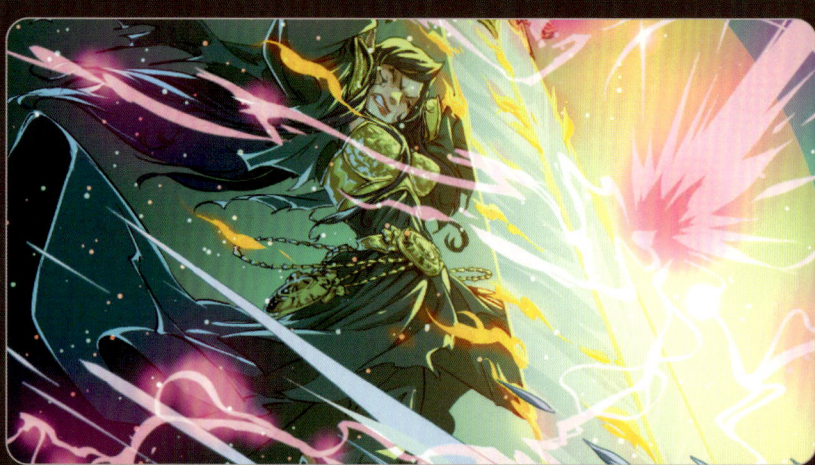

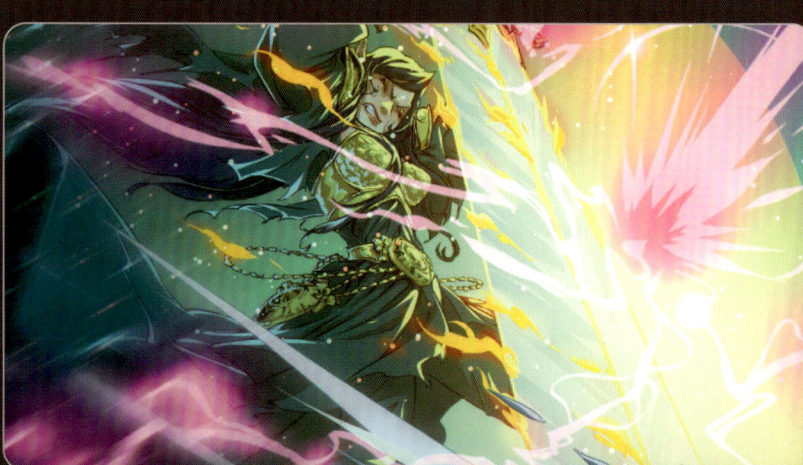

21 As a final touch, add more dynamism and movement by applying different types of Motion Blur and Blur to the panels. First, hide the layers containing the text bubbles, onomatopoeia, and panel borders. Next, create a copy of the entire webtoon by clicking on one of the layers and selecting Merge Visible to New Layer. This will create a new layer containing all of the artwork. Reserve this layer by duplicating it and hiding the original.

Use the Selection Area > Lasso sub tool to select panel by panel and apply different types of filters until you find the one that best creates your desired look. For example, for the ruins panel, try blurring the background slightly to create a greater sense of depth. If you apply Gaussian Blur to the entire panel, you will notice that the character also becomes blurred. To avoid this, add a mask to the layer and use the Airbrush tool to erase the parts that you don't want to appear blurred. This will reveal the crisp original layers underneath the edited duplicate.

< Experiment with using different types of
Motion Blur filters at various intensities –
each filter will produce unique results

22 Finish by adding the text to each speech bubble. Using the Text tool, click on the speech bubble and write in the dialogue. In the Tool Property palette, choose one of Clip Studio Paint's fonts. Maintain a consistent font size for all of the regular dialogue in the webtoon, and if the speech bubble is too small or large, use the Operation tool to adjust its size. For important or shouted lines, increase the font size. You can also use different fonts to represent characters speaking in different languages or tones, such as a staccato robotic voice or a quiet whisper. Adding colour to the dialogue can also reflect the tone of a character's voice. All of these options can be found in the Tool Property palette of the Text tool.

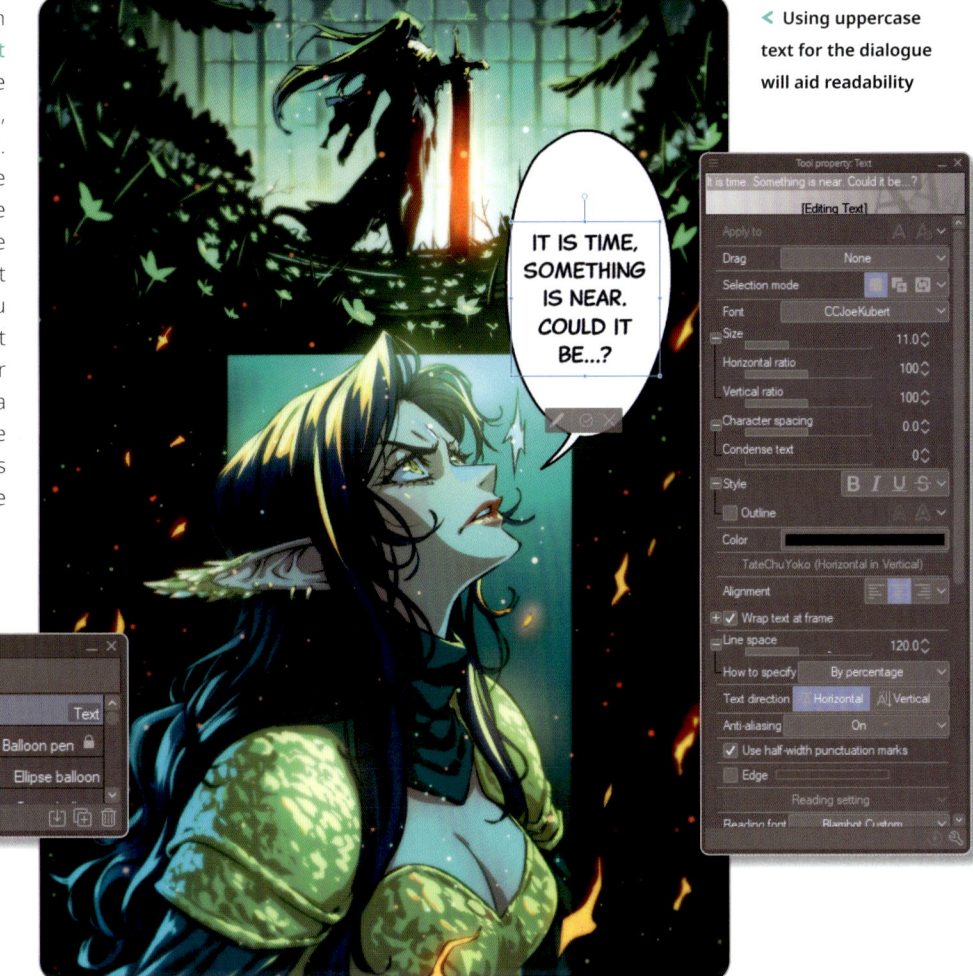

< Using uppercase text for the dialogue will aid readability

IT IS TIME, SOMETHING IS NEAR. COULD IT BE...?

Conclusion

You now know how to create a series of panels for a manga webtoon! In this case, the tense moments leading up to an epic battle between two formidable enemies. By following the steps in this tutorial, you've learnt how to create a smooth reading experience that invites the viewer to keep reading and leaves them eager to know what happens in the next chapter. These simple techniques can achieve powerful results and can be applied to any other theme or genre of webtoon, such as romance, detective, or sci-fi. Keep experimenting with Clip Studio Paint's range of tools and techniques as you continue your webtoon journey!

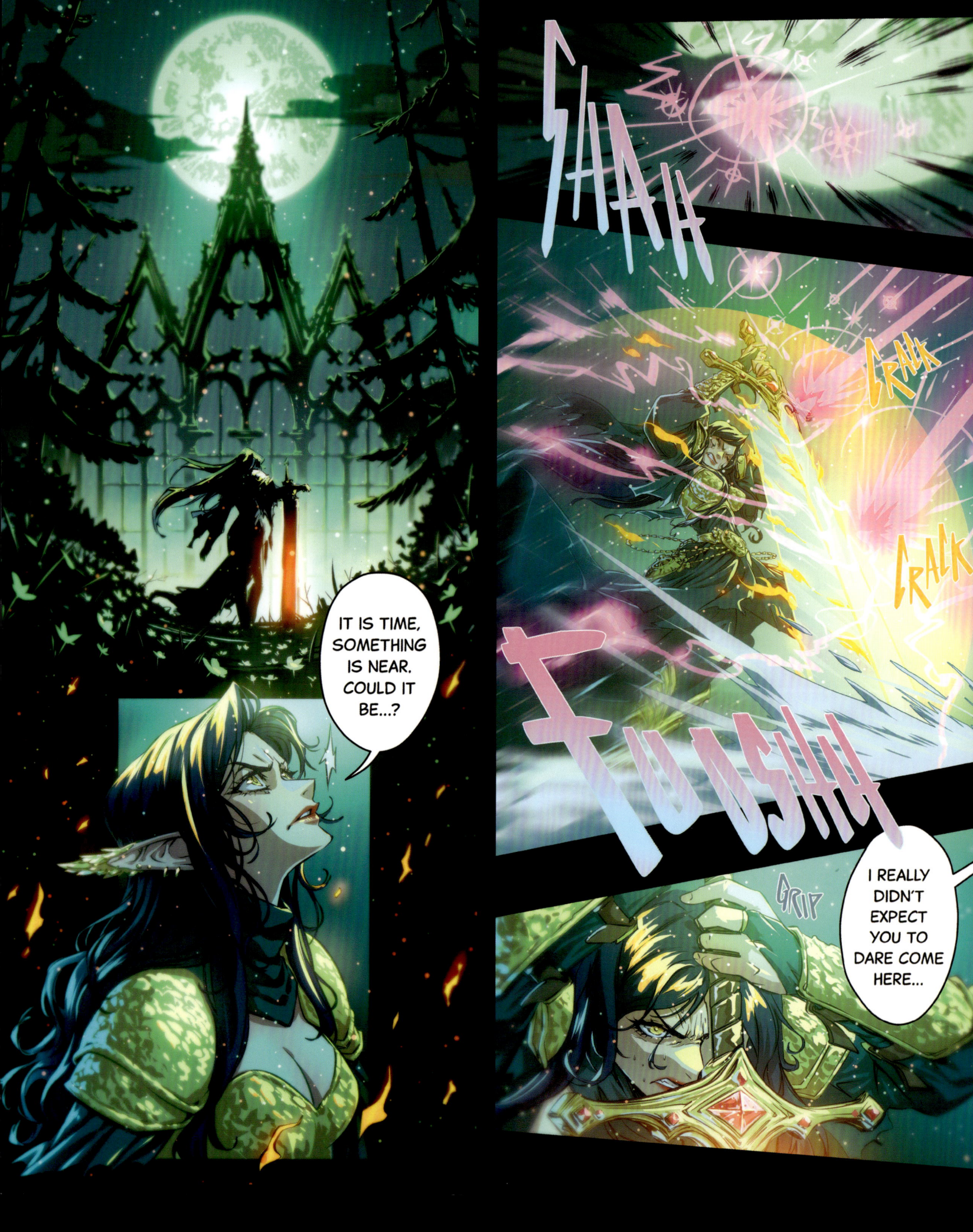

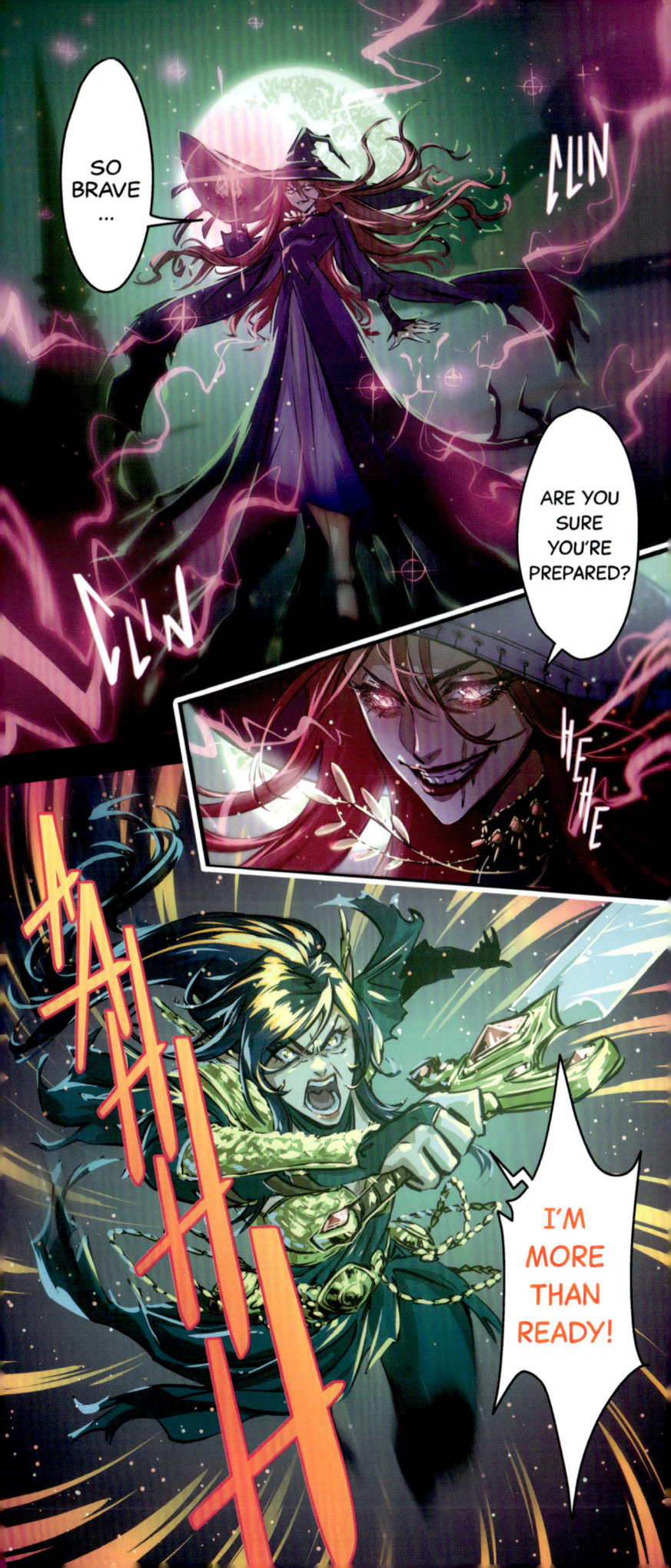

Gallery

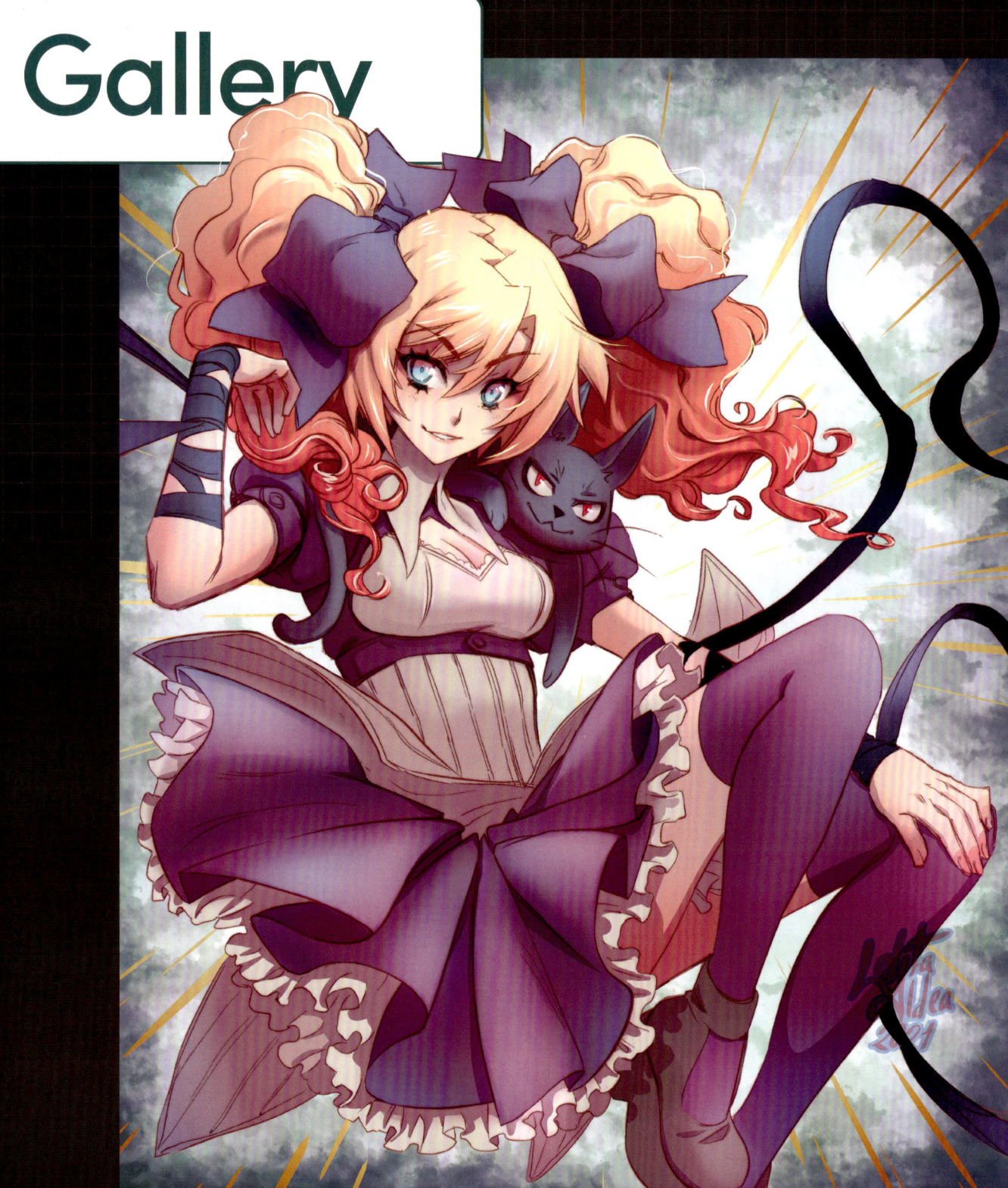

Charlotte © Lolita Aldea

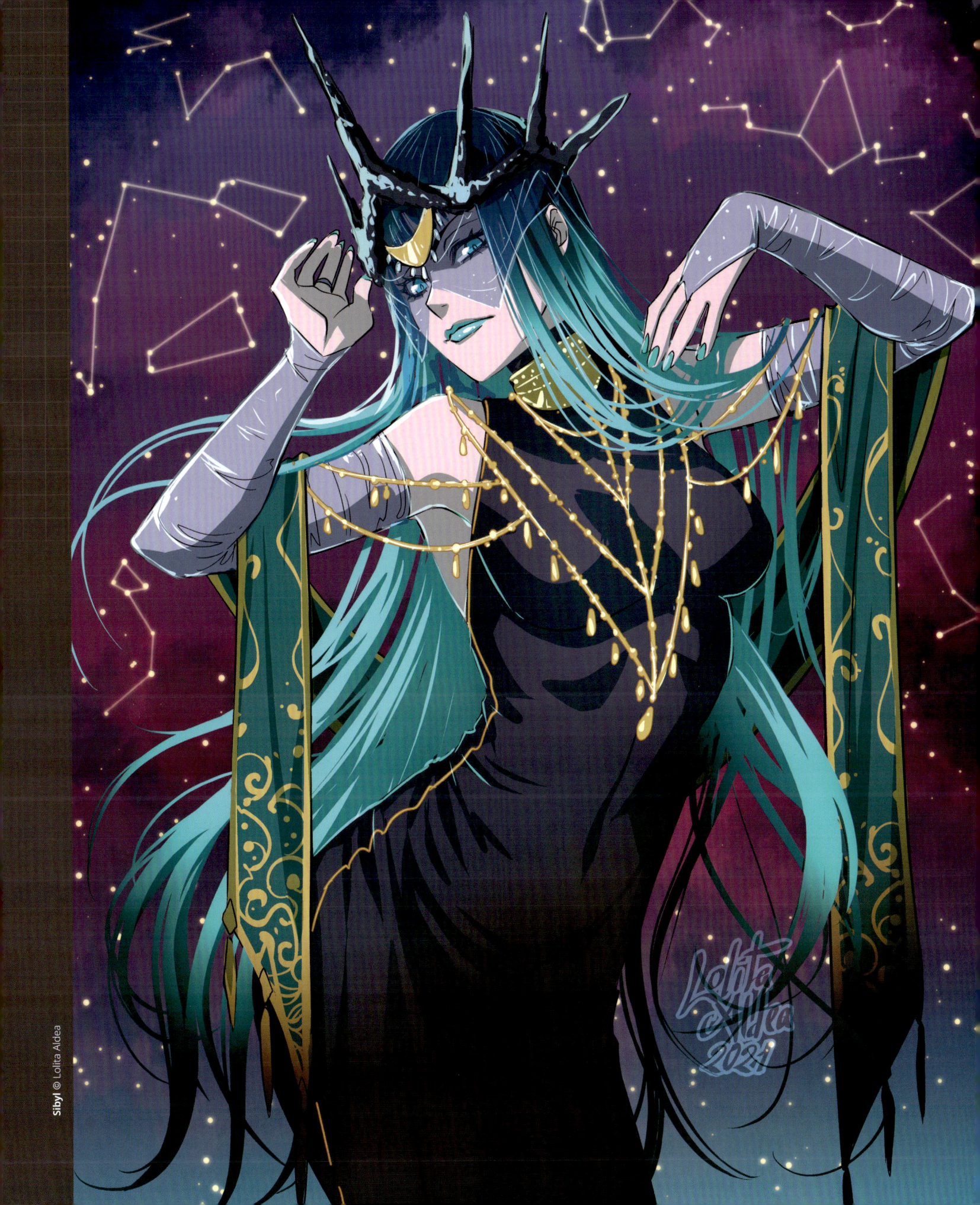

Glossary

ASSET
Pre-existing elements that can be added to your artwork or used as a reference. These materials include textures, brushes, 3D models, animation, and more.

BACKGROUND
The elements furthest away from the viewer in a painting, typically behind the focal point or objects of interest.

BLUR
A filter effect that blurs an image.

CANVAS
The area or space on which you can create an artwork.

COLOUR MODEL
A system used to identify colours by numerical values. There are numerous colour models, but the most used are: RGB (Red, Green, Blue), CMYK (Cyan, Magenta, Yellow, Black), and HSL (Hue, Saturation, Lightness).

CSP
Abbreviation of the Clip Studio Paint software.

DIALOGUE BOX
A small window that appears when executing a function or command. It may have spaces to fill and buttons to configure, or might simply be an information message.

EXPORT
To save a file in a format different to the original format so that it can be opened in another programme or software.

FILE
All of the information of your artwork is stored in a file. The file format typically used for Clip Studio Paint is CLIP (*. clip).

FOREGROUND
The elements that are placed at the front of a painting, closest to the viewer.

HUE/SATURATION/LIGHTNESS (HSL)
A colour model that defines colours by controlling the hue, saturation, and lightness. It can also be used as a tonal correction in CSP and other painting software.

IMPORT
To transfer a file from an external source into the software you're currently using.

LAYER
In digital painting, layers are like a stack of transparent sheets. You can arrange, paint on, and modify them to build up an artwork layer by layer, while keeping the elements on each layer separate and editable.

LINE ART
This drawing technique consists of only lines and strokes, without any shadow, shading, or tone.

MASK
Typically applied to a layer, a mask serves to flexibly hide or reveal parts of an element or image.

MENU
A list, whether dropdown or static, of options that can be selected to execute a function. There may be further submenus within it.

OPACITY
The level of transparency or translucency a layer or tool has. The less opacity, the more translucent it is. The more opacity, the less translucent it is.

PALETTE
A small window that can be docked, or sit freely, on the interface. Each palette has different tools, functions, and settings.

PERSPECTIVE
A technique used to create depth and three-dimensionality in drawings based on vanishing points and angled lines.

RASTER LAYER
This type of layer supports the creation of a pixel-based image, as opposed to a vector.

RESOLUTION
The level of sharpness and pixels of an image.

RGB
A common colour model that defines an image according to its red, green, and blue colour information.

SHORTCUT
A key or combination of keys that can be pressed to execute a function without needing to navigate menus.

TEXTURE
A type of material used to create an illusion of depth or complexity in an artwork. Texture can be used as an image or applied to a brush.

UNDO & REDO
Functions used to reverse (undo) or repeat (redo) an action in a digital software.

VECTOR GRAPHIC
Digital elements created from lines, points, and curves. They are generated mathematically, rather than by pixels, so can be easily scaled without losing their resolution (unlike raster images).

Downloadable resources

The following resources are available to download to help you complete the tutorials, as well as the Getting Started chapters. They can be downloaded from store.3dtotal.com/tools/downloadables/resources. We recommend that you download the resources before starting the tutorials.

GETTING STARTED
- Timelapse video

FANTASY SCENE
DEVIN ELLE KURTZ
- Timelapse video
- Line art

MANGA CHARACTER DESIGN
MMUMECHII
- Timelapse video
- Line art

CREATURE DESIGN
STEFAN KOSTIC
- Timelapse video
- Line art

CHARACTER DESIGN
DADOTRONIC
- Timelapse video
- Line art

REAL-WORLD SCENE
ORENJIKUN
- Timelapse video
- Line art
- Greyscale&ValueCheck.icc file

PICTURE-BOOK ILLUSTRATION
SIMONE FERRIERO AKA SIMZ
- Timelapse video
- Line art

COMIC
DYLAN TEAGUE
- Timelapse video
- Brushes
  - Ray Frendon_Teague's Folly (RNG)
  - Dylan Teague brush_pencil hatch 2
  - Dylan Teague brush_Soft 2

MANGA WEBTOON
LOLITA ALDEA
- Timelapse videos
- Line art

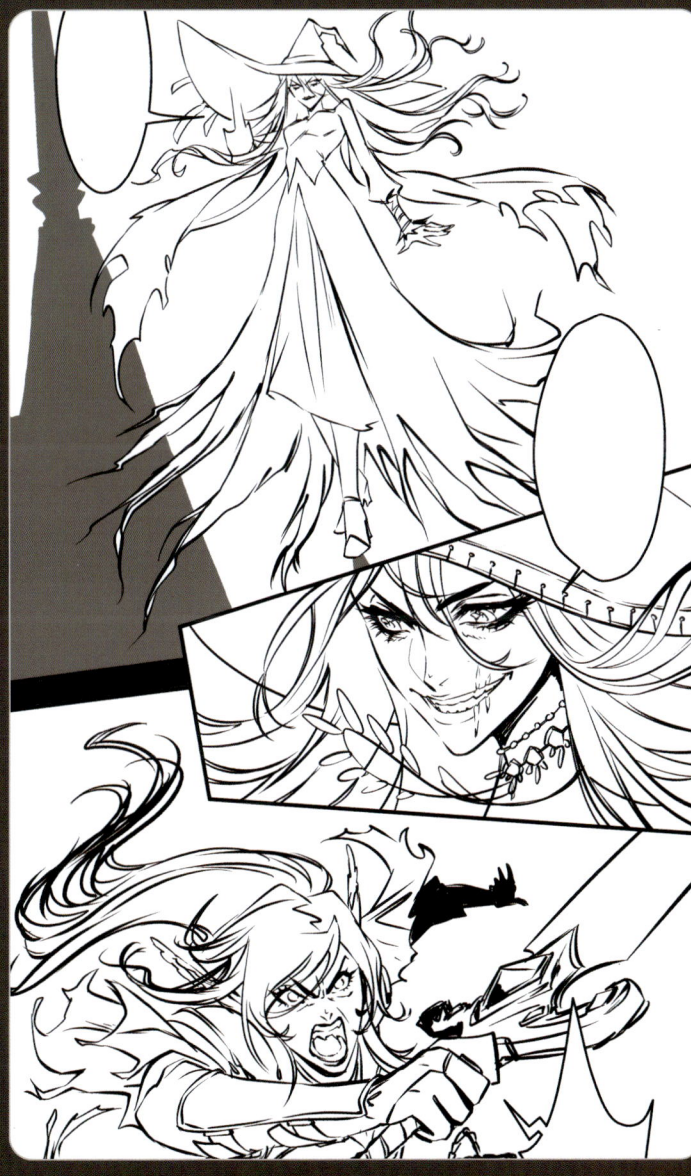

> An example of line art for a webtoon

Contributors

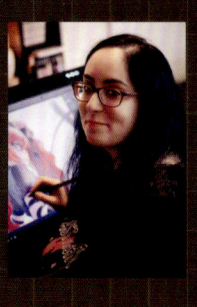

LOLITA ALDEA
Illustrator, comic-book artist & character designer

Lolita has been creating manga professionally since 2012. Her books have been translated into English, German, Korean, and Spanish. She also has experience working in animation, advertising, TTRPGs, video games, and concept art for miniatures.
lolitaaldea.com

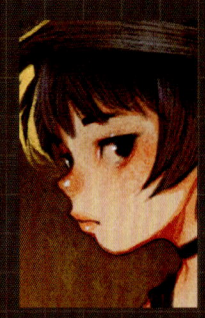

SIMONE FERRIERO AKA SIMZ
Artist

Based in Italy, Simone creates illustrations and comics and is well known for his paintings of witches and ghost cats. He works as an independent artist as well as on video games and other industry projects.
instagram.com/simz.art

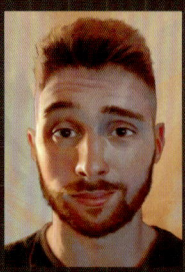

STEFAN KOSTIC
Freelance illustrator

Based in Belgrade, Serbia, Stefan is a self-taught digital artist and illustrator. He has worked on a variety of projects, including card illustrations, posters, and book covers.
kostart.carrd.co

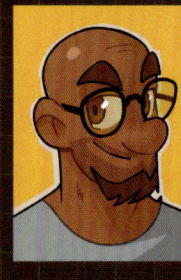

DADOTRONIC
2D artist

DADOtronic's artwork is inspired by imagery from the 1990s and 2000s, as well as retro games, graffiti, and tech. He has been creating artwork with Clip Studio Paint since its origin.
dadotronic.com

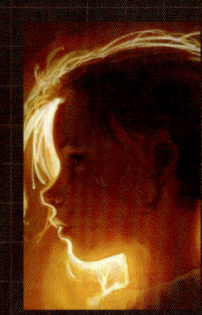

DEVIN ELLE KURTZ
Illustrator

Devin lives in the Los Angeles area, where she began her career in the animation industry. She currently illustrates for books, including her children's book *The Bakery Dragon* published by Random House Kids.
devinellekurtz.com

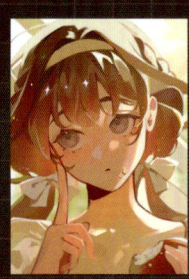

MMUMECHII
Digital illustrator

Mmumechii is an illustrator from Malaysia who enjoys drawing food and anime art. Sharing her illustrations and drawing process with others is her greatest joy!
mumechi.carrd.co

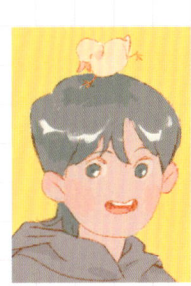

ORENJIKUN
Background artist & illustrator

Orenjikun is a self-taught background artist based in Australia. He primarily works in the animation industry for TV and film for clients such as Disney, Paramount, and LEGO.

instagram.com/orenjikun

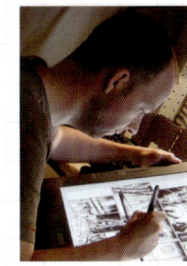

DYLAN TEAGUE
Freelance illustrator

Dylan has worked as a professional illustrator for around thirty years. He specializes in comics but also works on storyboarding, concept design, and advertising artwork.

instagram.com/dylbot2099

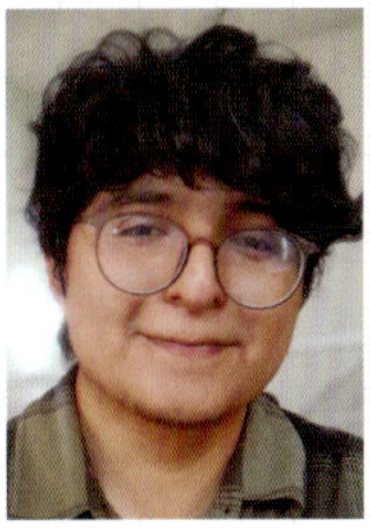

NIMROD VILLAR
Illustrator, concept artist & storyboard artist

Nimrod is a digital artist with over eight years' experience in illustration, concept art, and storyboarding. He has worked on private commissions, films, animated series and music videos, and video games. His interests include dark fantasy, futurism, and cartoon aesthetics.

instagram.com/nimrodarte

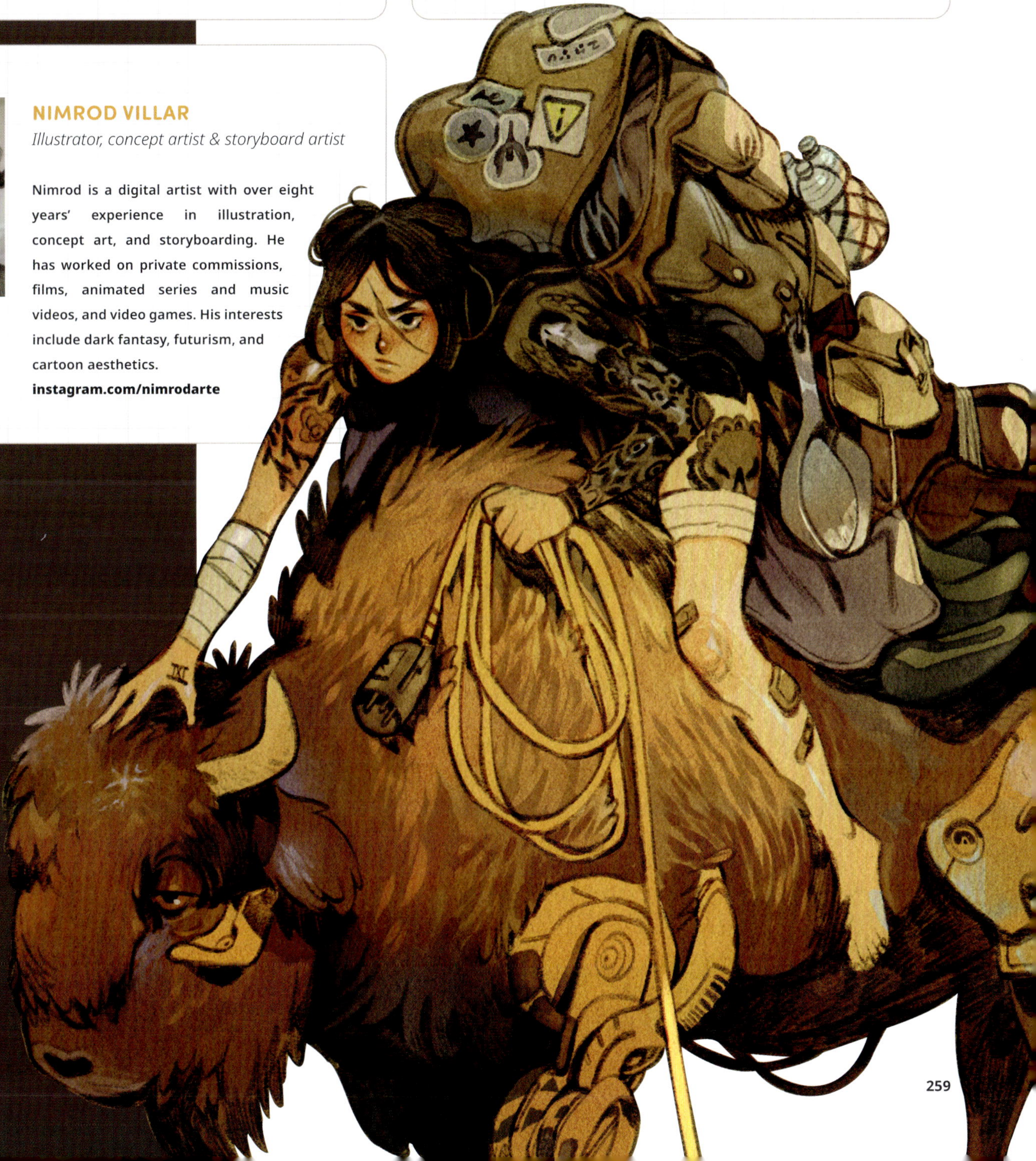

Bison © Simone Ferriero
AKA Simz

Index

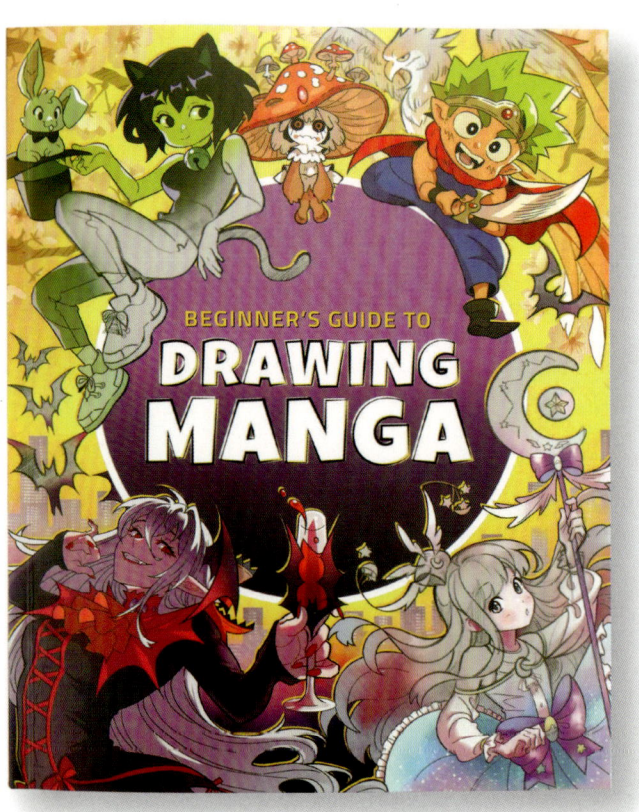

DRAWING MANGA

Dive into the exciting world of chibi fantasy heroes, martial arts fighters, and magical girls in this dynamic introduction to creating manga drawings. Learn how experienced artists turn their ideas into believable concepts, with easy-to-follow steps, helpful tips, and art theory made simple.

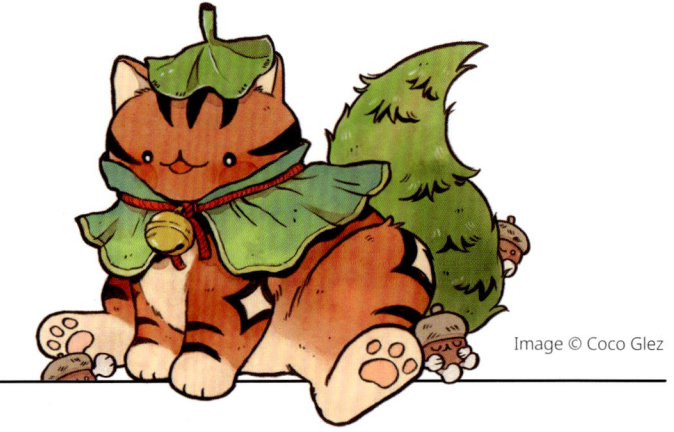

Image © Coco Glez

Sketching from the Imagination: ANIME & MANGA

Following on from the success of previous titles in this popular series, *Sketching from the Imagination: Anime & Manga* presents a wide array of interpretations of the Japanese art forms. You will have an access-all-areas pass to the sketchbooks of 50 accomplished artists from various fields as they share their inspiration, favourite tools, and techniques. Discover hundreds of sketches accompanied by the insightful narratives and advice of these illustrators, concept artists, and 3D artists alike. From action and adventure to the paranormal, horror, science fiction, and more, every corner of the anime genre is re-imagined for artists of all disciplines and skill levels.

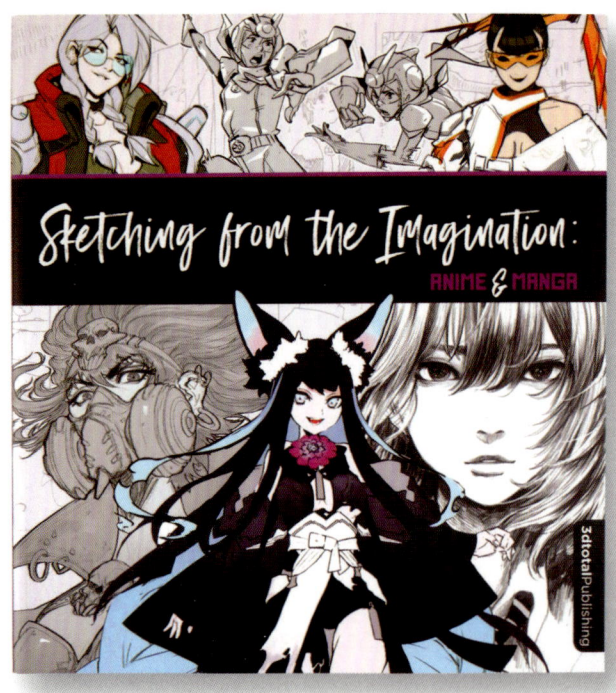

AVAILABLE AT STORE.3DTOTAL.COM

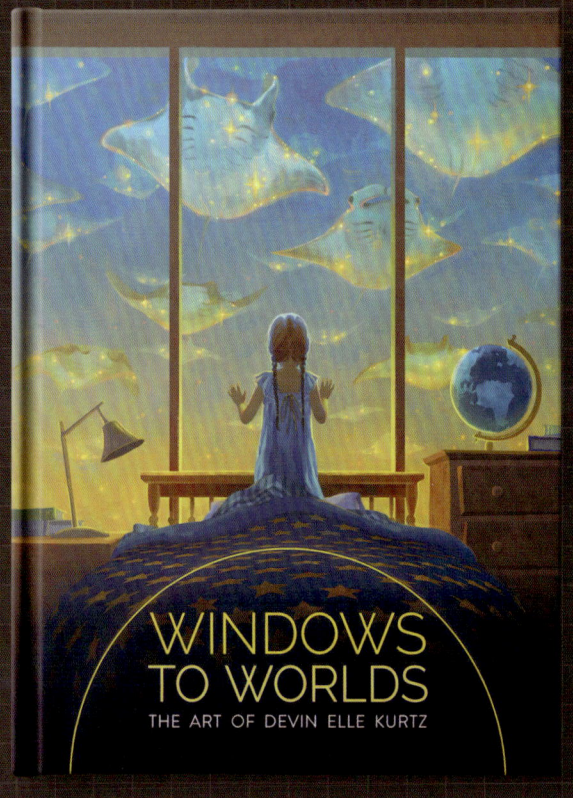

WINDOWS TO WORLDS
THE ART OF DEVIN ELLE KURTZ

Industry professional Devin Elle Kurtz inspires and instructs in her debut artbook. In addition to sharing how she has navigated the various paths of education, work, and personal struggles, Devin reveals her approach to colour, light, and storytelling. Readers will find theory, techniques, and favourite tools sitting alongside step-by-step tutorials as Devin helps them unlock their own artistic potential.

The Art of Simz

This magical debut from Simz is filled to the brim with the digital artist's much-loved modern witches and their spectral cat companions. Alongside an incredible collection of his work to date, readers will learn more about his creative journey, discover his coveted tools and techniques, and unveil how to find their own unique illustrative style. *The Art of Simz* is an enchanting addition to any bookshelf or coffee table, and will be sure to charm both artists and art lovers alike.

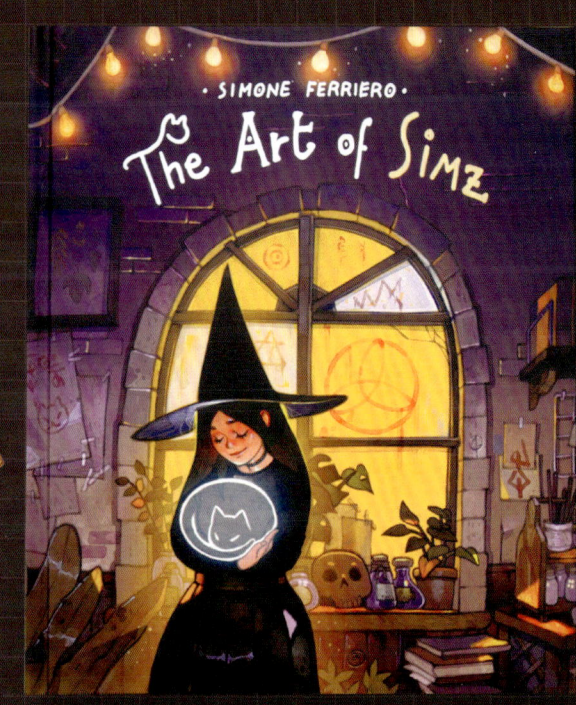

© Simone Ferriero AKA Simz

AVAILABLE AT STORE.3DTOTAL.COM

3dtotalPublishing

3dtotal Publishing is a trailblazing, creative publisher specializing in inspirational and educational resources for artists.

Our titles feature top industry professionals from around the globe who share their experience in skilfully written step-by-step tutorials and fascinating, detailed guides. Illustrated throughout with stunning artwork, these bestselling publications offer creative insight, expert advice, and essential motivation. Fans of digital art will enjoy our comprehensive volumes covering Adobe Photoshop, Procreate, and Blender, as well as our superb titles based around character design, including *Fundamentals of Character Design* and *Creating Characters for the Entertainment Industry*. The dedicated, high-quality blend of instruction and inspiration also extends to traditional art. Titles covering a range of techniques, genres, and abilities allow your creativity to flourish while building essential skills.

Well-established within the industry, we now offer over 100 titles and counting, many of which have been translated into multiple languages around the world. With something for every artist, we are proud to say that our books offer the 3dtotal package:

LEARN • CREATE • SHARE

Visit us at store.3dtotal.com

3dtotal Publishing is part of 3dtotal.com, a leading website for CG artists founded by Tom Greenway in 1999.